PRINCIPLES OF
POLYMER SYSTEMS

Fourth Edition

PRINCIPLES OF POLYMER SYSTEMS

Fourth Edition

Ferdinand Rodriguez

Professor of Chemical Engineering
Cornell University
Ithaca, New York

Taylor & Francis
Publishers since 1798

USA	Publishing Office:	Taylor & Francis 1101 Vermont Avenue, N.W., Suite 200 Washington, D.C. 20005-3521 Tel: (202) 289-2174 Fax: (202) 289-3665
	Distribution Center:	Taylor & Francis 1900 Frost Road, Suite 101 Bristol, PA 19007-1598 Tel: (215) 785-5800 Fax: (215) 785-5515
UK		Taylor & Francis Ltd. 1 Gunpowder Square London EC4A 3DE Tel: 171 583 0490 Fax: 171 583 0581

PRINCIPLES OF POLYMER SYSTEMS, Fourth Edition

1 2 3 4 5 6 7 8 9 0   BRBR   9 8 7 6

This book was set in Times Roman by Graphic Composition, Inc. The editors were Lynne Lackenbach and Holly Seltzer. Cover design by Michelle Fleitz. Prepress supervisor was Miriam Gonzalez. Printing and binding by Braun-Brumfield, Inc.

A CIP catalog record for this book is available from the British Library.
⊗ The paper in this publication meets the requirements of the ANSI Standard Z39.48-1984 (Permanence of Paper)

Library of Congress Cataloging-in-Publication Data

Rodriguez, Ferdinand,
    Principles of polymer systems/Ferdinand Rodriguez.—4th ed.
        p. cm.
    Includes bibliographical references and index.

    1. Polymers. I. Title.
    TP 156.P6R62 1996
    668.9—dc20   95-40393
                                                                                    CIP

ISBN 1-56032-325-6

CONTENTS

PREFACE

For this fourth edition, many of the generalizations made in previous editions continue to hold true. Most undergraduate students in science and engineering still will take at most one or two courses dealing with polymers. The American Chemical Society has proposed that polymers should be taught in a series of courses dealing with the chemistry and physics of polymers together with engineering aspects. Some specialized textbooks have been written to implement this concept. However, even at the graduate level, only a few universities have been willing to devote the time and the faculty resources to make such a curriculum a reality. These generally have been developed in the context of a field of polymer science and engineering separate from the traditional disciplines of chemistry and engineering.

This text offers a particular, integrated view of polymer science and engineering. Much of the material can be covered in one semester. Worked-out examples have been added in order to emphasize the importance of problem solving as a tool to understanding principles. Also, a list of keywords has been appended to each chapter to aid the student in reviewing topics. Recycling and resource recovery now is regarded as an essential part of polymer technology and has been treated in a separate chapter.

With the death of Herman Mark in 1992 at age 96, one of the last major links to the beginnings of polymer science in the 1920s has passed from the scene. His career in polymers reached all the way back to the controversies of the "macromolecular hypothesis" and continued actively into our present (still imperfect) understanding of polymer structure and function.

As with previous editions, I thank my students and coworkers for their added insights and continued stimulation. The comments from teachers who use the book have helped guide some of the revisions in this edition. I would like especially to thank Claude Cohen and Lee Vane for reviewing specific portions. As always, thanks go to my wife, Ethel, for her encouragement and patience.

Ferdinand Rodriguez

Equivalents for selected SI quantities

Quantity	SI unit	Value in cgs units	Value in British units
Mass	kg	1.000×10^3 g	2.200 lb_m
Lengh	m	1.000×10^2 cm	3.281 ft
Area	m^2	1.000×10^4 cm^2	10.76 ft^2
Volume	m^3	$1.000 \times 10^6 cm^3$, or	35.31 ft^3, or
		1.000×10^3 liter	264.2 gal (U.S.)
Force	N (newton)	1.000×10^5 dyn	0.2248 lb_f
Energy	J (joule), N·m	1.000×10^7 erg, or	0.7376 ft·lb_f, or
		0.2389 cal	0.9481×10^{-3} Btu
Pressure, stress	Pa (pascal), N·m^2	10.00 dyn	1.450×10^{-4} lb_f/in^2, psi
Viscosity	Pa·s	10.00 P (poise)	6.72 lb_m/ft·s
Power	W (watt), J/s	1.000×10^7 erg/s, or	1.341×10^{-3} hp, or
		14.34 cal/min	3.413 Btu/h
Specific heat	J/kg·K	2.389×10^2 cal/g· °C	2.389×10^2 Btu/lb_m· °F
Heat transfer coefficient	W/m^2·K	0.2389 cal/m^2·s· °C	0.1761 Btu/ft^2·h· °F
Impact strength	J/m		1.873×10^{-2} ft·lb_f/in

PREFACE TO THE FIRST EDITION

A man was asked the question, "Do you have trouble making decisions?" He thought a while and then finally answered, "Well, yes and no." The engineer or scientist who is asked to make generalizations about polymers often finds himself in the same position. In the interests of organizing the body of information about polymers which has accumulated since Baekeland, Staudinger, Mark, Carothers, and other pioneers started their work, there is a tendency to overgeneralize. The road of polymer discovery is strewn with the bones of absolute statements. In the older literature one finds such pronouncements as, "All polymer crystals are submicroscopic," "Five-membered rings are too stable to be opened to form linear polymers," "Maleic anhydride cannot be homopolymerized," and "Stereoregular polymers can be made only with an optically active catalyst." All of these have been disproved or qualified. In this book, any generalizations that are encountered are subject to the following *caveat*: "All generalizations are partially untrue, except this one."

It has been the author's aim in this book to relate the behavior of polymer systems whenever possible to examples that are part of everyday experience. With polymers the job should be simple, since many of the things we use—our clothing, our food, and our bodies—are made up of polymer systems.

It scarcely needs saying that this introductory text cannot treat any one subject exhaustively and must, perforce, omit some. However, the student who wants to learn more about polymers can easily, like Leacock's distraught young lord, jump on his horse and ride wildly off in all directions. Not only are there many journals devoted to polymers, the journals themselves divide and grow in yeasty fashion. Journals devoted to reviews of selected topics in polymers have made their appearance along with a flood of monographs. The list in Appendix 3 [in the first edition] includes many of the English-language sources. The futility of trying to present a complete picture of polymer systems in one text is best illustrated by reference to the massive *Encyclopedia of Polymer Science and Technology* (Interscience) which began to be published in 1964.

Despite the great leaps forward made in the last several decades toward understanding polymers, frontiers remain. The mechanisms of flame retarding by halogens and phosphates, drag reduction by dilute polymer solutions, filler reinforcement of rubber and plastics, and the freeze-thaw stabilization in human blood are incompletely understood. Significantly, in each case, the polymer does not have an isolated existence, but is part of a *system*. Perhaps the most challenging frontier of all involves the human system. The use of polymers to stabilize whole blood during extended storage at liquid nitrogen temperatures is but one example of the way in which polymer systems can be applied in the interests of humanity.

I gratefully acknowledge the permissions granted by journals and industrial firms to reproduce material originally appearing in their publications. Extensive comments on the original manuscript by Professor M. C. Williams of the University of California at Berkeley were very helpful. The assistance and encouragement from students, fellow teachers, and co-workers in industry have been important in making this book possible. I want to thank Professor Charles C. Winding especially for his constructive and friendly guidance. And, of course, I thank my wife and family for the understanding and patience they have shown at all times, but most conspicuously during the writing of this book.

Ferdinand Rodriguez

INTRODUCTION

1.1 POLYMERS

The term *polymer* denotes a molecule made up by the repetition of some simpler unit, the *mer* or *monomer*. It is almost impossible to define the term further to include the many species with which we shall deal and yet to exclude metals, ceramics, and crystalline forms of smaller molecules. In the broadest sense, these are polymers too.

Two mers may be combined to form a *dimer,* three to form a *trimer,* and so on. The number of repeating mer units is referred to as the *degree of polymerization* (see also Sec. 4.4). For low-molecular-weight polymers, the term *oligomer* (*oligo* being a prefix of Greek origin meaning few) is often used. Typically, oligomers might encompass degrees of polymerization of 2 to 20. The analogous term *oligosaccharide* usually includes molecules with 2 to 6 sugar units (see Sec. 15.3).

The term *macromolecule,* big molecule, also is used. Large molecules of complex structure can be covered better by this name than by the name "polymer," since the latter carries with it the connotation of a simple repeating unit. An example of a naturally occurring macromolecule is insulin, a protein hormone that occurs in the pancreas. It is best known as an agent to lower blood sugar in diabetic patients. The structure has repeating units with amide linkages:

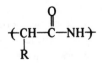

A macromolecule of insulin has a total of 51 such units with 16 variations of R. Proteins, nucleic acids (DNA, RNA), and enzymes generally are complex macromolecules. On the other hand, some naturally occurring polymers have a simple repeating unit. Starch, cellulose, and natural rubber are examples of polymers which are chains of a simple unit repeated thousands of times. Cellulose, in the form of cotton, is useful as a textile fiber without chemical modification. The polymer-based industries all started with such naturally occurring materials. A second stage in industrial development has been the modification of a natural polymer to make it more useful. All industries have found that entirely synthetic polymers, built up from small molecules such as ethylene and propylene, offer the broadest spectrum of properties. Obviously, the natural materials have a firm place in technology. However, when the idea arises of "tailoring" a polymer to certain end use requirements, the wholly synthetic polymers often provide an economical solution.

1.2 APPLICATIONS OF POLYMERS

To the general public, plastics, fibers, rubber, adhesives, paints, and coatings all are familiar as consumer products. The fact that all of these are based on polymers, and that often the same polymer is used for more than one application, is a much less familiar concept.

In Table 1.1 the change in usage of polymers in three major categories is summarized for two decades. Only the category of plastics has shown consistent growth. Of course, in every category there are some specialty polymers that go into specific "niche" markets representing areas of growth and profit. The commodity polymers may dominate the statistics, but they do not always represent the items of greatest value to a particular company. For example, the thermoplastic elastomers (see Sec. 4.5) have been able to challenge the older vulcanizable (cross-linkable) rubbers such as styrene-butadiene (SBR), neoprene, butyl, and nitrile.

Over the decades, the major uses of some materials have not changed much. Most fibers, both synthetic and natural, still go into clothing and house furnishings, just as they did over a century ago. Automobile parts including tires use over two-thirds of the rubber consumed in the United States, just as they did a half-century ago. Protective and decorative coatings and the adhesives sold in hardware stores still resemble, at least superficially, the products sold for many decades. However, it is in the area of plastics that diversity of materials and of applications has grown most conspicuously.

New or expanded applications are the main reason for the fivefold growth of plastics in the two decades (1974–1994) as seen in Table 1.1. In that period, the use of poly(vinyl chloride) for pipe, conduit, and siding in construction became common. The soft drink bottles made of poly(ethylene terephthalate) which now dominate grocery shelves have almost completely displaced glass bottles. It would come as a surprise to most consumers to find that soft drink bottles, video and audio tapes, and polyester apparel all are made from the same polymer. Likewise,

Table 1.1 Polymer consumption in the United States (in units of 10^6 kg)*

Category	1974		1984		1994	
All plastics (sales)	12,932		20,154		34,734	
Thermosets (total)	2,542		3,610		5,771	
Alkyd		370		109		(with other)
Epoxy		109		171		274
Phenolic		590		1,137		1,465
Polyester (unsaturated)		404		559		1,333
Polyurethane (mainly foams)		624		1,003		1,707
Urea and melamine		445		630		993
Thermoplastics (total)	10,390		16,544		28,963	
ABS		(with styrene)		499		677
Acrylic		246		240		(with other)
Cellulosics		76		50		(with other)
Nylon		83		177		419
Polyacetal		32		47		97
Polycarbonate		51		130		316
Polyester (thermoplastic)		10		604		1,564
Polyethylene (high density)		1,217		2,704		5,280
Polyethylene (low density)		2,729		3,825		6,394
Polyphenylene-based alloys		(with other)		(with other)		109
Polypropylene and copolymers		1,001		2,189		4,433
Polystyrene and copolymers		2,183		2,246		2,729
Poly(vinyl chloride) and vinyls		2,555		3,552		5,056
Thermoplastic elastomers		(with other)		201		394
Others		207		80		1,495
Rubber (production figures)						
Natural rubber (consumption)		738		751		997
Synthetic rubber (total)	2,498		1,987		2,099	
Styrene-butadiene (SBR)		1,466		888		851
Butadiene		310		433		495
Ethylene-propylene		126		185		263
Nitrile		88		65		84
Neoprene		163		93		75
Butyl		164		(with other)		(with other)
Synthetic isoprene		93		(with other)		(with other)
Other		88		323		331
Fibers (mill consumption figures)						
Natural						
Cotton		1,416		1,038		2,190
Wool		50		64		56
Synthetic	2,931		3,644		4,535	
Rayon (regenerated cellulose)		75		175		230
Cellulose acetate		163		90		(with rayon)
Nylon		917		1,086		1,250
Polyester		1,271		1,541		1,769
Acrylic		270		303		206
Polyolefin		235		450		1,081
Synthetic polymers (all categories)	18,361		25,785		41,369	

*Sources: U.S. Dept. of Commerce, *Chem. Eng. News, Mod. Plastics, Fiber Organon, Rubber Stat. Bull.*

the polypropylene now used for almost all automobile batteries is also used for indoor-outdoor carpeting, lawn furniture, and "polyolefin" intimate apparel.

1.3 POLYMER-BASED INDUSTRIES

Until the 1920s and 1930s, the several industries that depend on polymeric materials grew rather independently of one another and were based on natural or modified natural materials. These industries can be identified in the broad sense as

Plastics
Rubber
Fibers
Coatings
Adhesives

In fact, it was not until after World War II that the dividing lines became so blurred as to make it difficult to classify some companies.

Plastics

The word *plastic* denotes a material that can be shaped. Polymers that can be formed repeatedly by application of heat and pressure are called *thermoplastics.* Those which can be formed only once are called *thermosets.*

During Civil War days, daguerreotype cases were molded out of compositions containing shellac, gutta-percha, or other natural resins along with fibrous materials. The compression molding machinery and dies that were used were quite similar in principle to those used a hundred years later. Hyatt's discovery (1868) that nitrated cellulose mixed with camphor could be molded under pressure to give a hard, attractive material suitable for billiard balls and detachable collars is often cited as the start of the plastics industry. Others date it from 1907, when Baekeland patented a wholly synthetic material based on the reaction of phenol and formaldehyde. The industry has expanded continuously with scarcely a pause since 1930 (Fig. 1.1). The capacity for styrene production as a result of the synthetic rubber program after World War II was an important factor. The need for polyethylene and poly(vinyl chloride) during the war also accelerated industrial know-how. By 1960, each of these three polymers was being produced at the rate of over a billion pounds per year.

In general, the wholesale price pattern for plastics and resins parallels that of the Producer Price Index (PPI) for industrial commodities (Fig. 1.2). The more familiar measure of inflation is the Consumer Price Index (CPI), also shown in Fig. 1.2. One component of the CPI which makes a big difference from the other indexes is the cost of medical care, which more than doubled from 1982 to 1992.

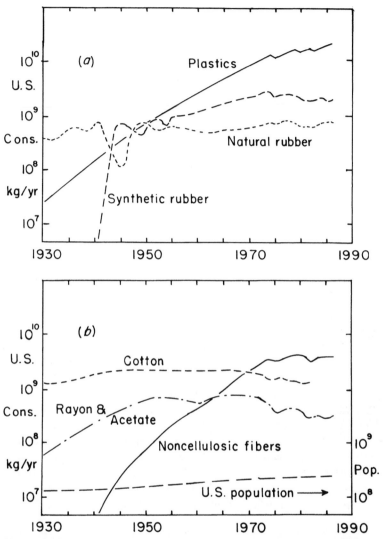

Figure 1.1 Polymer consumption can be compared with the gradual growth of the population of the United States, which was estimated to be 2.63×10^8 in 1995: (*a*) plastics and rubber; (*b*) fibers

It is to be expected that increased volume for any manufactured product will lead to economies of scale. This is the case, although there is no firm rule that applies. When a large number of diverse polymers are compared, there is a discernible trend which can be expressed as

$$\text{Selling price} = k_a(\text{production scale})^{-a} \qquad (1.1)$$

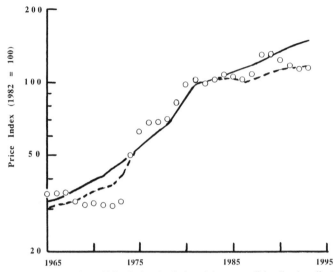

Figure 1.2 Producer Price Index for industrial commodities (broken line) and for plastics and resins (circles), compared with Consumer Price Index (solid line). The base year is 1982. (Source: U.S. Dept. of Commerce)

where a is between 0.1 and 0.4 and k_a increases from year to year. A widely used rule of thumb is that the cost of equipment increases with the production capacity to the 0.6 power. If all other factors follow the same trend, then, obviously, the cost of production would, indeed, decrease with the −0.4 power.

Another way of looking at prices is by a "learning" or "experience" curve. If price is plotted against time or against cumulative production volume, there should be a gradual decrease reflecting economies from debottlenecking, incremental increases in capacity with small investment, and, eventually, lowered plant depreciation. Even if volume per unit time remains constant, experience should lower price. Unfortunately, inflation causes the "baseline" of such a plot to drift, and attempts to correct for this (to plot in "constant" dollars) involve many assumptions.

Rubber

Even in pre-Columbian times, the natives of South and Central America made use of the latex obtained by puncturing the bark of certain trees. By coagulating the latex, a rubber ball could be produced that was used in the national sport of the Mayan Indians. MacIntosh and Hancock in England and Goodyear in the United States (1839) discovered that mixing natural rubber with sulfur gave a moldable composition that could be heated (vulcanized) and converted to a useful, nontacky, stable material suitable for raincoats, waterproof boots, and solid tires for carriages. The general progress of the rubber industry soon became associated with the automobile. Even today, about 70% of all types of rubber ends up in tires and a good part of the remainder is used in gaskets, grommets, and other parts of the modern auto. The pneumatic tire (Dunlop, 1888) and the use of carbon black as a reinforcing filler and of organic accelerators for the sulfur cross-linking reactions were achieved with natural rubber. Since World War I, most natural rubber has been cultivated in Malaysia and Indonesia rather than the Americas. Between World Wars I and II the development of synthetic rubbers was pursued, especially in Germany and the United States. In Germany, emphasis was on general-purpose rubber to reduce dependence on far-flung sources of supply. In the United States, more emphasis was placed on specialty rubbers, such as Thiokol rubber and neoprene, which would be more resistant to swelling by solvents than natural rubber. During World War II, Germany and Russia had developed synthetic rubbers so as to be self-sufficient. In the United States, a crash program was instituted that was to affect all other polymer-based industries eventually. Production of a styrene-butadiene copolymer (GR-S, later called SBR) went from zero in 1941 to over 700 $\times$ 10^6 kg in 1945. In Fig. 1.1, the drop in natural rubber consumption and the development of SBR rubber during World War II (1941 to 1946) can be seen. Despite a brief resurgence of interest in natural rubber after the war, the synthetic material remained in use. The primary results of the SBR program were:

1. Tremendous production capacity for styrene and butadiene monomers
2. Independence of American industry from rubber imports for many uses
3. Availability of synthetic rubber latex for other uses
4. New insight into polymer production and characterization

Not the least of the accomplishments was the overnight conversion of many scientists and engineers into polymer engineers, polymer chemists, polymer physicists, etc. After the war, their interest remained in the field, so that the rather thin prewar ranks of textile, rubber, paint, etc., researchers were suddenly swelled by large numbers of highly trained people who were not wedded to any one industrial viewpoint. The exchange of technology between industries that now is quite common can be credited in great measure to the proper recognition of the principles of polymer science and technology as the underlying basis for all these fields.

Since the early 1970s overall consumption of rubber in the United States has fluctuated around a total of about 2.1 $\times$ 10^9 kg/yr. About a third of this is natural rubber. SBR production peaked at 1.5 $\times$ 10^9 kg in 1973. Various factors have contributed to keep consumption from increasing, despite greater numbers of automobiles on the road. The trends to smaller automobiles, lower speed limits, and higher-priced fuel all softened the demand for rubber in transportation. Moreover, the change in construction of tires had several effects. Belted-radial tires use a higher ratio of natural rubber and *cis*-polybutadiene to SBR than the older bias-ply tires. The radial tires often wear longer, which helps to make up for their higher initial cost. The net result is a decreased demand for replacement tires. In the

1950s, a life of 20,000 mi (32,000 km) was considered acceptable for a bias-ply tire. Many radial tires now can be expected to last at least twice that long.

Fibers

Natural polymers still are an important part of the fiber world. Regenerated cellulose (rayon) and cellulose acetate were introduced around the turn of the century, but nylon was the only wholly synthetic polymer produced in quantity before World War II. In the 1960s, cotton and wool usage decreased while polyester and nylon production boomed. Since about 1970, rayon and acetate have diminished in importance (Fig. 1.1, Table 1.1). Pollution problems inherent in rayon manufacture caused some plants to close down in the 1970s. A factor favoring synthetic fibers has been the fashion trend to "easy-care" fabrics and garments. Cotton and wool require chemical treatment to compete on this basis. Synthetic fibers also have had a price advantage in recent years due partly to labor costs. Cotton and wool are "labor-intensive" materials compared to the synthetics.

Although cotton consumption in U.S. mills fluctuates from year to year, several points should be taken into account. A large fraction (as much as 50% in some years) of cotton production in the United States is exported rather than processed domestically into textiles. Also, a large amount of cotton is imported in the form of finished apparel without being included in the U.S. mill consumption figure. In the United States, annual imports of silk from 1979 to 1993 averaged 0.5 to 1 million kg and usage of wool averaged about 75 million kg, both of which are small compared to the 1,000 to 2,000 million kg of cotton in the same period. On a worldwide basis in 1993, demand for fibers (in millions of kg) was: cotton, 18,966; wool, 1,658; silk, 68; synthetic cellulosics, 2,246, and synthetic noncellulosics, 16,182 [1].

Coatings

Decorative and protective coatings have been based on unsaturated oils (linseed oil, tung oil) and natural resins (shellac, kuari gum) for centuries. In the 1930s synthetic resins (alkyd resins), which really are modified natural oils, became important. The SBR rubber program had its repercussions in the coatings industry, since styrene-butadiene latexes were found to be useful film formers. Latex paints based on vinyl acetate and acrylic resins now have been developed for outdoor as well as indoor applications. Such water-thinnable coatings represent most of the interior coatings and about half the exterior paints for houses. In 1994, total shipments of coatings amounted to about 1 billion gal, a little over half of which were "architectural" sales and the remainder were industrial finishes (to manufacturers of products such as automobiles and appliances) and other special-purpose coatings. Legislation in most states now limits the solvents that can be used in commercial coating applications, both in type and in amount. Water-thinned and powdered coatings have been substituted for solvent-based paints and lacquers in some instances.

Adhesives

Since adhesives often must be specifically designed to bond to two different surfaces, an even greater diversity of products has been developed than in coatings. In similar fashion to the coatings industry, the adhesives industry has made use of the latex technology developed in World War II. Many natural polymers are still used, but almost every new plastic and rubber developed has spawned a counterpart adhesive. The development of the highly advertised "super glues" based on the cyanoacrylates (see Sec. 12.4, Adhesives) was atypical in that the materials were used as adhesives first and not adapted from some prior application as plastics or coatings.

The starch and sodium silicate adhesives used to assemble paperboard materials still dominate the overall picture. Among the synthetics, phenolic and urea resins are widely used in plywood and particleboard manufacturing and in the surface treatment of fibrous glass insulation. Fibrous glass insulation has grown in importance as the United States has become more energy conscious. The phenolic resins used as binders for the glass give the characteristic pink color to the insulation.

1.4 FEEDSTOCKS FOR POLYMERS

Even when considered only as the monomer for the homopolymer, ethylene is a very important commodity. More than that, it is an essential ingredient in vinyl chloride and styrene, the other two largest volume monomers for plastics (Fig. 1.3). Polymer-grade propylene and butadiene are made in large measures as by-products of ethylene manufacture. As seen in Fig. 1.3, these two raw materials enter into the production of virtually all the other major plastic, rubber, and fiber polymers. Since polyethylene and polypropylene accounted for about half of the plastics produced in 1994, it is reasonable that the production of ethylene parallels rather closely the production of polymers in general and of polyethylene in particular (Fig. 1.4).

A complete olefins unit is a complex combination of operations. The Cain Chemical plant, which covers 14 acres in Chocolate Bayou, Texas, was started up in 1980 (Fig. 1.5). The unit can process various liquid feedstocks including gas field condensates, naphthas, and gas oils. There are nine cracking furnaces (shown in the background of Fig. 1.5). Also prominent are the 21 distillation columns. Not seen, but equally impressive to anyone who has replaced the washer in a dripping kitchen faucet, are 44,000 valves of all sizes. Ethylene (450×10^9 kg/yr) and propylene (300×10^9 kg/yr) are the major products. Coproducts are hydrogen, fuel gas, butadiene, benzene, toluene, pyrolysis gasoline, and tar.

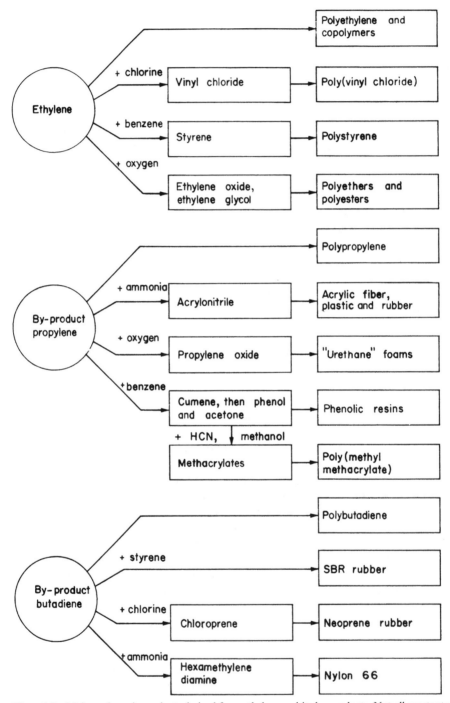

Figure 1.3 Major polymeric products derived from ethylene and its by-products. Not all reactants or products are shown

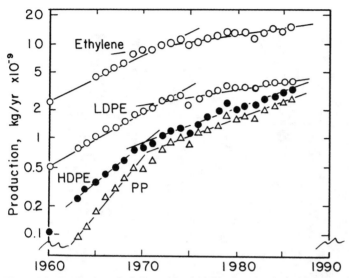

Figure 1.4 Production of ethylene and polyolefins (low-density polyethylene, branched or linear, high-density polyethylene, and polypropylene)

Building a new ethylene-producing plant is no small undertaking. A facility completed in 1985 to produce 500×10^6 kg/yr reportedly cost $580 million. That particular plant, at Mossmorran, Scotland, was a joint venture of two international petrochemical companies, Exxon and Shell.

Ethylene can be made from a variety of starting materials. Ethane recovered from natural gas is a "clean" feed, in that few by-products occur (Table 1.2). Natural gas was regarded as a diminishing resource in the 1970s, so almost all plants built then were made to use a variety of heavier feeds such as heavy naphtha (boiling range of 100 to 220°C) or atmospheric gas oil (AGO; boiling range of 200 to 350°C). However, the 1980s saw a renewed interest in light feeds including ethane. Many factors enter in, including government regulations regarding natural gas prices; this is a touchy political matter since many voters heat their homes with natural gas. Actually, almost any hydrocarbon source is a possible feed, but matters of aromaticity and sulfur content change the desirability from one refinery stream to another. The heavier feeds (Table 1.2) yield a greater variety of by-products that must be sold or used to make the process economical. Some chemical companies have been exploring the direct production of ethylene from crude oil, by-passing the usual oil refinery. By recycling various streams, the yield of ethylene can be optimized without having to market gasoline or fuel oil in competition with the traditional oil refiners. No full-scale plants have yet been built using a crude oil feed. The product distribution can differ from that listed in Table 1.2 for various reasons. High-*severity* operation may involve higher temperatures and longer resi-

Figure 1.5 The Cain Chemical olefins unit produces ethylene and propylene from naphthas and gas oils. The plant, designed by Lummus-Crest and built by Brown and Root, was started up in 1980. (Photo courtesy of Cain Chemical Inc., Subsidiary of Occidental Petroleum Corp., Alvin, Texas.)

Table 1.2 Typical product profile for production of 1000 parts by weight of ethylene from various feedstocks*

Product	Feedstock				
	Ethane	Propane	Naphtha	Atm. gas oil	Crude oil
Ethylene	1000	1000	1000	1000	1000
Acetylene	–	–	–	–	200
Propylene	24	375	420	557	135
Butadiene	30	69	141	187	90
Other C$_4$'s and gasoline	24	158	848	1019	435
Fuel oil	–	–	195	729	600
Residue gases	172	659	521	457	–
Pitch	–	–	–	–	583
Total feed	1253	2260	3126	3951	3050

*After S. C. Stinson, *Chem. Eng. News*, 57:32 (May 28, 1979).

dence time during cracking, which increases conversion per pass. Moderate-severity operation usually produces more by-products but lessens the amount of feed wasted in less salable materials such as coke.

With any feed, the hydrocarbon source mixed with steam is pyrolized at 850°C or higher for up to 1 s by passage through a tube set in a *cracking* furnace. The steam is added to decrease formation of coke on the tube walls. The hot effluent gases are *quenched* with water or oil to stop the reaction. Separations using distillation and absorption are used to obtain pure products as well as to recycle some materials. The relative merits of various feedstocks and cracking processes have been the topics of numerous journal articles. As a measure of energy usage in the United States, it may be pointed out that the entire synthetic polymer industry (plastics, rubber, and fibers) produced about 41 billion kg in 1994 while almost 8 times that weight of hydrocarbon was burned as fuel in passenger automobiles.

1.5 POLYMER SCIENCE AND TECHNOLOGY

The chemistry and physics of polymers lagged behind technology for many years. In 1920, Staudinger proposed his macromolecular hypothesis, saying that substances such as natural rubber were not colloidal, physical associations of small molecules but were truly long-chain molecules of extremely high molecular weight. Since such materials were not easily characterized by the methods used with small molecules, most chemists regarded polymer research as a rather undignified occupation. However, the studies of Emil Fischer on proteins, Meyer and Mark on cellulose, and Carothers on polycondensation had established a basis for acceptance of Staudinger's ideas by the 1930s. Once again, World War II must be recognized as a time when research accelerated sharply and theories started to catch up with practice. Debye's work on light scattering, Flory's on viscous flow (as well as other areas), and Harkins' on a theory of emulsion polymerization grew out of government-sponsored projects. The discovery by Ziegler of synthetic catalysts that would give ordered polymers and the extension of such systems by Natta in the 1950s changed many fundamental ideas of polymer science, as did the discovery of large single crystals of high polymers a few years later.

A capsule chronology of polymer technology and science is given in Table 1.3. Winfield [2] has presented an annotated retrospective look at the history of plastics, including many of the developments noted in Table 1.3. Another source of information on the history of polymers is the Beckman Center for the History of Chemistry in Philadelphia, which has a continuing program of receiving new material and publishing reviews.

1.6 TRENDS

The worldwide picture of plastics consumption indicates that the use of polymers is distributed throughout the developed and developing nations (Table 1.4). The rate of growth in Latin America, Asia/Oceania, and Africa/Mideast over the period 1982–1993 has been greater than that of North America and Europe.

Another way of looking at the future of polymers in the U.S. economy is to look at established trends. The change in plastics consumption can be compared with the change in the Gross Domestic Product (GDP) in constant dollars (Table 1.5). It is clear that plastics have been an increasingly important factor in the GDP. There seems to be no reason to expect the trend to be reversed.

Predictions of economic phenomena are notoriously inaccurate. There are various known factors that will continue to affect the polymer industry in the next decade. The large-volume plastics will find ever-increasing use in automobiles, furniture, and housing. Polymers for special applications in electronics, medicine, and other fields will continue to develop. This is true for rubber, fibers, coatings, and adhesives as well as for plastics. Fabrication techniques will be made more efficient.

The global picture of polymer production almost certainly will be changed by the building of polymerization facilities in countries with access to raw materials (such as polyethylene plants in Saudi Arabia). Building a polymer industry is high on the list of goals for most developing countries. Still, the average consumption of polymers in the world has a long way to go to match the usage in the industrialized nations (Table 1.4). Political and social developments may alter some patterns. Importation of textiles and apparel into the United States has been a major factor in the slowdown of domestic production of synthetic fibers. Tariff barriers may protect some industries and hurt others.

Table 1.3 A selected chronology of polymer science and technology

1770	Priestley is said to have given rubber its name because it can erase pencil marks
1806	Gough (England) experiments with elasticity of natural rubber
1838	Regnault (France) polymerizes vinylidene chloride via sunlight
1839	Goodyear (U.S.), MacIntosh and Hancock (England), vulcanization (cross-linking) of natural rubber
1859	Joule (England) demonstrates thermodynamic principles of elasticity of rubber
1860s	Molding of natural plastics such as shellac and gutta-percha
1868	Hyatt (U.S.), celluloid (cellulose nitrate molded articles)
1891	Chardonnet (France), regenerated cellulose via nitrate
1893 to 1898	Cross, Bevan, Beadle, Stearn (England), viscose rayon fibers
1907	Baekeland (U.S.), phenol-formaldehyde resins
1910	First rayon plant in United States
World War I	Cellulose acetate solutions ("dope") for aircraft; laminated plywood and fabric construction for aircraft fuselages
1884 to 1919	Emil Fischer (Germany) establishes formulas of many sugars and proteins
1920	Staudinger (Germany) advances macromolecular hypothesis
1928	Meyer and Mark (Germany) measure crystallite sizes in cellulose and rubber
1929	Carothers (U.S.) synthesizes and characterizes condensation polymers
1920s	Cellulose nitrate lacquers for autos
1924	Cellulose acetate fibers
1926	Alkyd resins from drying oils for coatings
1927	Poly(vinyl chloride)
	Cellulose acetate plastics
1929	Polysulfide (Thiokol) rubber
	Urea-formaldehyde resins
1930 to 1934	Kuhn, Guth, and Mark (Germany) derive mathematical models for polymer configurations; theory of rubber elasticity
1931	Poly(methyl methacrylate) plastics
	Neoprene (Duprene) synthetic rubber
1936	Poly(vinyl acetate) and poly(vinyl butyral) for laminated safety glass
1937	Polystyrene
	Styrene-butadiene (Buna S) and acrylonitrile-butadiene (Buna N) rubbers (Germany)
1938	Nylon 66 fibers
1939	Melamine-formaldehyde resins
	Poly(vinylidene chloride)
1940	Butyl rubber (U.S.)
1941	Polyethylene production (England)
1942	Unsaturated polyesters for laminates
World War II	Debye, light scattering of polymer solutions; Flory, viscosity of polymer solutions; Harkins, theory of emulsion polymerization; Weissenberg, normal stresses in polymer flow
	Silicones, fluorocarbon resins, polyurethanes (Germany), styrene-butadiene rubber in United States, latex-based paints
1944	Carboxymethyl cellulose
	Filament winding

Table 1.3 A selected chronology of polymer science and technology (*Continued*)

1945	Cellulose propionate
	Dielectric heating of polymers
	Fiber-reinforced boats
1946	Screw-injection molding
	Automatic transfer molding
1947	Epoxy resins
1948	ABS polymers
1950	Polyester fibers
1948 to 1950	Acrylic fibers
1949	Blow-molded bottles
1950	Poly(vinyl chloride) pipe
1950s	Ziegler, coordination complex polymerization; Natta, tacticity in polymers; Swarc, living polymers; interfacial polycondensation; Hogan and Banks, crystalline polypropylene; fluidized bed coating; Staudinger receives Nobel Prize (1953)
1954	Polyurethane foams in United States
	Styrene-acrylonitrile copolymers
1955	Williams-Landel-Ferry equation for time-temperature superposition of mechanical properties
1956	Linear polyethylene, acetals [poly(oxymethylene)]
1957	Polypropylene, polycarbonates
1957	Single crystals of polyethylene characterized, Keller, Till
1958	Rotational molding
1959	Chlorinated polyether
	Synthetic *cis*-polyisoprene and *cis*-polybutadiene rubbers
1960	T. Smith, the failure envelope
1960	Ethylene-propylene rubber
	Spandex fibers
1962	Phenoxy resins, polyimide resins
1965	Poly(phenylene oxide), polysulfones, styrene-butadiene block copolymers
1960s	Cyanoacrylate adhesives
	Aromatic polyamides, polyimides
	Silane coupling agents
	Thermoplastic elastomers (block styrene-butadiene copolymers)
	NMR applied to polymer structure analysis
	Orthogonal rheometer, Maxwell
	GPC analysis for mol. wt. distribution, Moore
	Differential scanning calorimetry
	Polybenzimidazoles, Marvel
	Torsional braid analysis, Gilham
	Ethylene-vinyl acetate copolymers
	Ionomers
	Natta and Ziegler share Nobel Prize (1963)
	Polysulfone
	Parylene coatings
	Polybenzimidazoles
	HDPE fuel tanks

Table 1.3 (*Continued*)

1970s	Polybutene (isotactic)
	Poly(butyl terephthalate)
	Thermoplastic elastomers based on copolyesters
	Poly(phenylene sulfide)
	Polynorbornene (rubber)
	Polyarylates
	Polyphosphazenes
	Soft contact lenses
	Reaction injection molding
	Interpenetrating networks
	High-performance liquid chromatography
	Submicrometer lithography for integrated circuits
	Polyester beverage bottles
	Structural foams
	Kevlar aromatic polyamide
	Polyethersulfone
	Reaction injection molding for auto fascia
	Unipol gas-phase polymerization
	Linear low-density polyethylene
	Polyarylates
	High-density polyethylene grocery bags
	Flory receives Nobel Prize (1974)
1980s	Polysilanes
	Liquid crystal polymers
	High-modulus fibers
	Polyether ether ketone
	Conducting polymers
	Polyetherimide
	Poly(methylpentene)
	Pultrusion
	Replacement of fluorocarbon blowing agents
	Group transfer polymerization
	Differential viscometry
1990s	Metallocene catalysts
	Large-scale recycling of polyester and HDPE bottles
	Poly(ethylene-2,6-naphthalene dicarboxylate)
	Living cationic polymerization (Kennedy)

Since 1980 the U.S. chemical industry has undergone some restructuring. In general, there are fewer producers of each commodity polymer than there were a decade ago. On the other hand, almost every company would like to participate in the specialty polymer market, where the profits are likely to be higher.

Table 1.4 Worldwide plastics consumption, 1993

Region	Plastics consumption		Est'd. population, millions
	Millions of kg	kg/person	
World	101,400	19	5,423
North America	30,620	—	—
United States	28,440	111	257
Canada	2,180	80	27
Western Europe	30,700	—	—
Austria	965	122	8
Belgium	1,720	172	10
France	3,910	68	57
Germany	8,930	111	80
Italy	4,490	78	58
Netherlands	940	62	15
Spain	2,240	57	39
Sweden	1,065	124	9
United Kingdom	3,515	61	58
Other	2,925	—	—
Asia/Oceania	25,700	—	—
Australia	1,080	61	18
China	2,650	2	1,170
Japan	12,200	98	124
South Korea	3,570	81	44
Taiwan	3,250	156	21
Other	2,950	—	—
Eastern Europe	7,560	—	—
Former Czechoslovakia	950	61	16
Hungary	390	38	10
Poland	690	18	38
Romania	360	16	23
Former USSR	4,460	15	300
Other	710	—	—
Latin America	4,950	—	—
Argentina	430	13	33
Brazil	1,720	11	158
Mexico	1,240	13	92
Other	1,560	—	—
Africa/Mideast	1,870	—	—

*Source: L. Young, *Mod. Plast.*, 71(12):A-16 (1994).

Table 1.5 Growth comparison, plastics and gross domestic product

Year	Gross Domestic Product, value Constant $(1987) \times 10^{-9}$	Ratio	Plastics sales, volume kg $\times 10^{-9}$	Ratio
1970	2,874	1.00	8.9	1.00
1975	3,206	1.11	10.3	1.16
1980	3,776	1.31	13.9	1.57
1985	4,280	1.49	18.5	2.08
1990	4,897	1.70	28.0	3.15
1993	5,136	1.79	31.3	3.52

KEYWORDS FOR CHAPTER 1

Section

1. Polymer
   Monomer
   Dimer
   Oligomer
   Macromolecule
3. Thermoplastic
   Thermoset
   Producer Price Index (PPI)
   Consumer Price Index (CPI)
4. Cracking
   Severity
6. Gross Domestic Product (GDP)

PROBLEMS

1.1 What are some large-scale uses of plastics and rubber in the home construction industry that are obvious to the layman?

1.2 What are some obvious uses of rubber and plastics in the automobile (in addition to tires)?

1.3 What are some drawbacks of ordinary cellulose-based paper for books and newspapers compared with what might be expected from a material based on polyethylene? What disadvantages might the plastic have (in addition to the initial one of price)? What is the potential market for "synthetic" newsprint?

1.4 If every one of the 10 million or so automobiles produced in the United States in a single year suddenly were equipped with a thermoplastic polycarbonate windshield, what new selling price might be expected for the general-purpose material?

1.5 What factors other than volume produced will affect the selling price of a polymer?

1.6 When plastics are used for a molded article, a specified volume of material must be used to achieve a certain effect. However, prices for plastics usually are given on a unit weight basis. If acetal resin is to compete on a cost per volume basis with polypropylene in a given application, what must the price ratio be on a weight basis (see Table A3.1)?

1.7 The price of the polymer is only one factor in the selling price of a finished article. Some other items might include other ingredients, production labor, decorating, packaging, and distribution. Rank the following items in the order of increased dependence on polymer cost:

Polyethylene garbage can   Tinted cellulose acetate sunglasses
Nylon panty hose   Nylon pocket comb
Nylon guitar string   Automobile tire

What factors are important for each item?

1.8 Using data from any recent issue of *Chemical Marketing Reporter* (Schnell Publishing Co.), plot the lowest polymer price listed versus monomer price for ethylene, propylene, styrene, vinyl chloride, and bisphenol A (polycarbonate). Should the ratio of polymer to monomer price increase or decrease with monomer price? Why?

REFERENCES

Production figures: "Statistical Abstract of the United States," issued annually by the Bureau of the Census, Department of Commerce, through the U.S. Government Printing Office, Washington, D.C. Summaries usually appear in *Chemical and Engineering News* (first week in June) and *Modern Plastics* (January). Prices of most plastics are reported in *Plastics Technology* and of many polymers and additives in *Chemical Marketing Reporter.*
1. *Fiber Organon,* 65:113 (1994).
2. Winfield, A. G.: *Plastics Eng.,* 48(5):32 (1992).

General References

History
Craver, J. K., and R. W. Tess (eds.): "Applied Polymer Science," ACS, Washington, D.C., 1975.
Dubois, J. H.: "Plastics History—U.S.A.," Cahners, Boston, 1971.
Elliott, E.: "Polymers & People: An Informal History," CHOC, Philadelphia, 1986.
Flory, P. J.: "Principles of Polymer Chemistry," chap. 1, Cornell Press, Ithaca, N.Y., 1953.
Morawetz, H.: "Polymers: The Origins and Growth of a Science," Wiley, New York, 1985.
Morris, P. J. T.: "Polymer Pioneers: A Popular History of the Science and Technology of Large Molecules," CHOC, Philadelphia, 1986.
Mossman, S. T. I., and P. J. T. Morris (eds.): "The Development of Plastics," CRC Press, Boca Raton, Fla., 1994.

Seymour, R. B., and T. Cheng (eds.): "History of Polyolefins," (Reidel Holland), Kluwer Academic, Norwell, Mass., 1985.

Seymour, R. B., and R. D. Deanin (eds.): "History of Polymeric Composites," VNU Science, Utrecht (Netherlands), 1987.

Seymour, R. B., and G. S. Kirshenbaum (eds.): "High Performance Polymers: Their Origin and Development," Elsevier Applied Science, New York, 1986.

Seymour, R. B., and H. F. Mark (eds.): "Organic Coatings: Their Origin and Development," Elsevier, Amsterdam, 1989.

Tess, R. W., and G. W. Poehlein (eds.): "Applied Polymer Science," 2d ed., ACS, Washington, D.C., 1985.

Introductory Textbooks and Handbooks

Aggarwal, S. L., and S. Russo (eds.): "Comprehensive Polymer Science," Pergamon, New York, 1992.

Allcock, H. R., and F. W. Lampe: "Contemporary Polymer Chemistry," Prentice-Hall, Englewood Cliffs, N.J., 1981.

Allen, G., and J. C. Bevington (eds): "Comprehensive Polymer Science," 7 vols, Pergamon, New York, 1988.

Ash, M., and I. Ash (eds.): "Encyclopedia of Plastics, Polymers, and Resins," 4 vols., Chemical Publishing, New York, 1987.

Baijal, M. D. (ed.): "Plastics Polymer Science & Technology," Wiley, New York, 1982.

Batzer, H., and F. Lohse: "Introduction to Macromolecular Chemistry," Wiley, New York, 1979.

Bawn, C. E. H.: "Macromolecular Science," Butterworth, Woburn, Mass., 1975.

Benoit, H., and P. Rempp (eds.): "Macromolecules," Pergamon, New York, 1982.

Billmeyer, F. W., Jr.: "Textbook of Polymer Science," 3d ed., Wiley, New York, 1984.

Birley, A. W., R. H. Heath, and M. J. Scott: "Plastics Materials, Properties and Applications," 2d ed., Chapman & Hall, New York, 1988.

Bovey, F. A., and F. H. Winslow (eds.): "Macromolecules, An Introduction to Polymer Science," Academic Press, New York, 1979.

Boyd, R. H., and P. J. Phillips: "The Science of Polymer Molecules," Cambridge Univ. Press, New York, 1994.

Brandrup, J., and E. H. Immergut (eds.): "Polymer Handbook," 3d ed., Wiley, New York, 1989.

Braun, D., H. Cherdron, and W. Kern: "Practical Macromolecular Organic Chemistry," Gordon & Breach, New York, 1984.

Brydson, J. A.: "Plastics Materials," 4th ed., Butterworth, Woburn, Mass., 1982.

Carley, J. F. (ed.): "Whittington's Dictionary of Plastics," 3d ed., Technomic, Lancaster, Pa., 1993.

Chanda, M., and S. K. Roy: "Plastics Technology Handbook," Dekker, New York, 1987.

Cheremisinoff, N. P. (ed.): "Handbook of Polymer Science and Technology," multivolume, Dekker, New York, 1994.

Coleman, M. M., and P. C. Painter: "Fundamentals of Polymer Science," Technomic, Lancaster, Pa., 1994.

Cowie, J. M. G.: "Polymers," 2d ed., Chapman & Hall, New York, 1991.

Driver, W. E.: "Plastics Chemistry and Technology," Van Nostrand-Reinhold, New York. 1979.

DuBois, J. H., and F. W. John: "Plastics," 5th ed., Van Nostrand-Reinhold, New York, 1974.

Elias, H.-G.: "Macromolecules, Structure and Properties," 2d ed., Plenum, New York, 1984.

Elias, H.-G.: "Macromolecules, Synthesis, Materials, and Technology," 2d ed., Plenum, New York, 1984.

Gowariker, V., N. V. Viswanathan, and J. Sreedhar: "Polymer Science," Wiley, New York, 1986.

Harper, C. A. (ed.): "Handbook of Plastics, Elastomers, and Composites," 2d ed., McGraw-Hill, New York, 1992.

Heath, R. J., and A. W. Birley: "Dictionary of Plastics Technology," Chapman & Hall, New York, 1993.

Hiemenz, P. C.: "Polymer Chemistry," Dekker, New York, 1984.

Jenkins, A. D.: "Polymer Science," vols. 1 and 2, Elsevier Applied Science, New York, 1972.

Jenkins, A. D., and J. F. Kennedy (eds.): "Macromolecular Chemistry," vol. 3, Royal Soc. Chem., London, 1985.

Kaufman, H. S., and J. J. Falcetta (eds.): "Introduction to Polymer Science and Technology," Wiley, New York, 1977.

Kleintjens, L. A., and P. J. Lemstra (eds.): "Integration of Fundamental Polymer Science and Technology," Elsevier, New York, 1986.

Kroschwitz, J. I. (ed.): "Encyclopedia of Polymer Science and Engineering," 2d ed., 17 vols., Wiley, New York, 1985–1990.

Kroschwitz, J. I. (ed.): "Polymers, an Encyclopedic Sourcebook of Engineering Properties," Wiley, New York, 1987.

Mark, J. E., A. Eisenberg, W. W. Graessley, L. Mandelkern, and J. L. Koenig: "Physical Properties of Polymers," ACS, Washington, D.C., 1984.

Moncrieff, R. W. (ed.): "Man-Made Fibres," Wiley-Halsted, New York, 1975.

Nicholson, J. W.: "The Chemistry of Polymers," CRC Press, Boca Raton, Fla., 1991.

Ravve, A: "Principles of Polymer Chemistry," Plenum, New York, 1995.

Rosen, S. L.: "Fundamental Principles of Polymeric Materials," 2d ed., Wiley, New York, 1993.

Rubin, I. I. (ed.): "Handbook of Plastic Materials and Technology," Wiley, New York, 1990.

Rudin, A.: "The Elements of Polymer Science and Engineering," Academic Press, New York, 1982.

Saunders, K. J.: "Organic Polymer Chemistry," 2d ed., Chapman & Hall, New York, 1988.

Seymour, R. B., and C. E. Carraher, Jr.: "Polymer Chemistry," 3d ed., Dekker, New York, 1992.

Sperling, L. H.: "Introduction to Physical Polymer Science," 2d ed., Wiley, New York, 1992.

Stevens, M. P.: "Polymer Chemistry," 2d ed., Oxford Univ. Press, New York, 1990.

Tess, R. W., and G. W. Poehlein (eds.): "Applied Polymer Science," 2d ed., ACS, Washington, D.C., 1985.

Ulrich, H.: "Raw Materials for Industrial Polymers," Hanser (Oxford Univ. Press), New York, 1988.

Ulrich, H.: "Introduction to Industrial Polymers," 2d ed., Hanser-Gardner, Cincinnati, Ohio, 1992.

Young, R. J., and P. A. Lovell: "Introduction to Polymers," 2d ed., Chapman & Hall, New York, 1991.

BASIC STRUCTURES OF POLYMERS

2.1 CLASSIFICATION SCHEMES

The study of any subject as vast and complex as polymers is simplified by gathering together the many thousands of examples that are known into a few categories about which generalized statements can be made. By speaking about polymers as a special topic, we already are categorizing materials as low-molecular-weight or as high-molecular-weight, i.e., polymers. However, some more useful classifications are by:

1. *Structure.* We can ask whether the polymer exists as a mass of separable, individual molecules or as a macroscopic network. Also, is it branched or linear? Is it a succession of randomly oriented units or does it have some preferred spatial orientation? We shall pursue this approach later in this chapter.
2. *Physical state.* Polymer molecules may be partially crystalline or completely disordered. The disordered state may be glassy and brittle, or it may be molten with the viscosity characteristic of a liquid or the elasticity we associate with a rubbery solid. We shall see that the distinctions depend on temperature, molecular weight, and chemical structure.
3. *Reaction to environment.* Within an industry, or within some other grouping, there may be an important difference in processing or end use behavior. In any application one can differentiate between low-cost, general-purpose materials and high-cost, specialty materials. For engineers this is a very logical categorization as a first step to specifying a material for a given use.

In the plastics industry an important factor in economic usage and in end use stability of a product is the behavior of the material at high temperatures. The term *thermoplastic* is applied to materials that soften and flow upon application of pressure and heat. Thus, most thermoplastic materials can be remolded many times, although chemical degradation may eventually limit the number of molding cycles. The obvious advantage is that a piece that is rejected or broken after molding can be ground up and remolded. The disadvantage is that there is a limiting temperature for the material in use above which it cannot be used as a structural element. The *deflection* temperature, formerly termed the *heat distortion* temperature, is measured by a standard loading at a standard rate of temperature rise.

The term *thermoset* is applied to materials that, once heated, react irreversibly so that subsequent applications of heat and pressure do not cause them to soften and flow. In this case, a rejected or scrapped piece cannot be ground up and remolded. On the other hand, the limiting upper temperature in use often is considerably higher than the molding temperature. Once again, chemical stability becomes a limiting factor.

4. *Chemical.* The elemental composition of a polymer, the chemical groups present (ether, ester, hydroxyl, etc.), or the manner of synthesis (chain propagation, transesterification, ring opening, etc.) may be used as a means of classifying polymers. A person or company attempting to exploit a unique raw material or process may profitably use such an approach.

5. *End use.* As has been mentioned before, the various polymer-consuming industries tend to think of a new material as an adhesive, a fiber, a rubber, a plastic, or a coating, even though the material may be adaptable to all these applications.

2.2 BONDING

Polymers generally are held together as large molecules by covalent bonds, while the separate molecules, or segments of the same molecule, are attracted to each other by *intermolecular forces,* also termed "secondary" or "van der Waals" forces. Although ionic and other types of bonds can occur in polymeric systems, they are not considered here. Covalent bonds are characterized by high energies (35 to 150 kcal/mol), short interatomic distances (0.11 to 0.16 nm), and relatively constant angles between successive bonds. Some important examples are given in Table 2.1. An apparent exception is the Si—O—Si bond, for which values from 104 to 180° have been reported. The flexibility of this bond, coupled with the ability of methyl groups attached to silicon to rotate about the Si—C bond, is responsible for poly(dimethylsiloxane) having the lowest glass-forming temperature of any rubber, −110°C [1]. Secondary forces are harder to characterize because they operate between molecules or segments of the same molecule rather than being localized to a pair of atoms. These forces increase in the presence of polar groups and decrease with increasing distance between molecules. One can see qualitatively the effect of

Table 2.1 Covalent bonds

Dimensions and energies[*]

Bond	Typical bond length, Å [2]	Typical average bond energy, kcal/mol [3]	Some typical bond angles [2]
C—C	1.54	83	
C=C	1.34	147	
C≡C	1.20	194	
C—H	1.09	99	
C—O	1.43	84	
C=O	1.23	171	
C—N	1.47	70	
C=N	1.27	147	
C≡N	1.16	213	
C—Si	1.87	69	
Si—O	1.64	88	
C—S	1.81	62	
C=S	1.71	114	
C—Cl	1.77	79	
S—S	2.04	51	
N—H	1.01	93	
S—H	1.35	81	
O—H	0.96	111	
O—O	1.48	33	
Si—N	1.74	–	

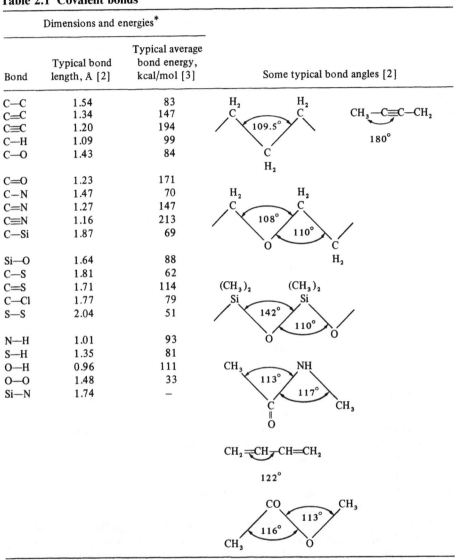

[*]1 Å = 0.1 nm; 1 kcal = 4.185 kJ.

polarity on intermolecular attraction in the boiling points of the series methane (CH_4, bp −161°C), methyl chloride (CH_3Cl, bp −24°C), and methyl alcohol (CH_3OH, bp 65°C). Because the energy of interaction varies inversely with the sixth power of the distance, small differences in structure can have large effects. For example, compare the boiling points of neopentane and *n*-pentane:

Table 2.2 Hydrogen bonds [4]

Bond	Typical bond length, A	Typical bond energy, kcal/mol
O—H- -O	2.7	3 to 6
O—H- -N	2.8	
N—H- -O	2.9	4
N—H- -N	3.1	3 to 5
O—H- -Cl	3.1	
N—H- -F	2.8	
N—H- -Cl	3.2	
F—H- -F	2.4	7

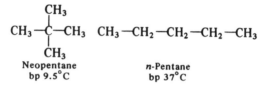

Neopentane
bp 9.5°C

n-Pentane
bp 37°C

Intermolecular spacings can be estimated from crystal lattice dimensions in polymers and from the average density of amorphous polymers. Distances of 0.2 to 1.0 nm and energies of 2 to 10 kcal/mol predominate in small molecules. In a large molecule the energy is better characterized by the cohesive energy density, which measures the energy per unit volume rather than per molecule (see Sec. 2.5). The hydrogen bond deserves special mention because its effect is localized. Tables 2.1 and 2.2 summarize some typical values of bond energies and dimensions. Hydrogen bonding has a great influence on the properties of cellulose (cotton) and polyamides (nylon and proteins), among others. The multiplicity of strong hydrogen bonds in dry cotton gives cotton many of the properties associated with a covalent-bonded *network* leading to insolubility and infusibility despite the fact that the truly covalent bonds give only a linear structure.

In general, covalent bonds govern the thermal and photochemical stability of polymers. Bond strength can be used as a clue to degradation mechanisms. For example, sulfur-vulcanized rubber is more likely to degrade at the comparatively weak —S—S— bonds than at the strong —C—C— bonds, both of which occur in the structure.

On the other hand, secondary forces determine most of the physical properties we associate with specific compounds. Melting, dissolving, vaporizing, adsorption, diffusion, deformation, and flow involve the making and breaking of intermolecular "bonds" so that molecules can move past one another or away from each other. In polymers the forces play the same role in the movements of individual segments of long-chain molecules.

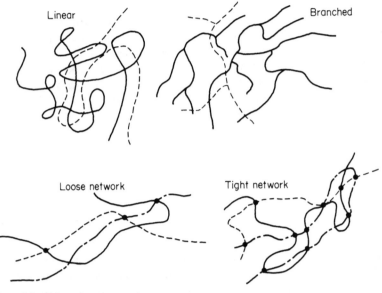

Figure 2.1   Polymer arrangements

The *arrangement* of the covalent structure in space leads to a convenient method of classification that helps explain polymer properties. Basically there are two such arrangements. The polymer may be present as *single* molecules, albeit of large individual size (i.e., mol wt of 10^7), or as an *infinite network*. The distinction is important because only the separate molecules can exhibit plastic flow and solubility. Accustomed as we are to thinking of the molecular scale as submicroscopic, it comes as something of a jolt to realize that the major portion of polymer in a tire or a bowling ball is really only one molecule. This is because all the separate molecules in the tire were connected to one another by sulfur cross-links during "vulcanization." (Incidentally, the molecular weight of a bowling ball polymer on this basis is about 10^{28}!)

One can calculate that a single molecule with molecular weight 10^6 and a density of 1 g/cm^3 should form a sphere 14.7 nm in diameter. This is large enough to be clearly visible by electron microscopy, and indeed such single molecules have been distinguished [5].

2.3 SINGLE MOLECULES

The single molecules may be *linear* or *branched* (Fig. 2.1). An important consequence is that branching interferes with the ordering of molecules, so that crystallinity decreases. Also, melt flow of branched molecules is more complicated by elastic effects.

If the polymer chain contains carbon atoms with two different substituents, the carbon is not symmetrical since the two parts of the chain to which it is connected are also different. That is, if we take the mirror image (the *enantiomer*) of such a substituted carbon, it cannot be superposed on the original without breaking covalent bonds. We refer to this as a *chiral* group and the carbon as the *chiral center*. In polymers such as polyethylene and polyisobutylene (with two methyl groups on every second chain carbon), there is a plane of symmetry, so the groups are *achiral*. *Optical isomerism* is familiar in the organic chemistry of sugars, amino acids, and many other biologically important molecules. As an example, consider the case of polypropylene, with the repeating unit

Every other carbon is a chiral center. Three structures can result. These are best visualized by looking at the main polymer chain in an extended planar zigzag conformation. In general, changes in structure caused by rotations about single bonds are termed *conformations,* and isomers that cannot be interchanged without bond breaking are termed *configurations.*

With polypropylene in the planar zigzag conformation, each pendant methyl group may be on one side of the chain, that is, all *d* or all *l*, using the terminology of stereochemistry. Using the terminology coined by Natta [6], this is an *isotactic structure* (Fig. 2.2). The regularity of such an arrangement facilitates the molecular ordering necessary in crystalline materials. A regular alternation of pendant groups is called a *syndiotactic structure.* Random placement of the methyl group gives an *atactic structure.* Until the advent of coordination complex polymerization, it was difficult to produce synthetically (with an optically inactive catalyst) any structure except the atactic. However, these structures and more complex variants have since been produced with a variety of catalyst systems. In the amorphous, random-coil state, *tacticity* may be a minor factor in determining properties, but in crystal lattice formation it often assumes a predominant role.

Poly(propylene oxide) is an example of a polymer that has a three-atom repeat unit in the chain. It has been made as all *d* or all *l* isotactic polymers, which are optically active, as the *d-l* isotactic mixture, and as the atactic polymer. In the planar zigzag presentation the isotactic form has the pendant methyl group aligned in *two* rows (Fig. 2.2*a*).

A second type of stereoisomerism is of the cis-trans variety familiar in small ethenic molecules. Because rotation is not free about a double bond, the substituents on either side can assume two attitudes (Fig. 2.2*b*). When the chain parts are on opposite sides, we have a trans conformation, and when they are on the same side, cis. In this case also, the catalyst systems that permit production of all-cis or all-trans polymers were developed in the 1950s [6].

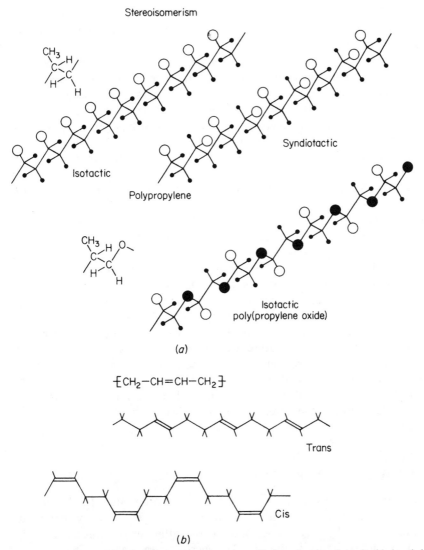

Stereoisomerism

Isotactic

Syndiotactic

Polypropylene

Isotactic
poly(propylene oxide)

(a)

$\{CH_2-CH=CH-CH_2\}$

Trans

Cis

(b)

Figure 2.2 (a) Stereoisomers from asymmetric carbon. (b) Stereoisomers from double bonds in chains of 1,4-polybutadiene

2.4 NETWORK MOLECULES

A loose polymer network can result from cross-linking a linear or branched polymer. Natural rubber consists mostly of a linear polymer that can be cross-linked to a loose network by reaction with 1 to 3% sulfur. The same polymer reacted with

40 to 50% sulfur is "hard rubber," a "tight" network polymer used for pocket combs and bowling balls (Fig. 2.1). Networks also may be formed directly, without the intermediate linear stage.

It was pointed out earlier (Sec. 2.1) that *thermoplastic* polymers soften without chemical change when heated and harden when cooled, whereas *thermosetting* polymers, once formed, do not soften upon heating below their decomposition temperatures. We can identify the thermoplastic polymers with the linear and branched structures. The thermosets generally form network polymers upon being heated the first time. As with all generalizations in the field of polymers, there are numerous exceptions. Cellulose is essentially a linear polymer (cotton, rayon). However, because of the strong hydrogen-bonded structure, softening does not occur below the decomposition temperature. Therefore it cannot be molded without breaking bonds.

A more complex arrangement results when two polymers form networks that overlap in space. One way of making such an *interpenetrating network* (IPN) is to swell a cross-linked polymer with a monomer. The IPN comes about when the monomer is polymerized into a second network, even though the two networks may have no covalent bonds in common [7–9]. Despite the fact that two polymers are not compatible when mixed as linear molecules, the IPN may exhibit no macroscopic phase separation and thus can be much stronger than a mechanical mixture.

Another kind of mixed network comes about when two linear polymers that are incompatible at room temperature are simultaneously cross-linked at a high temperature at which they are compatible. Phase separation on cooling may be inhibited by the covalent bonds, which limit movement of polymer chains. The term *semi-IPN* can refer to a linear polymer trapped in a network of another polymer. In the older literature these were sometimes called by the colorful term *snake-cage* systems.

2.5 COHESIVE ENERGY DENSITY

It has been mentioned before that many of the unique properties of polymers can be attributed to the fact that polymer segments are held together by covalent bonds in one direction but by secondary bonds in the other two. A measure of the strength of secondary bonds is given by the *cohesive energy density* (CED):

$$CED = \frac{\Delta E_v}{V_l} \tag{2.1}$$

where ΔE_v is the molar energy of vaporization and V_l is the molar volume of the liquid. For small molecules, ΔE_v can be measured by conventional means. The energy necessary to separate molecules from one another from the close spacing typical of the liquid state to the distant spacing of the vapor is evaluated directly by calorimetric methods or indirectly by measuring the vapor pressure p as a function of absolute temperature T. The Clapeyron equation can be used to extract an enthalpy of vaporization ΔH_v:

$$\frac{dp}{dT} = \frac{\Delta H_v}{T(V_g - V_l)} \tag{2.2}$$

where V_g and V_l are the molar volume of the compound in the vapor and liquid at temperature T. The internal energy of vaporization ΔE_v is related by

$$\Delta E_v = \Delta H_v - p(V_g - V_l) \tag{2.3}$$

Solubility Parameters (One-Dimensional)

Another useful parameter is the square root of the CED, the *solubility parameter* δ. In quoting values of these terms, the dimensions must be given. In the older literature, a widely used dimension for the solubility parameter is $(cal/cm^3)^{1/2}$, called the hildebrand. Other units used are $(J/cm^3)^{1/2}$ and $(MPa)^{1/2}$, which are, of course, identical. One hildebrand is the equivalent of 2.046 $(MPa)^{1/2}$. Solubility parameters for some common liquids are listed in Table 2.3.

Example 2.1

What is the solubility parameter for *n*-hexane at 25°C? Since V_g is much larger than V_l, the second term in Eq. (2.3) can be replaced by RT (assuming an ideal gas).

Data: Enthalpy of vaporization $\Delta H_v = 7540$ cal/mol; molar volume $= 86.2/0.654 = 131.8$ cm³/mol.

Solution:

$$CED = \frac{7540 - 1.987 \times 298}{131.8} = 52.7 \text{ cal/cm}^3$$

Solubility parameter $= \delta = 7.26$ hildebrands $= 14.9$ $(MPa)^{1/2}$

Liquids with like solubility parameters are apt to dissolve the same solutes and to be mutually compatible. This leads to an indirect method of measuring δ for a polymer. The dissolving of a polymer in a low-molecular-weight liquid causes the random coil to expand and occupy a greater volume than it would in the dry, amorphous state. If the polymer is composed of single molecules, viscous flow can occur, and the viscosity will be increased as the polymer expands. It is expected that when the polymer and solvent have the same δ, the maximum expansion will occur and therefore the highest viscosity (for a given concentration) will be obtained. By measuring the viscosity of solutions of the same polymer at the same concentration (usually dilute) in a variety of solvents, one should be able to deduce a consistent value of δ_2 for the polymer. If the polymer comprises a cross-linked network, solution cannot occur, but individual parts of the polymer chains, i.e., polymer segments, can solvate to give a swollen gel. Once again, it is expected that the maximum swelling will take place when δ_2 matches δ of the solvent.

Although the foregoing statements are borne out qualitatively by experiment, specific interactions and differences in molar volume of solvent make the estimation of δ_2 more complicated. An example is the swelling of a cross-linked butyl rubber [11] in 10 solvents (Fig. 2.3). In this figure, v_2 is the volume fraction of rubber in the swollen sample. It is obvious that solvents with δ between 8.0 and 9.5 swell butyl rubber more effectively than do solvents with δ below 8.0 or above 9.5.

The free-energy change on mixing solvent and polymer, ΔG_m, can be written as

$$\Delta G_m = RT[N_1 \ln(1 - v_2) + N_2 \ln v_2 + \chi N_1 v_2] \tag{2.4}$$

where N_1 and N_2 are the moles of solvent and polymer, respectively, v_2 is the volume fraction of polymer in the mixture, R is the gas constant (Boltzmann's constant per mole), and T is the absolute temperature. The *polymer-solvent interaction parameter* χ characterizes the interaction energy per mole of solvent divided by RT for a specific solvent-polymer pair. In a swollen gel there is also a change in free energy due to the elastic deformation of the polymer network.

At equilibrium the partial molar free energy of the solvent in the gel equals that of pure solvent. Using this fact and noting that the elastic energy term depends on N, the number of polymer chains per unit volume, and on V_1, the molar volume of the solvent, Flory [12] derives an equation that equates the mixing term (on the left) with the elastic term (on the right):

$$\ln(1 - v_2) + v_2 + \chi v_2^2 = -NV_1 \left(v_2^{1/3} - \frac{v_2}{2} \right) \tag{2.5}$$

While this equation is not a fundamental definition of χ, it has been used to derive experimental values of χ from swelling data. Orwoll [13] has reviewed the theoretical basis for χ and the experimental procedures for its measurement. Osmotic pressure, vapor sorption, gas-liquid chromatography, and several other methods can be used. Although it may not be the best absolute measure of χ, equilibrium swelling is a very practical method and relates directly to an important application, namely, the selection of materials for use in the presence of solvents.

The calculation of v_2 from known values of the other variables is simplified by recourse to a nomograph (Fig. 2.4). It is customary to calibrate the value of χ for a given solvent-polymer pair by using a sample that has had its cross-link density measured by an absolute method such as stress-strain behavior together with the rubber elasticity equation (Sec. 8.2). Some values of χ are listed in Table 2.4 for polymers in "good" solvents.

To connect Eq. (2.4) with δ, several semiempirical approaches have been tried. One approach is [11]:

$$\chi = \beta_1 + \frac{V_1}{RT} (\delta_1 - \delta_2)^2 \tag{2.6}$$

Table 2.3 Characteristic parameters for some solvents at 25°C [10]*

Liquid	Formula weight, g/mol	Density, g/cm³	Solubility parameter, $(MPa)^{1/2}$				H-bonding index γ_c
			δ_d	δ_p	δ_h	δ_{total}	
Acetic acid	60.1	1.044	13.9	12.2	18.9	26.5	—
Acetone	58.1	0.785	13.0	9.8	11.0	19.7	9.7
Acetonitrile	41.1	0.776	10.3	11.1	19.6	24.8	13.0
Acrylonitrile	53.1	0.801	10.6	12.5	14.0	21.6	12.7
n-Amyl alcohol	88.2	0.811	14.8	9.1	14.7	22.7	—
Benzene	78.1	0.874	16.1	8.6	4.1	18.7	0.0
Bromobenzene	157.0	1.486	18.4	8.2	0.0	20.1	—
n-Butanol	74.1	0.806	15.0	10.0	15.4	28.7	18.7
Carbon disulfide	76.1	1.256	10.9	16.6	4.3	20.3	0.0
Chlorobenzene	112.6	1.098	17.4	9.4	0.0	18.7	1.5
Chloroform	119.4	1.477	11.0	13.7	6.3	18.7	1.5
Cyclohexane	84.2	0.774	16.5	3.1	0.0	16.8	0.0
Cyclohexanone	98.2	0.942	15.6	9.4	11.0	21.3	8.4
n-Decane	142.3	0.725	15.8	0.0	0.0	15.8	0.0
Diacetone alcohol	116.2	0.934	10.7	11.4	12.6	20.0	13.0
Dibutyl phthalate	278.4	1.042	15.9	9.5	8.1	20.2	—
Diethyl ether	74.1	0.714	14.5	2.9	5.1	15.8	13.0
N,N-Diemthylformamide	73.1	0.944	17.4	13.7	11.3	24.8	11.7
Dimethyl sulfoxide	78.1	1.096	18.4	16.4	10.2	26.7	7.7
1,4-Dioxane	88.1	1.028	16.3	10.1	7.0	20.7	9.7
Ethanol	46.1	0.785	12.6	11.2	20.0	26.1	18.7
Ethyl acetate	88.1	0.894	13.4	8.6	8.9	18.2	8.4
Ethylbenzene	106.2	0.862	16.5	7.4	0.0	18.1	1.5
Ethylene dichloride (1,2-dichloroethane)	99.0	1.246	14.2	11.2	9.0	20.2	1.5
Ethylene glycol (EG)	62.1	1.110	10.1	15.1	29.8	34.9	20.6
EG, monobutyl ether (or 2-butoxyethanol)	118.2	0.896	13.3	7.9	13.0	20.2	13.0

Table 2.3 Characteristic parameters for some solvents at 25°C [10]* (Continued)

Liquid	Formula weight, g/mol	Density, g/cm³	Solubility parameter, (MPa)$^{1/2}$				H-bonding index γ_c
			δ_d	δ_p	δ_h	δ_{total}	
EG, monoethyl ether (or 2-ethoxyethanol)	90.1	0.925	13.0	9.1	15.2	21.9	13.0
EG, monoethyl ether, acetate ester	132.2	0.968	14.4	9.0	8.9	19.1	9.4
Glycerol	92.1	1.258	9.3	15.4	31.4	36.2	—
n-Heptane	100.2	0.679	15.3	0.0	0.0	15.3	0.0
n-Hexane	86.2	0.654	14.9	0.0	0.0	14.9	0.0
1-Hexene	84.2	0.668	14.4	3.9	0.0	15.0	—
Isophorone	138.2	0.917	16.4	9.4	3.2	19.2	8.6
Isopropyl acetate	102.1	0.866	14.3	8.4	5.7	17.6	8.6
Isopropyl alcohol	60.1	0.781	14.0	9.8	16.0	23.4	18.7
Methanol	32.0	0.786	11.6	13.0	24.0	29.7	18.7
Methylene chloride (dichloromethane)	84.9	1.316	13.4	11.7	9.6	20.2	1.5
Methyl ethyl ketone (2-butanone)	72.1	0.800	14.1	9.3	9.5	19.3	7.7

Methyl isobutyl ketone (3,3'-dimethyl-2-butanone)	100.2	0.796	14.4	8.1	5.9	17.6	7.7
Methyl methacrylate	90.1	0.930	13.5	10.1	8.5	18.9	—
N-Methyl pyrrolidone	99.1	1.020	16.5	10.4	13.5	23.7	—
Nitrobenzene	123.1	1.190	17.6	14.0	0.0	22.5	2.8
Nitroethane	75.1	1.045	19.0	13.0	0.0	23.0	2.5
1-Nitropropane	89.1	0.996	18.1	11.2	0.0	21.3	2.5
2-Nitropropane	89.1	0.985	16.5	10.4	6.6	20.6	2.5
n-Pentane	72.2	0.621	14.4	0.0	0.0	14.4	0.0
Perchloroethylene (tetrachloroethene)	165.9	1.611	11.4	15.2	0.0	19.0	—
Pyridine	79.1	0.978	17.6	10.1	7.7	21.7	18.1
Styrene	104.2	0.901	16.8	9.1	0.0	19.1	1.5
Tetrahydrofuran	72.1	0.882	13.3	11.0	6.7	18.5	—
Toluene	92.1	0.862	16.4	8.0	1.6	18.3	4.5
1,1,2-Trichloroethylene	131.4	1.455	11.7	14.0	4.4	18.7	—
Vinyl acetate	86.1	0.926	12.9	10.0	8.7	18.5	—
Water	18.02	0.997	12.2	22.8	40.4	48.0	39.0
p-Xylene	106.2	0.856	16.5	7.0	2.0	18.1	4.5

*Solubility parameters: Barton [10], pp. 94ff. H-bonding Index: Barton [10], pp. 186ff.

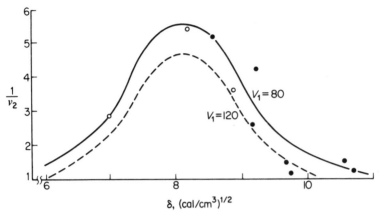

Figure 2.3 Swelling data for butyl rubber [11] in solvents with $100 < V_1 < 120$ (○) and with $80 < V_1 < 100$ (●). Curves for Eqs. (2.5) and (2.6) with $\beta_1 = 0.45$, $N = 1.0 \times 10^{-4}$, and $\delta_2 = 8.1$

where δ_1 and δ_2 relate to the solvent and polymer, respectively, R is the gas constant, and β_1 is a lattice parameter, usually 0.35 ± 0.1. Combination of Eqs. (2.5) and (2.6) gives the curves drawn in Fig. 2.3. Representative values of δ_2 (Table 2.5) obtained from swelling data show an increase with the polarity of the polymer. Another use of solubility parameters is in choosing compatible polymers for use in blends in surface coatings and adhesives. Maximum compatibility is expected to occur between polymers with the same δ_2.

Example 2.2

A cross-linked sample of polymer P swells to 4.55 times its original volume when immersed in solvent S at 27.0°C. Addition of a small amount of ethanol causes the swelling to decrease. If the polymer-solvent interaction parameter, χ, for S and P is 0.500, what is the solubility parameter of P? The lattice parameter, β, is 0.400.

Data for solvent S: FW = 216 g/mol, density = 1.00 g/cm³, and internal energy of vaporization = 63.2 kJ/mol.

Solution:

$$\delta_S = (\text{CED})^{0.5} = \left(\frac{\Delta E_v}{V_1}\right)^{0.5} = \left(\frac{63,200}{216}\right)^{0.5} = 17.1 \ (\text{MPa})^{0.5}$$

Using Eq. (2.6):

$$0.50 = 0.40 + \left(\frac{216}{8.31} \times 300\right)(17.1 - \delta_p)^2$$
$$\delta_P = 17.1 \pm 1.1$$

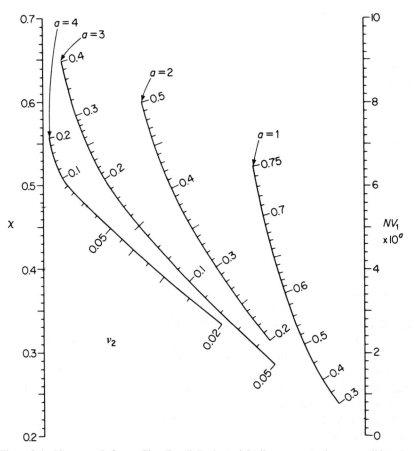

Figure 2.4 Nomograph for swelling Eq. (2.5). A straight line connects the compatible values of the three variables on their corresponding scales. Note that the exponent a varies with the scale selected for v_2. For example, when $\chi = 0.40$ and $v_2 = 0.3$, $NV_1 = 4.0 \times 10^{-2}$

Since δ for ethanol > 17, δ_P must be less than δ_S. Thus,

$$\delta_P = 16.0 \; (MPa)^{0.5}$$

Solubility Parameters (Two-Dimensional)

Because the variation of V_1 is not enough to reduce the scatter in plots like Figure 2.3, other molecular parameters have been sought to categorize solvent behavior further. One such approach is to add a second factor, the hydrogen-bonding index, γ_c. The change in hydrogen bonding that occurs when a solvent is mixed with deuterated methanol in specific proportions compared with the bonding when ben-

Table 2.4 Polymer-solvent interaction parameters at 25°C [11]

Polymer	Solvent	Interaction parameter χ
cis-Polyisoprene	Toluene ($V_1 = 106$)*	0.391
	Benzene ($V_1 = 89.0$)	0.437
Polyisobutylene	Toluene	0.557
	Cyclohexane ($V_1 = 108$)	0.436
Butadiene-styrene	Benzene	0.442
(71.5 : 28.5)	Cyclohexane	0.489
Butadiene-acrylonitrile		
82 : 18	Benzene	0.390
70 : 30	Benzene	0.486
61 : 39	Benzene	0.564

*V_1 is in cubic centimeters per mole. See references 13, 14, and 15 for more examples.

zene is the solvent can be measured by the shift of a peak in infrared spectrum away from a value of 2681 cm^{-1}. By using a scale for the hydrogen-bonding parameter that ranges from 0 for benzene and other hydrocarbons to 18.7 for ethanol and up to 39.0 for water, swelling and compatibility of polymers and solvents can be better predicted than by using δ alone. A two-dimensional plot for the swelling of a sample of cross-linked fluorocarbon rubber (Fig. 2.5) has contour lines at constant swell (swell = $1/v_2$). Values of the relative hydrogen-bonding index for various solvents appear in Table 2.3. The parameter locations for some major groups of solvents are indicated in Fig. 2.6. Such a map has great value in selecting solvents for a coating system or in judging probable resistance to swelling for a particular application.

Table 2.5 Three-dimensional solubility parameters for selected polymers (MPa)$^{1/2}$ [16]

Polymer	δ_d	δ_p	δ_h	$\delta_{total} = \delta_2$
Cellulose acetate	18.6	12.7	11.0	25.1
Polyisobutylene	14.5	2.5	4.7	15.5
Polyisoprene (cis)	16.6	1.4	−0.8	16.7
Nylon 66	18.6	5.1	12.3	22.9
Poly(acrylonitrile)	18.2	16.2	6.7	25.3
Poly(ethylene terephthalate)	19.4	3.5	8.6	21.5
Poly(methyl methacrylate)	18.6	10.5	7.5	22.7
Polystyrene	21.3	5.7	4.3	22.5
Polysulfone (bisphenol A)	19.0	0.0	7.0	20.3
Poly(vinyl acetate)	20.9	11.3	9.7	25.7
Poly(vinyl chloride)	18.8	10.0	3.1	21.5

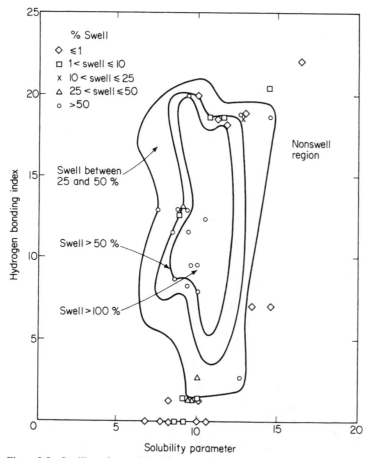

Figure 2.5 Swelling of cross-linked fluorocarbon rubber [17]. Solubility parameter is in hildebrands

Solubility Parameters (Three-Dimensional)

Another approach to generalizing solvent behavior is to break down the total solubility parameter into three components representing nonpolar, polar, and hydrogen-bonding contributions to the cohesive energy density:

$$\delta_t^2 \quad = \delta_d^2 \quad + \delta_p^2 \quad + \delta_h^2 \tag{2.7}$$

Total   Nonpolar   Polar   Hydrogen bonding

Values for the three components have been estimated by various means, most of which yield similar but not identical numbers. Both experimental and calculational methods have been employed. When the total CED is estimated from the experimental enthalpy of vaporization, the polar and hydrogen-bonding parameters may

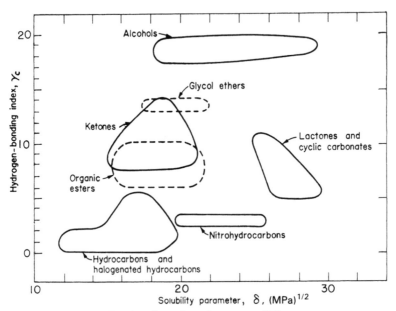

Figure 2.6   Parameter locations for major solvent groups [18]

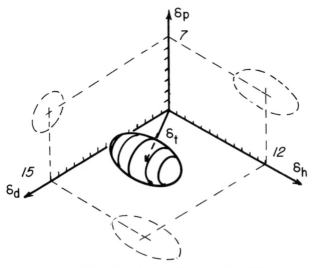

Figure 2.7   Solubility volume in three-dimensional solubility parameter space for a hypothetical situation where solubility is a maximum at the coordinates indicated and the polymer is insoluble outside the limits of the skin of the football-shaped volume. The total solubility parameter is represented by the length of the arrow from the origin to the center of the volume

be calculated using bond contribution methods. Values of these *Hansen parameters* are included in Table 2.3. In order to represent the solvent interaction with a polymer, a three-dimensional map is needed. An idealized representation is shown in Fig. 2.7 for a hypothetical situation in which the solubility volume is centered at values of $\delta_d : \delta_p : \delta_h = 15 : 7 : 12$. The total solubility parameter δ_t is about 20.4. Also shown are the projections of the surface of the solubility volume on each of the three planes. In practice, the volume containing the good solvents may resemble an irregular sausage rather than a spheroid. However, the correlation often is more satisfactory, as one might expect with the additional parameters available. On the other hand, the behavior of mixtures of solvents is not always predictable, and the molecular volume still can affect the correlation.

When a dynamic property such as dissolution rate is measured [19–21], the results may not necessarily be predicted from equilibrium behavior. For example, when water, a nonsolvent, is added to 2-butanone, poly(methyl methacrylate) dissolves more rapidly despite the fact that the solvent mixture is thermodynamically poorer than 2-butanone by itself (Fig. 2.8).

2.6 THERMODYNAMICS OF BINARY SOLUTIONS

The thermodynamic quantities of interest upon mixing two components are the changes in entropy, ΔS_m, enthalpy, ΔH_m, and free energy, ΔG_m, which are related at temperature T by

$$\Delta G_m = \Delta H_m - T \Delta S_m \tag{2.8}$$

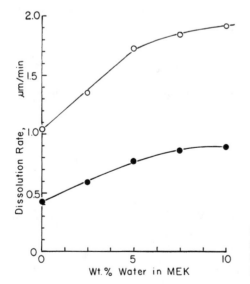

Figure 2.8 Dissolution rate for poly(methyl methacrylate) in methyl ethyl ketone (2-butanone) with various amounts of water at 27.5°C. $M_n = $ (●) 320×10^3; (○) 36×10^3 [20].

Entropy

Boltzmann's relationship states that the change in entropy of a system is proportional to the logarithm of the probability, that is, the number of ways the system can be arranged. When the two components are comparable in size, they can fit interchangeably in space, visualized as a lattice arrangement. The relationship, usually derived in elementary physical chemistry texts, reduces in the ideal case to

$$\Delta S_m = -k(N_1' \ln N_1 + N_2' \ln N_2) \tag{2.9}$$

where k is the Boltzmann constant (1.3807×10^{-23} J/K), and N' and N are the number of molecules and mole fractions of the two components, respectively.

When a polymer replaces component 2 in the same lattice, account must be taken of the fact that the volume occupied by the polymer is very much greater than that of the solvent. In the simplest approach, the number of lattice sites occupied by the polymer is increased in proportion to the volume fraction of polymer in the mixture. Thus, the mole fraction N is replaced in Eq. (2.9) by the corresponding volume fraction of each component:

$$\Delta S_m = -k(N_1' \ln v_1 + N_2' \ln v_2) \tag{2.10}$$

Enthalpy

The heat of mixing, ΔH_m, for ideal solutions is zero. For polymer solutions, the heat of mixing is the energy change involved in one contact between solvent and solute, ΔH_c, times the number of solvent-solute contacts. Assume that the number of contacts, N_c, is proportional to the volume fraction of solvent multiplied by the number of repeat units in the polymer, x_n, and Z, the number of effective contacts per segment. Then, for mixing N_2' molecules of polymer with N_1' molecules of solvent:

$$\Delta H_m = \Delta H_c N_2' N_c = \Delta H_c N_2' Z x_n v_1 \tag{2.11}$$

But the volume fractions are related by the number of lattice spaces taken up by each component:

$$\frac{v_1}{v_2} = \frac{N_1'}{(x_n N_2')} \tag{2.12}$$

Thus, substituting in Eq. (2.11):

$$\Delta H_m = \Delta H_c Z N_1' v_2 \tag{2.13}$$

When the equation is rewritten as

$$\Delta H_m = kT \chi N_1' v_2 \tag{2.14}$$

the symbol χ can be identified as the energy change in units of kT when a molecule of solvent is transferred from pure solvent to an infinite amount of polymer. Also, it is the energy in units of RT when a mole of solvent is transferred.

Free Energy

Combination of Eqs. (2.8), (2.10), and (2.14) yields:

$$\Delta G_m = kT(\chi N_1' v_2 + N_1' \ln v_1 + N_2' \ln v_2) \tag{2.15}$$

When k is replaced by R and the numbers of molecules are replaced by moles of molecules, Eq. (2.4) results. This brief treatment follows the early approach of Flory [22] and others. Experimentally, it is found that χ in Eq. (2.15) is not a simple function of $1/T$ as might be inferred from Eq. (2.14). The makeup of χ includes a constant term (entropic in nature) which corrects for the entropy expression. Also, χ is a function of concentration. Most listed values for χ are for the particular polymer in the given solvent at a stated temperature and at infinite dilution. A number of more rigorous expressions for ΔS_m and ΔG_m have been presented over the years [23]. The predictions of various theories have been compared with Monte Carlo simulations [24, 25].

KEYWORDS FOR CHAPTER 2

Section

1. Thermoplastic
   Thermoset
3. Conformation
   Configuration
   Tacticity
   Isotactic
   Syndiotactic
   Stereoisomerism
4. Interpenetrating network (IPN)
5. Cohesive energy density (CED)
   Solubility parameter, δ
   Polymer-solvent interaction parameter, χ

PROBLEMS

2.1 For a material with a density of 1.0 g/cm³, estimate the molecular weight of a sphere made up of one molecule with a diameter of (a) 100 Å, (b) 1.0 μm, and (c) 1.0 cm. Avogadro's number = 6.02×10^{23} molecules/mol.

2.2 A polyester has the following repeat unit:

$$\begin{array}{cccc} & & O & & & O \\ & & \| & & & \| \\ +CH_2-CHCl-CH_2-O-C-CH=CH-C-O+ \end{array}$$

What stereoisomers are possible?

2.3 Estimate the repeat distance between methyl groups in the same row of isotactic poly(propylene oxide) in the planar zigzag form (Fig. 2.2a). Repeat for the syndiotactic poly(propylene oxide).

2.4 A cross-linked sample of polyisobutylene swells 10 times its original volume in cyclohexane. What volume will it swell to in toluene?

2.5 A polymer of propylene oxide, $+(CH_2\ CH(CH_3)O)+$, is cross-linked so that the average distance between cross-links is 5000 chain atoms. The polymer density after cross-linking is 1.20 g/cm³. What is the density of a swollen sample in a solvent for which $\chi = 0.40$ and solvent density is 0.80 g/cm³? Assume additivity of volumes. Molecular weight of solvent is 102.

2.6 The gamma radiation of many polymers results in cross-linking. For a new polymer, assuming that cross-link density $N/2$ is directly proportional to radiation dose, deduce the cross-link density at the highest dose and χ, the interaction parameter.

Dose, Mrad	v_2 in n-hexane at 25°C
1	0.090
2	0.110
4	0.138
8	0.177
16	0.225

2.7 Calculate the extended length (in nm) of a molecule with 1000 repeat units of

(a) $+Si(CH_3)_2-O+$ (c) $+CH_2-CH_2+$
(b) $+CH_2-CO-O+$ (d) $+CH_2-O+$

Neglect the end-group contributions to length.

2.8 A single molecule of isotactic polypropylene has a molecular weight of 2×10^6. Calculate:

(a) The length in cm when the chain is extended (planar zigzag form).
(b) The volume occupied if the molecule forms a single crystal with a density of 0.906 g/cm³.

2.9 Heating natural rubber with 0.100 mol of peroxide/1000 cm³ gives a network which swells to 4.35 times its original volume in toluene (at 25°C). Each peroxide molecule yields one cross-link. Heating the same rubber with

28.0 g of sulfur/1000 cm³ (instead of the peroxide) gives a network which swells to 3.45 times its original volume (at 25°C). If all the sulfur atoms react to form bridges with variable numbers of atoms between chains, how many sulfur atoms are there in an average crosslink? The atomic weight of sulfur is 32.0. Toluene: density = 0.867 g/cm³, FW = 92.14 g/mol.

2.10 A polymer P with a very high initial molecular weight (1×10^6) is cross-linked by agent A. Each molecule of A brings about 1 cross-link. The cross-link density is measured by swelling the sample in a "perfect"* solvent. If a 5.00-cm³ sample of P with 0.020 g of A/100 g of P swells to 75.0 cm³, what concentration of A will yield a 5.00-cm³ sample which swells to only 25.0 cm³? (*"Perfect" in that the solubility parameters of the solvent and the polymer are equal.) Lattice parameter of solvent = 0.475, V_1 = 83.0 cm³/mol. Agent A: FW = 205 g/mol.

2.11 A cross-linked rubber sample swells to 10 times its original volume when immersed in solvent A. Assuming that this is the best possible solvent for the polymer, what is the solubility parameter of solvent B, which is made up by adding a small amount of a lactone to solvent A? The sample swells only to 5 times its original volume in B. Both solvents A and B have a molar volume of 100 cm³ and the same lattice parameter β_1. All tests are made at 27°C. Solvent A: CED = 85.0 cal/cm³, χ = 0.33.

2.12 A mixture of benzoyl peroxide (0.5 g) with poly(vinyl ethyl ether) (100 g) is heated to 175°C. Assuming that all the peroxide decomposes and that each molecule of peroxide extracts two hydrogen atoms to produce a single cross-link, calculate:

(a) The expected chain density N at 25°C in the final network.
(b) The swollen volume of 1 cm³ of polymer in benzene at 25°C.

Benzoyl peroxide, MW = 242. PVEE, MW = 72 (repeat unit), ρ = 0.97 g/cm³. V_1 (benzene) = 89.0 cm³/mol. χ (PVEE, benzene) = 0.40.

2.13 A silicone rubber sample is cross-linked by exposure to gamma radiation. The average number of chain atoms between cross-links is determined by a mechanical test to be 850. What volume (cm³) will be occupied by 1 g of rubber when it is equilibrated in n-octane at 20°C? The formula weight for dimethyl siloxane $-\!(Si(CMe_3)_2O)\!-$ is 78. The solid rubber has a density of 1.00 g/cm³. The formula weight of n-octane is 114 and its density is 0.703 g/cm³. χ (20°C) is 0.49.

2.14 If the polymer sample described in Example 2.2 is placed in solvent Q, which has the same lattice parameter as S, to what extent will the sample swell at 27.0°C? Data for solvent Q: FW = 88.0 g/mol, density = 0.860 g/cm³, internal energy of vaporization = 29.9 kJ/mol.

2.15 A cross-linked polymer is swollen in three solvents. What is the predicted swollen volume in solvent C? Assume that the lattice parameter is 0.300 for all three solvents. All data are taken at 27°C.

Solvent	Molar volume, cm³/mol	Swollen volume,* cm³	Solubility parameter, hildebrands
A	100	4.00	7.00
B	100	4.00	8.80
C	250	?	8.20

*Swollen volume of gel starting with 1.00 cm³ of original polymer.

2.16 A network polymer is produced by reacting 2 mol of a diisocyanate with 1 mol of pentaerythritol (the R group does not contain nitrogen):

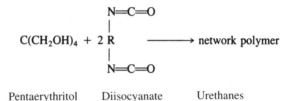

| Pentaerythritol | Diisocyanate | Urethanes |

The network polymer has a density of 1.05 g/cm³ and, according to an elemental analysis, 2.10 g/liter of nitrogen. What swollen volume can one expect from 1.00 cm³ of network polymer when it is equilibrated with chlorobenzene (density 1.10 g/cm³, MW = 112.5), for which system the polymer-solvent interaction parameter is 0.430?

2.17 A tetrafunctional telechelic polymer (possessing reactive end groups; see Sec. 4.5) is made by ring scission polymerization of propylene oxide (PO) on pentaerythritol (PE).

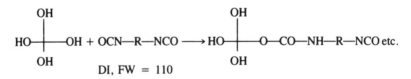

PE, FW = 136   PO, FW = 58.0

This is followed by network formation with the addition reaction of a low-molecular-weight diisocyanate (DI).

$$HO\!\!-\!\!\overset{\displaystyle OH}{\underset{\displaystyle OH}{\vert}}\!\!-\!\!OH + OCN\!\!-\!\!R\!\!-\!\!NCO \longrightarrow HO\!\!-\!\!\overset{\displaystyle OH}{\underset{\displaystyle OH}{\vert}}\!\!-\!\!O\!\!-\!\!CO\!\!-\!\!NH\!\!-\!\!R\!\!-\!\!NCO \text{ etc.}$$

DI, FW = 110

Given the following information:

200 mol of PO and 2 mol of DI are used per mole of PE.
Reaction is complete in each step.

Density of unswollen network polymer is 1.06 g/cm³.

The network polymer swells 22.0 times its original volume in toluene (at 27.0°C), which has a solubility parameter of 8.90 hildebrands, an FW of 92.14, and a density of 0.862 g/cm³.

what is the *solubility parameter* of the network polymer if the lattice parameter β_1 is 0.300?

2.18 A copolymer of allyl alcohol has some pendant OH groups. It is cross-linked through the OH groups by reaction with a negligible volume of a dicarboxylic acid:

$$P_1—OH + HO—CO—R—CO—OH + HO—P_2 \rightarrow$$
$$P_1—O—CO—R—CO—P_2 + HOH$$

The ester-cross-linked sample is then exposed to gamma radiation, which introduces covalent C—C cross-links in addition to the ester cross-links formed previously. When 10.0 cm³ of this *doubly cross-linked* polymer is placed in water, it swells rapidly to 55.6 cm³. However, after an extended time in water, all the ester bonds are hydrolyzed and now the same sample swells to 83.3 cm³. *What fraction of the cross-links* present in the doubly cross-linked sample (before hydrolysis) *were due to ester bonds?*

REFERENCES

1. Barry A. J., and H. N. Beck: chap. 5 in F. G. A. Stone and W. A. G. Graham (eds.), "Inorganic Polymers," Academic Press, New York, 1962.
2. Bowen, H. J. M., et al.: "Tables of Interatomic Distances and Configurations in Molecules and Ions," Spec. Publ. 11, Chem. Soc. (London), 1958.
3. Pauling, L.: "The Nature of the Chemical Bond," 3d ed., pp. 85, 189, Cornell Univ. Press, Ithaca, N.Y., 1960.
4. Pimentel, G. C., and A. L. McClellan: "The Hydrogen Bond," pp. 212, 224, 292, Freeman, San Francisco, 1960.
5. Richardson, M. J.: *Proc. Roy. Soc. (London)*, A279:50 (1964).
6. Natta, G., et al.: *J. Polymer Sci.*, 51:156 (1961). This number contains many articles on the subject. See also *SPE Trans.*, 3:(1963).
7. Sperling, L. H.: *Macromol. Revs.*, 12:141 (1977).
8. Sperling, L. H. (ed.): "Recent Advances in Polymer Blends, Grafts, and Blocks," Plenum, New York, 1974.
9. Manson, J. A., and L. H. Sperling: "Polymer Blends and Composites," Plenum, New York, 1976.
10. Barton, A. F. M.: "Handbook of Solubility Parameters and Other Cohesion Parameters," pp. 94, 186, CRC Press, Boca Raton, Fla., 1983.
11. Bristow, G. M., and W. F. Watson: *Trans. Faraday Soc.*, 54:1731 (1958).
12. Flory, P. J.: "Principles of Polymer Chemistry," chap. 13, Cornell Univ. Press, Ithaca, N.Y., 1952.
13. Orwoll, R. A.: *Rubber Chem. Technol.*, 50:451 (1977).
14. Bandrup, J., and E. H. Immergut (eds.): "Polymer Handbook," 3d ed., Wiley-Interscience, New York, 1989.
15. Sheehan, C. J., and A. L. Bisio: *Rubber Chem. Technol.*, 39:149 (1966).

16. Grulke, E. A.: in J. Bandrup and E. H. Immergut (eds.), "Polymer Handbook," 3d ed., p. 556, Wiley, New York, 1989.
17. Beerbower, A., L. A. Kaye, and D. A. Pattison: *Chem. Eng.,* 74:118 (Dec. 18, 1967).
18. "Solvent Formulating Maps for Elvacite Acrylic Resins," PA 12-770, E. I. Du Pont de Nemours and Co., 1971.
19. Rodriguez, F., P. D. Krasicky, and R. J. Groele: *Solid State Tech.,* 28(5):125 (1985).
20. Cooper, W. J., P. D. Krasicky, and F. Rodriguez: *J. Appl. Polym. Sci.,* 31:65 (1986).
21. Krasicky, P. D., R. J. Groele, J. A. Jubinsky, F. Rodriguez, Y. M. N. Namaste, and S. K. Obendorf: *Polym. Eng. Sci.,* 27:282 (1987).
22. Flory, P. J.: "Principles of Polymer Chemistry," chap. 12, Cornell Univ. Press, Ithaca, N.Y., 1952.
23. Schweitzer, K. S., and J. G. Curro: *Adv. Polym. Sci.,* 116:319 (1994).
24. Dickman, R., and C. K. Hall: *J. Chem. Phys.,* 85:3023 (1986).
25. Binder, K.: *Adv. Polym. Sci.,* 112:181 (1994).

General References

"Atomistic Modeling of Physical Properties," Springer-Verlag, Berlin, 1994.
Barton, A. F. M.: "Polymer-Liquid Interaction Parameters and Solubility Parameters," CRC Press, Boca Raton, Fla., 1990.
Barton, A. F. M.: "Handbook of Solubility Parameters," 2d ed., CRC Press, Boca Raton, Fla., 1991.
Coleman, M. M., P. C. Painter, and J. F. Graf: "Specific Interactions and the Miscibility of Polymer Blends," Technomic, Lancaster, Pa., 1991.
Elias, H.-G.: "Macromolecules, Structure and Properties," 2d ed., Plenum, New York, 1984.
Forsman, W. C. (ed.): "Polymers in Solution," Plenum, New York, 1986.
Han, C. D. (ed.): "Polymer Blends and Composites in Multiphase Systems," ACS, Washington, D.C., 1984.
Harris, F. W., and R. B. Seymour (eds.): "Structure-Solubility Relationships in Polymers," Academic Press, New York, 1977.
Hopfinger, A. J.: "Conformational Properties of Macromolecules," Academic Press, New York, 1973.
Klempner, D., and K. C. Frisch (eds.): "Polymer Alloys," Plenum, New York, 1977.
Manson, J. A., and L. H. Sperling: "Polymer Blends and Composites," Plenum, New York, 1976.
Olabisi, O., L. M. Robeson, and M. T. Shaw: "Polymer-Polymer Miscibility," Academic Press, New York, 1979.
Paul, D. R., and S. Newman: "Polymer Blends," Academic Press, New York, 1978.
Paul, D. R., and L. H. Sperling (eds.): "Multicomponent Polymer Materials," ACS, Washington, D.C., 1986.
Rempp, P., and G. Weill (eds.): "Polymer Thermodynamics and Radiation Scattering," Hüthig & Wepf, Basel, Switzerland, 1992.
Seymour, R. B., and C. E. Carraher, Jr.: "Structure-Property Relationships in Polymers," Plenum, New York, 1984.
Šolc, K. (ed.): "Polymer Compatibility and Incompatibility," Gordon & Breach, New York, 1982.
Sperling, L. H. (ed.): "Recent Advances in Polymer Blends, Grafts, and Blocks," Plenum, New York, 1974.
Sperling, L. H., and D. R. Paul (eds.): "Multicomponent Polymer Materials," ACS, Washington, D.C., 1985.
Thomas, E. L. (ed.): "Materials Science and Technology. Structure and Properties of Polymers," VCH, New York, 1993.
Urban, M. W., and C. D. Craver (eds.): "Structure-Property Relations in Polymers: Spectroscopy and Performance," ACS, Washington, D.C., 1993.
van Krevelen, D. W.: "Properties of Polymers," Elsevier, Amsterdam, 1990.

THREE

PHYSICAL STATES AND TRANSITIONS

3.1 PHYSICAL STATES

The physical states in which a polymer can exist can be idealized by considering first a long, regular polymer chain consisting of a succession of single bonds. Polyethylene, poly(styrene), or poly(methylene oxide) might serve as an example. With free rotation around each bond, the chain can assume an infinite number of conformations in space. During these rotations the bond angles and distances remain fixed. Three extreme conditions are possible.

1. *Completely free rotation.* This keeps the molecules in continuous motion. The wriggling molecules can slip past one another rather easily. This is a polymer *melt.* The higher the temperature, the more intense is the molecular motion.
2. *No rotation.* At some sufficiently low temperature, rotation around single bonds becomes impossible because of the energy barriers that a substituent on one chain atom encounters when trying to move past the substituent on an adjacent chain atom. Even hydrogen atoms on adjacent carbons interfere with one another at low enough temperatures. The polymer molecules become trapped in a chaotic, disordered, entangled state. This is a *glass.*
3. *Packing.* Polymer molecules may fit together in such a way that intermolecular attractions stabilize the chains in a regular lattice even though there are negligible intramolecular barriers to rotation around single bonds. This is the *crystalline* state of "long-range" order. If the temperature is lowered so that intermolecular barriers to rotation become great, the crystal is even more stable.

Each of these is an idealization. Factors that modify the picture are:

1. The polymer chain may contain double or triple bonds or rings, which do not permit rotation at any temperature without actual bond breaking.
2. The polymer may be branched or cross-linked.
3. The polymer may be short or long.
4. The polymer may not be homogeneous; parts may be in various states.
5. The polymer may be dissolved in a lower-molecular-weight liquid or in another polymer.
6. A polymer that is stressed may be oriented and not completely isotropic in the melt. Rapid cooling of the melt may preserve the orientation even when the stress is removed.

Two major transition temperatures are the glass transition temperature, T_g, and the melting temperature, T_m. T_g is the temperature below which free rotations cease because of intramolecular energy barriers. Values of T_g most often listed for polymers correspond to the stiffening temperature. Simple bending of a rod might be the criterion. The time scale of the test is important. Even hindered rotation around single bonds may be sufficient to allow adjustment to new conformations over a long time. A plastic that is brittle when hit with a hammer may sag under its own weight over a period of weeks at the same temperature. It is a glass in the first test, but a melt (of sorts) in the second.

Although we have described the glass as being homogeneous, many researchers challenge the idea. The interpretation of some results with microscopy and diffusion experiments suggests instead that glasses are made up of tiny, ordered *domains* with dimensions on the order of 5 to 15 nm. However, application of other techniques such as neutron scattering and light scattering have not confirmed the existence of the domains. The subject remains a very controversial one [1, 2].

If the long polymer chains are interconnected by widely separated cross-links, the segments between the cross-links can still assume the three states mentioned. In the case of state I, a loosely cross-linked melt is called a *rubber*. The extent to which the original molecules can assume new conformations is limited by the finite extensibility of segments between cross-links. If the number of cross-links is increased, the segments become shorter. Finally, the segments may be so short that rotations around single bonds are no longer possible and the system resembles a permanent glass even at temperatures where an uncross-linked polymer would be a melt.

The difficulty of fitting an entire polymer into a lattice leads to a generalization that crystallinity is never perfect. In fact, many "crystalline" polymers may have amorphous contents of 20 to 50%. This is many orders of magnitude greater than the imperfections associated with common metals. The amorphous portion may be glass, melt, or rubber, depending on temperature, time scale of testing, or cross-linking.

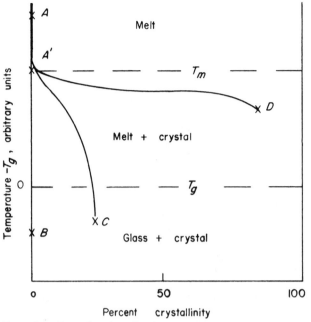

Figure 3.1 Examples of cooling from the melt (point A): All curves go through A' on cooling. B: Room-temperature state of polystyrene cooled at any rate from the melt. Also, typical state of nylon 66 that has been quenched from melt to a temperature below T_g. C: Typical final state of nylon 66 cooled at 1°C/s to a temperature below its T_g. D: Linear polyethylene cooled to room temperature (which is well above its T_g)

Another conceptual difficulty that may arise concerns the previously discussed phenomenon of tacticity (Sec. 2.3). Free rotation around single bonds does not imply interchangeability of two substituents on a chain atom. Thus an isotactic polymer (as in Fig. 2.2) may undergo an infinite number of rotations around each of the bonds in the main chain. However, when the chain is once again put in the planar zigzag form, all of the pendant groups will fall into exactly the same positions as they had before the rotations.

Before going on to discuss in greater detail the characteristics of these three basic states (glass, melt, and crystal), it might be well to describe some interrelations. Since no long-range molecular rearrangement is possible below T_g, it follows that if crystals are to form, crystallization will have to take place above that temperature. The highest temperature at which a crystal lattice is stable is the melting temperature, T_m. If no crystal lattice is stable, or if the material is quenched from the melt to a temperature well below T_g, behavior of the sort indicated by line A-B (Fig. 3.1) is possible. Ordinary atactic polystyrene and poly(methyl methacrylate) are two polymers that do not form crystals and, on cooling below their T_g's (both about 100°C), go directly from an easily deformed melt to a rigid, transparent,

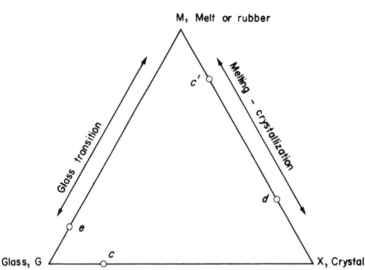

Figure 3.2 Composition diagram. Point *c* corresponds to *C* of Fig. 3.1, *d* to *D*, and *c'* to *C'*. Compositions *c* or *c'* may be rubbery even in the absence of covalent cross-links. Point *e* is possible only when
two portions of the same molecule have separate values of T_g

amorphous glass. Common nylon (nylon 66) quenched to room temperature might
follow the same path. However, if nylon is cooled at a constant rate (say, 1°/s) from
the melt, it might follow path *A-C*. Some crystals form when the temperature falls
below T_m. At a temperature below T_g, the crystals that form are surrounded by a
matrix of amorphous glass. If the glass is the major constituent, the material may
be quite brittle. When polyethylene is cooled to room temperature, it remains well
above its T_g. As indicated by line *A-D*, the polymer may be highly crystalline with
amorphous melt outside the crystalline regions. In response to a stress, the amorphous parts are rather easily deformed. Of course, above T_m there are no crystals.
Figure 3.2 is another attempt to interrelate the various physical states. Compositions *C* and *D* are near points *c* and *d,* respectively, in Fig. 3.2. It is possible to
have a melt with low crystallinity (point *c'*). If the crystals are large enough and
strong enough to act as massive cross-links connecting high-molecular segments,
the melt will act like a rubber as long as the temperature is between T_g and T_m. A
composition in which parts of molecules can be glassy while other parts remain in
the melt (point *e*) is achieved by the technique of block copolymerization (Sec.
4.9). Once again, rubbery behavior is attainable if the glassy domains act as cross-
links. Such a polymer is really two homopolymers tied together, each exhibiting its
own T_g. Because they are tied together, they cannot segregate macroscopically, but
can only do so on a microscopic basis. The thermoplastic elastomer based on styrene and butadiene is one commercial example of a system with two T_g's, one below
room temperature, the other above (see Sec. 4.5).

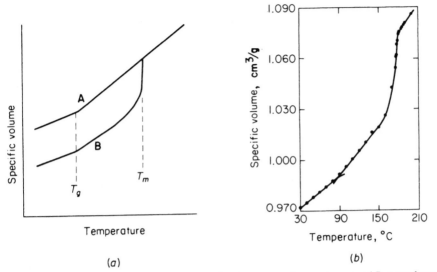

Figure 3.3 (a) Specific volume vs. temperature for A, an amorphous polymer, and B, a partly crystalline polymer. (b) Volume-temperature curve for pure poly(N,N'-sebacoyl piperazine), for which T_g is 82°C and T_m is 180 to 181°C [3]

3.2 AMORPHOUS POLYMERS

An amorphous polymer is easy to imagine. A bowl of spaghetti is a fair analogy. A bucket of worms is even better, because polymer molecules are constantly in motion. To be analogous to some common polymers, the worms would have to be 10^3 to 10^4 times longer than they are thick (that is, 5 mm diameter, 10 m long!). Obviously, the constant motion one notes in the bucket is not of whole worms at once, but of individual segments. The analogous *segmental Brownian motion* in polymers is very important in explaining flow and deformation. As previously mentioned, the segments themselves move by virtue of rotations around single bonds in the main chain of the polymer. The intensity of the motion increases with temperature. Below a certain temperature (T_g) the polymer segments do not have sufficient energy to move past one another. Rotations around single bonds become very difficult. If the material is stressed, the only reversible response can be for bond angles and distances to be strained, since no gross movements of segments can take place. Such a material is a *glass*. Only if the temperature is above T_g can the segments rearrange to relieve an externally applied stress. This important parameter, T_g, can be characterized also as an inflection point in the specific volume-temperature curve (Fig. 3.3). The *glass transition temperature* T_g resembles a second-order transition, because the change in volume is not discontinuous as it is with T_m. The abruptness of the transition can be rationalized by the concept of free volume (see Sec. 3.3). The segmental energy has to exceed a certain barrier value before a "hole" of segmental dimensions can be created for diffusion.

Whether or not T_g is a thermodynamic parameter is the subject of debate. A statistical thermodynamic model put forward by Gibbs and DiMarzio [4] predicts the existence and some of the characteristics of the glass transition. Still, attempts to verify the thermodynamic interrelation among changes in heat capacity, expansion coefficient, and compressibility at T_g have met with only occasional success [5]. Experimentally, T_g itself depends on the time scale of the experiment in which it is measured. As pointed out by Bueche [6], a 100-fold increase in heating rate when volume change is being measured increases the apparent T_g of polystyrene only about 5°C. Molecular weight affects T_g also. Theoretical equations of some complexity have been proposed [7]. An empirical correlation can be obtained in the form

$$T_g = T_{g,\infty} - \frac{K_g}{M_n} \tag{3.1}$$

where T_g for a finite molecular weight M_n is related to that for an infinitely long polymer, $T_{g,\infty}$. For example, the data of Beevers and White [7] for poly(methyl methacrylate) can be fitted by $K_g = 2.1 \times 10^5 °C \cdot mol/g$ and $T_{g,\infty} = 114°C$ when M_n exceeds 10^4. Likewise, polystyrene data are fitted by $K_g = 1.7 \times 10^5 °C \cdot mol/g$ and $T_{g,\infty} = 100°C$ down to molecular weights of less than 3×10^3 [6]. In both cases the maximum difference in T_g due to differences in M_n are small above $M_n = 5 \times 10^4$.

If we store energy in an amorphous polymer by stressing it and then attempt to recover the work we put in, there is always a certain amount of hysteresis or mechanical loss. This loss may be small above or below T_g, but it is always a maximum at T_g. In fact, this often is a more sensitive measurement than volume change on heating. The rationalization is not difficult. In the glass the stress is stored by bond distortions, which are easily recovered. Above T_g, polymer chains can be uncoiled readily from their random conformation by rotations about successive single bonds. Only at T_g is the uncoiling hindered by intermolecular forces. The discussion of deformation and flow of amorphous polymers is continued in Chaps. 7 and 8.

Segmental mobility is highly dependent on chain stiffness and somewhat dependent on intermolecular forces. It is expected that polymers with high polarity or high cohesive energy density should have higher transition temperatures than nonpolar materials. This is borne out approximately in Table 3.1. On the other hand, regularity, which is so important to crystallizability, counts for little in the amorphous transition. Since the segmental motion implies the concerted movement of many chain atoms (10 to 50 atoms), bulky side groups, which hinder rotation about single bonds, should raise T_g also. An interesting series in this respect is afforded by the vinyl ether polymers. As the polar ether linkage is diluted by longer alkyl side groups from methyl to butyl, T_g decreases. However, when the

Table 3.1 Transition temperatures for selected polymers [8–10]

Monomer unit	T_g, °C	T_m, °C	Total solubility parameter, $(MPa)^{1/2}$
Ethylene (linear)	−125	141	16.2
Propylene			
(isotactic)	−7	187	
(syndiotactic)	−9	150	
(atactic)	−10	—	
Butene-1 (isotactic)	−24	140	
Isobutylene	−73	—	15.5
4-Methylpentene-1 (isotactic)	30	166	
Styrene			
(isotactic)	100	240	
(syndiotactic)	—	270	
(atactic)	100	—	22.5
α-Methyl styrene	180	—	
Butadiene (1,4 polymer) (cis)	−102	12	
Butadiene (1,4 polymer) (trans)	−50	142	
Isoprene (1,4 polymer) (cis)	−70	39	16.7
Isoprene (1,4 polymer) (trans)	−50	80	
Acrylonitrile	125	—	
Chloroprene (1,4 polymer)	−50	80	18.8
Vinyl chloride (syndiotactic)	81/98	273	21.5
Vinyl acetate	32	—	19.2
Vinylidene chloride	−18	190	
Vinyl alcohol	85	265	
Vinyl methyl ether	−31	144	
Vinyl ethyl ether	−43	86	
Vinyl n-butyl ether	−55	64	
Vinyl iso-butyl ether	−20	115	
Methyl methacrylate	105	—	18.6
Ethyl acrylate	−24	—	
Acrylic acid	106	—	
Vinyl fluoride	41	200	
Vinylidene fluoride	−40	171	
Chlorotrifluoroethylene	100	220	
Tetrafluoroethylene	117	330	
Ethylene terephthalate	60/85	280	21.5
Caprolactone	−60	64	
Ethylene oxide	−41	69	
1,2-Propylene oxide	−72		
1,3-Propylene oxide	−78	75	
Formaldehyde (oxymethylene)	−82	184	
Acetaldehyde (isotactic)	−30	165	
Tetramethylene oxide	−84	57	
Phenylene sulfide	97	—	
Cellulose tributyrate	115	—	
Ethyl cellulose	43	—	21.1
Nylon 6	50/100	270	
Nylon 66	50	280	
Dimethyl siloxane	−127	−54	14.9

alkyl group is bunched up next to the chain, as with *tert*-butyl ether, rotation about the single bonds in the chain is made difficult and T_g is raised dramatically. Chain stiffness and, consequently, T_g are increased if single bonds in the polymer chain are replaced by multiple bonds about which there can be no rotation. Ring structures in the main chain contribute inflexibility as in poly(ethylene terephthalate) or cellulose derivatives. "Ladder" polymers have been synthesized in which there are no single bonds in the polymer chain. For example, pyrolysis of polyacrylonitrile gives an aromatic structure of high thermal stability:

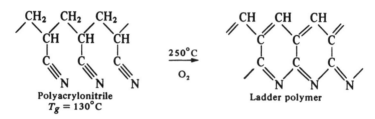

This particular material is so stable that it can be held directly in a flame in the form of woven cloth and not be changed physically or chemically [11]. When polyacrylonitrile is used as the precursor for carbon fibers, the ladder structure is often an intermediate [12]. The ladder polymer is subjected to carbonization (graphitization) typically at 1200 to 1400°C under nitrogen followed by additional heat treatment and special surface oxidation to improve adhesion. The carbon fibers so produced have stiffness moduli as high as 500 GPa (about 75×10^6 psi) and are widely used in high-performance composites.

For a copolymer in which the monomer units are distributed in a random fashion, polarity and chain stiffness of the copolymer are roughly the average of those for the individual homopolymers. Various thermodynamic approaches suggest rules for predicting the glass transition temperature of a copolymer in terms of the weight fraction of each monomer and the T_g of each homopolymer. However, an empirical equation due to Gordon and Taylor is widely used.

$$T_g(\text{copolymer}) = \frac{w_1(T_g)_1 + kw_2(T_g)_2}{w_1 + kw_2} \tag{3.2}$$

where k is a fitted constant [13]. When $k = 1$, a linear relationship results. When $k = (T_g)_1/(T_g)_2$ (using absolute temperatures), a reciprocal relationship will be derived:

$$\frac{1}{T_g(\text{copolymer})} = \frac{w_1}{(T_g)_1} + \frac{w_2}{(T_g)_2} \tag{3.3}$$

The relationship can be extended to systems with more than two monomers by adding another constant for each additional monomer.

Example 3.1

Monomers A and B are copolymerized with the following results:

Wt. fraction A in polymer	T_g, °C
1.000	108.0
0.850	75.0
0.260	28.0

What value of k fits these data to Eq. (3.2)? If B forms a copolymer with equal weights of B and C, and the same k applies, what T_g can be expected? The T_g of C is -37.0°C.

    Solution: Substituting in Eq. (3.2) and solving for $k(T_g)_B$:

$$75.0 = \frac{0.850(108.0) + 0.150k(T_g)_B}{0.850 + 0.150k}$$

$$28.0 = \frac{0.260(108.0) + 0.740k(T_g)_B}{0.260 + 0.740k}$$

$$k(T_g)_B = \frac{75.0(0.850 + 0.150k) - [0.850(108.0)]}{0.150} \quad \text{etc.}$$

$$k(T_g)_B = 75.0k - 187 = 28.0k - 31$$

so

$$k = 3.32 \quad \text{and} \quad (T_g)_B = 18.7°C$$

Then

$$(T_g)\text{new} = \frac{0.50(108.0) + 0.500(3.32)(-37.0)}{0.500 + 0.500(3.32)} = -3.4°C$$

3.3 PLASTICIZATION

The glass-transition temperature marks the onset of segmental mobility for a polymer. We can picture the movement of the segments due to a concentration gradient (diffusion) or due to a stress (viscous flow) as a jumping phenomenon. A segment moves or diffuses in a concerted manner into a vacancy or hole adjacent to it, leaving a hole of like dimensions behind. We know that the specific volume of a polymer increases as the temperature increases, as shown by the normal coefficient of expansion. However, the increase in volume is not uniform, and the local density on a molecular scale fluctuates to give an overall population of holes or "free volume." As previously mentioned, one experimental method of determining T_g is to note the temperature at which the coefficient of expansion changes. This greater

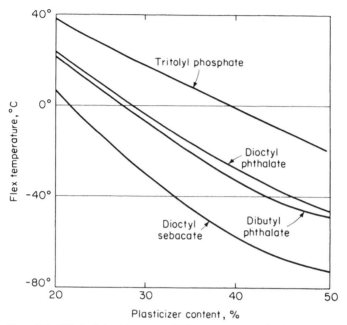

Figure 3.4 Effect of plasticizers on cold flex temperature of poly(vinyl chloride) compounds. The temperature is that at which torsional stiffness reaches some arbitrary value [14]

increase in volume with temperature above T_g represents the addition of free volume for segmental movement. In Sec. 7.5, a quantitative measure of "free volume" is given in the form of the Williams, Landel, and Ferry (WLF) equation.

If we mix two materials of different T_g's, we can imagine that each would contribute to the free volume of the system in proportion to the amount of material present. In this way we can rationalize that T_g should be a linear function of composition for copolymers. Likewise, if we mix a polymer with a solvent or another compatible polymer, we expect a T_g that is the weighted average of each component [Eq. (3.2) or (3.3)]. An important application of this principle is in plasticization. A plasticizer is a nonvolatile solvent that usually remains in the system in its ultimate use. Since solvent and polymer are not chemically bound to one another, the term "external plasticizer" is sometimes used to distinguish this case from copolymerization, where a comonomer of low T_g may be used to lower the T_g of the system (so-called internal plasticization). Poly(vinyl chloride) is very versatile because it is compatible with a variety of plasticizers and because the plasticized polymer remains quite stable both physically and chemically for long periods of time. The linearity of T_g with composition for several plasticizers is indicated by the behavior of the *flex temperature*, which is close to T_g for amorphous polymer

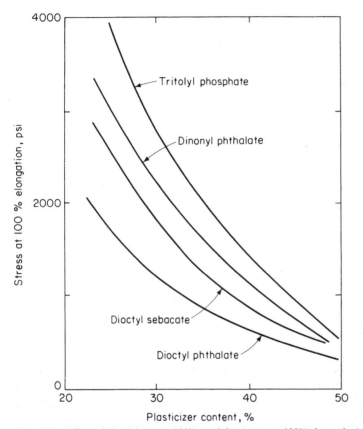

Figure 3.5 Effect of plasticizers on 100% modulus (stress at 100% elongation) of poly(vinyl chloride) compounds [14]

(Fig. 3.4). The efficiency in lowering T_g is not the only criterion for selecting a plasticizer. Cost, odor, biodegradability, high-temperature stability, and resistance to migration may be important in specific applications. Since stiffness of a polymer at temperature T decreases as $T - T_g$ increases, plasticization efficiency may be judged also by the lowering of the stiffness modulus at room temperature (Fig. 3.5).

Many commercial fibers are plasticized by small amounts of water. The familiar process of ironing a garment generally takes place between T_g and T_m so that creases can be controlled without destroying the overall structure due to crystallinity. In steam ironing, moisture lowers T_g, allowing a lower ironing temperature.

Another property that changes at the glass transition is the specific heat. The technique of differential scanning calorimetry (DSC) is often used to measure transitions (see Sec. 16.3). The method is convenient since it makes use of small

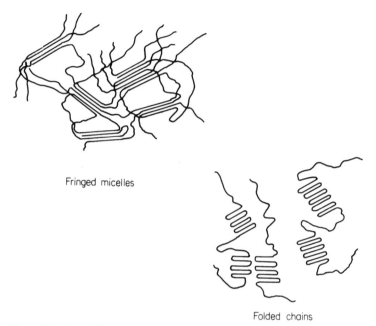

Fringed micelles

Folded chains

Figure 3.6 Crystallite arrangements

samples and is an automated procedure. Using the method of differential scanning calorimetry, the temperature range over which the transition occurs can be measured. Paul and co-workers [15] have observed that miscible polymer blends may be characterized by narrow or broad transitions. A broad transition does not mean that there is thermodynamic phase segregation. However, there are composition fluctuations possible in excess of the usual fluctuations in density in pure components. Thus, blends which are compatible in that only one glass transition is observed and which are transparent in both glassy and melt states may vary in the affinity of the components for each other without leading to phase separation. Most miscible polymer blends show phase separation on heating. The temperature at which this occurs is termed the *lower critical solution temperature* (LCST). As an example, a mixture with equal weights of poly(methyl methacrylate) and poly-(epichlorohydrin) is miscible at room temperature but becomes cloudy (indicating phase separation) at about 250°C [15].

3.4 CRYSTALLINITY

Polymers in the solid state can be completely amorphous, partially crystalline, or almost completely crystalline. Polymer crystals have the requirement that they must accommodate the covalent axis within an ordered structure. For some time

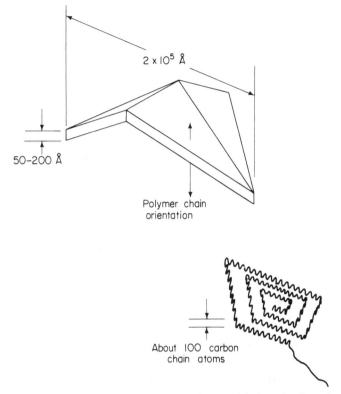

2×10^5 Å

50–200 Å

Polymer chain
orientation

About 100 carbon
chain atoms

Figure 3.7 A large single crystal of polyethylene might have the dimension indicated here. The growth mechanism involves the folding over of the planar zigzag chain on itself at intervals of about every 100 chain atoms. A single crystal may contain many individual molecules

the interpretation of x-ray diffraction patterns was that individual polymer molecules were partly crystalline and partly amorphous. The longest dimension of the crystallites in polycrystalline materials is usually about 5 to 50 nm, which is a small fraction of the length of a fully extended polymer molecule. A graphical representation of this once popular model is shown in Fig. 3.6. Here a long polymer chain wanders successively through disordered, random regions, through bundles of organized regions (micelles), again amorphous, again ordered, and so on. For a polymer that can be extended to a length of 5000 nm with crystal and amorphous domains averaging 10 nm, a single polymer thread might tie together a hundred or more crystallites. This "fringed micelle" model is inadequate in some ways. The abrupt change in density from crystal to amorphous region at the end of the micelle is unlikely. If some of the chains can fold on themselves, the transition to amorphous region can be accommodated. Such chain folding is clearly seen in single crystals.

For many years the statement could be made that all polymer crystals were submicroscopic. However, Till [16] and Keller [17] in 1957 produced single crystals of polyethylene up to 10 μm in one dimension. Subsequently, single crystals of many other polymers were obtained. A surprising feature of most of these crystals is that the covalent axis of the polymer chains is perpendicular to the longer dimensions of the crystals (Fig. 3.7). The explanation is that the polymer folds over on itself after about a hundred chain atoms have entered the lattice. The crystal thickness increases with the temperature of crystal formation or with subsequent annealing at a higher temperature. A polyethylene crystal with a thickness of 10 nm may be formed at 100°C. Heating the crystal at 130°C for several hours will increase the thickness to about 40 nm.

Both thermodynamic and kinetic arguments have been propounded to explain the spontaneous formation of the folded-chain structure [18]. The kinetic theory currently favored by some workers views the folding as a means of increasing the crystallization rate, even though crystals with completely extended chains might be more stable under equilibrium conditions. Indeed, polyethylene samples with molecular weights of less than 10,000 exhibit crystal thicknesses approximating the length of extended chains [19]. The four-sided hollow-pyramidal crystal is only one form found for polyethylene. Ridged crystals, truncated pyramids, and fibrillar forms are known also. Other materials show other features. Poly(oxymethylene) gives flat hexagonal crystals and polypropylene gives lathlike crystals. Polypropylene is an example where the folded chains are not derived from planar zigzag ribbons but from helices (see next section). The single-crystal evidence showing folded-chain structures for a variety of polymers leads one to think that the same structures occur in polycrystalline materials also.

The mechanical properties of crystalline materials can be viewed from two extreme positions. Materials of low crystallinity may be pictured as essentially amorphous polymer with the crystallites acting as massive cross-links, about 5 to 50 nm in diameter. The cross-links restrain the movement of the amorphous network just as covalent cross-links would. On the other hand, unlike the covalent bonds, the crystal cross-links can be melted or mechanically stressed beyond a rather low yield point. At the opposite end of the crystallinity spectrum, one can regard a highly crystalline material as a pure crystal that contains numerous defects such as chain ends, branches, folds, and foreign impurities. Mechanical failure of highly crystalline nylon, for example, bears a great resemblance to that of some metals, with deformation bands rather than the ragged failure typical of amorphous polymers.

X-ray and electron-diffraction studies of unoriented polymers allow the calculation of some interplanar distances in the crystalline domains. Diffraction by oriented fibers gives more information, since the axis of the fiber usually parallels one of the crystallographic axes. Both the conformation of single chains and the arrangement of the chains in the crystal lattice can be deduced, although the methods of analysis are neither simple nor unambiguous.

Isotactic polypropylene

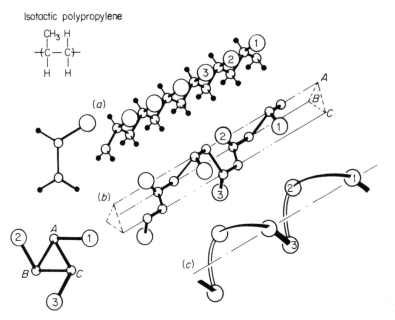

Figure 3.8 (a) Isotactic polypropylene. The large spheres represent the pendent methyl groups. (b) The H,3₁ helix with hydrogens omitted. (c) The pendent methyl groups as a helix [21]

3.5 CONFORMATION OF SINGLE CHAINS IN CRYSTALS

The planar zigzag conformation has been used to illustrate tacticity (Sec. 2.3). Some polymers assume this conformation in the crystal, e.g., polyethylene and 1,4-*cis*-polybutadiene. Isotactic polymers in which the pendant group is bulky often assume a helical form [20]. Polypropylene forms an H,3₁ helix, i.e., a helix in which three monomer units give one complete turn (Fig. 3.8). This simple helix can be generated by imagining the successive bonds being wrapped around a triangular mandrel, every other bond lying on a face and the alternating bonds lying on an edge. As seen in an end view (Fig. 3.8), the pendant methyl groups form a second triangular prism that contains the smaller one. Helices with more than three repeat units per turn are common. The α-helix assumed by many polymers of amino acids has about 18 units per five turns. This helical form for the single molecule may persist even in solution.

Other helices are favored by syndiotactic polymers and by some symmetrically substituted materials. A helix designated as H,2₁ can be generated by wrapping a chain around a mandrel with a square cross section (Fig. 3.9). Unlike the previous helices, the H,2₁ has bonds only on faces and not on the edges of the mandrel. It must be emphasized that only single-bond rotations and not bond bending or stretching are involved in achieving these conformations. Specific interactions be-

Syndiotactic
polypropylene

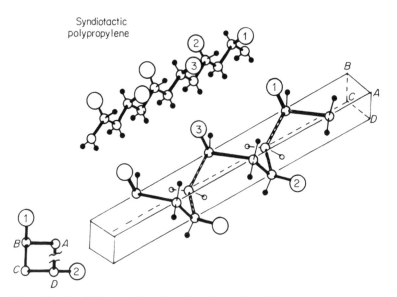

Figure 3.9 The H,2₁ helix of syndiotactic polypropylene [21]

tween pendant groups, such as hydrogen bonding in polymers of amino acids, or specific hindrances, such as repulsion between adjacent pendent methyl groups, often dictate the conformation with the most favorable free energy.

The fitting together of the chains in a lattice can take many forms. The unit cell of polyethylene (planar zigzag chains) is shown in Fig. 3.10. With some polymers, especially atactic, polar ones, crystallites may be formed that are lamellar. For example, they may have a helical form as individual chains and pack into a lattice in one direction, but not have a well-defined periodicity in the third dimension.

3.6 SPHERULITES AND DRAWING

Quiescent crystallization of a polymer from the melt or solution often results in a peculiar form of crystalline growth with a preferred chain orientation relative to a center (nucleus). For example, cooling poly(ethylene oxide) in a thin film results in the growth of circular areas emanating from nuclei, the growth stopping only when advancing fronts meet. When completely cooled, the areas may be as large as a centimeter in diameter (Fig. 3.11). Polarized light reveals that the polymer chains are oriented tangentially around each nucleus despite the fact that the area (a "spherulite," since it extends in three dimensions) consists of a multitude of crystallites and is not a single crystal. In most materials the spherulites do not exceed 100 μm in diameter.

When a highly crystalline material is stressed, a certain amount of energy can

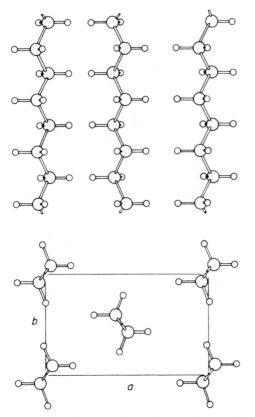

Figure 3.10 Model of the packing in the crystal structure of polyethylene in which $a = 7.4$ Å, $b = 4.94$ Å, and successive pendent atoms are 2.55 Å apart along the chain axis [22]

be stored by bond bending and stretching and other lattice distortions. Beyond the elastic limit (usually at a low elongation), rearrangement of the polymer chains often takes place so that chains are oriented in the direction of stress. This destroys the spherulitic pattern if it exists. This process of orientation by stretching (*drawing*) is used for most textile fibers (nylon, rayon, polyacrylonitrile). The oriented crystalline structure gives high strength and low elongation in the direction of the fiber axis. Drawing of crystalline materials often has another unusual feature, *necking*. If a sample of spherulitic nylon is stretched, elongation does not occur uniformly over the whole sample as it does when a rubber band is stretched. After a slight overall elongation (*ca.* 5%), a neck appears (Fig. 3.12) and grows if stretching is continued. Spherulites at the shoulder are pulled apart, and polymer chains are oriented in the direction of stress as they enter the neck. The orientation is permanent, since the new crystalline form is stable. Thus, it is not inconsistent that

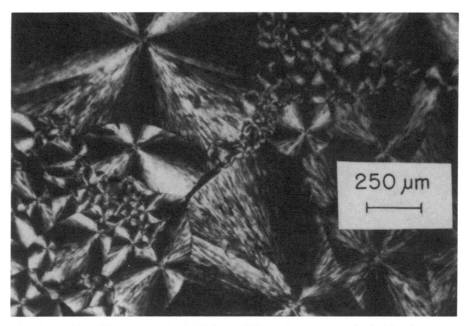

Figure 3.11 Spherulitic structure of poly(ethylene oxide) seen between crossed polarizing sheets

molded (spherulitic) nylon may be listed as having an elongation at break of about 300%, while drawn nylon fibers (oriented) have an ultimate elongation of only 20%. However, in drawn polyethylene, folded chains persist in oriented crystallites with tie molecules connecting the folded chain packets (Fig. 3.13).

The process just described involves going from random crystallites (short-range order, long-range disorder) to oriented crystallites (short- and long-range order). There are numerous polymers, which although amorphous in the unstressed state, on uniaxial stretching exhibit sharp x-ray patterns characteristic of oriented crystals. Usually the pattern disappears on release of the stress as the polymer goes back to the amorphous state. Characteristic repeat distances from x-ray patterns of these and stable crystalline materials allow estimation of unit cell dimensions, the crystal system, and also the ultimate density of perfectly crystalline material.

The organization of the crystals in a spherulite has been investigated for numerous materials. For example, single crystals have been observed to grow faster in one direction than another with branching so as to assume a sheaflike appearance and ultimately to form the nucleus of a spherulite [18]. The individual arms of the spherulite may have twists that lead to concentric rings or bands (Fig. 3.14). More commonly the nucleus of a spherulite is a solid impurity such as a catalyst residue. Such adventitious nuclei can be augmented or obscured by intentionally adding materials to act as nuclei. This can result in highly crystalline materials with

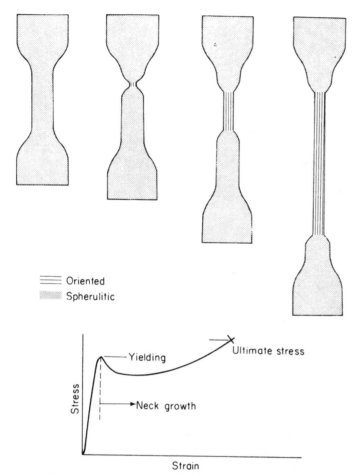

Figure 3.12   Successive stages in the drawing (elongation) of a spherulitic polymer

spherulites so small as not to scatter light appreciably. "Nucleated" materials are available commercially, where the mechanical advantages of crystallinity are desired without the opacity that comes from light scattering at the boundaries of large spherulites.

When crystallization occurs under stress, interesting new forms are seen. In particular, nucleation along flow lines in polyethylene results in a "shish kebab" structure, so-called because folded chain lamellae (the "kebabs") grow out at intervals from a center shaft of oriented polymer. While not all are as dramatic as the polyethylene case, "row structures" in which a central fibril acts as a contiguous source of nuclei are not uncommon when crystallization takes place in extrusion and molding, even under conditions of low melt stress [24].

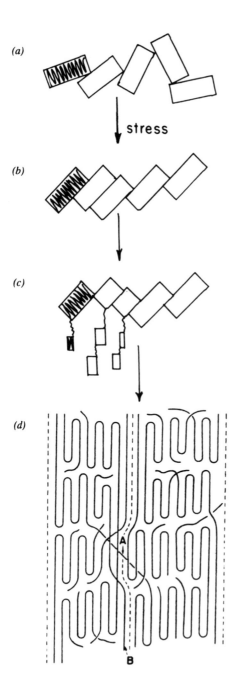

(a)

stress

(b)

(c)

(d)

Figure 3.13 The drawing process. (*a*) Twisted crystal lamellae that make up the spherulitic structure are pulled into (*b*) tilted prede-formed lamellae during the prenecking pe-riod. (*c*) The tilted lamellae are torn into blocks and tie molecules that make up the mi-crofibrils (*d*) with interfibrillar A and intrafi-brillar B tie molecules in the fully drawn state as proposed for polyethylene by Peterlin [23]

Figure 3.14 Slowly formed spherulites of linear polyethylene show a characteristic banded structure when observed between crossed polarizers

3.7 CRYSTALLIZATION

The tendency for a polymer to crystallize is enhanced by *regularity* and *polarity*. The molecules must fit together neatly, as well as attract one another, since most of the intermolecular energies vary inversely with the sixth power of the distance between molecules. Examples of the effect of *regularity* are given in Table 3.2.

The isotactic form is not always the most crystalline. In poly(vinyl alcohol), for example, the syndiotactic form crystallizes more readily than the isotactic because there are strong repulsive forces between adjacent hydroxyl groups [25] in the isotactic form.

Examples of the effect of polarity are:

Nonpolar   Polypropylene (atactic), noncrystalline
Polar   Poly(vinyl alcohol) (atactic), somewhat crystalline
Very polar   Polyamide (nylon 6), very crystalline

3.8 TRANSITION TEMPERATURES

Differential scanning calorimetry (DSC) has become the most popular analytical tool for determining transition temperatures as well as the degree of crystallinity

Table 3.2 Effect of Regularity on Polymer Crystallinity

Irregularity	Polymer	Typical percent crystallinity
Copolymer	Linear polyethylene	70
	Isotactic polypropylene	70
	Linear random copolymer	None
Tacticity	Isotactic polypropylene	70
	Atactic	None
Cis-trans	*trans*-1,4-Polybutadiene	40
	cis-1,4-Polybutadiene	30 (at 0°C-long times)
	Random cis and trans	None
Linearity	Linear polyethylene	70
	Branched polyethylene	40

(see Sec. 16.3). Small samples (5 to 15 mg) are sufficient, and a temperature scanning rate of 20°C/min is often feasible. Heat transfer is usually less efficient in dilatometers or in mechanical testing devices than in DSC. Both calorimetry and dynamic mechanical analysis are regarded as standard tests (ASTM D3418). Many other techniques have been applied to establish the upper temperature limit for dimensional stability, such as softening points (ASTM D1525) or microindenter penetration [26].

For most polymers, there is a single temperature (in an experiment with a given time scale) at which the onset of segmental motion occurs; it is termed the glass transition temperature T_g. Similarly, for those polymers which crystallize to any extent, there is a single melting temperature, T_m (which does depend to some extent on molecular weight). However, in both the amorphous and crystalline phases, additional rearrangements or relaxation processes can occur. Usually these do not result in obvious changes in properties. Sometimes they are observed as peaks in damping versus temperature plots resulting from oscillatory mechanical or electrical stresses (see Secs. 8.7 and 8.9).

Since crystallinity depends strongly on regularity, but T_g does not, it is conceivable that some copolymers might form glasses before they can crystallize. This is the case for ethylene-propylene copolymers of roughly equimolar proportions. When polymerized by a catalyst that gives highly crystalline, linear polyethylene or highly crystalline, isotactic polypropylene, a mixture of the monomers gives an amorphous, rubbery material that does not crystallize but only forms a glass on cooling to about −60°C. Another example is shown in Fig. 3.15. Here, adding about 20% of phenylmethylsiloxane is sufficient to destroy the crystallinity of poly(dimethylsiloxane) without substantially raising the T_g.

It is obvious from the foregoing that, for partially crystalline materials, T_m is always greater than T_g and that the difference is a maximum for homopolymers.

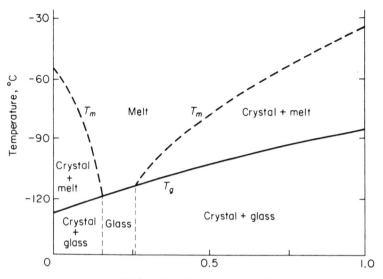

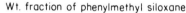

Wt. fraction of phenylmethyl siloxane

Figure 3.15 Loss of crystallinity by copolymerization of phenylmethyl siloxane with dimethyl siloxane [27]

An examination of these parameters for many homopolymers leads to the generalization that [28]

$$1.4 < \frac{T_m}{T_g} < 2.0 \tag{3.4}$$

The ratio generally is higher for symmetrical polymers such as poly(vinylidene chloride) than it is for unsymmetrical polymers such as isotactic polypropylene.

Some other generalizations can be made in light of what has been said so far. The rate of polymer crystallization increases as the temperature is lowered from T_m. However, no crystallization takes place effectively below T_g. Consequently, the maximum rate of crystallization occurs at a temperature between T_m and T_g (Fig. 3.16). The maximum rate occurs about halfway between T_g and T_m for natural rubber, but it can be much closer to T_m for some polymers such as poly(dimethylsiloxane) or polyethylene. If a sample of molten polymer is quenched rapidly to a temperature below T_g, a metastable glass may be obtained. Warming the sample just above T_g can increase crystallinity, making the sample stiffer. In the interesting situation that then arises, a glassy polymer is heated, becoming progressively softer as the temperature increases, suddenly stiffens, and then softens again as the T_m is approached (Fig. 3.17). See also Fig. 16.6 in Sec. 16.3.

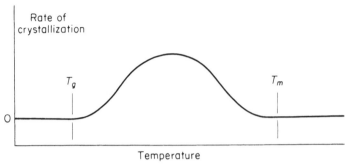

Figure 3.16 Rate of crystallization rises to a maximum between T_g and T_m

The *melting point* T_m of a polymer usually is more properly termed a melting range, because a single specimen consists of more than one molecular weight and more than one crystal size. Decreasing either molecular weight or crystal size lowers T_m somewhat. Besides the disappearance of opacity (seen by transmitted light) and polymer orientation (seen by transmitted polarized light), T_m can be characterized by the abrupt change in specific volume V that occurs (a first-order transition) (Fig. 3.18). Chain stiffness also has an effect on T_m. Two linear, nonpolar polymers are linear polyethylene and isotactic polystyrene. The bulky phenyl groups hinder rotation around single bonds in the polystyrene backbone. Since this rotation is the source of chain flexibility, it is not surprising that polystyrene is much stiffer than polyethylene. This shows up in melting points: polystyrene, $T_m = 230°C$; polyethylene, $T_m = 137°C$. Qualitatively, one can picture the effect of stiffness by comparing (1) the peeling of a leather strap from a wall to which it is nailed with (2)

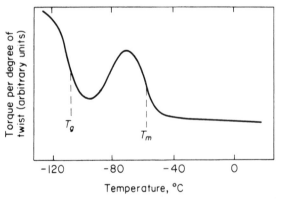

Figure 3.17 Quenched sample of silicone copolymer being heated from below T_g. Recrystallization between T_g and T_m increases stiffness, which then disappears when T_m is exceeded [27]

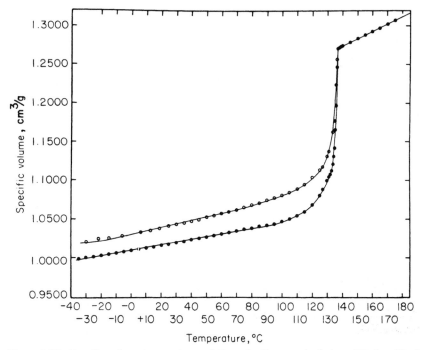

Figure 3.18 Specific volume-temperature relations for linear polyethylene (Marlex 50). Specimen slowly cooled from melt to room temperature prior to experiment (○) and specimen crystallized at 130°C for 40 days and then cooled to room temperature prior to experiment (●) [29]

the peeling of a stiff wooden board, also nailed. In the first, nails are stressed sequentially. In the second, all nails are stressed simultaneously. The molecular analogy for a nail is the intermolecular bond. Some typical values for T_m are summarized in Table 3.1.

Adding a solvent to a polymer decreases T_m in a manner predictable by the following equation [30]:

$$\frac{1}{T_m} - \frac{1}{T_m^0} = \frac{R}{\Delta H_u} \frac{V_u}{V_1} (v_1 - \chi v_1^2) \tag{3.5}$$

where V_u is the molar volume of the polymer repeat unit in the pure liquid state, ΔH_u is the heat of fusion per polymer repeat unit, V_1 is the molar volume of the solvent, v_1 is the volume fraction of the solvent, χ is the polymer-solvent interaction parameter, R is the gas constant, and T_m^0 is the melting point of the pure polymer (see Fig. 3.19, Table 3.3). Most of the time, the depression of T_m by a small amount of solvent is modest.

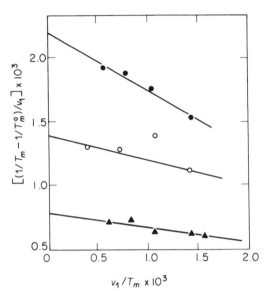

Figure 3.19 Temperature reduction as $(1/T_m - 1/T_m^\circ)/v_1$ plotted against v_1/T_m for cellulose tributyrate mixed with hydroquinone monomethyl ether (○), dimethyl phthalate (●) and ethyl laurate (▲) [31]. Linearity would not be greatly affected by replacing v_1/T_m by v_1

Example 3.2

If three volumes of poly(ε-caprolactone), $-(CH_2)_5-CO-O-$, are diluted with one volume of a good solvent for which $V_1 = V_u$ and $\chi = 0.400$, what is the depression in the melting point? Properties of the polymer are given in Table 3.3.

Solution:

$$\Delta H_u = 142 \times 114 = 16.2 \text{ kJ/mol}$$

$$\frac{1}{T_m} - \frac{1}{337 \text{ K}} = \left[\frac{8.314(\text{J/mol·K})}{16.2 \times 10^3 \text{ J/mol}}\right](0.25 - 0.4 \times 0.25^2) = 1.155 \times 10^{-4} \text{ K}^{-1}$$

$$T_m = 324 \text{ K} = 51°C$$

so

$$\Delta T_m = 64 - 51 = 13°C$$

Imposing a tensile strain on a polymer will cause a temporary orientation of chains in the direction of stress. Since this is a state of greater order, it represents a state of lower entropy, and therefore of higher free energy. An important consequence is that the melting temperature of crystals formed under stress will be higher than when unstressed. Careful measurements in which T_m is extrapolated to a limiting value for perfect crystallites have been made (Fig. 3.20). The available

Table 3.3 Heats of fusion for selected polymers [8–10]

Monomer unit	ΔH_u, J/g	T_m, °C	*Density, g/cm³ Amorph.	Cryst.	Repeat unit, g/mol
Ethylene (linear)	293	141	0.853	1.004	28.05
Propylene (isotactic)	79	187	0.853	0.946	42.08
Butene-1 (isotactic)	163	140	0.859	0.951	56.12
4-Methylpentene-1 (isotactic)	117	166	0.838	0.822	84.16
Styrene (isotactic)	96	240	1.054	1.126	104.15
Butadiene (1,4 polymer) (cis)	171	12	0.902	1.012	54.09
Butadiene (1,4 polymer) (trans)	67	142	0.891	1.036	54.09
Isoprene (1,4 polymer) (cis)	63	39	0.909	1.028	68.13
Isoprene (1,4 polymer) (trans)	63	80	0.906	1.051	68.13
Vinyl chloride (syndiotactic)	180	273	1.412	1.477	62.50
Vinyl alcohol	163	265	1.291	1.350	44.06
Ethylene terephthalate	138	280	1.336	1.514	192.18
ε-Caprolactone	142	64	1.095	1.194	114.14
Ethylene oxide	197	69	1.127	1.239	58.08
Formaldehyde (oxymethylene)	326	184	1.335	1.505	72.11
Nylon 6	230	270	1.090	1.190	113.16
Nylon 66	301	280	1.091	1.241	226.32

*Near 25°C.

theories are unsatisfactory in that they generally do not fit the observed T_m for unstressed polymer, although they can be used to correlate the changes in T_m at finite elongations.

A sizable increase in density with crystallization is characteristic of almost all materials. Only water, bismuth, and a few polymers are exceptions. The density of perfectly crystalline materials can be derived from x-ray measurements. The amorphous material is easily measured above its melting point and the density extrapolated to lower temperatures. Knowing these densities at any temperature together with the actual density of a sample gives a convenient estimate of the degree of crystallinity:

$$\text{Percent crystallinity} = \frac{\text{sample density} - \text{amorphous density}}{\text{perfect crystal density} - \text{amorphous density}} \times 100$$

(3.6)

Because chain atoms are involved in chain folding and in chain ends, it has been argued that even a single crystal could not be 100% crystalline on the basis described.

It should be obvious that characterization of a partly crystalline polymer is much more complex than mere specification of the fraction that is crystalline. Some factors that need to be taken into account have been listed by Magill [2] as:

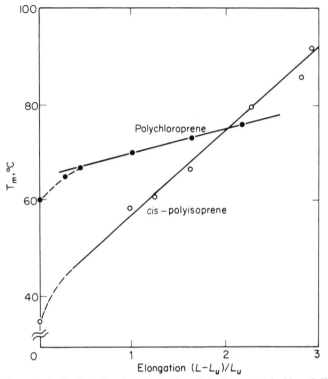

Figure 3.20   Strain-induced crystallization, L_u is the unstretched length [32]

1. Crystallite size, crystallite distribution, and crystallite perfection
2. Constraints on amorphous regions (matrix) that perturb it from its truly disordered condition
3. Defects or distortions in the paracrystalline sense
4. Presence of voids and surface stresses
5. Polymer chain chemistry, where induced chain irregularities prevent the system from attaining its lowest energy state
6. Transient and permanent entanglements in the amorphous regions
7. Inherent strains that give rise to nonrandom states
8. Chain length and chain ends
9. Nature of interzonal or surface layer with distance
10. Distribution of spherulite sizes
11. The relative sequence lengths for a block copolymer

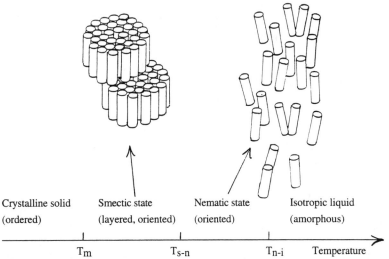

Crystalline solid   Smectic state   Nematic state   Isotropic liquid

(ordered)   (layered, oriented)   (oriented)   (amorphous)

T_m T_{s-n} T_{n-i} Temperature

Figure 3.21 Schematic visualization of possible mesomorphic transitions as temperature increases in a liquid crystal system showing the organization of mesogens (represented as cylinders) in the smectic and nematic states.

3.9 LIQUID CRYSTAL POLYMERS

Several intermediate states are possible between the three-dimensional crystal and the amorphous, isotropic liquid. These are the *thermotropic* (induced by temperature) *liquid crystalline,* and *mesomorphic* states [33, 34]. States can also be achieved by changes in concentration. For these, the term *lyotropic* (induced by solution) liquid crystals is used. In general, liquid crystalline behavior arises when molecules have rigidity and anisotropic polarizability. Some relatively simple molecules such as sodium stearate, a soap, exhibit varying degrees of liquid crystallinity. In fact, the name of one liquid crystal form, the *smectic* state, derives from the Greek word *smectos,* meaning soaplike. Another common liquid crystal form is the *nematic* state (from the Greek word *nematos,* meaning threadlike).

Using a tool such as differential scanning calorimetry (Sec. 16.3) or polarized-light microscopy, a progression can be seen for some materials going from the three-dimensional crystal to the two-dimensional smectic state or the one-dimensional nematic state and, finally, to the isotropic liquid state (Fig. 3.21). In the nematic phase, there is orientation alignment of the mesogens, but, unlike the smectic state, the centers of gravity of the mesogens are not located within layers [35]. There are several subgroups of the smectic phase. The nematic phase is the most common of the mesomorphic phases.

Rigid polymers often show liquid crystal behavior in moderately concentrated solutions. When a solution of a polyaramid (such as Kevlar) is extruded from a sulfuric acid solution, a highly oriented structure results that gives the recovered solid polymer a very high stiffness modulus and high tensile strength. The poly-aramid in this case is the polyamide based on terephthalic acid and *p*-phenylenediamine. Most *liquid crystal polymers* now reaching the market are linear, highly aromatic polyesters. They combine chemical stability with chain rigidity. They retain dimensional stability (remain as glasses) up to 200 or 300°C but are so chemically stable (resistant to oxidation) that they can be molded in the melt state when they are heated well above their glass transition temperatures. They are finding usage in electronic devices and automotive components.

Instead of the whole polymer consisting of a rigid chain, only part of the polymer may be a *mesogen* (liquid crystal-forming moiety). Blocks of stiff mole-cules may be located along the main chain or as pendent groups. They may be separated from each other by segments of flexible chains, so they will act as liquid crystals but remain part of the overall polymer. In these cases, the orientation corresponding to the smectic or nematic state may be induced by flow, changes in solvent, temperature, or the action of light or of a magnetic field. It is especially intriguing that cell wall membranes of living tissue are made of phospholipids that are segmented polymeric liquid crystals. The production of synthetic cells for use in biological systems is an example of the tailoring of polymers to perform highly specialized tasks.

KEYWORDS FOR CHAPTER 3

Section

1. Glass
   Crystallinity
   Melt
   Rubber
   T_f
   T_g
   ΔH_f
2. Segmental Brownian motion
   Ladder polymer
3. Free volume
   Plasticization
4. Spherulite
   Crystallite
   Fringed micelle
6. Drawing
7. Liquid crystal polymer
   Thermotropic

Lyotropic
Mesogen
Smectic
Nematic

PROBLEMS

3.1 When a saturated paraffin such as polymethylene $-(CH_2)-$ is chlorinated, chlorine replaces hydrogen at random. Invariably, small amounts of chlorine (10 to 50 wt% Cl) cause a lowering of softening point. Large amounts (*ca.* 70%) *raise* the softening point. Rationalize on the basis of intermolecular forces.

3.2 Polymers that are oriented on stretching may crystallize, since the molecules are then able to fit into a lattice more easily. How does this help explain why a rubber band heats up on stretching to near its breaking point? (Try this by holding a rubber band to your lips and suddenly stretching it.)

3.3 Copolymers of monomers A and B have the following stiffening temperatures:

Weight fraction of B	Stiffening temperature, °C
0.00	60
0.10	36
0.15	20.5
0.25	−8.0
0.35	−3.0
0.50	+5.0

Estimate T_g for homopolymers of B and of A. Explain your assumptions.

3.4 For the polyethylene of Fig. 3.18 (lower curve), calculate the fraction of crystalline material at 20°C assuming the coefficient of expansion for amorphous material is the same above and below T_m and that the density of perfect crystal can be obtained from Fig. 3.10.

3.5 Compare graphically the temperatures of Table 3.1 with Eq. (3.4). Also plot T_m and T_g versus δ_2. Rationalize extraordinary points.

3.6 From the data of Fig. 3.4, calculate the ratio of T_g for tritolyl phosphate to that for dioctyl sebacate.

3.7 What is wrong with describing a sample of polyisobutylene as "predominantly cis" or "syndiotactic"?

3.8 The following data are obtained when linear polyethylene is mixed with α-chloronaphthalene [36].

T_m, °C	v_1, vol. frac. of solvent
137.5	0.00
134.5	0.06
131	0.16
125	0.32
120	0.52
115	0.75
110	0.95

Densities of amorphous polyethylene and solvent are approximately 0.8 and 1.1 g/cm³, respectively. Estimate the heat of fusion of polyethylene and the polymer-solvent interaction parameter.

3.9 What is the repeat distance between pendent methyl groups that form a row in the H,3₁ helix of isotactic polypropylene? What is the analogous distance in the H,2₁ helix of syndiotactic polypropylene?

3.10 When held at a constant temperature, spherulites grow at a constant radial velocity. Crystallization rate may be studied by following the overall volume of a large sample containing many spherulites, each growing from a separate nucleus. Derive a formula for the change in density ρ with time t for a sample in which no new nuclei are formed after crystallization starts and in which the constant radial growth rate is $(dr/dt)_0$. Assume that we follow the density only during the time in which spherulites do not grow large enough to touch each other. Let densities of crystalline and amorphous polymer be ρ_c and ρ_a, respectively.

3.11 What melting temperature can be expected for poly(ethylene oxide) mixed with water when the volume fraction of solvent is 0.01, 0.02, and 0.05? Assume that $\chi = 0.45$. See Table 3.3 for additional data.

3.12 The T_g of nylon 66 is lowered by water or methanol. Using Eq. (3.3), estimate T_g for the two liquids assuming that the temperature at which $E' = 2.0$ GPa is near T_g (Fig. 8.18).

3.13 How much plasticizer with $T_g = -80°C$ would have to be added to a film of nylon 66 in order for T_g to be 25°C? If $V_1 = 200$ cm³/mol and $\chi = 0.40$, what would be the effect on T_m of this amount of plasticizer? Data: $\Delta H_u = 301$ J/g $\rho = 1.091$ g/cm³, $T_m = 280°C$, $T_g = 50°C$.

3.14 Two copolymers of ethylene ($CH_2\!\!=\!\!CH_2$) and propylene ($CH_2\!\!=\!\!CHCH_3$) have the same ratio of monomers. However, one is rubbery at room temperature and does not stiffen until the temperature is lowered to about $-70°C$, while the other is rather stiff, tough, and opaque at room temperature. Explain the difference.

3.15 What kind of isomerism might you expect in polymers of 1-butene ($CH_2\!\!=\!\!CHCH_2CH_3$)? Which form is likely to be least crystalline? What kind of isomerism do you expect in 1,4 polymerization of chloroprene (CH_2=CClCH=CH_2)? Which isomer should be most crystalline?

3.16 Why does an injection-molded polystyrene protractor show a complicated pattern when viewed in polarized light?

3.17 Polycaprolactone has $T_g = -60°C$ and $T_m = +60°C$. In the study of biodegradability at 25°C it would be desirable to vary the degree of crystallinity holding all other variables constant. Why is this difficult with polycaprolactone? Why should it be easier with poly(ethylene terephthalate), which has $T_g = +60°C$ and $T_m = 250°C$?

3.18 The Gordon-Taylor equation for the glass transition temperature of a two-component system in terms of weight fractions of components is

$$T = (T_g)_{mix} = \frac{w_1(T_g)_1 + kw_2(T_g)_2}{w_1 + kw_2}$$

(a) Show that $[(T_g)_1 - T]w_1/w_2 = \phi = kT - k(T_g)_2$.

(b) Plot ϕ versus T to get k and $(T_g)_2$, assuming $(T_g)_1 = 113°C$ for the system described by Anderson, who determined transition temperatures for blends of poly(methyl methacrylate), 1, with poly(epichlorohydrin), 2 [37]:

Weight fraction of 2	Transition temp., °C
0.000	113
0.200	52
0.300	42
0.430	11
0.700	−3
1.000	−19

3.19 Styrene forms a homopolymer with a T_g of 100.0°C. The rubbery copolymer with butadiene commonly called SBR has a T_g of −60°C, whereas the copolymer used for coatings has a T_g of 28.0°C. If a copolymer is made with equal weights of butadiene and vinyl acetate (T_g of polyacetate = 28°C), what T_g is expected for this new copolymer? Assume the validity of the Gordon-Taylor equation with the same k for any pair of monomers as long as monomer 1 is butadiene. SBR rubber, 25.0 wt% styrene; copolymer (coatings), 25.0 wt% butadiene.

3.20 When 100 g of PVC is mixed with 25 g of plasticizer Z, the glass transition temperature is lowered from 87°C (for PVC) to 0°C. A mixture of 100 g of PVC with 50 g of Z shows a glass transition temperature of −25°C. What glass transition temperature is to be expected from a mixture of 100 g of PVC with 100 g of Z?

REFERENCES

1. Geil, P. H.: *Ind. Eng. Chem. Prod. Res. Dev.,* 14:59 (1975).
2. Magill, J. H., in J. M. Schultz (ed.): "Treatise on Materials Science and Technology," vol. 10A, Academic Press, New York, 1977.
3. Flory, P. J., and H. K. Hall: *J. ACS,* 73:2532 (1951).
4. Gibbs, J. H., and E. A. DiMarzio: *J. Chem. Phys.,* 28:373, 807 (1958); *J. Polymer Sci.,* 40:121 (1959).
5. Gordon, M., in P. D. Ritchie (ed.): "Physics of Plastics," chap. 4, published for the Plastics Institute by Van Nostrand, Princeton, N.J., 1965.
6. Bueche, F.: "Physical Properties of Polymers," chaps. 4 and 5, Wiley-Interscience, New York, 1962.
7. Beevers, R. B., and E. F. T. White: *Trans. Faraday Soc.,* 56:744 (1960).
8. Runt, J. P.: "Crystallinity Determination," in *Encycl. Polym. Sci. Tech.,* 4:487 (1986).
9. Brandrup, J., and E. H. Immergut (eds.): "Polymer Handbook," 3d ed., Wiley, New York, 1989.
10. Lewis, O. G.: "Physical Constants of Linear Homopolymers," Springer-Verlag, New York, 1968.
11. Sorenson, W. R., and T. W. Campbell: "Preparative Methods of Polymer Chemistry," 2d ed., p. 237, Wiley-Interscience, New York, 1968.
12. Riggs, J. P.: Carbon Fibers, *Encyc. Polym. Sci. Eng.,* 2d ed., 2:640 (1985).
13. Gordon, M., and J. S. Taylor: *J. Appl. Chem.,* 2:493 (1952).
14. Lannon, D. A., and E. J. Hoskins in P. D. Ritchie (ed.): "Physics of Plastics," chap. 7, published for the Plastics Institute by Van Nostrand, Princeton, N.J., 1965.
15. Fernandes, A. C., J. W. Barlow, and D. R. Paul: *J. Appl. Polym. Sci.,* 32:5481 (1986).
16. Till, P. H.: *J. Polymer Sci.,* 24:301 (1957).
17. Keller, A.: *Phil. Mag.,* 2:1171 (1957).
18. Keller, A.: *Rept. Progr. Phys.,* 31:623 (1968).
19. Anderson, F. R.: *J. Polymer Sci.,* part C (3):123 (1963).
20. Miller, M. L.: "The Structure of Polymers," chap. 10, Reinhold, New York, 1966.
21. Rodriguez, F.: *J. Chem. Educ.,* 45:507 (1968).
22. Natta, G., and P. Corradini: *Rubber Chem. Technol.,* 33:703 (1960).
23. Peterlin, A.: *Textile Research J.,* 42:20 (1972); also, W. G. Perkins and R. S. Porter: *J. Mater. Sci.,* 12:2355 (1977).
24. Clark, E. S., and C. A. Garber: *Intern. J. Polymeric Mater.,* 1:31 (1971).
25. Fujii, K., et al.: *J. Polymer Sci.,* A2:2347 (1964).
26. Rodriguez, F., and T. Long: *J. Appl. Polym. Sci.,* 44:1281 (1992); and in N. P. Cheremisinoff and P. N. Cheremisinoff (eds.): "Handbook of Advanced Materials Testing," chap. 18, Dekker, New York, 1995.
27. Polmanteer, K. E., and M. J. Hunter: *J. Appl. Polymer Sci.,* 1:3 (1959).
28. Boyer, R. F.: *Rubber Chem. Technol.,* 36:1303 (1963).
29. Mandelkern, L.: *Rubber Chem. Technol.,* 32:1392 (1959).
30. Flory, P. J.: "Principles of Polymer Chemistry," chap. 13, Cornell Univ. Press, Ithaca, N.Y., 1952.
31. Mandelkern, L., and P. J. Flory: *J. ACS,* 73:3206 (1951).
32. Krigbaum, W. R., J. V. Dawkins, G. H. Via, and Y. I. Balta: *J. Polymer Sci.,* part A2, 4:475 (1966).
33. Goodby, J. W.: *Science,* 231:350 (1986).
34. Ober, C. K., J.-I. Jin, and R. W. Lenz: *Adv. Polym. Sci.,* 59:103 (1984).
35. Ballauff, M., in R. W. Cahn, P. Haasen, and E. J. Kramer (eds.): "Materials Science and Technology, vol. 12, Structure and Properties of Polymers," chap. 5, VCH, New York, 1993.
36. Quinn, F. A., Jr., and L. Mandelkern: *J. ACS,* 80:3178 (1958).
37. Anderson, C. C., and F. Rodriguez: *Proc. ACS Div. Polym. Matls.: Sci. Eng.,* 51:609 (1984).

General References

"Atomistic Modeling of Physical Properties," Springer-Verlag, Berlin, 1994.

Baltà-Calleja, F. J., and C. G. Vonk: "X-Ray Scattering of Synthetic Polymers," Elsevier, New York, 1989.

Bassett, D. C.: "Principles of Polymer Morphology," Cambridge, London, 1981.

Bassett, D. C. (ed.): "Developments in Crystalline Polymers—1," Elsevier Applied Science, New York, 1985.

Bicerano, J.: "Computational Modeling of Polymers," Dekker, New York, 1992.

Bicerano, J.: "Prediction of Polymer Properties," Dekker, New York, 1993.

Birley, A. W., B. Haworth, and J. Batchelor: "Physics of Plastics: Processing, Properties and Materials Engineering," Hanser-Gardner, Cincinnati, Ohio, 1992.

Blumstein, A. (ed.): "Liquid Crystalline Order in Polymers," Academic Press, New York, 1978.

Blumstein, A. (ed.): "Polymeric Liquid Crystals," Plenum, New York, 1985.

Boyer, R. F. (ed.): "Technological Aspects of the Mechanical Behavior of Polymers," Wiley-Interscience, New York, 1975.

Carfagna, C. (ed.): "Liquid Crystalline Polymers," Pergamon, New York, 1994.

Chapoy, L. L. (ed.): "Recent Advances in Liquid Crystalline Polymers," Elsevier Applied Science, New York, 1985.

Craver, C. D., and T. Provder (eds.): "Polymer Characterization: Physical Property, Spectroscopic, and Chromatographic Methods," ACS, Washington, D.C., 1990.

Deanin, R. D.: "Polymer Structure-Properties-Applications," Cahners, Boston, 1971.

Elias, H.-G.: "Macromolecules, Structure and Properties," 2d ed., Plenum, New York, 1984.

El-Nokaly, M. (ed.): "Polymer Association Structures: Microemulsion and Liquid Crystals," ACS, Washington, D.C., 1989.

Finlayson, K. M. (ed.): "Advances in Polymer Blends and Alloys Technology," vol. 4, Technomic, Lancaster, Pa., 1993.

Finlayson, K. M. (ed.): "Advances in Polymer Blends and Alloys," vol. 5, Technomic, Lancaster, Pa., 1994.

Fischer, E. W., and M. Dettenmaier: "Structure of polymeric glasses and melts," *J. Non-Cryst. Solids,* 31:181 (1978).

Fleer, G. J., M. A. C. Stuart, J. M. H. M. Scheutjens, T. Cosgrove, and B. Vincent: "Polymers at Interfaces," Chapman & Hall, New York, 1993.

Geil, P. H.: "Polymer Single Crystals," Wiley-Interscience, New York, 1963.

Gelin, B. R.: "Molecular Modeling of Polymer Structures and Properties," Hanser-Gardner, Cincinnati, Ohio, 1994.

Gordon, M., and N. A. Plate (eds.): "Liquid Crystal Polymers II/III," Springer-Verlag, New York, 1984.

Gruenwald, G.: "Plastics: How Structure Determines Properties," Hanser-Gardner, Cincinnati, Ohio, 1992.

Hall, I. H. (ed.): "Structure of Crystalline Polymers," Elsevier Applied Science, New York, 1984.

Hartwig, G.: "Polymer Properties at Room and Cryogenic Temperatures," Plenum, New York, 1995.

Haward, R. N. (ed.): "The Physics of Glassy Polymers," Wiley-Halsted, New York, 1973.

Hemsley, D. A.: "The Light Microscopy of Synthetic Polymers," Oxford Univ. Press, New York, 1984.

Höcker, H., and W. Kern (eds.): "Molecular and Supermolecular Order in Polymers," Huethig & Wepf, Mamaroneck, N.Y., 1984.

Höcker, H., and W. Kern (eds.): "New Developments on Polymer Structure and Morphology," Huethig & Wepf, Mamaroneck, N.Y., 1984.

Hope, P. S., and M. J. Folkes (eds.): "Polymer Blends and Alloys," Chapman & Hall, New York, 1993.

Hunt, B. J., and M. I. James (eds.): "Polymer Characterization," Chapman & Hall, New York, 1993.

Keinath, S. E., R. L. Miller, and J. K. Rieke (eds.): "Order in the Amorphous 'State' of Polymers," Plenum, New York, 1987.

Koenig, J. L.: "Chemical Microstructure of Polymer Chains," Wiley, New York, 1980.

Kovarskii, A. L. (ed.): "High-Pressure Chemistry and Physics of Polymers," CRC Press, Boca Raton, Fla., 1994.

Kryszewski, M., A. Galeski, and E. Martuscelli (eds.): "Polymer Blends," Plenum, New York, 1984.

Ladik, J. J.: "Quantum Theory of Polymers as Solids," Plenum, New York, 1987.

Lenz, R. W., and R. S. Stein (eds.): "Structure and Properties of Polymer Films," Plenum, New York, 1973.

March, N., and M. Tosi (eds.): "Polymers, Liquid Crystals, and Low-Dimensional Solids," Plenum, New York, 1984.

Mark, J. E., A. Eisenberg, W. W. Graessley, L. Mandelkern, E. T. Samulski, J. L. Koenig, and G. D. Wignall: "Physical Properties of Polymers," 2d ed., ACS, Washington, D.C., 1993.

McArdle, C. B. (ed.): "Side Chain Liquid Crystal Polymers," Chapman & Hall, New York, 1989.

McArdle, C. B.: "Applied Photochromic Polymer Systems," Chapman & Hall, New York, 1991.

Miller, M. L.: "The Structure of Polymers," Reinhold, New York, 1966.

Mitchell, J., Jr. (ed.): "Applied Polymer Analysis and Characterization," Hanser-Gardner, Cincinnati, Ohio, 1992.

Nagasawa, N. (ed.): "Molecular Conformation and Dynamics of Macromolecules in Condensed Systems," Elsevier, New York, 1988.

Nevell, T. P., and S. H. Zeronian (eds.): "Cellulose Chemistry and Its Applications," Wiley, New York, 1985.

Ogorkiewicz, R. M. (ed.): "Thermoplastics: Properties and Design," Wiley-Interscience, New York, 1974.

Peterlin, A. (ed.): "Plastic Deformation of Polymers," Dekker, New York, 1971.

Platè, N. A. (ed.): "Liquid-Crystal Polymers," Plenum, New York, 1993.

Platè, N. A., and V. P. Shibaev: "Comb-Shaped Polymers and Liquid Crystals," Plenum, New York, 1987.

Porter, D.: "Group Interaction Modelling of Polymer Properties," Dekker, New York, 1995.

Provder, T. (ed.): "Computer Applications in Applied Polymer Science II: Automation, Modeling, and Simulation," ACS, Washington, D.C., 1989.

Riew, C. K. (ed.): "Rubber-Toughened Plastics," ACS, Washington, D.C., 1989.

Ritchie, P. D. (ed.): "Physics of Plastics," Van Nostrand-Reinhold, New York, 1965.

Roe, R. J. (ed.): "Computer Simulation of Polymers," Prentice Hall, Englewood Cliffs, N.J., 1991.

Samuels, R. J.: "Structured Polymer Properties," Wiley-Interscience, New York, 1974.

Sandman, D. J. (ed.): "Crystallographically Ordered Polymers," ACS, Washington, D.C., 1987.

Sawyer, L., and D. T. Grubb: "Polymer Microscopy," Chapman & Hall (Methuen), New York, 1987.

Serafini, T. T., and J. L. Koenig (eds.): "Cryogenic Properties of Polymers," Dekker, New York, 1968.

Seymour, R. B., and C. E. Carraher, Jr.: "Structure-Property Relationships in Polymers," Plenum, New York, 1984.

Schröder, E., G. Müller, and K.-F. Arndt: "Polymer Characterization," Hanser-Gardner, Cincinnati, Ohio, 1989.

Stroeve, P., and A. C. Balazs (eds.): "Macromolecular Assemblies in Polymeric Systems," ACS, Washington, D.C., 1992.

Struik, L. C. E.: "Internal Stresses, Dimensional Instabilities and Molecular Orientations in Plastics," Wiley, New York, 1990.

Tadakoro, H.: "Structure of Crystalline Polymers," Wiley-Interscience, New York, 1979.

Thomas, E. L. (ed.): "Materials Science and Technology. Structure and Properties of Polymers," VCH, New York, 1993.

Tobolsky, A. V.: "Properties and Structure of Polymers," Wiley, New York, 1960.

Urban, M. W., and C. D. Craver (eds.): "Structure-Property Relations in Polymers: Spectroscopy and Performance," ACS, Washington, D.C., 1993.

van Krevelen, D. W.: "Properties of Polymers," Elsevier, Amsterdam, 1990.

Utracki, L. A.: "Polymer Alloys and Blends," Hanser-Gardner, Cincinnati, Ohio, 1990.

Utracki, L. A. (ed.): "Two-Phase Polymer Systems," Hanser-Gardner, Cincinnati, Ohio, 1991.

Vîlcu, R., and M. Leca: "Polymer Thermodynamics by Gas Chromatography," Elsevier, Amsterdam, 1989.

Ward, I. M. (ed.): "Structure and Properties of Oriented Polymers," Applied Science, London, 1975.

Ward, I. M. (ed.): "Developments in Oriented Polymers—1," Elsevier Applied Science, New York, 1982.

Ward, I. M. (ed.): "Developments in Oriented Polymers—2," Elsevier Applied Science, New York, 1987.

Weiss, R. A., and C. K. Ober (eds.): "Liquid-Crystalline Polymers," ACS, Washington, D.C., 1990.

Woodward, A. E.: "Atlas of Polymer Morphology," Hanser (Oxford Univ. Press), New York, 1988.

Woodward, A. E.: "Understanding Polymer Morphology," Hanser-Gardner, Cincinnati, Ohio, 1995.

Wunderlich, B.: "Macromolecular Physics/Crystal Structure, Morphology, Defects," Academic Press, New York, 1973.

Wunderlich, B.: "Macromolecular Physics/Crystal Nucleation, Growth, Annealing," Academic Press, New York, 1976.

Wunderlich, B.: "Macromolecular Physics/Crystal Melting," Academic Press, New York, 1980.

FOUR

POLYMER FORMATION

4.1 POLYMERIZATION REACTIONS

Polymerization is the process of joining together small molecules by covalent bonds. Around the turn of the century, chemists were reluctant to accept the idea of rubber, starch, and cotton as long, linear chains connected by covalent bonds. A popular alternative was the idea of an "associated colloidal" structure. As a matter of fact, some small molecules do exhibit such behavior. The extraordinary thickening power of aluminum disoaps in organic solvents is due to the formation of macroscopic structures from small molecules. However, the effective molecular weight of such a structure varies with concentration and temperature, whereas the molecular weights of true polymers with covalent links do not.

The reactions by which some complex, naturally occurring macromolecules are formed are not completely understood. Even *cis*-polyisoprene, natural rubber, which can be made in a single step from isoprene in the laboratory, is synthesized by a rather complex series of reactions in the rubber tree.

The ultimate starting materials for synthetic polymers are few in number. Petroleum, natural gas, coal tar, and cellulose are the main sources. In recent years ethylene and propylene from natural gas and petroleum refining have been used as a starting point for a variety of monomers.

The process of building up polymers from simple repeating units (monomers) can proceed with many variations. We can classify some of these as follows:

1. By the number of bonds each monomer can form in the reaction used, the functionality
2. By the kinetic scheme governing the polymerization reaction, chain vs. stepwise reactions
3. By the chemical reaction used to produce new bonds, ethenic addition, esterification, amidation, ester interchange, etc.
4. By the number of monomers used to give homopolymer, copolymer, terpolymer, etc.
5. By the physical arrangement, bulk, solution, suspension, or emulsion system

4.2 FUNCTIONALITY

Monomer can be converted to polymer by any reaction that creates new bonds. Fundamental to any polymerization scheme is the number of bonds that a given monomer can form. Carothers defined this number of bonds as the *functionality* of a monomer in a given reaction. Examples are given in Table 4.1. It should be obvious that a functionality of 2 can lead to linear structure. The necessity for specifying the reaction is emphasized by a monomer such as butadiene, which can react in several ways:

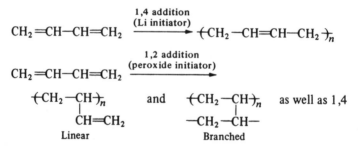

4.3 STEPWISE AND CHAIN POLYMERIZATIONS

We can envision two extremes of simplified behavior in the conversion of monomer to polymer. In a typical *stepwise polymerization,* each polymer formed can react further with monomer or other polymers. Each dimer, trimer, etc., is just as reactive as monomer. In a typical *chain polymerization* each polymer is formed in a comparatively short time, and then is "dead" and remains unchanged by the reaction of remaining monomer. Growing chains can add monomer, but neither monomer itself nor "dead" polymer can add monomer. Monomer, growing polymer, and "dead" polymer are quite different from one another in reactivity. Imagine a system of 100 monomer units (Fig. 4.1). We carry out a reaction that consumes 50 monomer units; i.e., half the original charge is unchanged while the other half is altered by having formed covalent bonds with other units.

Table 4.1 Functionality and structure

1. Functionality of 1 (A—)

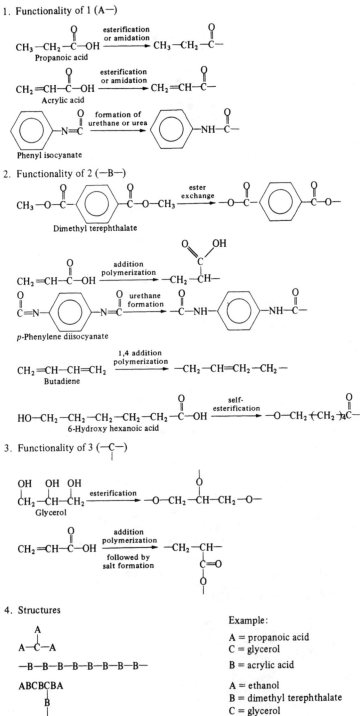

CH₃—CH₂—C(=O)—OH **Propanoic acid** → (esterification or amidation) → CH₃—CH₂—C(=O)—

CH₂=CH—C(=O)—OH **Acrylic acid** → (esterification or amidation) → CH₂=CH—C(=O)—

Phenyl isocyanate —N=C=O → (formation of urethane or urea) → —NH—C(=O)—

2. Functionality of 2 (—B—)

CH₃—O—C(=O)—⟨benzene⟩—C(=O)—O—CH₃ **Dimethyl terephthalate** → (ester exchange) → —O—C(=O)—⟨benzene⟩—C(=O)—O—

CH₂=CH—C(=O)—OH → (addition polymerization) → —CH₂—CH— (with pendant C(=O)OH)

O=C=N—⟨benzene⟩—N=C=O **p-Phenylene diisocyanate** → (urethane formation) → —C(=O)—NH—⟨benzene⟩—NH—C(=O)—

CH₂=CH—CH=CH₂ **Butadiene** → (1,4 addition polymerization) → —CH₂—CH=CH₂—CH₂—

HO—CH₂—CH₂—CH₂—CH₂—CH₂—C(=O)—OH **6-Hydroxy hexanoic acid** → (self-esterification) → —O—CH₂(CH₂)₄C(=O)—

3. Functionality of 3 (—C—)

OH OH OH
CH₂—CH—CH₂ **Glycerol** → (esterification) → —O—CH₂—CH(—O—)—CH₂—O—

CH₂=CH—C(=O)—OH → (addition polymerization followed by salt formation) → —CH₂—CH— (with pendant C=O, O—)

4. Structures

A—C(—A)(—A) (A above)

—B—B—B—B—B—B—B—B—

ABCBCBA (with B and A pendant below)

Example:

A = propanoic acid
C = glycerol
B = acrylic acid

A = ethanol
B = dimethyl terephthalate
C = glycerol

Stepwise   Chain

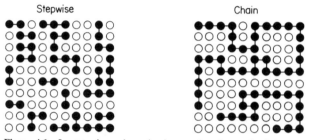

Figure 4.1 In stepwise polymerization, each polymer ($-\bullet_n$) is as reactive as monomer ($\circ$). In chain polymerization, each polymer is dead and can no longer react with monomer

The distribution of molecular weights is quite different for the two cases (Fig. 4.2). Further reaction of the stepwise polymer will gradually change the distribution toward higher values of x, the number of monomer units in a polymer. Further reaction in the chain polymer system will diminish the monomer concentration, but will also increase the amount of polymer at high values of x.

While these extremely simplified cases are met in practice, some very important exceptions occur that show the need for caution in making generalizations.

4.4 CHAIN POLYMERIZATION

Kinetic Scheme

Ethenic polymerization is an economically important class whose kinetics typify chain polymerization [1]. The terms "vinyl," "olefin," or "addition polymerization" are often used, although they are more restrictive. Usually three stages are essential to the formation of a useful high polymer:

Initiation is the creation of an "active" center such as a free radical, carbanion, or carbonium ion. For example, radicals from the thermal dissociation of benzoyl peroxide can initiate the chain polymerization of styrene.

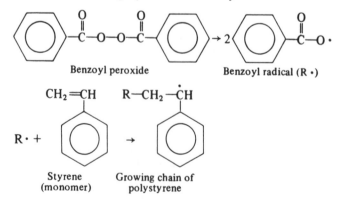

Benzoyl peroxide   Benzoyl radical (R ·)

Styrene
(monomer)

Growing chain of
polystyrene

Propagation is the addition of more monomer to growing chain end—usually very rapidly (for example, a molecular weight of 10^7 in 0.1 s) to final molecular weight.

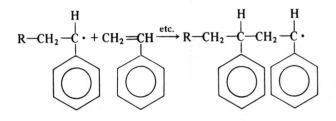

Termination is the disappearance of an "active" center.

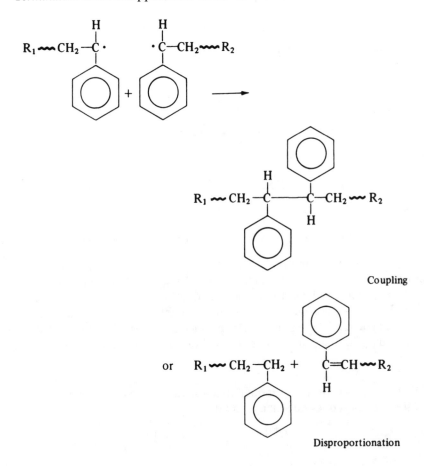

Coupling

or $R_1 \wedge CH_2-CH_2 +$ $C=CH \wedge R_2$

Disproportionation

Other propagating systems are described in Secs. 4.5 and 4.6.

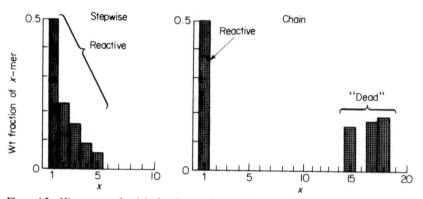

Figure 4.2 Histograms of weight fraction vs. degree of polymerization (x) for the examples of Fig. 4.1

A shorthand notation for these reactions is:
Initiation:

$$I \xrightarrow{k_i} 2R\cdot$$

$$R\cdot + M \xrightarrow{k_2} RM\cdot$$

Propagation:

$$RM\cdot + M \xrightarrow{k_p} RM_2\cdot \quad \text{or} \quad RM_n\cdot + M \xrightarrow{k_p} RM_{n+1}\cdot$$

Termination:

$$RM_n\cdot + RM_m\cdot \xrightarrow{k_t} R_2M_{n+m} \quad \text{or} \quad RM_n + RM_m$$

As seen here, each reaction is characterized by a rate constant. It is a cornerstone of polymerization theory that the reactivity of a growing polymer chain depends almost completely on the last unit added and not on previously added units or on the length of the chain. Therefore k_p is the same for a chain of any length.

The kinetic model consistent with this mechanism is simple, provided the following assumptions are valid (quantities in brackets are concentrations, e.g., moles per liter).

1. Reaction proceeds slowly enough that a steady state is reached where the radical population does not change rapidly with time

$$\left(\frac{d[R\cdot]}{dt} \quad \text{and} \quad \frac{d[M\cdot]}{dt} = 0 \right)$$

Material balances then give

$$2k_i[I] - k_2[R \cdot][M] = \frac{d[R \cdot]}{dt} = 0 \tag{4.1}$$

$$k_2[R \cdot][M] - 2k_t[M \cdot]^2 = \frac{d[M \cdot]}{dt} = 0 \tag{4.2}$$

2. The propagation reaction occurs so much more often than the others that it is effectively the only consumer of monomer.

$$\text{Rate of polymerization} = R_p = -\frac{d[M]}{dt} = k_p[M \cdot][M] \tag{4.3}$$

Under these conditions, we get first-order dependence on monomer and half-order dependence on initiator:

$$R_p = k_p \left(\frac{k_i}{k_t}\right)^{0.5} [M][I]^{0.5} \tag{4.4}$$

If only a certain fraction f of initiator fragments can successfully react with monomer, we have

$$R_p = k_p \left(\frac{f k_i}{k_t}\right)^{0.5} [M][I]^{0.5} \tag{4.5}$$

As a first approximation, f will be assumed here to be unity. Actual values often are in the range of 0.8 ± 0.2.

Often in experimental studies the monomer concentration is measured as a function of time for various starting concentrations of monomer and initiator and at various temperatures. When making high-molecular-weight polymer, relatively few initiator molecules will be consumed compared to the number of monomer molecules. If [I] can be assumed to be essentially constant, Eq. (4.4) can be integrated to give

$$2.303 \log_{10} \frac{[M]_0}{[M]} = k_p \left(\frac{k_i}{k_t}\right)^{0.5} [I]^{0.5} t \tag{4.6}$$

A semilog plot of concentration vs. time should give a line with an intersection for $t = 0$ at the initial concentration and a slope related to the rate constants and initiator concentration (Fig. 4.3). For various reasons, such as diffusion-controlled reaction or more complex mechanisms, the experimental order for a polymerization may be more or less than one. We shall see later (Sec. 5.6) that an emulsion polymerization may be zero order in overall monomer concentration. If conversion is followed only to 50% conversion (Fig. 4.3) the slight curvature presented by zero- to second-order reactions may not be readily seen. On the other hand, systematic

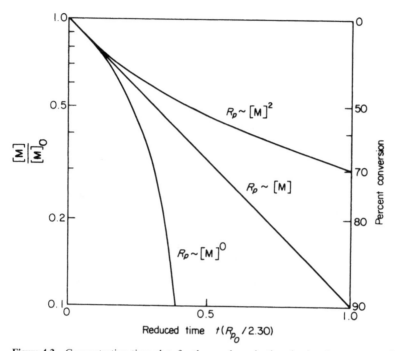

Figure 4.3 Concentration-time plots for three polymerizations having the same rate of monomer consumption, initially R_{p0}, but having a rate that depends on monomer concentration to the zero, first, or second power

errors in concentration analysis, heat transfer, or induction periods can produce curvature in such plots, which then relate to no fixed order. The deviations of many polymerizations from first-order kinetics (and half-order in initiator) often are a valuable clue to unraveling the sequence of reaction steps. Careful analysis and reasonable replication are essential. Another check on the order is to plot the polymerization rate, often the initial rate, against initial concentration of monomer and initiator (Fig. 4.4). The complication of simultaneous depletion of monomer and initiator is minimized here.

Example 4.1

In the laboratory, 100 g of monomer M and 1.00 g of peroxide Z are dissolved to make 1.000 liter of a solution in toluene. After 15.0 min at 70°C, the polymerization is stopped and the polymer is precipitated by the addition of methanol. After drying, the yield of polymer is 20.0 g. Assuming that the concentration of initiator

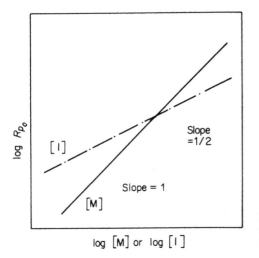

$$\log \ [M] \text{ or } \log \ [I]$$

Figure 4.4 Expected correlation of initial rate R_{po} with initial concentrations of initiator I or monomer M

and the density of the solution change negligibly, what percent conversion can be expected in 2.00 h for a plant batch which consists of 200 kg of monomer with 1.50 kg of peroxide in 2500 liters of solution in toluene at 70°C?

Solution: Let K = the rate constants lumped together and let FW = the formula weight of the peroxide. Then, in the laboratory preparation [using Eq. (4.6)]:

$$2.303 \log 1.25 = K\left(\frac{1.00}{FW}\right)^{1/2} 15.0$$

In the plant batch:

$$2.303 \log\left(\frac{[M]_0}{[M]}\right) = K\left(\frac{1.50}{2.50 \times FW}\right)^{1/2} 120$$

Thus,

$$\log\left(\frac{[M]_0}{[M]}\right)/\log 1.25 = \left(\frac{1.50}{2.50}\right)^{1/2} \frac{120}{15.0}$$

$$\frac{[M]_0}{[M]} = 3.99$$

and

$$\text{Percent conversion} = \frac{100([M] - [M]_0)}{[M]_0} = 74.9\%$$

Individual Rate Constants

The individual rate constants in Eq. (4.4) or (4.5) are not determined by the experiments described so far. The initiation rate k_i can be evaluated from nonpolymerization studies. The decomposition of a typical initiator is a first-order process:

$$-\frac{d[I]}{dt} = k_i[I] \tag{4.7}$$

The integrated form is

$$2.303 \log\left(\frac{[I]}{[I]_0}\right) = -k_i t \tag{4.8}$$

Another way of expressing the rate constant is by specifying the time it takes for the concentration of undecomposed initiator to fall to half its original value. The *half-life* is

$$t_{1/2} = \frac{0.693}{k_i} \tag{4.9}$$

As is the case with most rate constants, the temperature dependence of k_i can be correlated by an Arrhenius expression,

$$k_i = A \exp\left(\frac{-E_i}{RT}\right) \tag{4.10}$$

where E_i is the energy of activation for the decomposition process. Some typical values of $t_{1/2}$ as a function of temperature for organic peroxides are shown in Fig. 4.5. In Table 4.2 a few other radical sources are compared with two commonly used peroxides.

The integrated form of the polymerization equation [Eq. (4.6)] can be modified to take into account the time dependence of initiator concentration. Equation (4.8) is rewritten as

$$[I] = [I]_0 \exp(-k_i t) \tag{4.11}$$

This combined with Eq. (4.4) yields:

$$-\frac{d[M]}{dt} = \left(\frac{k_p^2}{k_t}\right)^{0.5} (k_i)^{0.5}[M][I]_0^{0.5} \exp\left(-\frac{k_i t}{2}\right) \tag{4.12}$$

The following equation is then obtained by integration between time zero and t (and letting $y = k_i t$):

$$2.303 \log\left(\frac{[M]}{[M]_0}\right) = \left(\frac{k_p^2}{k_t}\right)^{0.5} [I]_0^{0.5} \left(\frac{2}{y}\right)\left[1 - \exp\left(\frac{-y}{2}\right)\right] t \cdot k_i^{1/2} \tag{4.13}$$

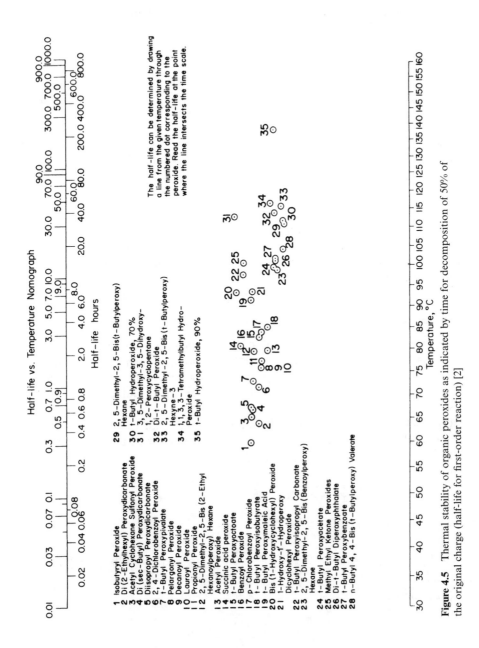

Figure 4.5 Thermal stability of organic peroxides as indicated by time for decomposition of 50% of the original charge (half-life for first-order reaction) [2]

97

Table 4.2 Typical rate constants for selected free-radical initiators [3, 4]

Name and formula	FW	T, °C	k_i, s^{-1}	E_i, kJ/mol
2,2'-azobisisobutyronitrile,	164.2	60	9.15×10^{-6}	129
Me$_2$C(CN)—N—N—C(CN)Me$_2$				
Benzoyl peroxide, (C$_6$H$_5$CO)$_2$O$_2$	242.2	60	2.76×10^{-6}	124
Dicumyl peroxide, [C$_6$H$_5$C(Me)$_2$]$_2$O$_2$	270.4	115	2.05×10^{-5}	159
*4,4'-azobis(4-cyanovaleric acid),				
HOOCCH$_2$CH$_2$—C(Me)(CN)—N				
‖				
HOOCCH$_2$CH$_2$—C(Me)(CN)—N	280.3	80	8.97×10^{-5}	141
*Potassium persulfate, K$_2$S$_2$O$_8$	270.3	30	8.83×10^{-10}	139

*Water soluble.

When y is very small, Eq. (4.13) reduces to Eq. (4.6). For example, when $y = 0.1$, the quantity $(2/y)[1 - \exp(-y/2)] = 0.975$. Although the equation is more exact than Eq. (4.6), it is not explicit in time. However, a trial-and-error solution for time usually will converge rapidly.

Example 4.2

When peroxide A is heated at 60°C in an inert solvent, it decomposes by a first-order process. An initial concentration of 5.00 mmol/liter changes to 4.00 mmol/liter after 1.00 h. Next, a solution of peroxide (0.400 mmol/liter) and monomer is held at 60°C. Additional data: $k_p = 1.8 \times 10^4$ liters/mol·s and $k_t = 1.45 \times 10^7$ liters/mol·s.

(a) What is the concentration of initiator after 10.0 min?
(b) What fraction of monomer should remain unconverted after 10.0 min?
Solution:
(a) From Eq. (4.8):

$$2.303 \log 0.800 = -k_i 3600 \qquad k_i = 62.0 \times 10^{-6} \text{ s}^{-1}$$

In the polymerization:

$$2.303 \log \frac{[I]}{4.00} = -62.0 \times 0^{-6} \times 600 = 0.0372 \ (= y) \ [I] = 0.385 \text{ mmol/liter}$$

(b) From Eq. (4.13):

$$\frac{2}{y}\left[1 - \exp\left(\frac{-y}{2}\right)\right] = 0.991$$

$$-2.303 \log\left(\frac{[M]}{[M]_0}\right) = (1.8 \times 10^4)\left(\frac{62.0 \times 10^{-6} \times 4.00 \times 10^{-4}}{1.45 \times 10^7}\right)^{1/2} 600 \times 0.991$$

Table 4.3 Selected propagation and termination rate constants [6]

Monomer	Temperature, °C	$k_p^2/k_t \times 10^3$, liters/mol·s	k_p, liters/mol·s	$k_t \times 10^{-6}$, liters/mol·s	E_p, kcal/mol	E_t, kcal/mol
Acrylamide	25	22,000	18,000	14.5		
Methyl acrylate	30	120	720	4.3	7	5
	60	460	2,090	9.5		
Methyl methacrylate	30	3.0	251	21	5	0.5
	60	10	515	25.5		
	80	21	800	30.5		
Styrene	50	0.38	209	115	7	2
Vinyl acetate	60	240	9,500	380	7	5
Vinyl chloride	25	35	6,200	1,100	4	4

$$-2.303 \log\left(\frac{[M]}{[M]_0}\right) = 0.443$$

$$\frac{[M]}{[M]_0} = 0.642 \qquad \text{(unconverted fraction of monomer remaining)}$$

A complication in using tabulated values for k_i is that not every radical produced initiates a chain. To measure the effective number of radicals, it is necessary to carry out an actual polymerization or a reaction with a compound that is known to react in the same way as monomer but more easily analyzed. If k_i is known with confidence from one monomer reaction, it can be used in others in order to separate a value of $k_p/(k_t)^{0.5}$ for the individual monomer. Some of the latter are listed in Table 4.3. A further separation of k_p from k_t is more difficult, but it can be attained by several methods. One of these is intermittent photoinitiation, which gives a measure of chain growth rate [5]. Another method is emulsion polymerization with a known particle population (Sec. 5.6). Some experimentally derived values for k_p and k_t are given in Table 4.3.

Molecular Size

The molecular weight that results from a chain polymerization with simple kinetics can be deduced from the scheme outlined so far. First, it is necessary to define the kinetic chain length and the degree of polymerization. The *kinetic chain length* v_n is the number of monomer units converted per initiating radical, so that

$$v_n = \frac{\text{rate of monomer consumption}}{-\text{ rate of radical formation}} = \frac{R_p}{2k_i[I]} \qquad (4.14)$$

Using Eq. (4.5) with $f = 1$, we get

$$v_n = \frac{k_p}{2}(k_i k_t)^{-0.5}[M][I]^{-0.5} \tag{4.15}$$

The number-average degree of polymerization x_n is

$$x_n = \frac{\text{total monomer units in system that have reacted}}{\text{total molecules in system}} \tag{4.16}$$

The corresponding molecular weight, M_n, is simply x_n multiplied by the formula weight of the repeat unit (the monomeric molecular weight).

It should be noted that, by convention, in stepwise polymerization, unreacted monomer is *included* in the numerator of Eq. (4.16). However, in chain polymerization, unreacted monomer usually is *not included*. Likewise, unreacted monomer molecules are counted in the denominator for stepwise polymerization but not for chain polymerization. The rationale for this is based on the usual discontinuity between monomer and polymer in chain polymerizations (Fig. 4.2). If termination is by disproportionation, x_n is the same as v_n; if by coupling, $x_n = 2v_n$. Both mechanisms are important in practice. For example, styrene terminates almost exclusively by coupling, while methyl methacrylate typically terminates 58% by disproportionation and 42% by coupling [7]. As polymerization proceeds, v_n, and therefore x_n, will decrease because monomer is being consumed. The distribution of molecular weights being produced then is caused by the random nature of termination together with the constantly changing average molecular weight, since each "dead" polymer increment retains its initial molecular weight.

Example 4.3

Assume that methyl acrylate terminates by coupling, $(k_p^2/k_t) = 0.460$ liter/mol·s at 60°C (Table 4.3). If an azo compound with a half-life of 10.0 h at this temperature is used to initiate polymerization of a solution of 260.0 g of monomer in toluene (to make 1.000 liter), what concentration of initiator will give an initial number-average molecular weight of 500,000? What will be the initial rate of conversion?

Solution:

$$x_n = \frac{500,000}{86.09} = 5,810 \text{ repeat units/molecule}$$

$$k_i = \frac{0.693}{36,000} = 19.3 \times 10^{-6}\,s^{-1} \qquad [M]_0 = \left(\frac{260.0}{86.09}\right) = 3.02 \text{ mol/liter}$$

In Eq. (4.15), $x_n = 2v_n$ and

$$5810 = \left(\frac{0.460}{19.3 \times 10^{-6}}\right)^{1/2} \times 3.02[I]_0^{-1/2} \qquad [I]_0 = 80.2 \text{ mmol/liter}$$

$$-\left(\frac{d[M]}{dt}\right)_0 = (0.460 \times 19.3 \times 10^{-6} \times 80.2 \times 10^{-3})^{1/2} \times 3.02$$

$$= 2.55 \times 10^{-3} \text{ mol/liter·s or } 9.18 \text{ mol/liter·h.}$$

Chain Transfer and Inhibition

It often happens that x_n is much smaller than v_n. A major reason for this is *chain transfer*. For example, dodecyl mercaptan ($C_{12}H_{25}SH$ or R′SH) is widely used as a *chain transfer agent* where it is desirable to decrease molecular weight in radical polymerization. The agent enters the propagation scheme by giving up a proton to the growing radical chain, terminating it, but possibly initiating a new chain.

$$\text{R′SH} + \text{M·} \xrightarrow{k_3} \text{MH} + \text{R′S·}$$

$$\text{R′S·} + \text{M} \xrightarrow{k_4} \text{R′SM·} \qquad \text{(equivalent to M·)}$$

If $k_3 \simeq k_4 \simeq k_p$, the interposing of such an agent will not change v_n, but it will decrease x_n, since more than one dead polymer chain has been formed from one initiating fragment (R·). In the commercial production of styrene-butadiene rubber, dodecyl mercaptan might be added at a ratio of 1 part of mercaptan to 200 parts of monomer, enough to reduce the average molecular weight severalfold. Of course, other species present during polymerization can act as chain transfer agents. Solvent, initiator, monomer, or polymer sometimes are involved. In the simple case of only one transfer reaction with constant k_3 and with k_4 large compared to k_3, the rate of radical termination leading to new molecules is increased by the number of times the radical is transferred:

$$\text{Rate of molecule generation} = Yk_t[\text{M·}]^2 + k_3[\text{R′SH}][\text{M·}] \qquad (4.17)$$

where Y has a value of 1 if termination is by coupling and a value of 2 if termination is by disproportionation. Then the degree of polymerization becomes

$$x_n = \frac{k_p[\text{M}][\text{M·}]}{\text{rate of molecule generation}} \qquad (4.18)$$

Rearranging and substituting (4.17) into (4.18):

$$(x_n)^{-1} = \frac{Yk_t[\text{M·}]}{k_p[\text{M}]} + \frac{k_3[\text{R′SH}]}{k_p[\text{M}]} \qquad (4.19)$$

or

$$(x_n)^{-1} = (x_n)_0^{-1} + \frac{C_s[\text{R}'\text{SH}]}{[\text{M}]} \tag{4.20}$$

where $(x_n)_0$ is the number-average degree of polymerization that would occur in the absence of any chain transfer. A plot of $(x_n)^{-1}$ vs. $[\text{R}'\text{SH}]/[\text{M}]$ should give a straight line with an intercept of $(x_n)_0^{-1}$ and a slope of C_s, the *chain transfer constant*. The chain transfer constant is specific for a given monomer and agent combination under particular conditions of solvent composition and temperature. In the more general case, chain transfer can take place between radical chains and many other species such as monomer, solvent, or even "dead" polymer. Any molecule from which a hydrogen atom can be abstracted to terminate one chain and to leave behind a radical of adding monomer is a possible chain transfer agent. A broader form of Eq. (4.20) for the general case is:

$$(x_n)^{-1} = (x_n)_0^{-1} + \sum \frac{C_i[\text{A}_i]}{[\text{M}]} \tag{4.21}$$

where C_i is the chain transfer constant for the agent A_i present in concentration $[\text{A}_i]$.

Example 4.4

Polymerization of a 0.500-molar solution of methyl methacrylate (FW $= 100.12$) in toluene is carried out at 60°C. The initial polymer formed has a number-average molecular weight of 500,000. What concentration (g/liter) of carbon tetrabromide will reduce the molecular weight to 100,000? The C_i for CBr_4 (FW $= 331.65$) is 0.270 [8].

   Solution:

$$(x_n)_0 = \frac{500,000}{100.1} = 5,000 \qquad (x_n) = \frac{100,000}{100.1} = 1,000$$

$$[\text{M}] = 0.500 \text{ mol/liter}$$

(Eq). 4.21: $\quad \dfrac{1}{1,000} = \dfrac{1}{5,000} + \dfrac{0.270[\text{A}_i]}{0.500}$

$$[\text{A}_i] = 1.48 \text{ mmol/liter} = 0.49 \text{ g/liter}$$

If $k_4 < k_p$, the overall rate will be decreased by chain transfer. This is "degradative" chain transfer. In fact, if k_4 is so small as to be negligible, chain transfer results in inhibition and the agent is a free-radical sink, or *inhibitor*. Almost all ethenic monomers are stored and shipped containing such a material to prevent adventitious polymerization. Hydroquinone and diphenylamine are compounds that at concentrations of 10 to 200 ppm are effective inhibitors.

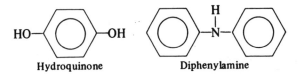

Hydroquinone   Diphenylamine

The radicals that result from these compounds are unreactive toward monomer. Before a polymerization is carried out, inhibitors can be removed by distillation or by extraction (e.g., dilute caustic in the case of hydroquinone in a water-insoluble monomer), or they can simply be overcome by an excess of initiator, which "uses up" the inhibitor during an "induction period."

Dissolved oxygen in a polymer system often acts as an inhibitor by reacting with radicals to give stable species. The variable induction periods that result can be annoying. Free-radical polymerizations usually are carried out under an inert atmosphere for this reason. Some early workers sought to obviate the necessity for deaeration in commercial reactors by adding reducing agents to act as scavengers for oxygen. The greatly increased rates that resulted when that was done were later explained by the fact that, while oxygen scavenging was not very efficient, the reducing agent-oxidizing initiator combination produced radicals much faster than the oxidizing agent could by itself. Such *redux couples* now are commonly employed to obtain fast rates at low temperatures by "reduction activation." For example, the initial rate of polymerization of acrylamide with oxidizing agent (persulfate) is increased sevenfold by a modest amount of reducing agent (thiosulfate) (Fig. 4.6).

In many systems a polymer can be dead to the usual addition of monomer, but alive in the sense that it can act by a different mechanism as a transfer agent. At high temperatures and pressures, growing polyethylene chains abstract protons from themselves ("backbiting") or from other dead polymer.

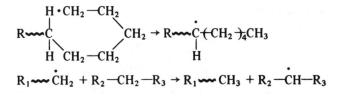

These two processes result in short and long branches, respectively, because each new radical can add to monomer. Monomer itself can be a chain transfer agent. It is hard to tell whether vinyl chloride (CH_2=CHCl) terminates by coupling or by disproportionation, since most chains are terminated by having a growing radical abstract a chlorine atom from a monomer molecule.

$$R-CH_2-\dot{C}HCl + CH_2=CHCl \rightarrow R-CH_2-CHCl_2 + CH_2=\dot{C}H$$

The resultant radicals are not thought to initiate chains but rather to combine to form butadiene which subsequently copolymerizes with vinyl chloride [10].

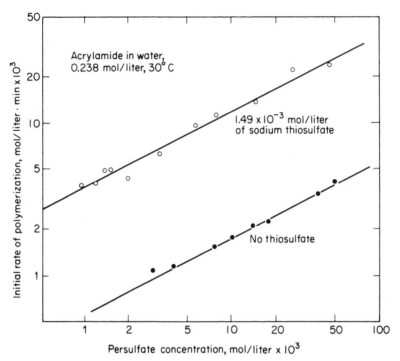

Figure 4.6 Increase in polymerization rate by addition of reducing agent, $Na_2S_2O_3$, to reaction initiated by $K_2S_2O_8$. Slope of log-log plot is $\frac{1}{2}$ in either case [9]

Temperature Dependence

The temperature dependence of the reaction rate constants k_i, k_p, etc., can be expressed as conventional Arrhenius equations:

$$k = A \exp\left(-\frac{E_a}{RT}\right) \tag{4.22}$$

where E_a is the *activation energy*, A is the *collision frequency factor*, R is the gas constant, and T is the absolute temperature. The temperature dependence of the overall polymerization rate R_p follows from Eq. (4.4):

$$k_p\left(\frac{k_i}{k_t}\right)^{1/2} = A_p\left(\frac{A_i}{A_t}\right)^{1/2} \exp\left(\frac{E_i/2 - E_p - E_t/2}{RT}\right) \tag{4.23}$$

As an example, E_i for persulfate decomposition in water is 33.5 kcal/mol and $E_p - E_t/2$ for acrylamide polymerization is 1.5 kcal/mol. The calculated activation energy for R_p, 18.25 kcal/mol, is in fair agreement with a measured value of 16.9 kcal/mol [9].

Generally, the overall rate of polymerization increases with temperature. However, the variation of degree of polymerization with temperature follows from Eq. (4.15):

$$\frac{d(\ln x_n)}{dT} = \frac{d(\ln v_n)}{dT} = \frac{E_p - E_t/2 - E_i/2}{RT^2} \tag{4.24}$$

Example 4.5

Solution polymerization of monomer A initiated by peroxide P at two temperatures (using the same concentrations of A and P) gives the following results:

	35.0°C	53.0°C
Conversion in 30 min	25.0%	65.0%
Initial molecular weight	500×10^3	300×10^3

What is the energy of activation, E_i, for the decomposition of P?

Solution: We can abbreviate Eqs. (4.6) and (4.15) as

$$\ln\left(\frac{[M]_0}{[M]}\right) = K(k_i)^{1/2}[I]^{1/2}t \quad \text{and} \quad M_n = M_0 x_n = \frac{M_0 K}{(k_i)^{1/2}[I]^{1/2}}$$

where M_0 is the formula weight of a repeat unit. Only K and k_i depend on temperature.

At 35.0°C (condition A):

$$\ln\left(\frac{1}{0.75}\right) = K_A(k_i)_A^{1/2}[I]^{1/2}(30) \quad \text{and} \quad 500,000 = \frac{M_0 K_A}{(k_i)_A^{1/2}[I]^{1/2}}$$

so that

$$\ln\frac{(1/0.75)}{500,000} = \frac{(k_i)_A(30)}{M_0}$$

At 53.0°C (condition B):

$$\ln\left(\frac{1}{0.35}\right) = K_B(k_i)_B^{1/2}[I]^{1/2}(30) \quad \text{and} \quad 300,000 = \frac{M_0 K_B}{(k_i)_B^{1/2}[I]^{1/2}}$$

so that

$$\ln\frac{(1/0.35)}{300,000} = \frac{(k_i)_B(30)}{M_0}$$

The ratio $(k_i)_B/(k_i)_A = 6.08$ (all other terms cancel out), so

$$\ln\left[\frac{(k_i)_B}{(k_i)_A}\right] = 1.805 = (E_i/8.3145 \text{ J/mol·K})(1/308 \text{ K} - 1/326 \text{ K})$$

And $E_i = 83.7$ kJ/mol.

Conjugated Diene Monomers

This discussion of chain polymerization has centered on free-radical polymerization of an ethenic monomer. Conjugated dienes such as 1,3-butadiene often polymerize as bifunctional monomers with 1,4 addition rather than as tetrafunctional monomers.

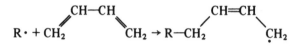

One double bond remains in the main chain for each monomer that reacts. Free-radical polymerization can also produce a 1,2 form of addition at the same time, so that the polymer that results may be a kind of copolymer with some unsaturated side groups and perhaps even some network formation. Ionic and coordination complex schemes of polymerization, described in the next section, not only can give almost all 1,4 addition with butadiene, they also can be tailored to give all cis or all trans polymer as well. Free-radical polymerization usually gives a mixture of isomeric forms.

4.5 IONIC POLYMERIZATION

The reactivity of ethenic monomers to polymerization by radicals, ions, and complexing agents varies with the structure in a manner that can be correlated though not always quantitatively predicted. A convenient classification is that by Schildknecht (Table 4.4). It can be seen that for the vinyl monomer ($CH_2{=}CHX$), cationic initiation is favored when X is electron-donating and anionic when X is electron-withdrawing. In radical polymerization, the steric hindrance of the methyl group next to the double bond causes methacrylates, $CH_2{=}C(CH_3)(COOR)$, to react much more slowly than the corresponding acrylates, $CH_2{=}CH(COOR)$.

A major difference between radical polymerization and the various ionic methods is that, in the latter, the incoming monomer must fit between the growing chain end and an associated ion or complex. The growing radical chain, on the other hand, has no such impediment at the growing end.

Cationic Polymerization

Cationic polymerizations tend to be very rapid even at low temperatures. The polymerization of isobutylene with $AlCl_3$ or BF_3 is carried out commercially at $-100°C$

Table 4.4 Type of chain polymerization suitable for common monomers [11]

Cationic only

Isobutylene and derivatives, $CH_2=C(CH_3)_2$,
$CH_2=C(CH_3)R$
Alkyl vinyl ethers, $CH_2=CHOR$, and related ether types,
$CH_2=C(R)OR$, $CH_2=C(OR)_2$,
$CH_2OCH=CHOCH_3$
Coumarone, indene
Derivatives of α-methylstyrene,
$CH_2=C(CH_3)$

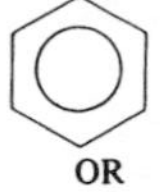

OR

Free radical only

Halogenated vinyls,
$CH_2=CHX$
$CH_2=CX_2$
$CF_2=CFX$
$CF_2=CX_2$
where X = halogen, hydrogen
(but $CH_2=CH_2$ not included)
Vinyl esters,
$CH_2=CHOCOR$

Cationic or free radical

N-vinylcarbazole,

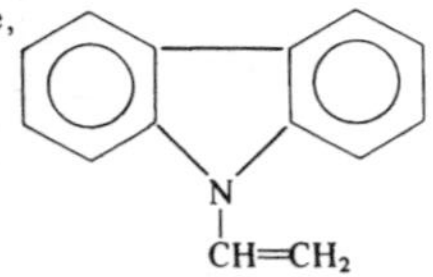

N-vinylpyrrolidone,

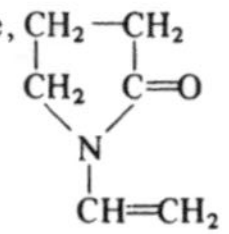

Anionic only

Vinylidene cyanide,
$CH_2=C(CN)_2$, and related
cyano derivatives, $CH_2=C(CN)Y$,
where Y = SO_2R, CF_3, COOR
Nitroethylenes,
$CH_2=C(NO_2)R$

Free radical or anionic

Acrylic and methacrylic esters
Vinylidene esters, $CH_2=C(COOR)_2$
Derivatives of acrylonitrile, $CH_2=CRCN$
$CH_2=CRCONH_2$

Cationic, free radical, or anionic

Ethylene, $CH_2=CH_2$: butadiene, $CH_2=CH—CH=CH—CH=CH$; styrene, $CH_2=CH$; α-methylstyrene, methyl vinyl ketone

Coordination or supported metal oxide catalyst

α-Olefins including ethylene, dienes, alkyl vinyl ethers

Table 4.5 Some chain polymerization systems

Carbocation (cationic polymerization) [13]

Initiation:

$$BF_3 \cdot H_2O + CH_2 = \overset{\underset{\displaystyle CH_3}{|}}{\underset{\underset{\displaystyle CH_3}{|}}{C}} \rightarrow CH_3 - \overset{\underset{\displaystyle CH_3}{|}}{\underset{\underset{\displaystyle CH_3}{|}}{C^+}} \; [BF_3 \cdot OH]^- \quad (\text{gegen ion})$$

Propagation:

$$CH_3 - \overset{\underset{\displaystyle CH_3}{|}}{\underset{\underset{\displaystyle CH_3}{|}}{C^+}} \; [BF_3 \cdot OH]^- + CH_2 = \overset{\underset{\displaystyle CH_3}{|}}{\underset{\underset{\displaystyle CH_3}{|}}{C}} \rightarrow CH_3 - \overset{\underset{\displaystyle CH_3}{|}}{\underset{\underset{\displaystyle CH_3}{|}}{C}} - CH_2 - \overset{\underset{\displaystyle CH_3}{|}}{\underset{\underset{\displaystyle CH_3}{|}}{C^+}} \; [BF_3 \cdot OH]^-$$

Termination:

$$\sim\!\!\sim\!\! CH_2 - \overset{\underset{\displaystyle CH_3}{|}}{\underset{\underset{\displaystyle CH_3}{|}}{C^+}} \; [BF_3 \cdot OH]^- \rightarrow \sim\!\!\sim\!\! CH_2 - \overset{\underset{\displaystyle CH_3}{|}}{C} = CH_2 + H^+[BF_3 \cdot OH]^-$$

Carbanion (anionic polymerization) [11]

Initiation:

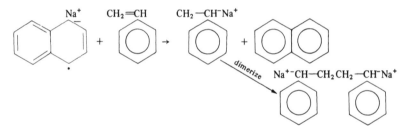

Propagation:

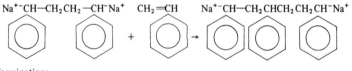

Termination:

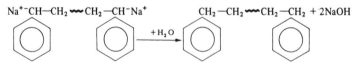

Table 4.5 (*Continued*)

Insertion polymerization (Ziegler catalyst) [15]

Initiation and propagation:

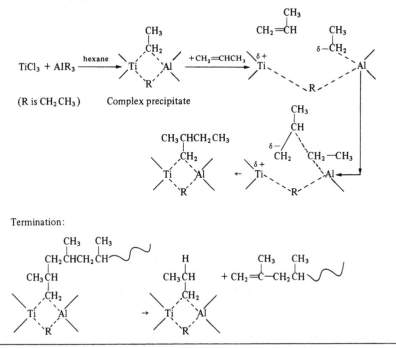

Termination:

(see Sec. 14.3). An estimate of the lifetime of a growing chain of isobutylene in this case is about 10^{-6} s [12]. This is much shorter than the usual lifetime of a free-radical chain such as vinyl acetate, which may be several seconds. If ionic catalysis is used in a heterogeneous system—e.g., insoluble catalyst, soluble monomer, and polymer—the rate of diffusion of monomer to and polymer from the surface may control the overall rate. In such cases, intensity of agitation, catalyst particle size, and viscosity of solvent can become very important. Usually a cocatalyst is involved. For example, water is the cocatalyst with BF_3 (Table 4.5) and it is $(BF_3 \cdot OH)^-$ that forms the "gegen" ion (also called the "counter ion") at the growing end of the chain. The fitting of the incoming monomer unit between chain and gegen ion can lead to stereoregularity, as when vinyl alkyl ethers are polymerized at very low temperatures [16].

As recently as 1975, it seemed unlikely that cationic polymerization would lead to a "living" system. A "living" polymerization can be defined as one with no termination mechanism. The growing end of the living polymer remains reactive

Table 4.6 Synthesis of a triblock polymer using cationic polymerization at −80°C [18]

		Concentration (M = mol/liter)	
Ingredient	Function	Step 1	Step 2
DCE, dicumyl ether	Initiator	1.6 mM	—
DMA, N,N-dimethylacetamide	Electron donor	1.6 mM	1.6 mM
DtBP, 2,6-di-tert-butylpyridine	Proton trap	1.6 mM	1.6 mM
TiCl$_4$, titanium tetrachloride	Activator	0.0256 M	0.0128 M
IB, isobutylene	Monomer	1.45 M	—
In, indene	Monomer	—	0.37 M
MCHx, methylcyclohexane	Solvent	40% by vol.	
CH$_3$Cl, methyl chloride	Solvent	60% by vol.	
MeOH, methanol	Quencher	—	—

Characterization (molecular weights obtained using column chromatography with PIB standards):

	M_n	M_w/M_n
PIB midsegment	54,250	1.19
PIn-b-PIB-b-PIn triblock	93,400	1.39

after all the monomer (say, monomer A) has been added. If a new monomer, B, is introduced into the system, it may add to the living end and result in a block of polymer B being covalently attached to the previously existing block of A. Cationic initiation was long regarded as having "built-in" termination and chain transfer mechanisms that would not allow keeping the end group reactive. However, it has been found that certain combinations of cationic initiators and modifiers will indeed allow the formation of living polymers [17].

Kennedy [18] has described the formation of a triblock polymer with a center block based on isobutylene and end blocks based on indene. By using an initiator which is difunctional, the center block has two living ends so that both end blocks are added simultaneously. In one example (see Table 4.6 for definitions), DCE, DMA, DtBP, IB, and solvents were charged to a flask held at −80°C. To start the first step, the indicated increment of TiCl was added rapidly. Conversion proceeded according to approximately first-order kinetics in monomer (about 50% conversion in 20 min, 80% in 50 min). After 90 min, commencing the second step required the addition of a solution of In, DMA, and DtBP (also at −80°C). The second step also was carried out for 90 min. At this point, a few milliliters of MeOH were added to quench the end groups. It can be seen (Table 4.6) that the molecular weight obtained is close to the theoretical value predicted from the initial concentrations of DCE, IB, and In. If each molecule of DCE generates one molecule of polymer, the molecular weight of the midsegment should equal the molar ratio of IB to DCE times the formula weight of IB:

$$M_n(\text{predicted}) = 56.1 \times \frac{[\text{IB}]}{[\text{DCE}]} = 56.1 \times \frac{1.45}{0.0016} = 51,000$$

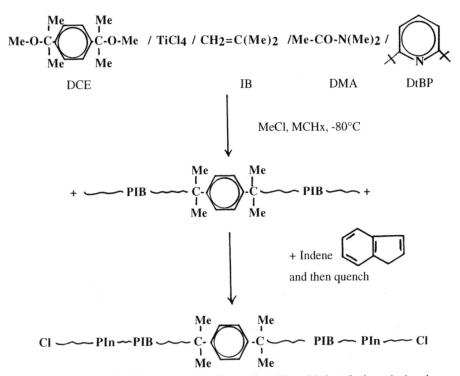

Figure 4.7 Synthesis of a triblock copolymer of isobutylene, IB, and indene, In, by cationic polymerization (See also Table 4.6) [18]

A similar calculation predicts M_n = 78,000 for the triblock, which is somewhat lower than the measured value. As will be seen later (see Sec. 6.2), a ratio of M_w/M_n much less than 2 is indicative of a narrow molecular-weight distribution and is typical of living polymers. The steps are summarized in Fig. 4.7.

Other kinds of polymerizations such as those which involve opening rings can be initiated by cationic mechanisms. Photogenerated acids can be used to open oxirane rings, for example.

Anionic Polymerization

Although there are several mechanisms by which living polymers can be prepared, anionic polymerization probably represents the most successful commercial application. The particular usefulness of lithium alkyls has been with dienes to give *cis*-polysioprene and *cis*-polybutadiene, although they can yield isotactic or atactic polymers of styrene and methyl methacrylate. It is a peculiarity of polymerizations with lithium alkyl that there is no termination step. The rate of polymerization depends on the amount of initiator and monomer present [19].

Initiation:

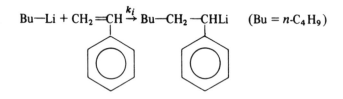

Propagation:

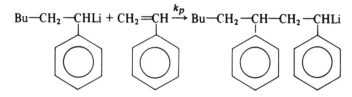

If the rate of initiation is much faster than the rate of propagation, each initiator should start one polymer chain. If all these start at time zero and grow to consume all the monomer, a narrow molecular-weight distribution (Poisson) results (see Sec. 6.2). The number-average degree of polymerization is given by $x_n = [M]/[I]$, where [M] is the initial monomer concentration and [I] is the initiator and, therefore, also the polymer concentration in moles per unit volume.

When butyl lithium is added to a monomer solution, the rate of initiation and propagation both depend on monomer concentration. If, however, the initiator is "seeded" by adding some monomer and then added to the remaining monomer, only the propagation step is observed. Under these circumstances the reaction is first order in monomer (Fig. 4.8), the straight line corresponding to

$$-\frac{d[M]}{dt} = k_p[M]\,[R_s] \tag{4.25}$$

or

$$\ln\frac{[M]_0}{[M]} = k_p t[R_s] \tag{4.26}$$

where [M] is the monomer concentration at time t and $[M]_0$ at time $t = 0$ and $[R_s]$ is the concentration of growing polymer "seeds." The rate is proportional to the initiator concentration to the first power in tetrahydrofuran and to the second power in hydrocarbon solvents. This is because the lithium is associated in nonpolar solvents as a dimer. The final step is decomposition of the lithium alkyl by addition of water or alcohol. Various end groups can be attached to the chains by the reagent employed to "kill" the polymer. Since the polymer chains are "alive," a second monomer can be grown on a "seed" made from a different monomer.

Commercial copolymers based on styrene and butadiene have been prepared

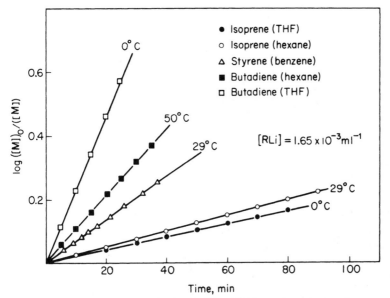

Figure 4.8 First-order rate plots for "seeded" butyl lithium polymerizations for monomers in benzene, hexane, or tetrahydrofuran [19]

by this method. Both the random (both monomers present) and block (first one monomer added then the next) are available.

In one example [20], styrene (S) (60 g) in benzene (1400 g) is polymerized using 0.003 mol of *sec*-butyl lithium at 40°. Then isoprene (I) (450 g) is added and polymerized. Finally, an additional 60 g of styrene (S) is added and polymerized. Alcohol can then be added to cap the chain end and to precipitate the polymer. The chain thus made has the structure SIS, where each letter in the abbreviation represents a *block* of monomer. A somewhat different block copolymer can be made by coupling together "living polymers" instead of merely capping them. Careful removal of moisture and air is necessary in either case. To make such a *radial* block copolymer, ordinary glass beverage bottles may be used as reactors [21]. Cyclohexane (40 cm³) is added followed by styrene (16 g) and *sec*-butyl lithium. After 40 min at 70°C in a tumbling rack, the bottle is cooled to 40°C and butadiene (B) (24 g) is added. At this point the orange-colored polystyryl anion is converted to the colorless polybutadienyl anion. Another hour at 70° gives essentially complete conversion. The final step is to add a coupling agent that will join the living polymers together. If a difunctional agent such as methylene iodide is used, a linear block copolymer is formed.

$$SBLi + CH_2I_2 \rightarrow SB\!-\!CH_2\!-\!BS + 2LiI$$

Methyl trichlorosilane is a trifunctional agent.

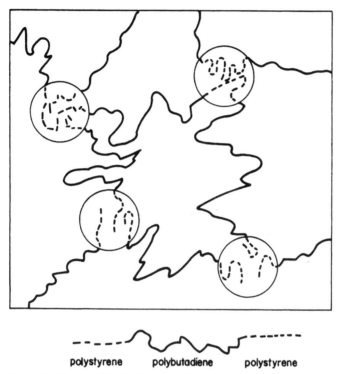

polystyrene   polybutadiene   polystyrene

Figure 4.9   Typical triblock thermoplastic elastomer. Circles are glassy domains of polystyrene acting as temperature-sensitive cross-links holding the polybutadiene segments in place

$$\text{SBLi} + \text{Cl}_3\text{Si}-\text{CH}_3 \rightarrow \text{SB}-\overset{\overset{\textstyle \text{CH}_3}{\textstyle |}}{\underset{\underset{\textstyle \text{BS}}{\textstyle |}}{\text{Si}}}-\text{BS} + 3\,\text{LiCl}$$

In either case the coupling is highly efficient, over 90% of the SBLi chains being incorporated into the block copolymer. The examples cited here are useful as elastomers.

At room temperature these thermoplastic elastomers act as though they were covalently cross-linked, exhibiting high resilience and low creep. However, they flow when heated above 100°C, like true thermoplastics. The reason is that the large polystyrene blocks form glassy aggregates that function as massive cross-links (Fig. 4.9). The independent behavior of the polybutadiene portion is evident in the behavior of dynamic loss with changing temperature (Fig. 4.10). Block copolymers can be made with tensile strengths of over 3000 psi (20 MPa) and high elongations. In contrast, conventional SBR rubber cannot be made to exceed 1 or 2 MPa whether cross-linked or not unless a reinforcing filler is added.

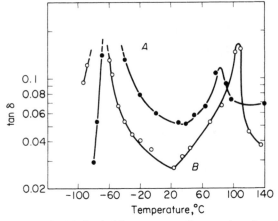

Figure 4.10 Mechanical loss tangents of styrene-butadiene block copolymers that have been cross-linked. *A* has 6% styrene as blocks and 19% as random units, while *B* has all 25% styrene in blocks of about 350 units. Test run at a frequency of 0.1 Hz [22]

When the initiator is based on the reaction product of sodium and naphthalene (Table 4.5), the growing chain is difunctional. It is possible to put various functional groups on both ends of a single chain in a single step. Low-molecular-weight polymers (oligomers) of styrene, a-methlystyrene, and diene monomers have been produced with end groups as diverse as carboxyl, hydroxyl, and amine. These then can be used as blocks in other polymers by appropriate reactions such as esterification or amide formation.

Polar monomers present a problem with the sodium and lithium initiators described above. For example, methyl methacrylate does not polymerize satisfactorily due to side reactions such as cyclization. A number of modified initiating systems have been found that do result in high-molecular-weight, narrow-distribution polymers of the methacrylates and acrylates. Moreover, there is some control over tacticity. Workers at ICI described the polymerization of methacrylates using a modified lithium initiator [23]. The living nature of the growing polymer chain was demonstrated unequivocally in one experiment in which monomer was added in separate portions. Between additions the molecular weight was measured.

One part of the initiator, Al(BHT)(iB$_2$), was produced by reacting BHT (commonly called butylated hydroxytoluene) with trisiobutylaluminum. The actual initiating structure (Fig. 4.11) is the reaction product of Al(BHT)(iB$_2$) with *tert*-butyllithium (BuLi). The molecular size of the resulting polymer depends only on the concentration of the BuLi as long as the ratio of Al(BHT)(iB$_2$) to BuLi exceeds unity. In the experiment, 2.46 mmol of Al(BHT)(iB$_2$) was added to 80 ml of dry toluene. After cooling the mixture to $-10°$C, 1.23 mmol of BuLi was added as a solution in pentane. The first monomer addition consisted of 75 mol of methyl

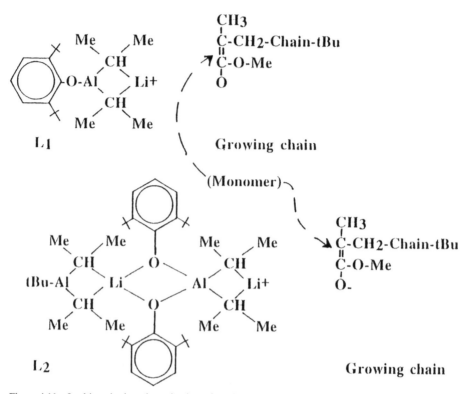

Figure 4.11 In this anionic polymerization of methyl methacrylate, monomer is inserted between the propagating center L_1 or L_2 and the growing end of the chain [23]

methacrylate for each mol of BuLi. The reaction was complete within 5 min at 0°C. The M_n measured on a very small sample removed from the reaction mixture was 11,700. After waiting 60 min, an identical increment of monomer added to the system increased M_n to 23,200. Additional increments at widely spaced time intervals increased M_n to 33,900 and then to 44,100.

It is obvious that all the polymer chains remained "alive" throughout the entire process. Also, M_w/M_n remained at about 1.1 throughout the experiment. The fact that the measured molecular size is around 1.5 times the ratio of monomer to initiator indicates that not all the lithium was available for initiation. The explanation given is that there are two forms of complex in equilibrium, L_1 and L_2 (Fig. 4.11). In the example cited above, equal molar quantities of L_1 and L_2 should result in the observed molecular size since the effective number of initiation sites are decreased by a factor of 2/3. The polymer produced in this experiment was about 70% syndiotactic and 3% isotactic. Block and random copolymers can be prepared with this initiating system.

Other workers [24] have used modified lithium initiators to polymerize acrylates in addition to methacrylates in a living fashion. The polymerizations are complete in 10 min even though the temperature is typically under −80°C.

4.6 INSERTION POLYMERIZATION

The term *coordination complex polymerization* was used for some time to embrace many systems that could give some control over the stereochemistry of the product. However, *insertion polymerization* is a more general description of polymerizations that include a "hindered propagation site" [25].

Ziegler-Natta Catalysts

Something of a revolution occurred in the realm of polymer chemistry in 1955 when Giulio Natta demonstrated conclusively that stereospecific polymers could be produced with synthetic catalysts [26]. The systems he employed were primarily α-olefins polymerized by Ziegler catalysts—the discovery of Karl Ziegler. In recognition of this work, Natta and Ziegler shared the Nobel Prize in chemistry for 1963. Since the early work, isotactic, syndiotactic, and cis and trans polymers have been made by using ionic and even some free-radical initiators at low temperatures with various types of monomers.

Many systems other than free-radical or ionic have been investigated for the production of polymers. Several have assumed commercial importance because they lead to structures that cannot easily be achieved by free-radical or ionic routes. Ziegler-type catalysts [27] originally involved the formation of a complex precipitate from aluminum triethyl and titanium tetrachloride:

$$AlEt_3 + TiCl_4 \xrightarrow{\text{hexane}} \text{complex, colored precipitate} + \text{some volatile by-products}$$

Ethylene bubbled into the suspension at room temperature polymerizes very rapidly to give a high-molecular-weight, linear polyethylene. The linearity leads to higher crystallinity than for branched polymer, so that the product differs significantly from high-pressure, free-radical polythylene. With propylene, an isotactic polymer can be produced. The mechanism may proceed as outlined in Table 4.5, the stereospecificity stemming from the peculiar topography of the catalyst at the site where monomer is inserted. For a time it was postulated that a heterogeneous catalyst would be needed for stereospecific activity, but this was refuted by subsequent work with similar catalysts, which, though complex, were soluble.

With diolefins such as butadiene and isoprene, both *cis*- and *trans*-1,4 polymers can be obtained. Variations in transition halide and/or ratio of alkyl to halide give different structures. For example, $TiCl_4$—AlR_3 with Al/Ti > 1 gives 96% *cis*-polyisoprene and with Al/Ti < 1 gives 95% *trans*-polyisoprene [27].

It is instructive to see how the Ziegler-Natta catalyst has evolved with respect

to its use for the polymerization of propylene. Goodall [28] identifies three generations of catalysts. The first, dating to the 1950s, resulted from the reaction of TiCl
with triethylaluminum in cold hydrocarbons. The resulting slurry was slowly
heated to 160 to 200°C for several hours to convert β-$TiCl_3$ (brown) to the γ-$TiCl_3$
(purple) which had been shown to be essential for propylene polymerization. A
cocatalyst which alkylates the Ti atoms generating active centers (usually diethyl-
aluminum chloride) also should be present. Early on it was found that electron
donors such as ethers, esters, and amines would enhance the polymerization by
reacting with the closely related ethylaluminum dichloride, which, unlike the
monochloride, poisons the catalyst.

A second generation was developed in the 1970s. It was found that ether extraction of the brown precipitate to remove $AlCl_3$ followed by heat treatment at
60 to 100°C in the presence of excess $TiCl_4$ gave a catalyst with smaller, more
active crystallites.

In the third generation, during the 1980s, $MgCl_2$ was used as a support for the
catalyst. Since over 400 patents on catalyst preparation were granted in the decade
before 1985, it is difficult to identify one optimum process. Goodall describes a
"common" method as starting by ball-milling anhydrous MgCl with ethyl benzoate
to make a catalyst support to which is added $TiCl_4$ with a diluent such as hexane.
After washing the solid with more hexane, the "precatalyst" is dried in a vacuum.
The final polymerization system requires the presence of trialkylaluminum and a
Lewis base such as ethylbenzoate.

The three generations can be characterized by the fraction of isotactic polypropylene, PP, produced, and by the productivity, kilograms of PP per gram of Ti.
The first generation yielded PP which was only about 92% isotactic. A separate
solvent extraction of atactic polymer was required to give a suitable product. The
subsequent generations yielded 96% isotactic polymer, obviating the extraction
step. In the first two generations, productivities of 15 kg PP/g Ti were typical. With
the third generation, 1500 kg PP/g Ti can be obtained. In a laboratory preparation,
a third-generation-type catalyst yielded 25 kg PP/g Ti in 1 h at 63°C and 4 bar
pressure [29]. The PP was 93% isotactic.

Tait [30] points out that the original milled catalyst particles (perhaps 5 to 50
mm in diameter) are fragile and break easily. As the polymerization proceeds, the
catalyst-polymer particle fractures gradually from the outside and the number of
active centers increases. The final polymer particle that results does not have the
intact original catalyst particle at its center, but instead is multigrained with many
individual catalyst crystallites dispersed throughout.

Metallocene Catalysts

Although they have been described as a type of Ziegler-Natta catalyst, the *metallocenes* have given rise to polyolefins of such novel properties as to merit a separate
category. A simple metallocene is defined as the cyclopentadienyl derivative of a
transition metal. In practice, the metallocene is used in conjunction with a methyl-

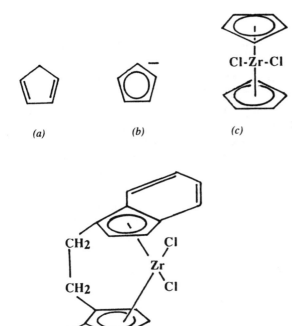

Figure 4.12 Structures of (*a*) cyclopentadiene, CP, (*b*) CP⁻, (*c*) CP₂ZrCl₂, and (*d*) rac-ethylene(indenyl)zirconium dichloride

aluminoxane, MAO. When cyclopentadiene loses a hydrogen from a singly bonded carbon in the ring, a very stable *p*-bonded structure similar to benzene is formed, but with an overall negative charge (Fig. 4.12). Kaminsky [31] reported the polymerization of propylene, ethylene, and various comonomers using Cp₂ZrCl₂ and similar soluble metallocenes with a large excess of coactivator MAO. The MAO is produced by the controlled partial hydrolysis of trimethylaluminum.

$$Al(CH_3)_3 + H_2O \rightarrow (CH_3)\{Al(CH_3)\!-\!O\}_n\!-\!Al(CH_3)_2$$

The structure of MAO with *n* in the range of 4 to 20 apparently varies with the method of preparation. Branched chains and rings have been postulated in addition to the linear form shown above. The typical ratio of Al in the MAO to transition metal in the metallocene is of the order of 100 to 10,000.

The dicyclopentadiene compound yielded atactic polypropylene as demonstrated by extraction with hydrocarbon solvents. However, when Kaminsky replaced the *p*-bonded cyclopentadienyl ligands by ethylene-bridged indenyl rings (Fig. 4.12), isotactic polymer was produced, and less than 1% of the polymer was extractable. Furthermore, the productivity at 21°C over an 8-h period was about 400 kg PP/g Zr.

Commercialization of metallocene technology has taken the form of several competing product lines. Exxon introduced a line of Exxpol resins made from metallocene/MAO catalysts with the addition of a noncoordinating anion such as tri(pentafluorophenyl) boron, said to provide charge balance and to stabilize the catalyst cation [32]. Because the metallocenes are soluble and well defined, they are termed "single-site" catalysts as opposed to the conventional heterogeneous Ziegler-Natta catalyst, which has a multiplicity of reaction sites differing in activity. Where the Ziegler-Natta-based polymers have broad molecular-weight distributions and uneven comonomer distributions (for ethylene-hexene copolymerization, for example), the metallocene-based polymers have narrow molecular-weight distributions and uniform comonomer distributions. Crystalline polymers of propylene and amorphous copolymers of ethylene with propylene have also appeared. In one example [33], a reaction mixture with MAO yielded 2.5 kg PP/g Ti in 1 h at 30°C.

In a somewhat confusing development, another U.S. company, Dow, was able to obtain a composition-of-matter patent [34] for "elastic substantially linear olefin polymers." The said polymers are defined in terms of shear sensitivity which arises from long chain branching. The confusion to the casual observer is that the catalysts mentioned in the Exxon patent [33] are cited as yielding the new materials. In their commercial literature Dow refers to their "Insite" process as using "constrained geometry" catalysts. Also mentioned in news releases is the possible commercial production of syndiotactic polystyrene using similar processes. At least a dozen other companies are actively involved with metallocene-based polyolefins.

Supported Chromium Catalysts

Metal oxides have been used to give linear polyolefins, although they have not been used for isotactic structures. Nickel, cobalt, chromic, or vanadium oxides supported on silica-alumina have been found effective for ethylene and α-olefin copolymers at low temperatures. In 1975 it could be said that most linear polyethylene being marketed was being produced using the catalyst developed by the Phillips Company (essentially chromium oxide on a silica-alumina base) [35]. The SiO_2/Al_2O_3 support is saturated with an aqueous CrO_3 solution. When this material is activated by heating in air at 500 to 800°C, chromate and dichromates form by reaction with support surface hydroxyl groups. Reduction of the chromium by monomer or hydrogen leads to a site where ethylene can attach to form a chromium-ethyl bond. Polymerization results from insertion of monomer units at the chromium-alkyl bond.

Marsden has reviewed industrial catalysts which typically contain 1% by weight of chromium on the surface of the silica support [36]. The preparation of the silica support is critical, since each catalyst particle creates a polymer particle much larger than itself. It is vital that the support fragment as polymerization proceeds so that access to active sites can continue. Surface areas can range from 50 to 1000 m^2/g, and pore volumes can range from 0.4 to 3 cm^3/g. High pore volume generally increases productivity. Active catalysts fracture rapidly to particles in the 7- to 10-μm range within the first few minutes of polymerization. Since as much as 5000 g of polymer can grow on 1 g of catalyst, it is obvious that the original surface would become inaccessible to incoming monomer after a short time if it were not for the fragmentation which disperses the catalyst throughout the polymer particle. Many patents have specified the addition of titanium, zirconium, aluminum, boron, or fluorides in order to enhance the metal oxide activity. Whether incorporated on the surface or in the bulk of the support, titanium, in particular, can affect the molecular weight of the polyethylene made and also the productivity of the catalyst.

 J. Paul Hogan and Robert L. Banks (Phillips Petroleum) had observed the formation of crystalline polypropylene as well as linear polyethylene in 1951 using their own metal oxide catalyst system. In fact, about a half-dozen companies filed patent applications in 1953 covering crystalline polypropylene. The consequent, technically complex legal battles for priority extended over a period of three decades. Finally, in 1983, the U.S. patent on crystalline polypropylene was issued to Phillips. Hogan and Banks received the Perkin Medal in 1987 in recognition of their achievement.

Group Transfer Polymerization

It has been mentioned that linear polymers of olefins, diolefins, and styrene with narrow molecular-weight distributions are possible when anionic "living polymer" systems are used. Anionic methods have only recently been successful with acrylic monomers. *Group transfer polymerization* is a method for producing linear, living polymer systems from acrylates and methacrylates using organosilicon initiators [37, 38]. The mechanism basically is an anion-catalyzed polymerization with the initiator consisting of a ketene silyl acetal. The insertion point is different from the usual metal-carbon bond site of other anionic systems. Typically, the starting material is methyl trimethylsilyl dimethyl ketene acetal (C), which adds a monomer such as methyl methacrylate (M) in the presence of a nucleophilic activator such as bifluoride ion (B):

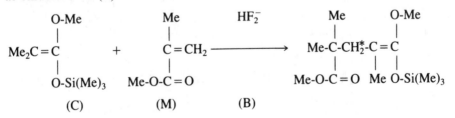

Propagation takes place by insertion of additional monomer at the position marked with an asterisk. In a clean system there is no termination, so a block of one monomer, say methyl methacrylate, can be formed followed by a block of a second monomer, say glycidyl methacrylate, and so on. Molecular-weight distributions can be very narrow. The termination step can be controlled to put a reactive site at one or both ends of the final polymer. In one example, two living chains are coupled by reaction with 1,4-bis(bromomethyl)benzene followed by "quenching" the silyl group with tetrabutylammonium bromide. The result is a single polymer molecule with a carboxyl group at each end. Polymers with reactive end groups have been called *telechelic,* based on the Greek words meaning far off (*tele*) and claw (*chele*). Such polymers are useful in subsequent chain-extension reactions. Although syndiotactic units tend to be favored in group transfer polymers, a great deal of random tacticity usually is also found. Brittain has reviewed the mechanism of group transfer polymerization, especially as it is used to produce random and block copolymers [39].

4.7 STEPWISE POLYMERIZATION

It is important to realize that in chain reactions, propagation to final molecular weight is very rapid. In the styrene polymerization, for example, we might start a bulk polymerization with a peroxide and stop it after only 1% of the monomer is converted to polymer. Analysis would show that the mixture consisted of 99% unreacted monomer and 1% of a high polymer. Conversion of the rest of the monomer to polymer would not affect the already formed material, which, with the exception of some "living" polymer systems, is dead. We cannot ordinarily make a low-molecular-weight polymer by addition polymerization and then increase its molecular weight by more of the same reaction. This is a major point of difference between chain and stepwise polymerization.

Most reactions other than olefin addition reactions lead to polymer chains containing atoms *other* than aliphatic carbon in the backbone. Another distinction is that usually polymers formed at early stages of conversion are not dead but can react as easily as monomers. The principle of equal reactivity regardless of molecular weight is fundamental to stepwise polymerization also.

Some reactions involve condensation in that some part of the reacting system is eliminated as a small molecule (Table 4.7). A good example is esterification, where water is eliminated between an acid and an alcohol. Because there is an equilibrium between reactants and products, the rate of conversion can be controlled by the rate of removal of one of the products, namely, water.

$$(n + 1)\text{HO}-\text{CH}_2\left(\text{CH}_2\right)_6\overset{\text{O}}{\underset{\|}{\text{C}}}-\text{OH} \xrightarrow{\text{H}^+} \text{HO}\left(\text{CH}_2\left(\text{CH}_2\right)_6\overset{\text{O}}{\underset{\|}{\text{C}}}-\text{O}\right)_{n+1}\text{H} + n\text{H}_2\text{O}$$

If we symbolize reactive hydroxyl by A, carboxyl by B, and the ester group by E, we can write the formulas for monomer, dimer, trimer, etc., as

Table 4.7 Stepwise polymerization systems

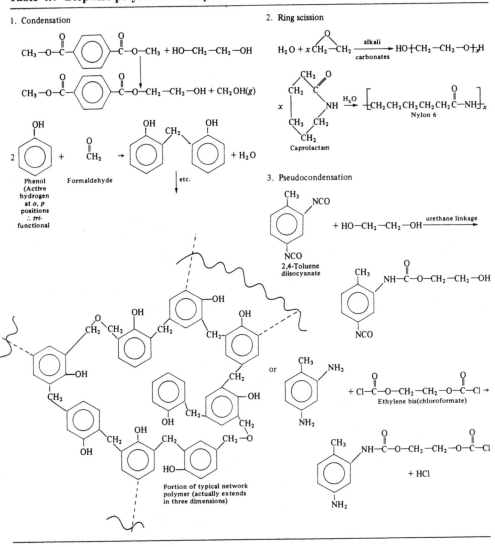

1. Condensation

2. Ring scission

Phenol (Active hydrogen at *o*, *p* positions ∴ *tri*-functional)

Formaldehyde

etc.

Caprolactam

Nylon 6

3. Pseudocondensation

2,4-Toluene diisocyanate

urethane linkage

Ethylene bis(chloroformate)

+ HCl

Portion of typical network polymer (actually extends in three dimensions)

or

Monomer A—B   dimer A—E—B   trimer A—E—E—B

x-mer A—$(E)_{x-1}$B

It is apparent that each successive member of the series has the same reactive groups as the monomer despite its larger size. With the principle of equal reactivity for all molecular sizes, a further condensation reaction is as likely to involve an

x-mer as it is monomer. Starting with N_A molecules and forming N_E ester groups, we can define the extent of polymerization p as

$$p = \frac{N_E}{N_A} \qquad (4.27)$$

This parameter is zero for pure monomer, and it reaches a value of 1 when every end group has been reacted. The number of molecules left in the system when N_E groups have been formed, N_R, is

$$N_R = N_A - N_E \qquad (4.28)$$

The number-average degree of polymerization $\bar{x}_n$ becomes

$$\bar{x}_n = \frac{N_A}{N_R} = \frac{1}{1-p} \qquad (4.29)$$

Later (Sec. 6.2) we shall use this example to derive a distribution of molecular weights. Where polyfunctional monomers are involved, successively higher conversion of monomer to polymer increases the probability of forming a network. It is important to be able to predict the maximum degree of conversion to which it is safe to go before network or gel formation occurs. A simple condensation between a trifunctional molecule T, which does not react with itself but does react with a bifunctional molecule D, has been analyzed [40]:

$$HO-\overset{\overset{O}{\|}}{C}-CH_2\,CH_2\,CH_2\,CH_2-\overset{\overset{O}{\|}}{C}-OH$$
Adipic acid, D

$$HO-CH_2-\underset{\underset{OH}{|}}{CH}-CH_2-OH$$
Glycerol, T

$$D + T \rightarrow D-T--D-T-D-T-$$

with D and –T–D–T– branches

etc.

If the initial ratio of hydroxyl groups to carboxyl groups is r and the number of hydroxyl groups and carboxyl groups initially is n_T and N_D, respectively, incipient gelation will take place when $(N_E)_g$ ester groups have been formed, where

$$(N_E)_g = N_T(2r)^{-0.5} \qquad (4.30)$$

At this point the probability that each polymer molecule is connected to another one becomes 1; that is, an infinite network has been formed. However, only part of the system is in this network. A substantial amount of polymer remains free and soluble. Further reaction rapidly reduces the soluble, extractable portion. For a monomer of functionality f reacting with another of functionality 2, the fraction

Table 4.8 Stepwise reactions with polyfunctional monomers

	Functionality f_i	Initial moles of reactant	Moles of functional group	ρ_i
1. Two reactants, $r = 1.1$				
Pentaerythritol	4	1.1	4.4	1.0
Adipic acid	2	2.0	4.0	1.0
2. Two alcohols, $r = 1.0$				
Sorbitol	6	0.33	2	0.5
Ethylene glycol	2	1	2	0.5
Adipic acid	2	2.0	4.0	1.0

of reactive higher-functionality groups that have been consumed at the gel point p_F is given by

$$\left(\frac{1}{p_F}\right)^2 = r(f - 1) \tag{4.31}$$

For example, take the case of the esterification of 1.1 mol of pentaerythritol with 2 mol of adipic acid:

$$\begin{array}{cc}
\underset{\substack{|\\ \text{CH}_2-\text{OH} \\ | \\ \text{HO}-\text{CH}_2-\overset{|}{\underset{|}{\text{C}}}-\text{CH}_2\text{OH} \\ | \\ \text{CH}_2-\text{OH}}}{} &
\underset{\substack{\text{O} \qquad\qquad\qquad \text{O} \\ \| \qquad\qquad\qquad \| \\ \text{HO}-\text{C}-\text{CH}_2\text{CH}_2\text{CH}_2\text{CH}_2\text{C}-\text{OH}}}{} \\
\text{Pentaerythritol} & \text{Adipic acid} \\
f = 4 & f = 2
\end{array}$$

The ratio of reactive groups r is $4.4/4.0 = 1.1$. Equation (4.31) indicates that if more than 55% of the hydroxyl groups are reacted ($p_f \geq 0.55$), the system will form a gel. Since r is constant, the fraction of carboxyl groups reacted p_B is simply rp_F, or 0.605. In actual experiments the predictions based on this equation are somewhat conservative because some of the cross-links formed are intramolecular and do not contribute to the total network formation. When there are N kinds of polyfunctional reactants in the same system, an average functionality can be used, where ρ_i is the ratio of the reactive groups on reactant i of functionality f_i to all groups that react the same way:

$$\bar{f} = \sum_1^N f_i \rho_i \tag{4.32}$$

In this way $1/3$ mol of sorbitol, f_i = six hydroxyl groups per molecule, with one of ethylene glycol, f_i = two hydroxyl groups per molecule, would be the equivalent of 1 mol of pentaerythritol, both in number of available hydroxyl groups and in average functionality (Table 4.8).

Stepwise reactions can offer special problems in controlling reaction rate and molecular weight. In *ring scission polymerization* there may be a thermodynamic equilibrium that varies with temperature, as in making nylon 6 from caprolactam (see next section).

In many isocyanate systems, reactions can be so rapid as to present special problems in control. Isocyanate foam production involves the following reactions:

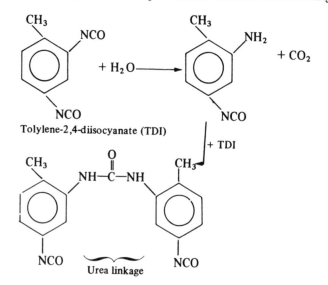

Also:

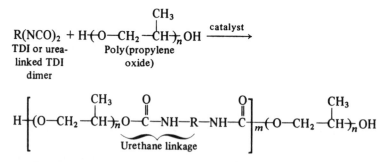

These reactions are so rapid at room temperature that special machinery had to be developed to bring the ingredients (plus others) together in a matter of seconds, giving a continuous mat of foam product, i.e., the urethane polymer indicated. The important point here is that despite the condensation with ejection of a small molecule, the reaction is not easily controlled.

Stepwise polymerization can sometimes be stopped at an early stage and completed in a separate piece of equipment. Such two-stage polymerizations are quite common. In the isocyanate example above, the poly(propylene oxide) is a low-molecular-weight ($\bar{x}_n = 15$), ring scission polymer which then becomes part of the isocyanate foam by a different reaction. Network polymers often are made in two stages, which may use the same reaction. Phenol-formaldehyde condensation is carried out in two stages (Table 4.7). The first produces a mixture of low-molecular-weight linear and branched polymers that are still fusible and soluble in organic solvents. This reaction may be carried out in a large glass-lined reactor. Final production of the cross-linked network uses the same reaction carried out concurrently with the final forming operation, i.e., molding. This is necessary, because a thermoset material like this cross-linked network cannot be remolded (nor can it be easily removed from a 5000-gal reactor if the first stage proceeds too far too fast).

Many commercially important polyesters and polyamides are made by combining a *dyad* consisting of two dissimilar monomers of functionality 2, A and B, such that A will react only with B and not with itself, and vice versa. The result is something like a copolymer. However, since it is perfectly alternating, it is more convenient to think of it as a dyadic homopolymer. Poly(ethylene terephthalate) and nylon 66, for example, exhibit the regularity and crystallinity typical of homopolymers and are useful as fibers. If one were to use a mixture of diacids rather than a single one in making a polyamide, the result would no longer be a perfect dyadic polymer. The regularity would be disrupted and the tendency to crystallize diminished as in normal random copolymerization.

4.8 RING SCISSION POLYMERIZATION

When a ring is opened to form a linear polymer, the propagation step may resemble either chain or stepwise polymerization. It was pointed out in the previous section that an equilibrium may be set up between open-chain and ring structures when the monomer (that is, the ring) is as reactive as growing polymer. In making nylon 6 from caprolactam, no condensation is involved because the product has the same composition as the reactants (Table 4.7). At 220°C with 1 mol of water to 10 mol of monomer initially, an equilibrium is reached rather rapidly in which the reaction mixture contains about 5% unchanged monomer, even though the number-average degree of polymerization is nearly 100 [41]. The residue of monomer can present a problem in purification. It may be necessary to inactivate the equilibration catalyst and then remove the unreacted monomer by vacuum-stripping or by solvent extraction.

Usually only small rings are involved in such equilibria. The silicones are an exception in that a continuous series of rings of increasing sizes are formed. Silicone rings can be isolated with up to 80 ring atoms (see Sec. 15.7). The polymerization of cyclopentene with a tungsten catalyst also leads to an equilibrium among large rings. A catalyst can be formed when WCl_6 with tetraalkyltin modified by ethyl ether. The combination bears some resemblance to the Ziegler system (Sec.

Table 4.9 Ring scission by olefin metathesis [42]

A. Polymerization of cyclopentene

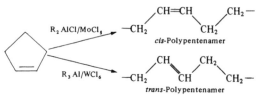

B. Olefin metathesis

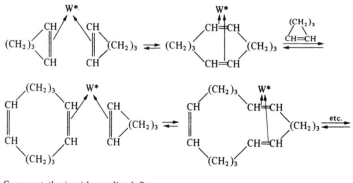

C. Cross metathesis with acyclic olefin

$$\left[-[(CH_2)_3-CH=CH]_X-(CH_2)_3-CH \quad CH-R_1 \right.$$
$$\left. \qquad\qquad\qquad\qquad\qquad\qquad -CH \quad CH-R_2 \right. \rightleftharpoons$$
$$R_2-CH=CH[(CH_2)_3-CH=CH]_X(CH_2)_3-CH=CH-R_1$$

4.6). The reactions are summarized in Table 4.9. The term *olefin metathesis* has been applied to this reaction [42]. When an acyclic olefin is introduced, the large rings are interrupted and made into open-chain polymers. In one example, monomer is completely converted to polymer in 30 min at 0°C with a ratio of monomer to tungsten of about 10,000. The low temperature is needed to keep the equilibrium concentration of cyclopentene down. In general, higher temperatures favor rings over open-chain polymers.

Another common feature of most ring scission schemes is the low heat of polymerization. Polybutadiene has a very similar structure to polypentenamer, but the isothermal conversion of butadiene to polymer requires removal of 74.9 kJ/mol, while conversion of cyclopentene to polymer requires removal of only 18.4 kJ/mol. A commercial polyalkenamer is described later (Sec. 14.4).

For strained rings with only three or four atoms, ring scission is easy and the heat of polymerization is comparatively large. In going from liquid monomer to amorphous, solid polymer, the heat of polymerization is 94.5 kJ/mol for ethylene oxide and 105 kJ/mol for a cyclobutane ring. The reactions of epoxy resins containing the three-membered oxirane ring are described in Sec. 12.4 both as coatings and as adhesives. Much newer are thermosetting resins containing the cyclobutane ring [43]. The typical experimental diketone resin shown here has a formula weight of 338, so the heat released on ring opening is about 620 kJ/g. To polymerize the monomer (T_m = 152°C), it is first melted by heating at 160°C. When the temperature is raised to 200°C, polymerization occurs rapidly to form a network structure with no added catalyst. Most often the resin is used in composites with large amounts of glass fibers or other reinforcement.

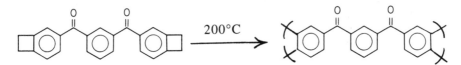

Experimental diketone-bis-benzocyclobutene resin [44].

4.9 COPOLYMERIZATION

When more than one monomer is polymerized at the same time, a variety of structures can result. In the simple case of two monomers A and B in a chain polymerization, one can derive relationships that will help answer some questions about the degree of heterogeneity of the products made under various conditions:

1. What is the composition of a polymer made from the limited conversion of a mixture of two monomers?
2. How will two monomers that never have copolymerized before interact?

The first is answered by introducing relative reactivities. The second is approached by the Alfrey-Price Q-e scheme. To express the relationship of polymer composition to the monomer composition from which it is being formed, we go back to the kinetic scheme of Sec. 4.4. If we can assume that we achieve a "steady-state" population of chain radicals that grow to high molecular weight, the material balances for various species are few and simple for a binary system.

Assumptions

1. Over any short time interval, the concentration of free radicals does not change appreciably with time; that is, $d[\text{radicals}]/dt = 0$.
2. A growing polymer chain's reactivity is determined solely by the last monomer unit added. This reactivity is independent of the molecular weight.

3. The only monomer-consuming reactions taking place, together with their rate constants, are

$$A\cdot + A \xrightarrow{k_{11}} A\cdot$$

$$A\cdot + B \xrightarrow{k_{12}} B\cdot \qquad A, B = \text{monomers A and B}$$

$$B\cdot + B \xrightarrow{k_{22}} B\cdot \qquad A\cdot, B\cdot = \text{growing chains ending in groups A and B}$$

$$B\cdot + A \xrightarrow{k_{21}} A\cdot$$

That is, a growing chain whose last added unit was monomer A has reactivity determined by A only. This chain can add another A, in which case reactivity is unchanged; it may also react with B, in which case the new reactivity is determined by B.

4. Propagation is the only reaction of importance, since it is repeated many times for each initiation or termination step.

Material Balances

1. Rate of creation of radicals ($t = \text{time}$):

$$\frac{d[\text{A·}]}{dt} = k_{21}[\text{B·}]\,[\text{A}] - k_{12}[\text{A·}]\,[\text{B}] = 0 \qquad \text{(assumption 1)} \qquad (4.33)$$

or

$$\frac{[\text{A·}]}{[\text{B·}]} = \frac{k_{21}[\text{A}]}{k_{12}[\text{B}]}$$

2. Rate of consumption (disappearance) of monomers A and B:

$$-\frac{d[\text{A}]}{dt} = k_{11}[\text{A·}]\,[\text{A}] + k_{21}[\text{B·}]\,[\text{A}] \qquad (4.34a)$$

$$-\frac{d[\text{B}]}{dt} = k_{12}[\text{A·}]\,[\text{B}] + k_{22}[\text{B·}]\,[\text{B}] \qquad (4.34b)$$

We define the relative reactivities as

$$r_1 = \frac{k_{11}}{k_{12}} \qquad \text{The ratio of reactivity of monomer 1 (A) toward itself to the reactivity of monomer 1 toward monomer 2 (B)}$$

$$r_2 = \frac{k_{22}}{k_{21}}$$

Now at the instant when monomer concentrations are [A] and [B] (moles per liter), the rate at which monomer is entering a growing polymer chain is

$-d[A]/dt$ and $-d[B]/dt$. The mole fraction of monomer A being added to the growing polymer at this instant, F_1, then is

$$F_1 = \frac{d[A]/dt}{d[A]/dt + d[B]/dt} \tag{4.35}$$

Of course,

$$\frac{F_1}{F_2} = \frac{d[A]}{d[B]} \tag{4.36}$$

Now we let

$$f_1 = \frac{[A]}{[A] + [B]} \quad \text{and} \quad \frac{f_1}{f_2} = \frac{[A]}{[B]} \tag{4.37}$$

Combining Eqs. (4.33) through (4.37) gives

$$\frac{F_1}{F_2} = \frac{(r_1 f_1/f_2) + 1}{(r_2 f_2/f_1) + 1} \tag{4.38}$$

or

$$\frac{F_1}{1 - F_1} = \frac{r_1 f_1/(1 - f_1) + 1}{r_2(1 - f_1)/f_1 + 1}$$

Thus we have related the "instantaneous" copolymer composition F_1 to the "instantaneous" monomer composition f_1 by two parameters r_1 and r_2. We say "instantaneous" because if A and B are consumed at different rates so that $F_1 \neq f_1$, the value of f_1 will change as monomer is converted to polymer in a batch operation.

Special Cases

Several cases simplify Eq. (4.38). If k_{12} and k_{21} are nil, no copolymer is formed, only a mixture of homopolymers. If k_{11} and k_{22} are nil, only perfectly alternating copolymer is formed ($F_1 = F_2 = 0.5$). If $r_1 = r_2 = 1$, then $F_1 = f_1$. An interesting special case is $r_1 r_2 = 1$, when Eq. (4.38) becomes

$$\frac{F_1}{1 - F_1} = \frac{r_1 f_1}{1 - f_1} \tag{4.39}$$

This resembles the vapor y-liquid x composition at constant relative volatility α_r

$$\frac{y}{1 - y} = \frac{\alpha_r x}{1 - x} \tag{4.40}$$

In such "ideal" copolymerization the F_1-f_1 curve never crosses the F_1-f_1 diagonal, whereas when $r_1r_2 \neq 1$, there can be a point where they cross to give an "azeotrope" at which $F_1 = f_1$ (Fig. 4.13). Equation (4.38) can be rewritten as:

$$Y = r_1X \frac{1 + r_1X}{r_1r_2 + r_1X} \tag{4.41}$$

where $Y = F_1/(1 - F_1)$, $X = f_1/(1 - f_1)$, and r_1 and r_2 are the relative reactivities. Terpolymers (three monomers) require six relative reactivities and can be handled conveniently only in special cases or by computer methods.

Example 4.6

When monomers A and B are copolymerized, an azeotrope (polymer composition same as feed) is formed at a ratio of 1 mol of A to 2 mol of B. If monomer A is known not to homopolymerize:

(a) What polymer composition (mol fraction of A) will result initially when a mixture of 1 mol of A with 9 mol of B is polymerized?

(b) Will polymer formed after 50% conversion from an initial mixture of 4 mol of A with 6 mol of B contain more or less of the same amount of A compared to that formed at 1% conversion?

Solution:

(a) The azeotrope corresponds to $F_1 = f_1 = 1/3$ and $r_1 = 0$.
From Eq. (4.38):

$$\frac{1/3}{2/3} = (1 + 2r_2)^{-1} \quad \text{so} \quad r_2 = 0.500$$

When $f_1 = 0.100$,

$$\frac{F_1}{1 - F_1} = \frac{1}{1 + 0.500 \times 9} = \frac{1}{5.5} \quad \text{and} \quad F_1 = 0.154$$

(b) When $f_1 = 0.400$,

$$\frac{F_1}{1 - F_1} = \frac{1}{1 + 0.500 \times 3/2} = \frac{1}{1.75} \quad \text{and} \quad F_1 = 0.364$$

So polymer has less A than the monomer feed and B is consumed faster than A. Both f_1 and F_1 must increase with conversion. Thus polymer at 50% conversion must be richer in A than at 1% conversion.

For the simple case of only two monomers, the problem of calculating F_1 as a function of degree of conversion and initial monomer composition is analogous to the situation in differential distillation of a binary mixture. Monomer is converted

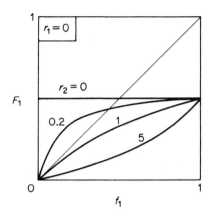

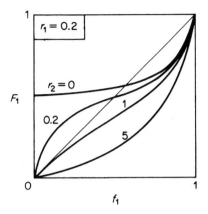

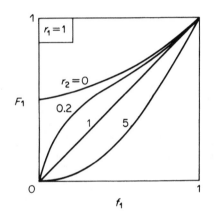

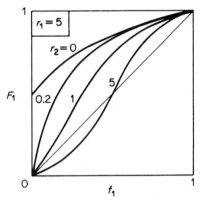

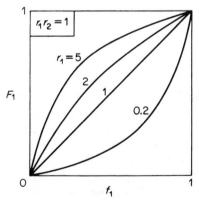

Figure 4.13 F_1, mole fraction of monomer 1 in polymer formed when f_1 is the mole fraction of monomer 1 in the feed for various combinations of the relative reactivity ratios

to polymer irreversibly just as volatile liquids are distilled out from a pot irreversibly. A material balance gives the Rayleigh equation [45] (also called the Skeist equation):

$$\ln \frac{n}{N_0} = \int_{(f_1)_0}^{f_1} \frac{df_1}{F_1 - f_1} \tag{4.42}$$

where N/N_0 is the mole fraction of all monomer present initially that is still unreacted and $(f_1)_0$ and f_1 are the mole fractions of component A initially and at N/N_0. This can be integrated graphically or analytically, although the analytic result is not tidy except when $r_1 r_2 = 1$. For that case:

$$\left(\frac{N}{N_0}\right)^{r_1-1} = \frac{f_1}{(f_1)_0} \left[\frac{1 - (f_1)_0}{1 - f_1}\right]^{r_1} \tag{4.43}$$

Also

$$\frac{N_1}{(N_1)_0} = \left[\frac{N_2}{(N_2)_0}\right]^{r_1} \tag{4.44}$$

where N_1 and N_2 are the number of moles of monomers A and B still in the monomer phase at any time. In the general case [46]:

$$\frac{N}{N_0} = \left[\frac{f_1}{(f_1)_0}\right]^{\alpha} \left[\frac{f_2}{(f_2)_0}\right]^{\beta} \left[\frac{(f_1)_0 - \delta}{f_1 - \delta}\right]^{\gamma} \tag{4.45}$$

where $\alpha = \dfrac{r_2}{1 - r_2}$

$\beta = \dfrac{r_1}{1 - r_1}$

$\gamma = \dfrac{1 - r_1 r_2}{(1 - r_1)(1 - r_2)}$

$\delta = \dfrac{1 - r_2}{(2 - r_1 - r_2)}$

At $(f_1)_0 = \delta$, an azeotrope is formed.

Some results for an equimolar monomer feed of vinyl acetate and vinyl laurate are shown in Fig. 4.14. Although the average polymer composition F_1 is never changed much and eventually has to equal the original monomer ratio, the polymer being made at high conversions may approximate pure vinyl laurate. This is a real problem in polymerization methods that attempt complete conversion. Partial conversion with recycling of unreacted monomer plus makeup of the more reactive monomer is one method of regulation. Another is to add the more reactive monomer during the reaction to maintain the same monomer and therefore the same

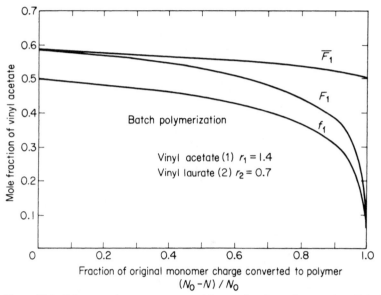

Figure 4.14 Polymer and monomer compositions as functions of conversion. F_1 is the mole fraction of monomer 1 in the polymer being formed at a point where the feed composition has shifted to f_1. $\overline{F}_1$ is the average composition of all the polymer generated up to that point

polymer composition [47]. Some examples (Table 4.10) are typical in that $r_1 r_2$ seldom exceeds unity. A comprehensive set of relative reactivity ratios has been published by Alfrey and Young [48]. However, it is desirable to have some parameters associated with a single monomer that would allow prediction of r_1 and r_2 for various combinations with that monomer just as bond contribution and atomic radii are used to estimate bond strengths and interatomic distances. According to the Alfrey-Price relationship [48, 49], each monomer can be characterized by values of Q and e that are related to relative reactivities by

$$\ln r_1 = \ln \left(\frac{Q_1}{Q_2}\right) - e_1(e_1 - e_2) \quad \text{and} \quad \ln r_1 r_2 = -(e_2 - e_1)^2 \quad (4.46)$$

The parameter Q_1 is mainly affected by the relative stability of the polymer chain radical resulting from the addition of monomer 1 to the growing end. As a point of reference, styrene is assigned values of $Q = 1$ and $e = -0.80$. Resonance-stabilized monomers such as 1,3-butadiene are expected to have high Q values, whereas non-conjugated monomers such as ethylene ($Q = 0.015$, $e = -0.20$) have low Q values. The parameter e is expected to reflect the polarity of the monomer and the polymer radical resulting from the addition of the monomer. An electron-donating substituent on the end of a polymer radical, as in p-methoxystyrene ($Q = 1.36$, $e = -1.11$),

Table 4.10 Relative reactivities and copolymerization parameters [48]

Radical	1,3-Butadiene	Maleic anhydride	Styrene	Vinyl chloride	Methyl methacrylate	Acrylonitrile	Q	e
1,3-Butadiene	1.0	–	1.4	8.8	0.53*	0.35	2.39	-1.05
Maleic anhydride	–	1.0	0.0†	0.008‡	0.03†	0†	0.23	2.25
Styrene	0.5	0.01†	1.0	35	0.50†	0.37	1.00	-0.80
Vinyl chloride	0.035	0.296‡	0.077	1.0	0†	0.074	0.044	0.20
Methyl methacrylate	0.06*	3.5†	0.50†	12.5†	1.0	23†	0.74	0.40
Acrylonitrile	0.0	6†	0.070	3.7	0.0	1.0	0.60	1.20

All data at 50°C except as noted: *5°C, †60°C, ‡75°C.
Example: For styrene (i_j) and acrylonitrile (2), $r_1 = 0.37$ and $r_2 = 0.070$.

decreases e, while an electron-withdrawing substituent, as in p-introstyrene ($Q = 1.63$, $e = 0.39$), increases e. The Q-e scheme is analogous to, but not identical with, the linear free-energy relationship, the Hammett equation, so important to the physical organic chemistry of small molecules. In general the scheme is only moderately successful, but it is useful where experimental data are sparse. Numerous average values have been calculated and tabulated by Young in the previously cited article.

Other Copolymers

The copolymers discussed so far are *random* in that we polymerize a mixture of monomers. Even in this case, long runs of one monomer or the other may occur, especially when one monomer is present in higher proportion than the other. An extreme case is the *block* copolymer. Sequential addition of butadiene, then styrene, then butadiene to lithium alkyl will give a structure with long runs or blocks of each monomer (see Sec. 4.5, Anionic Systems).

$$\text{SBBSSSBSBSSBSBBS} \qquad \text{S}(\text{S})_m\text{S}-\text{B}(\text{B})_n\text{B}-\text{S}(\text{S})_x\text{S}$$

Random copolymer	Block copolymer

$$\text{S = Styrene, B = Butadiene}$$

If an ethylene oxide (E) polymer terminating in hydroxyls is used to start a propylene oxide (P) polymerization, a block polymer results with hydrophilic (E) portions and hydrophobic (P) portions. The products at molecular weights of 1000 to 2000 are useful as surface-active agents.

$$\text{HOE(E)}_x\text{EH} \xrightarrow{\text{P}} \text{HOP(P)}_z\text{P}-\text{E(E)}_x\text{E}-\text{P(P)}_y\text{PH}$$

$$E = CH_2 \overset{O}{\diagup\diagdown} CH_2 \qquad P = CH_2 \overset{O}{\diagup\diagdown} \underset{\underset{CH_3}{|}}{CH}$$

or

$$(CH_2-CH_2-O) \qquad \text{or} \qquad (CH_2-\underset{\underset{CH_3}{|}}{CH}-O)$$

A third variety is the *graft* copolymer. In this case, branches of one monomer are grown on a main stem of a previously formed polymer molecule. For example, polyethylene can be irradiated in air with gamma rays or accelerated electrons, which leave peroxides or free radicals "trapped" on the polymer backbone. Exposed to a reactive monomer such as acrylonitrile ($CH_2=CHCN$), polymerization is initiated at the free-radical sites, and branches of poly(acrylonitrile) grow on the polyethylene stem [50].

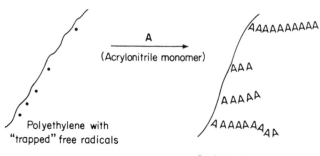

Graft copolymer

4.10 BIOSYNTHESIS OF POLYMERS

Polymers produced by biological routes invariably are made by some kind of stepwise polymerization rather than a chain polymerization. An interesting example is afforded by *cis*-1,4-polyisoprene. Isoprene can be made from propylene through a sequence of dimerization, isomerization, and steam demethanization [51]:

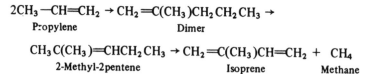

$$2CH_3 —CH\!=\!CH_2 \rightarrow CH_2\!=\!C(CH_3)CH_2\,CH_2\,CH_3 \rightarrow$$

Propylene   Dimer

$$CH_3\,C(CH_3)\!=\!CHCH_2\,CH_3 \rightarrow CH_2\!=\!C(CH_3)CH\!=\!CH_2 + CH_4$$

2-Methyl-2pentene   Isoprene   Methane

The monomer is easily converted to the *cis*-1,4 polymer by using a butyl lithium initiator in a hydrocarbon solvent (Sec. 4.5, Anionic Systems). The identical structure is produced in the plant tissues of the rubber tree *Hevea brasiliensis*. Although over 2000 species of plant produce polyisoprenes, only the one cited is utilized on a large commercial scale, mainly in the warm climates of Malaysia, Indonesia, and Liberia. Brazil, original home of the plant, no longer contributes significantly to the world's supply. Like other plants, the tree contains tubes or vessels that convey water and salts from the roots to the leaves and a second system that carries soluble sugars down from the leaves in the sap. Unlike most other plants, the rubber tree has a third system of vessels in which the sugars are converted into a latex, a stabilized emulsion of rubber particles along with the other compounds necessary to the process such as enzymes, sterols, and lipids. The function of the rubber in the tree is unknown. Apparently it does not serve as a food reserve in the way that starch does in other systems. The primitive raw material used in the latex formation is acetic acid [52, 53]. Many intermediate products have been isolated or have been demonstrated by incorporation of ^{14}C-labeled compounds in the growing plant. A number of enzymes and other catalysts are necessary. Some stages are indicated in the sequence of Table 4.11. The manner in which the pyrophosphate adds to the growing chain and the specific enzyme that is involved are not known with certainty.

Table 4.11 Intermediates in the biosynthesis of rubber

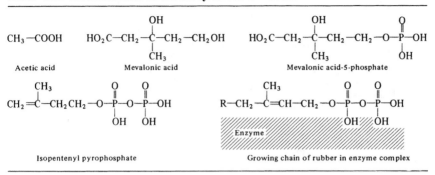

Acetic acid Mevalonic acid Mevalonic acid-5-phosphate

Isopentenyl pyrophosphate Growing chain of rubber in enzyme complex

Although most commercially important polysaccharides are recovered from plants and trees, progress has been made in producing some by controlled fermentations in conventional plant equipment. Raw cotton contains over 85% cellulose, which is the linear polymer of β-D-glucopyranose (Fig. 4.15). The formal indication of structure for the pyranose unit is somewhat misleading, since the rings are not planar and the bond angle at the oxygen connecting successive rings is distorted. The important feature to note is that all of the ring carbons are asymmetric and therefore give rise to possible stereoisomers. The changes in physical properties can be great. For example, the unbranched component of starch, amylose, differs only in that the carbon marked with an asterisk in Fig. 4.15 has the oxygen below the ring rather than above as in cellulose. However, amylose dissolves in water but cellulose does not. As in the case of rubber, phosphates appear to be intermediates in the biosynthesis of polysaccharides. Glucose 1-phosphate in the presence of an enzyme extracted from potato juice can be converted to amylose. The energy for carrying out polymerization is obtained by the oxidation of some of the starting monosaccharide. A variety of microorganisms can produce cellulose from sugar in the proper culture medium. However, cellulose from such an operation cannot compete economically with polymer from cotton or wood.

The biopolymers discussed here have a relatively simple structure in which a single or a very few repeat units are involved. The enzymes that catalyze these reactions are complex proteins. A large number of laboratories work today on the structure of proteins and other polymeric components of living organisms such as ribonucleic acid (RNA) and deoxyribonucleic acid (DNA). The importance of nucleic acids (combinations of phosphoric acid, sugars, and nitrogenous bases) is that they, in cooperation with proteins, are the main constituents of viruses and genes and occur in all living cells. Synthesis of these materials is remote but conceivable. The production of simpler polymers is much more likely with the present state of the art and certainly less fraught with philosophical, moral, religious, and legal significance than the creation of "test tube life."

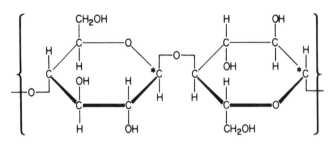

(a)

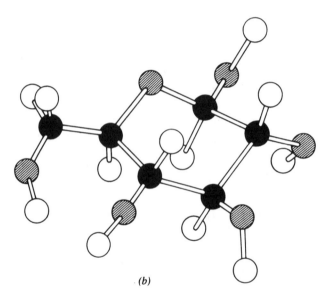

(b)

Figure 4.15 (a) Cellulose repeating structure, β-D-glucopyranose unit; (b) D-Glucose.

A close approximation to test tube creation of a biologically active substance was made at Stanford University [54]. Using a tritium-labeled (+) DNA as a template, workers were able to make an artificial (−) DNA from modified monomer units. The resulting (−) DNA was different from that usually found in nature, but it was able to act as a reverse template to reproduce (+) DNA identical with the starting material. This experiment does not demonstrate total synthesis of living matter from the elements. It does show that genetic modification by chemical means is a distinct possibility. Total syntheses of enzymes have been reported (see Sec. 15.4).

4.11 POLYMER MODIFICATION

Whenever a polymer is formed in several stages, each subsequent stage is a kind of polymer modification. All cross-linking reactions (vulcanizations) of rubber would fall in this category, together with the molding of network polymers such as the phenolformaldehyde resins. However, the term "polymer modification" is used here to mean changes wrought by the polymer manufacturer in an operation separate from the final molding, spinning, or formation. Modifications are feasible when a polymeric raw material is available that can have its value enhanced substantially by the operation. Two naturally occurring polymers in this category are α-cellulose and natural rubber. In the form of cotton, α-cellulose is already valuable as a fiber. Even with cotton, the chemical treatment to give wash-and-wear properties may amount to a polymer modification. But cotton linters, the short fibers remaining from cotton ginning, and also α-cellulose from wood pulping are not easily utilized. The hydrogen-bonded structure of cellulose prevents it from melting or dissolving below its decomposition temperature. Permanent reaction of these hydroxyl groups by esterification or etherification results in materials that are soluble and fusible—and therefore capable of being molded or cast in useful forms.

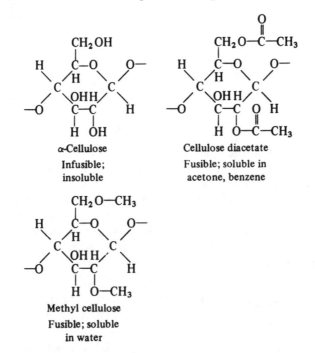

Sometimes several purposes can be served at once. Chlorination of natural rubber decreases the flammability of the polymer. At the same time it raises the T_g so that the polymer can be used as a binder for traffic paints. Nitration of cellulose

Table 4.12 Typical polymer modifications

Purpose of modification	Initial polymer	Chemical reactions	Example of final form
Change physical form	α-Cellulose	Regeneration via xanthate	Rayon, cellophane
Change solubility	α-Cellulose	Esterification	Cellulose acetate Cellulose nitrate

		Etherification	Hydroxy ethyl cellulose
			Carboxy methyl cellulose
			    Ethyl cellulose
			    Methyl cellulose

	Poly(vinyl acetate)	Hydrolysis	Poly(vinyl alcohol)
	Poly(vinyl alcohol)	Acetal formation	Poly(vinyl butyral) Poly(vinyl formal)
Introduce cross-linking sites	Butyl rubber	Bromination	Brominated butyl rubber
	Poly(vinyl alcohol)	Esterification	Poly(vinyl cinnamate)
Increase flammability	α-Cellulose	Esterification	Cellulose nitrate (guncotton)
Decrease flammability	Natural rubber	Chlorination	Chlorinated rubber
Change mechanical	Natural rubber	Graft copolymerizations of methyl methacrylate	Graft copolymer

allows the native cotton to be plasticized and molded. Flammability is enhanced at the same time. This may be a disadvantage to makers of molded articles for decoration and apparel. However, the munitions maker regards the flammability and propellant power of guncotton, cullulose nitrate, as its most important advantage. Most often a polymer modification is economically justifiable only when the original polymer occurs naturally or as a by-product. Sometimes modification is the only good way to make a material. A case in point is the photosensitive polymer poly(vinyl cinnamate):

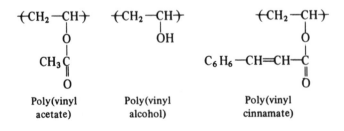

An attempt to polymerize vinyl cinnamate would give a cross-linked network, since the functionality is 4. Even poly(vinyl alcohol) cannot be made directly, because the monomer is unknown (isomeric with acetaldehyde). Thus it is necessary first to polymerize vinyl acetate, then to hydrolyze the polymer to the polyalcohol. Finally the alcohol is esterified by cinnamic acid (functionality of 1 in an esterification). Exposure of a film of poly(vinyl cinnamate) to ultraviolet light causes rapid formation of a cross-linked, insoluble network (see Sec. 14.5). It is used as a "photoresist" to establish etched patterns on lithographic plates. Some other typical modifications are listed in Table 4.12.

KEYWORDS FOR CHAPTER 4

Section

2. Functionality
3. Stepwise polymerization
   Chain polymerization
4. Initiation
   Initiator
   Propagation
   Termination
   Coupling
   Disproportionation
   Free-radical polymerization
   Half-life
   Kinetic chain length
   Degree of polymerization
   Chain transfer
   Chain transfer agent
   Inhibition
   Inhibitor
   Redox initiation
   Activation energy
5. Anionic polymerization
   Cationic polymerization
   Gegen ion
   Living polymer
   Block copolymer
   Thermoplastic elastomer
6. Coordination complex polymerization
   Insertion polymerization
   Group transfer polymerization
   Ziegler-Natta catalysis

Telechelic
Metallocene
7. Dyad
8. Ring scission polymerization
   Olefin metathesis
9. Random copolymer
   Graft polymer
   Q-e scheme
   Relative reactivity
   Azeotrope
10. Biosynthesis
11. Polymer modification

PROBLEMS

4.1 What structures can occur in the polymer when chloroprene, CH_2=CCl—CH=CH_2, is polymerized?

4.2 Consider the monomer

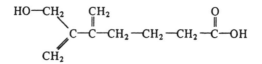

(*a*) What is the functionality of this monomer in a free-radical or ionic polymerization? Name and describe the isomeric forms expected in linear polymers formed by such reactions.

(*b*) What is the functionality of this monomer in a polyesterification? Name and describe the isomeric forms expected in linear polymers formed by such a reaction.

(*c*) Which would you expect to have the higher T_g, linear polymer from (*a*) or (*b*)? Why?

4.3 What is the energy of activation for the decomposition of di-*tert*-butyl peroxide (Fig. 4.5)? For bis(1-hydroxy cyclohexyl) peroxide?

4.4 Two monomers are polymerized in homogeneous solution with the following results:

Series 1	Monomer A	Monomer B
Initial monomer concentration, mol/liter	0.100	0.200
Time to convert 5% of original monomer charge to polymer, h	0.100	0.300
Initiator concentration, mol/liter	0.0397	0.0397

In a second series at the same temperature, it is desired that both polymers produced initially have the same degree of polymerization. It can be assumed that both polymers terminate exclusively by coupling. In this second series both monomers initially are 0.300 mol/liter. What should be the ratio of concentration of initiator for monomer B compared with that for monomer A in this second series?

4.5 The following data are obtained for the polymerization of a new monomer. Predict the time for 50% conversion in run D. Calculate the energy of activation for the polymerization. $R = 1.987$ cal/mol·K.

Run	Temperature, °C	Conversion, %	Time, min	Initial monomer conc., mol/liter	Initial initiator conc., mol/liter
A	60	50	500	1.00	0.0025
B	80	75	700	0.50	0.0010
C	60	40	584	0.80	0.0010
D	60	50	—	0.25	0.0100

4.6 This table gives the initial rate of polymerization $(-d[M]/dt)_0$ for each combination of monomer and peroxide in mmol/liter·min. In each case, initial concentrations of monomer and initiator are 0.100 and 0.00100 mol/liter, respectively. Calculate rates (*a*) and (*b*).

Monomer	Peroxide I		Peroxide II	
	30°C	80°C	30°C	80°C
A. Styrene	1.00	2.00	0.500	1.50
B. Methyl methacrylate	2.00	5.00	(*a*)	(*c*)
C. Vinyl acetate	3.00	(*b*)	(*d*)	7.50

4.7 Using a k_i based on Fig. 4.5 and $k_p/(k_t)^{0.5}$ from Table 4.3, calculate the time needed to convert half of a charge of methyl methacrylate to polymer using benzoyl peroxide as the initiator in benzene solution at 60°C. What number-average degree of polymerization do you expect initially? Initial charge per 100 ml of solution: 10 g methyl methacrylate and 0.1 g benzoyl peroxide. What fraction of the initiator has been used after 50% conversion of monomer?

4.8 When peroxide P is held at 70°C for 10 h, 90% of the original peroxide remains undecomposed. Now if a 5% solution of ethyl mermanate containing 0.000100 mol/liter of peroxide P is polymerized at 70°C, 40% of the original monomer charge is converted to polymer in 1 h. How long will it take the polymerize 90% of the original monomer charge in a solution containing 10% ethyl mermanate (initially) with 0.0100 mol/liter of peroxide P?

4.9 Initiator B has twice the half-life that initiator A has at 70°C. Monomer C polymerizes three times as fast as monomer D at 70°C when initiator A is used and all the concentrations are the same. If both C and D terminate by coupling, what is the ratio of degree of polymerization for case I to that of case II?

	Case I	Case II
Monomer	C	D
Initiator	A	B
Concentration of monomer	1 mol/liter	2 mol/liter
Concentration of initiator	0.001 mol/liter	0.005 mol/liter

4.10 When peroxide A is heated to 60°C in an inert solvent, it decomposes by a first-order process. An initial concentration of 5.0 mmol/liter of peroxide changes to 4.0 mmol/liter after 1.0 h.

(a) In the following system what fraction of the monomer should remain *unconverted* after 10 min at 60°C?

(b) What is the concentration of initiator after 10 min? *System parameters* (all at 60°C):

$k_p = 1.8 \times 10^4$ liter/mol·s
$k_t = 1.45 \times 10^7$ liter/mol·s
Initiator concentration $= 4.0 \times 10^{-4}$ mol/liter

4.11 Methyl acrylate (1.0 mol) is polymerized in benzene (1.0 liter of solution) using succinic acid peroxide (1.0×10^{-3} mol) at 60°C. How long will it take to convert 10% of the monomer to polymer? How long will it take to convert 90% of the monomer to polymer? If the polymerization were carried out adiabatically, how much would the temperature have risen after 10% conversion? Assume that the specific heat of the benzene solution is 370 cal/°C-liter.

4.12 When monomer M is polymerized at 50°C using 0.0200 mol/liter of initiator and 2.00 mol/liter of monomer, the molecular weight formed initially is 100,000 g/mol. The same molecular weight is produced at 63°C with the same amount of initiator but with a monomer concentration of 4.55 mol/liter. What concentration of initiator should be used at 70°C to get a molecular weight of 150,000 g/mol if the monomer concentration is 6.00 mol/liter?

4.13 Solution polymerization of monomer A initiated by peroxide P at two temperatures (using the same concentrations of A and P) gives the following results:

Fig 4.5 gives both $t_{1/2}$ POLYMER FORMATION **147**

	35.0°C	53.0°C
Conversion in 30 min	25.0%	65.0%
Initial molecular weight	500,000	300,000

What is the energy of activation, E_a, for the decomposition of P?

4.14 The decomposition of benzoyl peroxide is characterized by a half-life of 50 h at 60°C and an activation energy of 27.3 kcal/mol. How much peroxide (mol/liter) is needed to convert one-third of the original charge of vinyl monomer (original concentration of 2 mol/liter) to polymer in 85 min at 70°C? Data: $[k_p^2/k_t] = 128$ liter/mol·h at 70°C.

4.15 Monomer M (FW = 156) is polymerized using peroxide P. Termination is by disproportionation. The peroxide has a half-life of 1.00 h at 90°C and an activation energy of 30.0 kcal/mol. In a run at 60°C where the initial charge to the reactor is 1.00×10^{-3} mol of P and 2.00 mol of M in a total volume of 2,500 ml, the reaction is stopped after 35.0 min, at which time 12.5 g of polymer has been formed. What number-average molecular weight is expected of the polymer?

4.16 Monomer M is to be polymerized using a peroxide initiator, I. The peroxide has a half-life of 6.00 h at 70°C. The polymerization is to be carried out with the following requirements:

Polymerization time:	2.00 min
Polymerization temperature:	70°C
Conversion:	2.00%
Initial degree of polymerization:	1200

If the efficiency (f) = 1.00, and termination is by coupling, what concentration of peroxide should be used (in mol peroxide/mol of monomer)?

4.17 Peroxide G has twice the half-life that peroxide H has at the same temperature, 65°C. Monomer K polymerizes three times as fast as monomer L at 65°C when initiator H is used with either monomer and all the concentrations are the same. If both K and L homopolymers terminate by coupling, what is the ratio of the initial degree of polymerization for case I to that of case II as described in the following table? That is, what is $(x_n)_I/(x_n)_{II}$?

	Initial concentrations of reactants, polymerizations at 65°C	
	Case I	Case II
Monomer	[K] = 1.00 mol/liter	[L] = 2.00 mol/liter
Initiator	[H] = 0.00100 mol/liter	[G] = 0.00500 mol/liter

4.18 An aliphatic acrylate was polymerized by a free-radical mechanism and terminated by coupling. A peroxide characterized by a half-life of 1.0 h at 90°C and an activation energy of 30.0 kcal/mol was used for initiation. A polymer with a number-average degree of polymerization of 10,000 was produced by polymerization at 70°C using an initiator concentration of 3.5×10^{-4} mol/liter. If 2.00% of the initial monomer present was converted to polymer in 45.0 min, what was the initial monomer concentration?

4.19 When radicals are produced by photoinitiation, the rate of radical production can be independent of temperature. Thermal decomposition of a peroxide is, of course, temperature dependent. When styrene is photopolymerized in two experiments in which only the temperature is varied, the time it takes to convert 20.0% of the original charge of monomer to polymer is 20.0 min at 60°C and 18.0 min at 70°C. On the other hand, when an organic peroxide is the thermal initiator, the corresponding times for 20.0% conversion are 40.0 min at 70°C and 23.0 min at 80°C. If the half-life of the peroxide at 90°C is 10.0 h, what is its half-life at 60°C?

4.20 When monomer M is polymerized using peroxide Q at 57.0°C, 5.00% of the monomer is converted to polymer after 3.00 h. At 82.0°C, 15.00% is converted after only 1.00 h. The molecular weight initially produced at the higher temperature has exactly one-half the value of that from the lower temperature. If the energy of activation of the termination reaction is negligible, what is the energy of activation for the propagation reaction? The same starting concentrations of monomer and peroxide are used at both temperatures.

4.21 If one reacts butyl lithium (LiBu) with a small amount of monomer, a seed of living polymer is produced.

$$\text{LiBu} + 2M \rightarrow \text{Li—M—M—Bu} = S$$

Now we mix 10^{-3} mol of S with 2 mol of fresh monomer M and find that the reaction is first order in M. Half the monomer, 1 mol, is converted to polymer in 50 min at 25°C. *Estimate* k_p. Total volume of system is 1 liter. There is no termination reaction.

4.22 A vinyl monomer (CH_2=CHY, MW = 213) is polymerized by a free-radical initiator in the presence of dodecyl mercaptan ($C_{12}H_{23}SH$). Analysis of the purified polymer shows:

Number average degree of polymerization = 430
Sulfur content = 5.45×10^{-6} g-atoms/g
Terminal unsaturation = C mol/g

If half the kinetic chains are terminated by coupling and half by disproportionation, what is the expected value of C?

4.23 Show how the copolymer equation can be rearranged to give:

$$\frac{Y-1}{X} = r_1 - \left(\frac{Y}{X^2}\right) r_2$$

Derive r_1 and r_2 from a plot of these data by Dainton [55] for acrylamide (1): methacrylamide (2).

Compositions		
Feed $f_1/(1 - f_1)$ $\frac{f_1}{f_2}$		$\frac{F_1}{F_2}$ Polymer $F_1/(1 - F_1)$
0.125		0.150
0.250		0.358
0.500		0.602
1.000		1.33
4.000		4.72
8.000		10.63

Plot F_1 versus f_1 and compare the data with the copolymer equation on these coordinates.

4.24 Draw curves (on the same plot) of polymer composition (F_1) versus monomer composition (f_1) for the following systems:

(a) Butadiene (1), styrene (2), 60°C; $r_1 = 1.39$, $r_2 = 0.78$.
(b) Vinyl acetate (1), styrene (2), 60°C; $r_1 = 0.01$, $r_2 = 55$.
(c) Maleic anhydride (1), isopropenyl acetate (2), 60°C; $r_1 = 0.002$, $r_2 = 0.032$.

Assuming that one starts with an equimolar mixture of monomers in each case, which will contain more of monomer (1), the polymer formed first or the polymer formed later?

4.25 Compare the experimental values of r_1 and r_2 from Table 4.10 with those calculated by using Q and e values from the same table for several pairs of monomers.

4.26 Acrylonitrile, 1 mol, is copolymerized with methyl vinyl ketone, 2 mol, at 60°C. Regarding acrylonitrile as monomer 1, $r_1 = 0.60$ and $r_2 = 1.66$. Calculate the mole fraction of acrylonitrile in polymer made initially. What is the mole fraction of acrylonitrile in the polymer being formed when 2 mol of monomer have been converted to polymer?

4.27 In the copolymerization of monomers 1 and 2, $r_2 = 1.0$ and $r_2 = 0.5$. Initially, $f_2 = 2f_1$.

(a) Which monomer predominates in the polymer formed initially?
(b) After 10% of the monomer charge is converted to polymer, does the polymer being made have more or less of monomer 1 than the polymer that was made initially? Show your calculations and reasons.

4.28 Using the data given below, select concentrations of initiator and chain transfer agent that will give poly(vinyl acetate) with an initial molecular weight (assuming coupling) of 15,000. The polymerization (at 60°C) should be 50% complete in 8.00 min.

	FW	Initial conc., g/liter
Monomer: vinyl acetate	86.09	500.0
Initiator: 2,2'-azo-bis-2-isobutyl- 4-methylvaleronitrile	332	—
Transfer agent: benzyl mercaptan	124.21	—

Solvent: methanol (to make 1 liter)
$k_i = 3.78 \times 10^{-4} \text{ s}^{-1}$, $k_p = 15 \times 10^3$ liter/mol·s, $k_t = 600 \times 10^6$ liter/mol·s
$C_s = 0.885$ (dimensionless) (Data from *Polymer Handbook,* Wiley)

4.29 The following chain transfer constants are reported for the polymerization of styrene at 60°C (*Polymer Handbook,* Sec. II, Wiley):

Reagent	Chain transfer constant, (degree of polym.)$^{-1}$	Mol. wt., g/mol	Density, g/cm^3
Toluene	Essentially zero	92.1	0.866
Styrene	0.60×10^{-4}	104.1	0.903
Acetone	1.5×10^{-4}	58.08	0.792
Isopropyl alcohol	3.1×10^{-4}	60.09	0.789
Carbon tetrachloride	92.0×10^{-4}	153.84	1.595

Under a certain set of conditions, one can produce polystyrene with a degree of polymerization (x_n) of 5000. The conditions are that a peroxide-type initiator is used at 60°C and the initial concentration of monomer is 2.00 mol/liter with toluene as the only solvent. What concentration* of additive (g/liter) would reduce the x_n to 800? Calculate separately for each of the three additives. Assume additivity of volumes.

4.30 The molecular weight of polymer when vinyl butyrate is polymerized in an inert solvent is 500,000. With all other conditions the same, addition of 5 g/liter of 1-dodecanethiol decreases the molecular weight to 150,000. What concentration of 1-dodecanethiol will give a molecular weight of 275,000 (other conditions remaining the same)?

4.31 When monomers A and B are copolymerized, an azeotrope (polymer composition same as feed) is formed at a ratio of 7 mol of A to 3 mol of B. If monomer B is known *not* to homopolymerize:

*Concentration is in terms of the additive taking the place of part of the toluene and keeping the initial monomer concentration at 2.00 mol/liter.

(a) What polymer composition (mole fraction of A) will result initially when a mixture of 9 mol of A with 1 mol of B is polymerized?

(b) Will polymer formed after 50% conversion from an initial mixture of 4 mol of A with 6 mol of B contain:
1. More A than polymer formed at 1% conversion?
2. Less A than polymer formed at 1% conversion?
3. Same fraction of A as polymer formed at 1% conversion?

4.32 A random copolymer of vinyl chloroacetate (1) and vinyl acetate (2) is made with $F_1 = 0.00500$ and a density of 1.04 g/cm^3. Then each chlorine is removed by reaction with metallic zinc to form zinc chloride (removed by a water wash) and a carbon-carbon cross-link (one for each pair of chlorines). Assuming a complete network with no loose ends, what swollen volume will 2.80 g of cross-linked copolymer occupy in a solvent with a molar volume of 82.5 cm^3/mol and $\chi = 0.520$?

$$-CH_2-CH- \qquad \text{vinyl acetate, } Y = H, \text{ FW} = 86.09$$

$$\mid$$

$$O \qquad\qquad \text{vinyl chloroacetate, } Y = Cl, \text{ FW} = 120.53$$

$$\mid$$

$$O=C-CH_2Y$$

4.33 Homopolymer A has a T_g of 120°C. Homopolymer B has a T_g of −40°C. A and B are copolymerized (to very low conversion) and the glass transition temperatures of the copolymers are measured. For a monomer feed of 50:50 (mole ratio), the T_g is 74°C. For a monomer feed of 25:75, A:B, T_g is 36°C. What is the expected T_g when the monomer feed is 85:15, A:B? Assume:

(a) There is a linear relationship between T_g and weight fraction of A in the polymer.

(b) Density does not change with composition.

(c) Monomers A and B have the same molecular weight.

4.34 When monomers A and B are copolymerized in a mole ratio of 2 of A to 5 of B, the polymer formed (initially) has a composition exactly the same as the monomer feed. However, when the monomer ratio is 5 of A to 1 of B, the polymer has a ratio of 5 mol of A to 3 of B. What monomer feed composition (mole fraction of A) will yield a polymer (initially) with equal moles of A and B?

4.35 Monomers A and B copolymerize to give an azeotropic composition in which the weight fraction of B is 0.750. A copolymer with a weight fraction of A is 0.750 can be made when the monomer feed has a weight fraction of A equal to 0.810. What mole fraction of A will be contained in a copolymer made from a monomer mixture with 10 mol each of A and B?

Monomer A: FW = 104   Monomer B: FW = 171

4.36 Monomers A, B, and C are polymerized together. Illustrate all the possible propagation reactions with equations and constants. Define the relative reaction rates (r_i). What set of values for r_i describes production of mixtures of homopolymers?

4.37 Monomer A and monomer B are copolymerized under various conditions. Predict the composition (mol% of A) initially formed at the start of a copolymerization with an equimolar feed.

Previous data

Run I: Feed: 99 mol A, 1 mol B. Initial polymer: 0.50 mol%B. *instantaneous*
Run II: Feed: 1 mol A, 99 mol B. Initial polymer: 2.00 mol%A.

4.38 An equimolar mixture of methyl acrylate and styrene is copolymerized. It is found that polymer radicals based on methyl acrylate add methyl acrylate monomer at a rate that is only 18% of the rate at which they add styrene. Polymer radicals based on styrene add methyl acrylate at a rate 133% of the rate at which they react with styrene. For what feed composition will a copolymer be produced that is uniform in composition (does not change with conversion)? →azeotrope !

4.39 The reaction of 1.00 mol of monomer A with 4.00 mol of monomer B yields an azeotrope with a T_g of 15.0°C. It is also established that a polymer made from a monomer feed of equimolar quantities of A and B has a T_g of 30.0°C. A nonazeotropic polymer is made from a monomer feed with 4.00 mol of A and 1.00 mol of B. Given that the T_g of A is 125.0°C and the T_g of B is 0.0°C, what is the T_g of the nonazeotropic polymer? Monomers A and B have the same formula weight and density.

4.40 Monomers A and B yield homopolymers with glass transition temperatures of 210.0°C and 20.0°C, respectively. When they are copolymerized, an equimolar feed yields a polymer with a mole fraction of A = 0.800 and a glass transition temperature of 95.0°C. Monomer A has a formula weight of 124 and monomer B has a formula weight of 102. The copolymerization is "ideal" in that $r_1 r_2 = 1$. If a copolymer is made from a feed with a mole fraction of A = 0.300, what glass transition temperature is expected for the polymer?

4.41 The reaction rate of an isocyanate with a primary alcohol may be given by

$$-\frac{d[B]}{dt} = -\frac{d[A]}{dt} = k[A][B]$$

where [B] and [A] are concentrations of isocyanate and hydroxyl groups, respectively, in moles per liter. For the following mixture, estimate the time required for gel to form. Assume $k = 0.50$ liter/mol·h

3,3′ Dimethylphenylmethylmethane- 4,4′-diisocyanate($C_{17}H_{14}N_2O_2$)	2 mol
2-Ethylhexanol ($C_8H_{18}O$)	1.5 mol
Trimethylol propane ($C_6H_{14}O_3$)	1 mol
Benzene	To make 1 liter

4.42 Polymer A is mixed with a small amount of peroxide P in order to carry out a cross-linking operation. It can be assumed that each peroxide molecule that decomposes produces two radicals that, together, introduce one cross-link into the polymer. Thus decomposition of one molecule of peroxide brings about one cross-link. The polymer/peroxide mixture is divided into three portions, and cross-linking is carried out at various temperatures for various times. In order to characterize the reaction, samples are swollen in a solvent for which the *polymer-solvent interaction parameter* is 0.400. From the data furnished below, calculate the activation energy for peroxide decomposition. An additional assumption is that the peroxide decomposes at the same rate that it would decompose in the absence of polymer.

Cross-linking temp., °C	Time, h	v_2, volume fraction of polymer in swollen gel
110	0.500	0.340
110	1.000	0.390
135	0.500	0.380

REFERENCES

1. Flory, P. J.: "Principles of Polymer Chemistry," pp. 106–132, Cornell Univ. Press, Ithaca, 1953.
2. "Lucidol Organic Peroxides," Lucidol Division, Wallace and Tiernan, Inc., Buffalo, N.Y., 1968.
3. Brandrup, J., and E. H. Immergut (eds.): "Polymer Handbook," 3d ed. Wiley, New York, 1989.
4. Kolthoff, I. M., and I. K. Miller: *J. ACS,* 73:3055 (1951).
5. Flory, P. J.: "Principles of Polymer Chemistry," pp. 148–158, Cornell Univ. Press, Ithaca, N.Y., 1953; M. S. Matheson, E. E. Auer, E. B. Bevilacqua, and E. J. Hart: *J. ACS,* 71:497 (1949).
6. Bandrup, J., and E. H. Immergut: "Polymer Handbook," 2d ed., pp. II-47 to II-52, Wiley-Interscience, New York, 1975.
7. Bevington, J., and H. Mellville: *J. Polymer Sci.,* 12:449 (1954).
8. Brandrup, J., and E. H. Immergut (eds.): "Polymer Handbook," 3rd ed., p. II-109, Wiley, New York, 1989.
9. Riggs, J. P.: "The Aqueous Phase Polymerization of Acrylamide: A Kinetics and Mechanism Study," Ph.D. thesis, Cornell University, Ithaca, N.Y., 1964.
10. Razuvayev, G. A., et al.: *Polymer Sci. USSR,* 1020 (1962).
11. Billmeyer, F. W., Jr.: "Textbook of Polymer Science," p. 292, Wiley-Interscience, New York, 1962.
12. Kennedy, J. P., and R. M. Thomas: *J. Polymer Sci.,* 49:189 (1961).
13. Flory, P. J.: "Principles of Polymer Chemistry," p. 220, Cornell Univ. Press, Ithaca, 1953.
14. Lenz, R. W.: "Organic Chemistry of Synthetic High Polymers," p. 418. Wiley-Interscience, New York, 1967.

15. Lenz, R. W.: "Organic Chemistry of Synthetic High Polymers," p. 623, Wiley-Interscience, New York, 1967.
16. Russell, K. E., and G. J. Wilson, in C. E. Schildknecht (ed.): "Polymerization Processes," Wiley-Interscience, New York, 1977.
17. Kennedy, J. P., and B. Ivan: "Designed Polymers by Carbocationic Macromolecular Engineering," Hanser, New York, 1992.
18. Kennedy, J. P., S. Midha, and Y. Tsunogae: *Macromolecules,* 26:429 (1993).
19. Morton, M.: *AIChE Symp. Polymer Kinetics Catalyst Systems,* preprint 30, December 1961.
20. Holden, G., and R. Milkovich: *U.S. Patent 3,265,765* (assigned to Shell Oil Company), Aug. 9, 1966.
21. Hsieh, H. L.: *Rubber Chem. Technol.,* 49:1305 (1976).
22. Zelinski, R., and C. W. Childers: *Rubber Chem. Technol.,* 41:161 (1968).
23. Ballard, D. G. H., R. J. Bowles, D. M. Haddleton, S. N. Richards, R. Sellens, and D. L. Twose: *Macromolecules,* 25:5907 (1992).
24. Bayard, P., J. P. Teyssie, S. Varshney, and J. S. Wang: *Polymer Bull.,* 33:381 (1994).
25. Pino, P., U. Giannini, and L. Porri: "Insertion polymerization," in *Enc. Polym. Sci. Eng. 2d ed.,* 8:147 (1987).
26. Natta, G., et al.: *J. Polymer Sci.,* 16:143 (1955).
27. Cooper, W., in J. C. Robb and F. W. Peaker (eds.): "Progress in High Polymers," vol. 1, p. 279, Heywood, London, 1961.
28. Goodall, B. L.: *J. Chem. Educ.,* 63:191 (1986).
29. Guyot, A., C. Bobichon, R. Spitz, L. Duranel, and J. L. Lacombe in W. Kaminsky and H. Sinn (eds.): "Olefin Polymerization," p. 13, Springer-Verlag, New York, 1987.
30. Tait, P. J. T., in W. Kaminsky and H. Sinn (eds.): "Olefin Polymerization," p. 309, Springer-Verlag, New York, 1987.
31. Kaminsky, W., in T. Keii and K. Soga (eds.): "Catalytic Polymerization of Olefins," p. 293, Elsevier, New York, 1986.
32. Montagna, A. A., and J. C. Floyd: *Hydrocarbon Proc.,* 73(11):57 (1994).
33. Canich, J. A. M.: *U.S. Patent 5,026,798,* June 25, 1991.
34. Lai, S. Y., J. R. Wilson, G. W. Knight, J. C. Stevens, and P. W. S. Chum: *U.S. Patent 5,272,236,* Dec. 21, 1993.
35. McMillan, F. M., in J. K. Craver and R. W. Tess (eds.): "Applied Polymer Science," p. 328, ACS, Washington, D.C., 1975.
36. Marsden, C. E.: *Plastics, Rubber, Comp. Proc. and Appl.,* 21:193 (1994).
37. Webster, O. W., W. R. Hertler, D. Y. Sogah, W. B. Farnham, and T. V. RajanBabu: *J. Am. Chem. Soc.,* 105:5706 (1983).
38. Sogah, D. Y., W. R. Hertler, O. W. Webster, and G. M. Cohen: *Macromolecules,* 20:1473 (1987).
39. Brittain, W. J.: *Rubber Chem. Tech.* 65:580 (1992).
40. Flory, P. J.: "Principles of Polymer Chemistry," chap. 9, Cornell Univ. Press, Ithaca, N.Y., 1953.
41. Tobolsky, A. V.: "Properties and Structure of Polymers," chap. 6, Wiley, New York, 1960.
42. Calderon, N., and R. L. Hinrichs: *Chem. Tech.,* 4:627 (1974).
43. Kirchhoff, R. A., and K. J. Bruza: *Prog. Polymer Sci.* 18:85 (1993).
44. "Experimental Benzocyclobutene for Aerospace Applications," The Dow Chemical Company, Midland, Mich., 1992.
45. McCabe, W., J. C. Smith, and P. Harriott: "Unit Operations of Chemical Engineering," 4th ed., p. 507, McGraw-Hill, New York, 1985.
46. Meyer, V. E., and G. G. Lowry: *ACS Polymer Div. Preprints,* 5:60 (1964): *J. Polym. Sci.,* A3:2843 (1965).
47. Hanna, R. J.: *Ind. Eng. Chem.,* 49:208 (1957).
48. Alfrey, T., Jr., and L. J. Young, in G. E. Ham (ed.): "Copolymerization," chap. 2, Wiley-Interscience, New york, 1964.
49. Alfrey, T., Jr., and C. C. Price: *J. Polymer Sci.,* 2:101 (1947).

50. Hoffman, A. S., and R. Bacskai, in G. E. Ham (ed.): "Copolymerization," chap. 6, Wiley-Interscience, New York, 1964.
51. *Hydrocarbon Process Petrol. Refiner,* 46(11):192 (November 1967).
52. Archer, B. L., et al., Biosynthesis of Rubber, in L. Bateman (ed.): "The Chemistry and Physics of Rubber-like Substances," Wiley, New York, 1963.
53. Jirgensons, B.: "Natural Organic Macromolecules," p. 116, Pergamon, New York, 1962.
54. Goulian, M., A. Kornberg, and R. L. Sinsheimer: *Proc. Natl. Acad. Sci., U.S.,* 58:2321 (1967).
55. Dainton, F. S., and W. D. Sisley: *Trans. Far. Soc.,* 59:1385 (1963).

General References

Aggarwal, S. L. (ed.): "Block Polymers," Plenum, New York, 1970.
Akelah, A., and A. Moet: "Functionalized Polymers," Chapman & Hall, New York, 1990.
Albright, L. F.: "Processes for Major Addition-Type Plastics and Their Monomers," McGraw-Hill, New York, 1974.
Allen, N. S. (ed.): "Photopolymerisation and Photoimaging Science and Technology," Elsevier, New York, 1989.
Allen, P. E. M., and C. R. Patrick: "Kinetics and Mechanisms of Polymerization Reactions," Halsted-Wiley, New York, 1974.
Bailey, F. E., Jr., and J. V. Koleske (eds.): "Alkylene Oxides and Their Polymers," Dekker, New York, 1991.
Bamford, C. H., and C. F. H. Tipper: "Comprehensive Chemical Kinetics," vol. 14, "Free Radical Polymerization," Elsevier, New York, 1976.
Benham, J. L., and J. F. Kinstle (eds.): "Chemical Reactions on Polymers," ACS, Washington, D.C., 1988.
Biederman, H., and Y. Osada: "Plasma Polymerization Processes," Elsevier, New York, 1992.
Boor, J., Jr.: "Ziegler-Natta Catalysts and Polymerizations," Academic Press, New York, 1979.
Brunelle, D. J. (ed.): "Ring-Opening Polymerization," Hanser-Gardner, Cincinnati, Ohio, 1993.
Butler, G. B.: "Cyclopolymerization and Cyclocopolymerization," Dekker, New York, 1992.
Carraher, C. E., and M. Tsuda (eds.): "Modification of Polymers," ACS, Washington, D.C., 1980.
Casale, A., R. S. Porter, and W. H. Sharkey, in H. J. Cantow (ed.): "Polymerization," Springer-Verlag, New York, 1975.
Cassidy, P. E.: "Thermally Stable Polymers—Synthesis and Properties," Dekker, New York, 1980.
Ceresa, R. J. (ed.): "Block and Graft Copolymerization," Wiley, New York, vol. 1, 1973; vol. 2, 1976.
Chien, J. C. W. (ed.): "Coordination Polymerization," Academic Press, New York, 1975.
Chum, H. L. (ed.): "Polymers from Biobased Materials," Noyes, Park Ridge, N.J., 1991.
Cowie, J. M. G. (ed.): "Alternating Copolymers," Plenum, New York, 1985.
Culbertson, B. M., and J. E. McGrath (eds.): "Advances in Polymer Synthesis," Plenum, New York, 1985.
Dickstein, W. H.: "Rigid Rod Star-Block Copolymers," Technomic, Lancaster, Pa., 1990.
Dragutan, V., A. T. Balaban, and M. Dimonie: "Olefin Metathesis and Ring-Opening Polymerization of Cyclo-Olefins," Wiley, New York, 1986.
Elias, H.-G.: "Macromolecules, Synthesis, Materials, and Technology," 2d ed., Plenum, New York, 1984.
Ellis, B. (ed.): "Chemistry and Technology of Epoxy Resins," Chapman & Hall, New York, 1993.
Erusalimskii, B. L.: "Mechanisms of Ionic Polymerization," Plenum, New York, 1986.
Fontanille, M., and A. Guyot (eds.): "Recent Advances in Mechanistic and Synthetic Aspects of Polymerization," (Reidel Holland), Kluwer Academic, Norwell, Mass., 1987.
Frisch, K. C., and S. L. Reegen (eds.): "Ring-Opening Polymerization," Dekker, New York, 1969.
Goethals, E. J. (ed.): "Telechelic Polymers: Synthesis and Applications," CRC Press, Boca Raton, FL, 1988.

Ham, G. E.: "Copolymerization," Wiley-Interscience, New York, 1964.

Ham, G. E. (ed.): "Vinyl Polymerization," Dekker, New York, part 1, 1967; part 2, 1969.

Haward, R. N. (ed.): "Developments in Polymerisation—1," Elsevier Applied Science, New York, 1979.

Haward, R. N. (ed.): "Developments in Polymerisation—2," Elsevier Applied Science, New York, 1979.

Haward, R. N. (ed.): "Developments in Polymerisation—3," Elsevier Applied Science, New York, 1982.

Hogen-Esch, T. E., and J. Smid (eds.): "Recent Advances in Anionic Polymerization," Elsevier Applied Science, New York, 1987.

Ivin, K. J., and T. Saegusa (eds.): "Ring-Opening Polymerization," 3 vols., Elsevier Applied Science, New York, 1984.

Jenkins, A. D., and A. Ledwith (eds.): "Reactivity, Mechanism, and Structure in Polymer Chemistry," Wiley-Interscience, New York, 1974.

Kaminsky, W., and H. Sinn (eds.): "Transition Metals and Organometallics as Catalysts for Olefin Polymerization," Springer, New York, 1988.

Kennedy, J. P.: "Cationic Polymerization of Olefins: A Critical Inventory," Wiley-Interscience, New York, 1975.

Kennedy, J. P., and B. Ivan: "Designed Polymers by Carbocationic Macromolecular Engineering Theory and Practice," Hanser-Gardner, Cincinnati, Ohio, 1992.

Ketley, A. D.: "The Stereochemistry of Macromolecules," Dekker, New York, vol. 1, 1967; vol. 2, 1967.

Kricheldorf, H. R. (ed.): "Handbook of Polymer Synthesis, Pts A & B," Dekker, New York, 1991.

Kucera, M.: "Mechanism and Kinetics of Addition Polymerizations," Elsevier, New York, 1991.

Lazar, M., T. Bleha, and J. Rychly: "Chemical Reactions of Natural and Synthetic Polymers," Prentice Hall, Englewood Cliffs, N.J., 1989.

Lenz, R. W.: "Organic Chemistry of Synthetic High Polymers," Wiley-Interscience, New York, 1967.

Lenz, R. W., and F. Ciardelli (eds.): "Preparation and Properties of Stereoregular Polymers," Reidel, Hingham, Mass., 1979.

Macromolecular Syntheses: A Periodic Publication of Methods for the Preparation of Macromolecules, Wiley-Interscience, New York, with various editors and years of issue, vol. 6, 1977.

McGrath, J. (ed.): "Ring-Opening Polymerization," ACS, Washington, D.C., 1985.

Meier, D. J. (ed.): "Block Copolymers," Gordon & Breach, New York, 1983.

Millich, F., and C. E. Carraher (eds.): "Interfacial Synthesis," Dekker, New York, 1973.

Morgan, P. W.: "Condensation Polymers by Interfacial and Solution Methods," Wiley-Interscience, New York, 1965.

Morton, M.: "Anionic Polymerization: Principles and Practice," Academic Press, New York, 1983.

Nonhebel, D. C., and J. C. Walton: "Free Radical Chemistry," Cambridge, New York, 1974.

Noshay, A., and J. E. McGrath: "Block Copolymers: Overview and Critical Survey," Academic Press, New York, 1976.

Odian, G.: "Principles of Polymerization," 3d ed., Wiley, New York, 1991.

Paleos, C. M. (ed.): "Polymerization in Organized Media," Gordon & Breach, New York, 1992.

Platzer, N. A. J. (ed.): "Polymerization Kinetics and Technology," ACS, Washington, D.C., 1973.

Platzer, N. A. J. (ed.): "Polymerization Reactions and New Polymers," ACS, Washington, D.C., 1973.

Platzer, N. A. J. (ed.): "Copolymers, Polyblends, and Composites," ACS, Washington, D.C., 1976.

Plesch, P. H., in J. C. Robb and F. W. Peaker (eds.): "Progress in High Polymers," vol. 2, Heywood, London, 1968.

Pomogailo, A. D., and V. S. Savost'yanov: "Synthesis and Polymerization of Metal-Containing Monomers," CRC Press, Boca Raton, Fla., 1994.

Quirck, R. P. (ed.): "Transition Metal Catalyzed Polymerization," Gordon & Breach, New York, 1983.

Rempp, P., and E. W. Merrill: "Polymer Synthesis," Huethig & Wepf, Mamaroneck, N.Y., 1986.

Roberts, A. D. (ed.): "Natural Rubber Science and Technology," Oxford Univ. Press, New York, 1988.

Sadhir, R. K., and R. M. Luck (eds.): "Expanding Monomers: Synthesis, Characterization, and Applications," CRC Press, Boca Rabon, Fla., 1992.

Saegusa, T., and E. Goethals (eds.): "Ring-Opening Polymerization," ACS, Washington, D.C., 1977.

Sandler, S. R., and W. Karo: "Polymer Synthesis," vol. 1, Academic Press, New York, 1974.

Sawada, H.: "Thermodynamics of Polymerization," Dekker, New York, 1976.

Schildknecht, C. E., and I. Skeist (eds.): "Polymerization Processes," Wiley-Interscience, New York, 1977.

Semlyen, J. A. (ed.): "Cyclic Polymers," Elsevier Applied Science, New York, 1986.

Soga, K., and M. Terano (eds.): "Catalyst Design for Tailor-Made Polyolefins," Elsevier, Amsterdam, 1994.

Solomon, D. H. (ed.): "Step-Growth Polymerizations," Dekker, New York, 1972.

Sorenson, W. R., and T. W. Campbell: "Preparative Methods of Polymer Chemistry," 2d ed., Wiley-Interscience, New York, 1968.

Stang, P. J. (ed.): "Vinyl Cations," Academic Press, New York, 1979.

Starks, C.: "Free Radical Telomerization," Academic Press, New York, 1974.

Swarc, M., and M. Van Beylens: "Ionic Polymerization," Chapman & Hall, New York, 1993.

Takemoto, K., Y. Inaki, and R. M. Ottenbrite (eds.): "Functional Monomers and Polymers," Dekker, 1987.

Tsurata, T., and K. F. O'Driscoll (eds.): "Structure and Mechanism in Vinyl Polymerization," Dekker, New York, 1969.

Uglea, C. V., and I. A. Negulescu: "Synthesis and Characterization of Oligomers," CRC Press, Boca Raton, Fla., 1991.

Ulrich, H.: "Raw Materials for Industrial Polymers," Hanser-Gardner, Cincinnati, Ohio, 1988.

Van der Ven, S.: "Polypropylene and Other Polyolefins: Polymerization and Characterization," Elsevier, New York, 1990.

Vandenberg, E. J., and J. C. Salamone (eds.): "Catalysis in Polymer Synthesis," ACS, Washington, D.C., 1992.

Vogl, O., and J. Furukawa (eds.): "Polymerization of Heterocyclics," Dekker, New York, 1973.

Yamashita, Y.: "Chemistry and Industry of Macromonomers," Hüthig & Wepf, Basel, Switzerland, 1992.

Yasuda, H.: "Plasma Polymerization," Academic, New York, 1985.

Yocum, R. H., and E. B. Nyquist (eds.): "Functional Monomers," 2 vols., Dekker, New York, 1973.

CHAPTER FIVE

POLYMERIZATION PROCESSES

5.1 DESIGN CRITERIA

Many of the problems that attend mechanical design of polymerization systems are common to ordinary organic reactions. Toxic and flammable monomers and catalysts, noxious odors, and sticky solids are dealt with routinely in chemical and petroleum plants. Some notable examples in the field of commercial polymer processes are:

Toxicity. Acrylonitrile ($CH_2{=}CH{-}CN$) has the toxicity of inorganic cyanides. A maximal concentration of 20 ppm in air is recommended for 8-h exposures.

Flammability. Many Ziegler catalysts involve aluminum triethyl, which is pyrophoric (bursts spontaneously into flames on exposure to air).

Odors. The lower acrylates can have odors that are penetrating and disagreeable even in low concentrations. Special catalytic systems may be installed to destroy traces of monomer in effluent gas streams.

In most cases the polymers themselves do not present these problems. On the other hand, unlike low-molecular-weight products, polymers are not generally subjected to purification by extraction, distillation, or crystallization after they have been formed. The ingredients present during formation often remain as part of the final product. Various techniques, bulk, solution, suspension, and emulsion, are used to produce polymers. Some of the criteria for use of these are connected with thermal effects.

Table 5.1 Heats of polymerization and changes in volume
(Liquid monomer to amorphous polymer in the range 20 to 25°C) [1, 2]

Monomer	Structural unit	FW, g/mol	$-\Delta Hp$ kJ/mol	$-\Delta Hp$ kJ/g	Spec. volume, cm³/g Monomer	Spec. volume, cm³/g Polymer
1. Ethylene	—CH₂—CH₂—	28.05	102	3.62	gas	0.853
2. Propylene	—CH₂—CH(CH₃)—	42.07	84	0.50	0.519	0.853
3. Isobutylene	—CH₂—C(CH₃)₂—	56.11	48	0.86	0.594	0.914
4. Butene-1	—CH₂—CH(C₂H₅)—	56.11	84	1.49	0.595	0.859
5. Isoprene	—CH₂—C(CH₃)=CH—CH₂—	68.12	75	1.10	0.681	0.909
6. Styrene	—CH₂—CH(C₆H₅)—	104.15	70	0.67	0.906	1.047
7. α-Methyl styrene	—CH₂—C(CH₃)(C₆H₅)—	118.18	35	0.30	0.908	1.066
8. Vinyl chloride	—CH₂—CHCl—	62.50	96	1.54	0.911	1.412
9. Vinyl acetate	—CH₂—CH(C₃H₃O₂)—	86.09	88	1.02	0.932	1.190
10. Acrylonitrile	—CH₂—CH(CN)—	53.06	78	1.46	0.806	1.184
11. Methyl methacrylate	—CH₂—C(CH₃)(C₂H₃O₂)—	100.12	56	0.56	0.944	1.190
12. Ethyl acrylate	—CH₂—CH(C₃H₅O₂)—	100.12	78	0.78	0.924	1.120
13. Methyl acrylate	—CH₂—CH(C₂H₃O₂)—	86.09	78	0.91	0.953	1.220
14. Acrylamide	—CH₂—CH(CONH₂)—	71.08	79	1.12	1.122	1.300
15. Tetrahydrofuran	—CH₂—CH₂—CH₂—CH₂—O—	72.11	23	0.32	0.886	0.985
16. 1,3-Butadiene	—CH₂—CH=CH—CH₂—	54.09	73	1.35	0.621	0.902
17. Chloroprene	—CH₂—C(Cl)=CH—CH₂—	88.54	68	0.77	0.958	1.243
18. Ethylene oxide	—CH₂—CH₂—O—	44.05	95	2.15	gas	1.127

Data sources: "Polymer Handbook," 3d ed., Secs. II and IV; J. P. Runt, *Enc. Polym Sci. Eng.*, 4:487 (1986).

The conversion of a double bond to a single bond is accompanied by an exothermic *heat of polymerization* of the order of 10 to 20 kcal/mol. With monomers of molecular weight of about 100 and specific heat of about 0.5 cal/°C·g, this means an adiabatic temperature rise of 200 to 400°C. Removal of this heat of polymerization often limits the rate at which the reaction can be carried out, especially because most monomers and polymers are poor conductors of heat. As pointed out earlier, a higher temperature often gives a lower molecular weight. Because of this, a variable temperature widens the distribution of molecular weights. Also, a rise in temperature increases the rate of reaction and the rate of heat generation. Highly substituted monomers have more steric repulsion in the polymer form. This is reflected in lower heats of polymerization. In Table 5.1, compare styrene with α-methyl styrene and acrylates with methacrylate.

As a monomer is converted to polymer in a homogeneous system, the viscosity can increase rapidly. In a highly viscous medium, small monomer molecules still can diffuse readily to growing chains so that k_p remains relatively constant but large growing chains cannot diffuse easily toward each other, so that k_t can decrease considerably. According to Eq. (4.5), then, the rate should increase. Also, according to Eq. (4.15), the degree of polymerization should increase. This sudden increase in rate, *autoacceleration* or the *Trommsdorff effect,* is pronounced when high-molecular-weight polymer is formed, since the viscosity of the solution increases in proportion to the molecular weight raised to somewhere between the second and tenth power in many cases. The transition from normal kinetics to autoacceleration can be quite sharp. It is aggravated by the higher rate of heat generation, which can raise the temperature and further increase the rate. The addition of large amounts of initiator to give a lower molecular weight or the presence of a solvent to keep the viscosity down can delay the onset of autoacceleration. On the other hand, the addition of a polyfunctional monomer will bring about branching and perhaps gel formation with early onset of the effect.

Polymers invariably are more dense than their monomers even when both are amorphous. The *shrinkage* on polymerization can be as much as 10 to 20%. In a batch polymerization, this often means a variable vapor volume above the reacting liquids. The agitation systems may be complicated, as in the suspension polymerization of styrene or methyl methacrylate, where monomer is lighter than water and must be drawn down into the agitated core, but polymer is heavier than water and must be pulled off the bottom of the reactor. This change in volume is the basis for kinetic measurements based on *dilatometry.* Even in dilute (1 to 5%) solutions of monomer in an inert solvent the increase in density is enough to be able to measure continuously the volume of the system and, therefore, the extent of reaction. Usually the density is directly proportional to conversion, simplifying calculations. In the dilatometer shown in Fig. 5.1, the bulb has a volume of about 50 ml and the capillary is 0.1 cm in diameter. When a dilute solution of acrylamide, $CH_2{=}CH{-}CO{-}NH_2$, say 20 g/liter, is polymerized, a 1% conversion of monomer to polymer can be seen as a change in capillary height of about 0.25 cm.

Example 5.1

Two cylinders are filled to a depth of 10.0 cm with (*a*) methyl methacrylate and (*b*) styrene. When both are polymerized completely, what will be the average depth of polymer in each cylinder (assuming no lateral shrinkage)?

Solution: Using data from Table 5.1,

$$\text{Methyl methacrylate:} \quad \frac{0.944 \times 10 \text{ cm}}{1.19} = 7.9 \text{ cm}$$

$$\text{Styrene:} \quad \frac{0.906 \times 10 \text{ cm}}{1.047} = 8.7 \text{ cm}$$

Certain classes of strained monomers do not shrink and may even expand on polymerization [4]. One of the first examples reported was a spiro ortho ester which was polymerized to a polymer with an estimated molecular weight of 25,000 using a boron trifluoride etherate initiator [5]. In this case there was no measurable shrinkage on polymerization. The monomer is made by reacting ethylene oxide with α-butyrolactone:

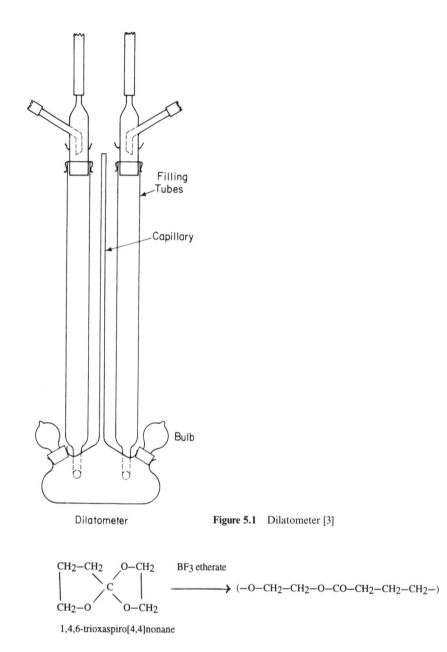

Dilatometer   **Figure 5.1**   Dilatometer [3]

CH$_2$–CH$_2$ O–CH$_2$ BF$_3$ etherate

(–O–CH$_2$–CH$_2$–O–CO–CH$_2$–CH$_2$–CH$_2$–)

CH$_2$–O   O–CH$_2$

1,4,6-trioxaspiro[4,4]nonane

    Polymerization may be carried out with monomer alone (*bulk*), in a solvent (*solution*), as an emulsion in water (*emulsion*), or as droplets, each one comprising an individual bulk polymerization, suspended in water (*suspension*). All four methods are commercially applied to radical-initiated chain polymers such as polystyrene. Most ionic and coordination complex systems are inactivated by water, so that only bulk or solution methods can be used. Also, rather few condensations are carried out in emulsion or suspension. However, ethylene dichloride and sodium polysulfide are condensed to give ethylene polysulfide rubber and sodium chloride in an aqueous emulsion. Gas-phase polymerization and interfacial condensation are special techniques, which are mentioned under solution polymerization.

5.2 BULK POLYMERIZATION

In bulk polymerization, also called *mass* or *block* polymerization, monomer and polymer (and initiator) are the only components. When only part of the monomer charge is converted to polymer, however, the problems encountered are more typical of the solution method, which is discussed next. We can differentiate between quiescent and stirred bulk polymerizations. Both methods are applied to systems where polymer is soluble in monomer and progressively increases in viscosity with conversion. In quiescent systems, gel formation, corresponding to infinite viscosity, can occur. Because of the heat of polymerization and autoacceleration, reaction rate is difficult to control. Heat removal is impeded by high viscosity and low thermal conductivity. The removal of traces of unreacted monomer from the final product is difficult because of low diffusion rates. Conversion of all monomer is difficult for the same reason.

Quiescent Bulk Polymerization (Monomer Casting)

Typically, nonstirred polymerizations are used to produce materials in useful shapes. Phenol-formaldehyde condensation carried out in a mold under pressure is an example that has been mentioned before. The casting of poly(methyl methacrylate) sheet has several advantages. For example, the Trommsdorff effect gives a large fraction of extremely high-molecular-weight molecules. These in turn make for a tough material with high melt viscosity that is ideal for sheet-forming operations. It would be hard to make this material in another form and then mold it as sheet because of this high melt viscosity. The major problems involved are (1) heat removal to prevent boiling of monomer, which would leave bubbles in the sheet; (2) conversion of all monomer; and (3) accommodation of the 21% shrinkage that occurs on polymerization. Problems 1 and 3 are alleviated somewhat by filling the mold with a degassed syrup of low-molecular-weight polymer (10 to 30% of syrup) and benzoyl peroxide (0.02 to 0.05%) in methyl methacrylate rather than with monomer alone. The mold consists of two glass plates separated by a flexible gasket that is also the confining wall of the mold (Fig. 5.2). After filling, the entire assembly is placed in an air oven. A typical time-temperature profile (Fig. 5.3) shows an exotherm due to autoacceleration after about 18 h. The shrinkage is accommodated by the flexible mold wall. Virtually complete conversion of monomer is assured by a 10-h "soak" at 85°C.

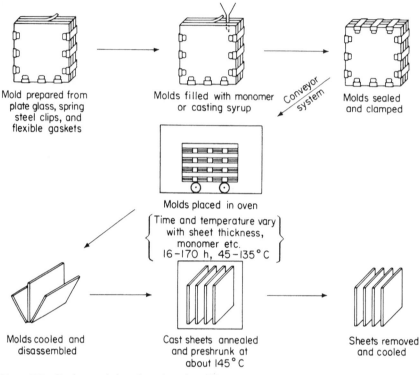

Mold prepared from plate glass, spring steel clips, and flexible gaskets

Molds filled with monomer or casting syrup

Conveyor system

Molds sealed and clamped

Molds placed in oven

Time and temperature vary with sheet thickness, monomer etc. 16–170 h, 45–135° C

Molds cooled and disassembled

Cast sheets annealed and preshrunk at about 145° C

Sheets removed and cooled

Figure 5.2 Casting methyl methacrylate sheet [6]

A continuous sheet-casting operation has obvious advantages over a batch process [8]. The casting syrup is confined between stainless-steel belts by using a flexible gasket as in the batch process. The belts run progressively through polymerization and annealing zones. It is important that pressure be maintained (by external rollers) so that monomer boiling will not occur. Since a higher temperature can be used, the total time for polymerization is decreased compared to the batch operation. The molecular-weight distribution that results from continuous polymerization will differ from that for the batch-cast polymer, so the products are not identical. Sheet thickness for the continuous casting process usually is limited to the range of 2 to 12 mm [9].

Stirred Bulk Polymerization

Although polystyrene and poly(methyl methacrylate) are commercially produced in stirred continuous-bulk systems, mechanical details of the processes are not readily available. In some cases, conversion is limited to less than 70% or so, and

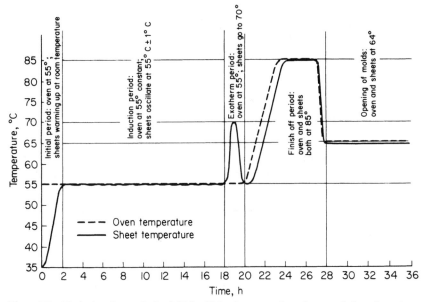

Figure 5.3 Typical curing cycle for 0.25-in-thick, clear, cast sheet from methyl methacrylate syrup [7]

the remaining monomer is stripped and recycled. These are, in effect, solution polymerizations and are treated in the next section. Equipment that will handle liquids progressively from monomer (*ca.* 0.01 poise) to polymer (*ca.* 10^5 poise) with efficient heat removal is usually designed specifically for a given installation. Conventional turbine- or propeller-agitated vessels can handle a limited degree of conversion. With low-molecular-weight condensation polymers, the completed polymer melt may be transferable by gear pumps or merely extruded from the reactor by application of moderate pressure. Nylon 66 is sufficiently low in viscosity above its melting temperature that it can be pumped through spinnerets (small orifices) directly from the polymerization vessel. In one design (Fig. 15.7) the continuous reactor consists of a series of tubes where nylon salt is first heated, water is flashed off, and finally the molten polymer is extruded directly from the reactor.

$$+ \, H_3N \text{\footnotesize +} CH_2 \text{\footnotesize)}_6 \, NH_3^+OOC \text{\footnotesize +} CH_2 \text{\footnotesize)}_4 \, COO^- \xrightarrow[-H_2O]{\text{heat}}$$

Nylon salt

$$\text{\footnotesize +} \text{\footnotesize (} CH_2 \text{\footnotesize)}_6 \, NH{-}CO \text{\footnotesize (} CH_2 \text{\footnotesize)}_4 \, CO{-}NH \text{\footnotesize +}$$

Nylon 66

High-molecular-weight materials need special equipment similar to screw extruders to convey the melt and remove heat simultaneously (Fig. 5.4). Usually in these continuous systems, high temperatures employed in the later stages are used in order to assure flow of the melt and complete reaction of monomer. This works in the opposite direction from the Trommsdorff effect as far as molecular weight is concerned, so that a lower average molecular weight is obtained than by quiescent polymerization.

5.3 SOLUTION POLYMERIZATION

The main advantage of a diluent is to take up heat of polymerization by a rise in temperature or by vaporization. We can subdivide solution polymerizations by the phases present. Invariably, monomer and diluent are miscible, so the combinations are:

	1	2	3	4
Monomer and diluent	Soluble	Soluble	Soluble	Soluble
Initiator	Soluble	Insoluble	Insoluble	Soluble
Polymer	Soluble	Soluble	Insoluble	Insoluble

Some physical schemes to carry out these reactions can be illustrated with specific examples.

1. *All components soluble.* In one process [11], ethylene containing a trace of oxygen (initiator) is compressed to 2000 atm and continuously pumped to a heated tube 5 mm ID and 20 m long (Fig. 5.5). With an average residence time of 45 s at 175°C, 22% of the ethylene is converted (3 kg polymer/h). At the end of the tube a valve is opened at intervals and the pressure drops slightly as polymer and monomer are ejected in a cyclic manner. The monomer flashes off, leaving polymer to be washed, dried, and packed. In this pilot-scale example, the monomer is the diluent. The extreme pressure favors the tube geometry. In a production unit the tube can be over a kilometer in length. Stirred autoclaves operating at 2000 to 3000 atm are sometimes used in place of the tubular geometry. Further details are given in Sec. 14.2.

2. *Insoluble initiator.* A chromia silica alumina catalyst (an initiator that can be regenerated) will convert ethylene to linear polymer at moderate pressures. One method [11] is to pass a dilute (2 to 4%) solution of ethylene in a saturated hydrocarbon over a fixed bed of catalyst at 150 to 180°C and 300 to 700 psi (2.1 to 4.8 MPa) (Fig. 5.6). Conversion may be high, but there is considerable volume of solvent to recover and recycle. Periodic reactivation of catalyst at about 50-h intervals is required.

3. *Insoluble initiator and polymer.* With a polymerizing system of ethylene and an agitated catalyst suspension at lower temperatures, polyethylene and catalyst may be insoluble and form a slurry which is removed continuously (Fig. 5.7). If productivity (ratio of polymer formed to catalyst charged) is high enough, simple drying may suffice. If not, washing out of catalyst may be needed.

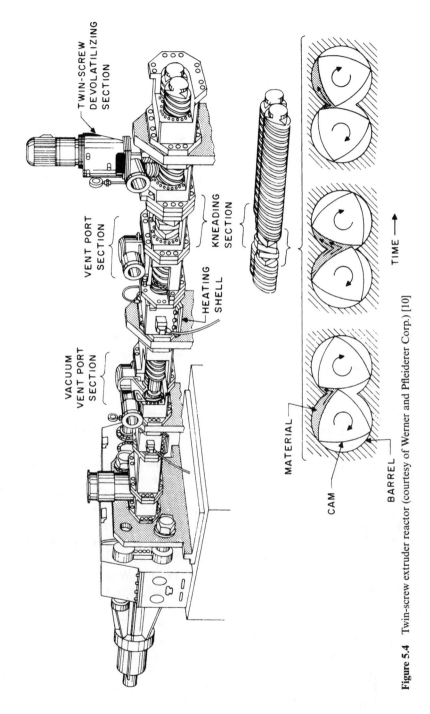

Figure 5.4 Twin-screw extruder reactor (courtesy of Werner and Pfleiderer Corp.) [10]

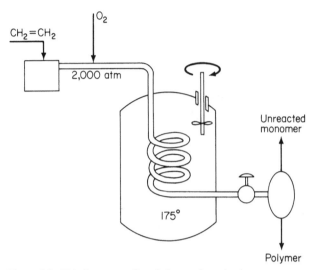

Figure 5.5 Tubular reactor for ethylene polymerization

A fluidized-bed, gas-phase polymerization bears some resemblance to the process just described. Fine catalyst particles are contacted with rapidly upward-moving ethylene vapors below the melting point of polyethylene. This has become an important commercial process with definite economic advantages over solvent-based systems (see Sec. 14.2).

4. *Insoluble polymer.* Heating an aqueous solution of acrylonitrile with persulfate as initiator at 80°C results in a stringy precipitate of polyacrylonitrile that can be filtered out and dried.

An interesting example of condensation polymerization in this category is *interfacial condensation* [12]. A diacid chloride is dissolved in a heavy, dense solvent such as tetrachlorethylene. As a stagnant layer above it is poured a solution of diamine in water. At the interface a layer of polyamide (nylon) is formed almost immediately. But the reaction stops because of the slow diffusion of reactants to each other through the polymeric interface. However, as the interface is pulled out, reactants contact each other and form more polymer (Fig. 5.8). Within limits, the rate of reaction is controlled by the rate of interface removal. With pure sebacoyl chloride and hexamethylene diamine, nylon 6-10 is formed from which fibers or films can be made.

$$H_2N-CH_2+CH_2)_4CH_2-NH_2 + Cl-\overset{\overset{\displaystyle O}{\|}}{C}+CH_2)_8\overset{\overset{\displaystyle O}{\|}}{C}-Cl \rightarrow$$

Hexamethylene diamine   Sebacoyl chloride

$$\left[-NH+CH_2)_6NH-\overset{\overset{\displaystyle O}{\|}}{C}+CH_2)_8\overset{\overset{\displaystyle O}{\|}}{C}-\right] + HCl$$

Nylon 6-10

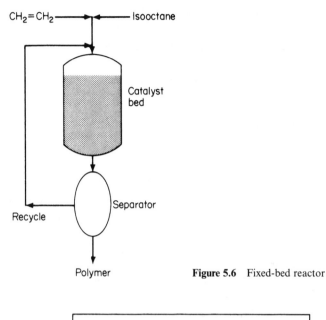

Figure 5.6   Fixed-bed reactor

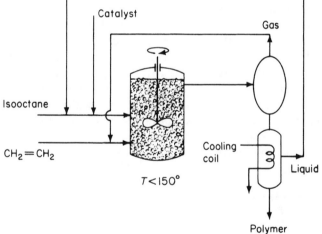

Figure 5.7   Slurry reactor

The demonstration based on the apparatus of Fig. 5.8 has been popularized under the name "the nylon rope trick" [12–14]. It has been used at countless science fairs and at all levels of sophistication from elementary school to university. On the other hand, it is not very convenient even for laboratory preparations. A typical preparative method starts with the same ingredients as the nylon rope trick. The aqueous phase is placed in a high-speed stirrer (blender) and the organic phase added while the system is stirred at high speed. Suspending agents or surfactants may be added to stabilize the polymer-solvent dispersion which results. Plant-scale operations also use stirred tanks rather than quiescent interfaces.

For all these systems a conventional glass-lined or stainless-steel reactor is suitable for batch operations when reaction pressures and product flow characteristics are within certain limits. Recently, scraped-surface, stainless-steel reactors have been built that will handle pressures from a full vacuum to 6 atm and viscosities of 1 mPa·s to 10 kPa·s for large batches.

5.4 SUSPENSION POLYMERIZATION

If the monomer is insoluble in water, bulk polymerization can be carried out in suspended droplets. The water phase becomes the heat transfer medium. Since it is the continuous phase, viscosity changes very little with conversion, so that heat transfer to the reactor walls can be efficient. The behavior inside the droplets is very much like the bulk polymerizations previously described. But because the droplets are only 10 to 1000 μm in diameter, more rapid reaction rates can be tolerated without boiling the monomer. To keep the droplets from coalescing as they proceed from liquid to solid states via a sticky phase, a protective colloid (suspending agent) and careful stirring are used. Poly(vinyl alcohol) dissolved in the aqueous phase is a typical suspending agent. Electrostatic charges are induced on the suspended monomer-polymer particles, which retard coalescence. They do not prevent the particles from rising (creaming) together if agitation is stopped. The particle size and size distribution are affected by the suspending agent and stirring rate. In the particle diameter range of 10 to 1000 μm, Fondy and Bates found that particle diameter varied inversely with impeller tip speed raised to the 1.8 power for a variety of impeller designs [15]. A typical vessel suitable for suspension or emulsion polymerization is shown in Fig. 14.16.

A simple example of suspension polymerization can be carried out in a glass flask with agitation.

Aqueous phase:	
Water	400 ml
Poly(vinyl alcohol)*	1 g
Oil phase:	
Methyl methacrylate	100 g
Benzoyl peroxide	1 g

*Copolymer with about 12% vinyl acetate.

Under a nitrogen atmosphere the aqueous phase is heated to 80°C. The oil phase is added with stirring at about 150 to 300 rpm with a paddle agitator. After an hour or so a slight exotherm, 5 to 10°C, is noted. Particles of the slurry exam-

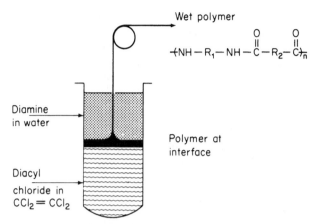

Figure 5.8 Interfacial condensation

ined before this time are sticky and usually agglomerate on cessation of stirring. After the exotherm, when most conversion has taken place, the particles are hard and do not agglomerate. Initially the monomer phase is lighter than water and agitation near the surface is essential. After the exotherm, enough polymer has been formed to make the oil phase heavier than water, so that settling out occurs if stirring is inadequate. Final recovery of the beads, which are 100 to 1000 μm in diameter, is by filtration and washing.

In most cases, polymerization is carried to high conversion and pearl-like beads of polymer result, especially if the polymer is amorphous and below its glass transition temperature. Two important exceptions are vinyl acetate and vinyl chloride. Vinyl acetate can be polymerized to give a stable suspension of about 50% solids content if the particle size is kept small (1 to 15 μm diameter). Plasticized with castor oil so that the polymer will be above T_g at room temperature, the suspension makes an excellent quick-setting adhesive for wood, paper, and cloth. Poly(vinyl chloride) is insoluble in its monomer. Suspension polymerization can be carried to 70 to 90% conversion under moderate pressure (6 atm) and temperature (80°C) by using an oil-soluble initiator such as lauroyl peroxide or azoisobutyronitrile. During this period, polymer precipitates in the droplets. Release of pressure and stripping of monomer by pulling a vacuum leaves the polymer portion of the droplets as porous particles that can be recovered by filtration and drying or directly by spray drying. Such porous particles are ideally suited for adsorbing plasticizers in "dry-blending" operations.

A combination of solution, suspension, and network polymerization has been used for a special-purpose material. Moore [16] describes the copolymerization of styrene with divinyl benzene as a 20% solution in toluene with peroxide initiator. All these ingredients are suspended in water. The swollen-gel droplets that result are well suited for gel permeation chromatography columns (Sec. 6.4). Ion-exchange resins can be made from cross-linked suspension-polymerized droplets (without the solvent), since the physical form does not have to be changed and the subsequent reactions can be performed on the beads. Sulfonation of styrene-divinyl benzene copolymer beads leads to a strong-acid ion exchanger (see Sec. 14.3 and Fig. 14.11).

5.5 EMULSION POLYMERIZATION [17]

Some important distinctions between emulsion and suspension polymerizations are:

1. Emulsions usually are comprised of small particles, 0.05 to 5 μm, compared with suspensions, particles 10 to 1000 μm diameter.
2. Water-soluble initiators are used rather than monomer-soluble ones.
3. The end product usually is a stable *latex*—an emulsion of polymer in water rather than a filterable suspension.

Because of these conditions, the mechanism of polymerization is basically different. The essentials of an emulsion polymerization system are monomer, a surface-active agent (surfactant), initiator, and water. Initially, the surfactant is in the form of *micelles,* spherical or rodlike aggregates of 50 to 100 surfactant molecules with their hydrophobic "tails" oriented inward and their hydrophilic "heads" outward. These micelles form whenever the concentration of surfactant exceeds a rather low *critical micelle concentration.* The manner in which the critical micelle concentration for a given surfactant is measured is instructive for our purposes. If a dilute aqueous solution of an oil-soluble dye such as eosin is titrated with a dilute soap solution, a soap concentration is reached where the color suddenly disappears. The explanation is that the dye has dissolved in or has been extracted by the micellar interiors. This is the same point at which surface tension stops decreasing rapidly with added soap. If monomer is added to a micellar dispersion, most of it remains as rather large droplets, but some of it dissolves in the micelles just as the dye does. Since they are smaller, the micelles present much greater surface area than do the droplets. Consequently, when free radicals are generated in the aqueous phase, the micelles capture most of them.

After a few percent conversion, the system consists of (1) stabilized, monomer-swollen polymer particles rather than micelles and (2) monomer, which is still mainly in droplets, although it constantly diffuses to replenish the swollen particles where polymerization continues. The result is, when monomer is quite water insoluble,

$$R_p = k_p[M] [M\cdot] \tag{5.1}$$

where [M] now is the concentration of monomer in the swollen polymer particles. Since this concentration, moles per liter of swollen particles, may remain constant from low conversions up to 70 or 80% conversion, R_p should be independent of the total concentration of monomer, i.e., moles per liter of emulsion. This means that the superficial reaction rate is pseudo-zero order in monomer. Now, it can be shown that a polymer particle cannot tolerate more than one growing chain at a time. Two radicals in the same particle mutually terminate rapidly. Therefore, on average, half the particles will contain radicals at any instant. As an analogy, picture a group of cups standing in the rain. Each cup has received either an odd or an even number of raindrops. With enough cups and enough rain, the chances are that half the cups have received an even number of drops and the other half an odd number. If the raindrops are initiating radicals and the cups are polymer droplets, only the half that have received an odd number of droplets are growing, the even cases having been stopped by mutual termination.

So [M·] is simply $N_p/2$, where N_p is the number of particles per unit volume of emulsion and

$$R_p = \frac{k_p[\mathrm{M}]N_p}{2} \tag{5.2}$$

Similar reasoning, disregarding chain transfer, and assuming termination by coupling, gives

$$x_n = \frac{k_p N_p[\mathrm{M}]}{d[\mathrm{M}\cdot]/dt} \tag{5.3}$$

where $d[\mathrm{M}\cdot]/dt$ is the rate of radical formation (not the *net* rate) and may be proportional to the square root of initiator concentration in the aqueous phase.

Example 5.2

Assume that during the emulsion polymerization of isoprene (FW = 68.12) with 0.10 potassium laurate at 50°C, during stage 2, the growing swollen particles contain 200 g of monomer per liter of swollen polymer particles. Assume also that the final latex has 400 g of polymer per liter with particles of 45.0 nm diameter. The polymer has a density of 0.90 g/cm. Using Fig. 5.9, estimate k_p (liters/mol·h).

Solution: The slope of the appropriate line in Fig. 5.9 is 100% conversion/ 30 h. Complete conversion amounts to 400/68.12 = 5.87 mol/liter of isoprene. Thus,

$$R_p = \frac{5.87}{30} = 0.196 \text{ mol/L·h}$$

In the particles, [M] = (200/68.12) = 2.94 mol/liter. The mass of each particle is

$$\frac{\pi D^3 \rho}{6} = \frac{\pi(45.0 \times 10^{-7} \text{ cm})^3\, 0.90\ (\text{g/cm}^3)}{6} = 4.29 \times 10^{-17} \text{ g}$$

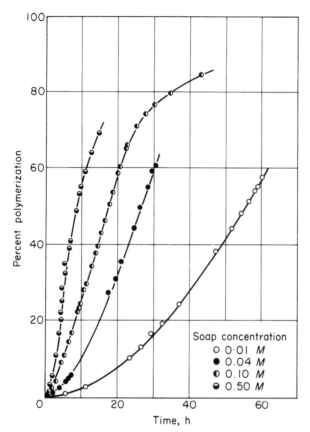

Figure 5.9 Polymerization of isoprene at 50°C for four concentrations of emulsifier (potassium laurate) [19]

So

$$N_p = \frac{400}{4.29} \times 10^{-17} = 9.3 \times 10^{18} \text{ particles/liter}$$

Divide N_p by Avogadro's number (6.02×10^{23}) to get moles per liter.

$$k_p = \frac{2R_p}{[M]N_p} = \frac{2 \times 0.196 \times (6.02 \times 10^{23})}{2.94 \times 9.3 \times 10^{18}}$$
$$= 8.6 \times 10^3 \text{ liters/mol·h}$$

(The literature value is 10×10^3 liters/mol·h [2])

The number of particles per unit volume, N_p, can be related to the concentrations of surfactant and initiator. The derivation can be complex, but the Smith-

Ewart approach [17, 18] introduces a number of simplifying assumptions. The entire process of establishing N_p takes place during the first part of polymerization, when the surfactant is still present in the form of micelles. Each effective radical generated from the initiator converts a micelle into a swollen polymer particle, which then grows at a constant rate ($\mu = dv/dt$). The capture of radicals by particles in competition with micelles is ignored for the moment. Every particle formed grows in area as well as in volume. Each increment of area requires a coating of surfactant, which is obtained from unreacted micelles. Thus micelles disappear for two reasons: Some are converted to polymer particles, and some supply surfactant to growing particles.

At some time t_1 all the surfactant in the system has coated the surface of particles, leaving none in the form of micelles. At that point, N_p no longer changes and the rate of polymerization is given by Eq. (5.2). If effective radicals are being generated constantly at a rate ρ_r and each one starts a particle, then at t_1,

$$N_p = \rho_r t_1 \tag{5.4}$$

Next, we can relate t_1 to the surface area of particles being formed.

1. A single particle is formed at time τ and then grows until time t_1. The volume of this particle (neglecting the initial volume of the micelle) at this time will be

$$v(t_1, \tau) = \mu(t_1 - \tau) \tag{5.5}$$

The surface area of the particle is related to the volume by geometry:

$$a(t_1, \tau) = (36\pi)^{1/3}[\mu(t_1 - \tau)]^{2/3} \tag{5.6}$$

2. The number of particles generated in the time interval $d\tau$ is $\rho_r d\tau$. Thus the total surface area of particles at t_1, using Eq. (5.6), is:

$$A_t = \int_0^{t_1} a(t_1, \tau)\rho_r\, d\tau = (36\pi)^{1/3}(0.6\rho_r)\mu^{2/3}t_1^{5/3} \tag{5.7}$$

3. The total area at t_1 is also given by

$$A_t = a_s[S] \tag{5.8}$$

where a_s is the surface area occupied by a mole of surfactant and [S] is the surfactant concentration (in mol/liter).

4. Combining Eqs. (5.4), (5.7), and (5.8), we can eliminate A_t and t_1:

$$N_p = 0.53 \left(\frac{\rho_r}{\mu}\right)^{0.4} (a_s[S])^{0.6} \tag{5.9}$$

The total number of particles per liter should vary with surfactant concentration to the 0.6 power and with initiator concentration (via ρ_r) to the 0.4 power. As noted earlier, at time t_1 (usually corresponding to a very small fraction of conversion) N_p no longer increases and the rate of polymerization reverts to Eq. (5.2).

This derivation can be modified to take into account radicals entering particles in competition with micelles. The only change is that the factor 0.53 becomes 0.37. More serious differences occur when particles are large, when initiator decomposition is very slow, or when the monomer is partly soluble in water or insoluble in polymer [18]. However, the form of Eq. (5.9) often holds even though the proportionality factor must be determined experimentally.

An increase in surfactant concentration increases N and, therefore, increases R_p and x_n (Fig. 5.9). High molecular weights are possible in emulsion polymerization because initiation is in one phase, water, and termination is in another phase, monomer-polymer. The combination of high molecular weight with high rate is one reason for the popularity of this method. *Seeded polymerizations* can be useful for making large-particle-size latexes. A completed "seed" latex is diluted to give the desirable value of N_p particles per liter of emulsion. No additional surfactant is added, so no new particles are formed. Monomer is fed in and initiator is added. Polymerization occurs in the previously formed particles, so that each one grows as monomer diffuses to it and is converted (Fig. 5.10). When the seed monomer and added monomer are different, graft copolymers may be formed provided the dead polymer in the seed can react by means of residual unsaturation or chain transfer.

A small amount of water solubility can alter the polymerization characteristics. Vinyl acetate dissolves in water only to the extent of 2%, but this is enough to give a combination of emulsion and solution polymerization when emulsion polymerization is attempted (Fig. 5.10). The rate also is proportional to the initiator concentration [20]. Another exceptional case is vinylidene chloride, where monomer is almost insoluble in polymer but is adsorbed on the surface of polymer particles. Under these circumstances, faster stirring can accelerate the reaction (Fig. 5.11) [21].

The latex that results from emulsion polymerization may be the desired form for the intended end use. Some paints and adhesives can use latexes directly. However, if massive polymer is desired, recovery may involve coagulation by heating. freezing, salt or acid addition, spray drying, or mechanical turbulence. Surfactants, coagulants, and initiator fragments often remain as impurities in the final product under these circumstances. In the production of styrene-butadiene rubber (SBR), sodium soaps of fatty acids are used as emulsifiers. Acidification of the final latex simultaneously destroys the surfactant and deposits free fatty acid in the coagulum, where it is a valuable adjunct to the sulfur cross-linking reaction used later on for vulcanization. Both styrene-butadiene rubber and polychloroprene rubber have been produced in continuous systems in which a number of stirred reactors are connected in series.

Several variants of suspension and emulsion polymerization can be used to produce particles with almost uniform particle size. Careful control of an emulsion polymerization in the absence of a separate surfactant can produce almost monodisperse (uniform size) latex particles with diameters as large as 0.5 μm. Swelling

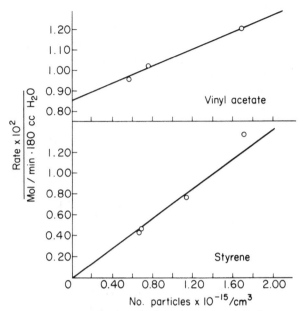

Figure 5.10 Rate of polymerization of styrene and vinyl acetate vs. number of particles using a seed of poly(vinyl acetate) latex [20]

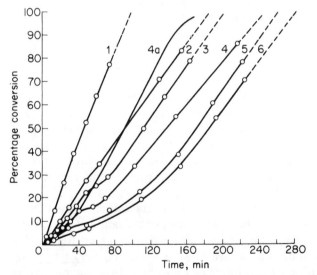

Figure 5.11 Effect of surfactant concentration on emulsion polymerization of vinylidene chloride at 36°C [21]. Initiator is 0.15 g $(NH_4)_2S_2O_8$ and 0.15 g $Na_2S_2OL_3$/100 g monomer. Sodium lauryl sulfate concentration is (1) 10.0 g, (2) 5.0 g, (3) 3.0 g, (4 and 4a) 2.0 g, (5) 1.0 g, (6) 0.50 g/100 g monomer. Stirring rate is 285 rpm except for 4a, which is 756 rpm.

such a "seed" latex with additional monomer and continuing the polymerization, but now with a monomer-soluble initiator, can produce much larger, monodisperse particles [22]. Another route to uniform particles is *dispersion* polymerization in an organic solvent. For example, polymerization of styrene with benzoyl peroxide as the initiator produces a precipitate when the medium (a mixture of alcohols) is not a solvent for the polymer. If hydroxypropyl cellulose is present as a stabilizer (such as the protective colloid in usual suspension polymerization), essentially monodisperse particles can be produced [23]. Perhaps the most elegant method of producing highly monodisperse particles with diameters of 3 to 30 μm involves sending a reactor into space [24]. The microgravity environment in a space shuttle craft has been used to make polystyrene latexes. The technique starts with a seed latex, which is swollen with additional monomer. The initiator is a combination of water-soluble persulfate and monomer-soluble azo compound. The particles actually produced in space have been marketed as microscopy standards by the U.S. National Bureau of Standards.

5.6 COMPARISON OF POLYMERIZATION METHODS

Bulk polymerization offers real hazards. The thermal conductivities of monomers and polymers are low, and viscosity buildup limits heat transfer by forced convection. Removal of unreacted monomer is difficult because of the low surface-to-volume ratio. On the other hand, the level of impurities can be held down by use of low initiator levels and diligent monomer removal.

Solution polymerization offers easier temperature control because of (1) added heat capacity of solvent and (2) lower viscosity. Removal of last traces of solvent and unreacted monomer can be difficult. Level of impurities can be very low, since the initiator residues can be washed out.

Temperature control is more convenient in *emulsion polymerization,* because the viscosity changes very little with conversion. Also, thermal conductivity and specific heat of water are higher than those for organic solvents. Removal of monomer can be accomplished without coagulating the latex. However, impurity levels are usually rather high because surfactant and coagulant residues are hard to remove. The agglomeration processes tend to give porous particles that trap some of the aqueous phase with its dissolved salts and surfactant.

Suspension polymerization is essentially a bulk polymerization carried out in droplets. Temperature control is complicated by the unstable nature of the suspension. Agitation is critical. Often, as the viscosity *within the beads* rises, the reaction rate increases suddenly (Trommsdorff effect). This leads to a surge in heat generation, which does not usually occur in solution or emulsion polymerization. On the other hand, the viscosity of the continuous phase (water) does not change during the reaction, so that control is easier than in bulk polymerization. Monomer recovery parallels that in emulsion polymerization. Polymer recovery is simple and usually leads to lower impurity levels than with emulsion polymers.

The differences in preparing polystyrene by several methods can be studied in Figs. 14.10 and 14.11.

Polymer processing after the reaction step varies with the solubility, the physical state, and the desired final form of the polymer, as well as with the extent of conversion of monomer to polymer. Coagulation of latexes and filtration of suspension polymers have been mentioned. Flow sheets for industrial processes in Chaps. 14 and 15 illustrate a wide variety of processing operations. Washing to remove catalyst residues, low-molecular-weight polymer, or atactic fractions may be practical for a polymer such as polypropylene, which can be handled as a slurry of relatively hard particles. Recrystallization is a common procedure for low-molecular-weight organic compounds but not for polymers. Purification by dissolving and precipitation is avoided because of the expense involved with copious quantities of solvents. Even deionized water can be ruinously expensive if 100 volumes are used for a single volume of polymer. Most polymers recovered by precipitation separate as slimy, sticky masses that are hard to handle. Heating, cooling, and pumping viscous melts and solutions require the same attention to power input and surface scraping as polymerization design. Final removal of liquids, whether solvent, diluent, or unreacted monomer, is complicated by the high viscosities and low diffusivities characteristic of most polymers. Vacuum evaporators, latex strippers, and vented extruders are used, as well as belt, pan, spray, and rotary kiln types of dryers.

KEYWORDS FOR CHAPTER 5

Section

1. Heat of polymerization, ΔH_p
   Autoacceleration
   Trommsdorff effect
   Shrinkage
   Dilatometry
2. Bulk polymerization
3. Solution polymerization
   Gas-phase polymerization
   Interfacial polymerization
4. Suspension polymerization
   Protective colloid
5. Emulsion polymerization
   Micelle
   Critical micelle concentration

Surfactant
Seeded polymerization
Latex
Dispersion polymerization

PROBLEMS

5.1   One kilogram of a 20% (by weight) solution of acrylamide in water at 30°C is polymerized adiabatically with a redox initiator. The peak temperature reached is 80°C. What is the molar heat of polymerization?

Molecular weight (monomer) = 71
Heat capacity (monomer and polymer) = 0.5 cal/g · °C
Heat capacity (water) = 1.0 cal/g · °C
Heat capacity of container, stirrer, etc. = 0.1 kcal/°C

5.2   In a particular adiabatic solution polymerization, the temperature, T, is found to vary with time, t, according to the expression:

$$\frac{1}{(T - T_0)} = \frac{A}{t} + B$$

For a run in which the initial monomer concentration is 0.800 mol/liter and the initial temperature is 50.0°C, $A = 12.5$ min/°C and $B = 0.0550$ (°C)$^{-1}$. If the specific heat of the solution is 0.650 cal/(°C, ml) and the volume of the solution remains constant with conversion, calculate:

(a) The heat of polymerization
(b) The time for 10% conversion of monomer to polymer

5.3   (a) In what basic way is emulsion polymerization similar to suspension polymerization? (b) What advantage does this give over solution polymerization? (c) In what way do emulsion and suspension polymerizations differ basically? (d) In an unstirred, bulk polymerization that proceeds to 100% conversion, how can the effects of shrinkage and heat of polymerization be handled?

5.4   Why would one expect the apparent order (in monomer) of an emulsion polymerization to rise from zero after 80 or 90% conversion? What maximum value should it reach?

5.5   What order of rate dependence would be expected in a suspension polymerization initially? Why might the rate change abruptly after 5 to 20% conversion?

5.6   According to Fig. 5.10, the rate of emulsion polymerization for styrene varies directly with the number of particles at 60°C. Taking k_p (at 50°C) from

Table 4.3, calculate the dynamic concentration of monomer in particles under these conditions. The experimental value of 5.2 mol/liter has been found by M. Morton et al. [25].

5.7 Company Y has available a 5% (by weight) latex of poly(methyl methacrylate) containing particles that average 0.40 μm in diameter. In order to grow these to a larger size, 8 kg of monomer will be fed into the latex per kilogram of polymer as polymerization proceeds without further addition of emulsifier. The reaction is carried on until all monomer is added to the reactor and the ratio of monomer to polymer in the reactor has decreased to 0.1. Then the unreacted monomer will be steam-stripped and recovered.

Estimate the time required for the reaction in hours, the rate of heat removal initially (W/m³ of original latex), and the final particle diameter.

Data:

Isothermal reaction at 70°C
$\Delta H_p = -13.0$ kcal/mol (54.4 kJ/mol)
Density (polymer) = 1.2 g/cm³
Density (monomer) = 0.9 g/cm³
$k_p = 640$ liter/mol · s
Dynamic solubility of monomer in polymer = 1 g monomer/2 g polymer

5.8 In an emulsion polymerization, all the ingredients are charged at time $t = 0$. The time to convert various amounts of monomer to polymer is indicated in the following table. Predict the time for 30% conversion. Polymerization is "normal" (no water solubility or polymer precipitation).

Fraction of original charge of monomer converted to polymer	Time (h)
0.05	2.2
0.12	4.0
0.155	4.9

5.9 Using the information in Fig. 5.9, estimate the power of the dependence of N_p on surfactant concentration. In plotting the estimates of the stage 2 rates, use error bars on the graph to indicate the range of uncertainty. Hint: What should be plotted and why?

5.10 Assume that during the emulsion polymerization of isoprene with 0.10 M potassium laurate at 50°C, during stage 2, the growing swollen polymer particles contain 20 g of monomer per 100 ml of swollen polymer. Assume also that the final latex has 40 g of polymer per 100 ml with particles of 450 Å diameter. The polymer has a density of 0.90 g/cm³. Using Fig. 5.9, estimate k_p (liter/mol · h).

5.11 The following ingredients are charged to a reactor:

Poly(vinyl acetate) particles (latex)	5.00 g, 2.00×10^{18} particles
Styrene monomer	100 g
Water	0.900 liter
Initiator	(sufficient)
Total volume about 1 liter	

(a) According to Fig. 5.10, how long will it take to convert 50.0 g of styrene to polymer?

(b) If the particle density is 1.05 g/cm³, what is the particle diameter of the original poly(vinyl acetate) seed?

(c) If the particle density does not change with further polymerization, what is the particle diameter after 50.0 g of styrene has been polymerized and the unreacted monomer has been vacuum-stripped off?

5.12 Monomer M can be polymerized using peroxide P in solution at 60°C. The half-life of P at that temperature is 5.00 h. When initial monomer concentration is 0.400 mol/liter and initial peroxide concentration is 4.00×10^{-2} mol/liter, 30.0% of the monomer is converted to polymer in 25.0 min. The conversion rate (at 60.0°C) for the same monomer in an emulsion of 1.00×10^{-6} mol of particles is 13.6 mol/h. The concentration of monomer in the particles is constant at 4.00 mol/liter. What is the termination rate constant?

5.13 (a) Estimate the gas velocity (m/s) in a fluidized-bed polymerization. Given:

Reactor diameter	2.50 m
Production rate	60×10^6 kg/8000 h
Conversion/pass	3.00%
Gas density (ethylene at 20 atm, 100°C)	16.0 kg/m³ $= \rho$

(b) What diameter particle (micrometers) will be settled out rather than carried around the reactor loop? Assume that Newton's law holds for the terminal velocity of a sphere in still medium [26]:

$$u_t^2 = \frac{(1.75)^2 g D_p (\rho_p - \rho)}{\rho}$$

where u_t is the terminal velocity, $g = 9.81$ m/s², ρ_p is the particle density (900 kg/m³), and ρ is the gas density as before.

5.14 Propylene is polymerized in a slurry reactor by an equimolar mixture of aluminum triethyl and titanium tetrachloride. The catalyst residue remains with the polymer in a hydrolyzed form. If a customer specifies a maximum ash content of 0.10 wt% in product, what productivity (moles of monomer converted per mole of catalyst) must be achieved? Assume the ash is entirely Al_2O_3 and TiO_2.

5.15 If 20,000 kg of vinyl chloride is polymerized in a 40-m³ jacketed reactor with 25 m² of heat transfer area, what overall rate of heat transfer is needed when the reaction takes place in 6 h? If the jacket-side coefficient is 6000 W/m² · °C and the temperature difference is 40°C, what inside coefficient is necessary? Neglect wall resistance.

5.16 In an adiabatic tubular reactor, styrene is converted partially to polymer at high pressure and the mixture of monomer and polymer is sprayed into a vacuum chamber with evaporation of monomer and recovery of polymer. If the heat of vaporization is 355 J/g and the monomer enters the tube at 50°C, what fraction can be converted to polymer per pass and still allow polymer recovery at 50°C?

5.17 Monomer A is polymerized at 80°C by two methods in two separate runs: (*a*) in an inert solvent (no chemical interactions), and (*b*) as an aqueous suspension. If the *initial* rate of polymerization for 1 liter of solution is 0.057 mol/h, what is the expected *initial* rate for 1 liter of suspension (in mol/h)?

	Solution (1 liter)	Suspension (1 liter)
Monomer	50 g (0.50 mol, 60 cm³)	same
Benzoyl peroxide	1.0×10^{-3} mol	1.0×10^{-4} mol
	(negligible volume)	(negligible volume)
Diluent	940 cm³ of benzene	940 cm³ of water
Additives	none	0.1 g of poly(vinyl alcohol)

5.18 In a solution polymerization, heat removal may be complicated by an increase in viscosity as polymerization proceeds. For the following situation, the heat of polymerization can be handled initially by a temperature difference between contents and jacket of 20°C. What temperature difference is required when half the charge has been converted to polymer? Conditions are:

(*a*) Initial charge of monomer is 100 kg/m³ of solution.

(*b*) The overall coefficient of heat transfer is 2000 W/m² · °C at the start of the reaction. It varies with viscosity of contents to the inverse one-third power.

(*c*) (Viscosity of solution)/(initial viscosity) = exp(0.05*c*), where *c* is the concentration of polymer in kg/m³.

(*d*) The rate of polymerization changes with concentration according to Eq. (4.4) with initiator concentration held constant.

5.19 For the conditions of the previous problem, if the monomer is acrylamide, the solvent is water, and the time for 50% conversion is 35 min, what heat transfer area is required per cubic meter of solution?

REFERENCES

1. Runt, J. P.: "Crystallinity Determination," in *Enc. Polym. Sci. Tech.,* 4:487 (1986).
2. Brandrup, J., and E. H. Immergut (eds.): "Polymer Handbook," 3d ed., Secs. II and IV, Wiley, New York, 1989.
3. Riggs, J. P., and F. Rodriguez: *J. Polymer Sci., A1,* 5:3151 (1967).
4. T. Takata and T. Endo: *Prog. Polymer Sci.,* 18:839 (1993).
5. Bailey, W. J., and R. L. Sun: *Polymer Preprints,* 13(1):281 (1972).
6. Riddle, E. H., and P. A. Horrigan, in P. H. Groggins (ed.): "Unit Processes in Organic Synthesis," 5th ed., p. 1015, McGraw-Hill, New York, 1958.
7. *Mod. Plastics,* 42(1A, Encycl. Issue):684 (1964).
8. Nauman, E. B.: "Bulk Polymerization," in J. I. Kroschwitz (ed.), *Enc. Polym. Sci. Eng.,* 2:500 (1985).
9. Harbison, W. C.: "Casting," in J. I. Kroschwitz (ed.), *Enc. Polym. Sci. Eng.,* 2:692 (1985).
10. Mack, W. A., and R. Herter: "Extruder Reactors for Polymer Production," Werner and Pfleiderer Corp., Waldwick, N.J. (AIChE Meeting, Boston, 1975).
11. Sittig, M.: "Polyolefin Resin Processes," Gulf Publishing, Houston, 1961.
12. Sorenson, W. R., and T. W. Campbell: "Preparative Methods of Polymer Chemistry," 2d ed., p. 92, Wiley-Interscience, New York, 1968.
13. Morgan, P. W., and S. L. Kwolek: *J. Chem. Ed.,* 36:182, 530 (1959).
14. McCaffery, E. L.: "Laboratory Preparation of Macromolecular Chemistry," McGraw-Hill, New York, 1970.
15. Fondy, P. L., and R. L. Bates: *AIChE J.,* 9:338 (1963).
16. Moore, J. C.: *J. Polymer Sci.,* A2:835 (1964).
17. Flory, P. J.: "Principles of Polymer Chemistry," pp. 203–217, Cornell Univ. Press, Ithaca, N.Y., 1953.
18. Blackley, D. C.: "Emulsion Polymerization," p. 99, Halsted-Wiley, New York, 1975.
19. Harkins, W. D.: *J. ACS,* 69:1428 (1947).
20. Patsiga, R., M. Litt, and V. Stannett: *J. Phys. Chem.,* 64:801 (1960).
21. Hay, P. M., et al.: *J. Appl. Polymer Sci.,* 5:23 (1961).
22. Ugelstad, J., and P. C. Mork: *Adv. Coll. Int. Sci.,* 13:101 (1980).
23. Ober, C. K., and M. L. Hair: *J. Polym. Sci. Chem. Ed.,* 25:1395 (1987).
24. Vanderhoff, J. W., M. S. El-Aasser, F. J. Micale, E. D. Sudol, C-M. Tseng, H-R. Sheu, and D. M. Kornfeld: *ACS Polymer Preprints,* 28(2):455 (1987).
25. Morton, M., P. P. Salatiello, and H. Landfield: *J. Polymer Sci.,* 8:279 (1952).
26. McCabe, W. L., J. C. Smith, and P. Harriott: "Unit Operations of Chemical Engineering," 4th ed., p. 143, McGraw-Hill, New York, 1985.

General References

Albright, L. F.: "Processes for Major Addition-Type Plastics and Their Monomers," McGraw-Hill, New York, 1974.
Athey, R. D., Jr.: "Emulsion Polymer Technology," Dekker, 1991.
Barrett, K. E. J. (ed.): "Dispersion Polymerization in Organic Media," Wiley-Interscience, New York, 1975.
Biesenberger, J. A.: "Devolatilization of Polymers," Macmillan, New York, 1983.
Biesenberger, J. A., and D. H. Sebastian: "Principles of Polymerization Engineering," Wiley, New York, 1983.
Bishop, R. B.: "Practical Polymerization for Polystyrene," Cahners, Boston, 1972.
Blackley, D. C.: "Emulsion Polymerization," Halsted-Wiley, New York, 1975.

Burgess, R. H. (ed.): "Manufacturing and Processing PVC," Elsevier Applied Science, New York, 1981.

Calvert, K. O. (ed.): "Polymer Latices and Their Applications," Elsevier Applied Science, New York, 1981.

Daniels, E. S., E. D. Sudol, and M. S. El-Aasser (eds.): "Polymer Latexes: Preparation, Characterization, and Applications," ACS, Washington, D.C., 1992.

El-Aasser, M. S., and J. W. Vanderhoff (eds.): "Emulsion Polymerisation of Vinyl Acetate," Elsevier Applied Science, New York, 1981.

Elias, H.-G.: "Macromolecules, Synthesis, Materials, and Technology," 2d ed., Plenum, New York, 1984.

Fontanille, M., and A. Guyot (eds.): "Recent Advances in Mechanistic and Synthetic Aspects of Polymerization," (Reidel Holland), Kluwer Academic, Norwell, Mass., 1987.

Guillot, I., and C. Pichot (eds.): "Emulsion Copolymerization," Huethig & Wepf, Mamaroneck, N.Y., 1984.

Guiot, P., and P. Couvreur: "Polymeric Nanoparticles and Microspheres," CRC Press, Boca Raton, Fla., 1986.

Henderson, J. N., and T. C. Bouton (eds.): "Polymerization Reactors and Processes," ACS, Washington, D.C., 1979.

McCrum, N. G., C. P. Buckley, and C. B. Bucknall: "Principles of Polymer Engineering," Oxford, New York, 1988.

McGreavy, C. (ed.): "Polymer Reaction Engineering," Chapman & Hall, New York, 1993.

Millich, F., and C. E. Carraher (eds.): "Interfacial Synthesis," Dekker, New York, 1973.

Nass, L. I., and C. A. Heiberger (eds.): "Encyclopedia of PVC, Resin Manufacture and Properties," 2d ed., Dekker, New York, 1986.

Paleos, C. M. (ed.): "Polymerization in Organized Media," Gordon & Breach, New York, 1992.

Piirma, I., and J. L. Gardon (eds.): "Emulsion Polymerization," ACS, Washington, D.C., 1976.

Platzer, N. A. J. (ed.): "Polymerization Kinetics and Technology," ACS, Washington, D.C., 1973.

Platzer, N. A. J. (ed.): "Polymerization Reactions and New Polymers," ACS, Washington, D.C., 1973.

"Polymer Manufacturing, Technology and Health Effects," Noyes Data, Park Ridge, N.J., 1986.

Reichert, K.-H., and W. Geiseler (eds.): "Polymer Reaction Engineering," Huethig & Wepf, Mamaroneck, N.Y., 1986.

Ryan, M. E. (ed.): "Fundamentals of Polymerization and Polymer Processing Technology," Gordon & Breach, New York, 1983.

Sadhir, R. K., and R. M. Luck (eds.): "Expanding Monomers: Synthesis, Characterization, and Applications," CRC Press, Boca Raton, Fla., 1992.

Schildknecht, C. E.: "Polymer Processes," Wiley-Interscience, New York, 1956.

Schildknecht, C. E., and I. Skeist (eds.): "Polymerization Processes," Wiley-Interscience, New York, 1977.

Schork, F. J., P. B. Deshpande, and K. W. Leffew: "Control of Polymerization Reactors," Dekker, New York, 1993.

Shen, M., and A. T. Bell (eds.): "Plasma Polymerization," ACS, Washington, D.C., 1979.

Smith, W. M. (ed.): "Manufacture of Plastics," Van Nostrand-Reinhold, Cincinnati, 1964.

van der Hoff, B. M. E., in K. Shinoda (ed.): "Solvent Properties of Surfactant Solutions," Dekker, New York, 1967.

CHAPTER SIX

THE MOLECULAR WEIGHT OF POLYMERS

6.1 AVERAGE MOLECULAR WEIGHT

The high molecular weight of polymers is responsible for many of the properties that make polymers valuable as a class of materials. Although there is no sharp dividing line, we can draw an imaginary one at a molecular weight of about 2000, since that is near the limit of convenient purification by distillation or extraction. Also, differences between members of a homologous series differing in steps of 100 or so (the molecular weight of a typical monomer) become so slight as to prevent clean separations. Some polymers, such as polyethylene, poly(ethylene oxide), and dimethyl silicones, are commercially available in sizes ranging almost continuously from monomer or dimer up to molecular weights in the millions. Except for the smaller members of each series, each of these products is a mixture of different-sized molecules having an *average molecular weight,* which we can define and measure in several ways, and a *distribution* of molecular weights. Distributions used to be rather tedious to measure and not as often specified as the average molecular weight. However, chromatographic techniques changed this situation.

If someone told you that he wanted to drop into your hand, from a height of 1 ft, a series of 1000 steel balls with an average diameter of 2.4 in, you might agree to help him in his playful experiment. After all, a 2.4-in diameter steel ball weighs only about 2 lb. However, if he said the average diameter was 23.6 in, your attitude might be less cooperative. Both numbers could refer to the same series of balls, the difference being in the manner in which the "average" diameter is calculated. If

Table 6.1 Hypothetical distribution of balls

Number of balls N_i	Diameter D_i, in	Length $N_i D_i$	Area ($\times 1/\pi$) $N_i D_i^2$	Volume ($\times 6/\pi$) $N_i D_i^3$
900	1	900	900	900
50	5	250	1,250	6,250
50	25	1250	31,250	781,250
$\Sigma N_i = 1000$		$\Sigma N_i D_i = 2400$	$\Sigma = 33,400$	$\Sigma = 788,400$

the population is as shown in Table 6.1, we can calculate an average diameter in several ways:

The average diameter $\overline{D}_L$, based on length (one dimension):

$$\overline{D}_L = \frac{\Sigma N_i D_i}{\Sigma N_i} = \frac{2400}{1000} = 2.4 \text{ in}$$

The average diameter $\overline{D}_A$, based on area (two dimensions):

$$\overline{D}_A = \frac{\Sigma N_i D_i^2}{\Sigma N_i D_i} = \frac{33,400}{2400} = 13.9 \text{ in}$$

The average diameter $\overline{D}_V$, based on volume (three dimensions):

$$\overline{D}_V = \frac{\Sigma N_i D_i^3}{\Sigma N_i D_i^2} = \frac{788,400}{33,400} = 23.6 \text{ in}$$

Although $\overline{D}_L$ reflects the preponderant number of small balls, the 1-in balls represent only about 0.1% of the total volume. $\overline{D}_V$ reflects the importance of the few large balls, which represent 99% of the volume (and weight) of the system. Incidentally, each 25-in ball weighs about 1 ton.

Let us now imagine a population of polymers with molecular weights distributed as in Table 6.1, where N is now the number of molecules of molecular weight $M = D$. The quantity we called $N_i D_i$ before is now $N_i M_i = W_i$, the weight of species with molecular weight M_i. We have used before the concept of a number-average molecular weight M_n, where

$$M_n = \frac{\text{total weight of system}}{\text{molecules in system}} \tag{6.1}$$

In terms of any population, then,

$$M_n = \frac{\Sigma N_i M_i}{\Sigma N_i} = \frac{\Sigma W_i}{\Sigma (W_i/M_i)} \tag{6.2}$$

As in the case of the steel balls, M_n is very sensitive to the concentration of low-molecular-weight species. The weight-average molecular weight M_w is defined as

$$M_w = \frac{\sum N_i M_i^2}{\sum N_i M_i} = \frac{\sum W_i M_i}{\sum W_i} \tag{6.3}$$

In correlating such important polymer properties as viscosity or toughness, M_w often is a more useful parameter than M_n. Higher averages are defined as

$$M_z = \frac{\sum N_i M_i^3}{\sum N_i M_i^2} = \frac{\sum W_i M_i^2}{\sum W_i M_i} \tag{6.4}$$

$$M_{z+1} = \frac{\sum N_i M_i^4}{\sum N_i M_i^3} = \frac{\sum W_i M_i^3}{\sum W_i M_i^2} \tag{6.5}$$

These are used less often than M_n and M_w. The ratio of M_w to M_n is a measure of the broadness of a distribution, since each is influenced by an opposite end of the population. The quantity M_w/M_n is the *polydispersity index* (PDI).

Example 6.1

A polymer shipment is to be made up by blending three lots of polyethylene, A, B, and C. How much of each lot is needed to make up a shipment of 50,000 kg with a weight-average molecular weight of 250,000 and a polydispersity index (PDI) of 3.65? Let $w_i = W_i/\sum W_i$

Lot	M_w	PDI
A	500,000	2.50
B	250,000	2.00
C	125,000	2.50

Solution:

$$M_w = \sum w_i(M_w)_i$$
$$250,000 = 500,000w_A + 250,000w_B + 125,000(1 - w_A - w_B)$$
$$2.50 = 5.00w_A + 2.50w_B + 1.25(1 - w_A - w_B)$$
$$w_A = 0.333 - 0.333w_B$$
$$\frac{1}{M_n} = \sum \frac{w_i}{(M_n)_i}$$

(For convenience, divide molecular weights by 1000.)

$$\text{Each } (M_n)_i = \left(\frac{M_w}{\text{PDI}}\right)_i$$

$$\frac{1}{68.5} = \frac{w_A}{200} + \frac{w_B}{125} + \frac{1 - w_A - w_B}{50}$$
$$w_A = 0.360 - 0.800 w_B$$

Solve simultaneously:

$$w_A = 0.058 \qquad w_B = 0.314 \qquad w_C = 0.628$$

Amounts needed are 2,900 kg A, 15,700 kg B, and 31,400 kg C.

6.2 THEORETICAL DISTRIBUTIONS

With the possible exception of some naturally occurring proteins, all polymers are mixtures of many molecular weights ("polydisperse"). This is a consequence of the random nature of polymerization reactions. For example, take the case analyzed by Flory [1] in which the monomer is an ω-hydroxy acid (see also Sec. 4.7).

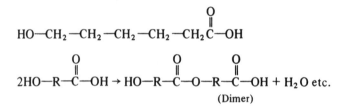

This is a typical stepwise polymerization in which each successively higher molecular weight, dimer, trimer, etc., is just as reactive as monomer. An important parameter is the fraction converted, p. If we have a population of such monomers, each can react to form an ester group. Each esterification eliminates one carboxyl and one hydroxyl group. With five monomers (Fig. 6.1) there would be a maximum of five esterifications ($p = 1$). Initially, p is 0. After one reaction, $p = 0.20$. Abbreviating the unreacted sites by the symbol U and the esterified sites by the symbol E, we can generate a distribution of molecular weights by random reaction of 50 out of 100 sites ($p = 0.5$). If we had alternated sites of reaction with unreacted sites, we would have generated 50 dimers and no other species. Another possible extreme would be to generate 49 monomers and 1 polymer molecule with a degree of polymerization x of 51. We have 50 molecules in either case. With random reaction (Fig. 6.1), monomers predominate in number of molecules N_x, although not necessarily in weight xN_x. If our sample population were much larger, as it is in any macroscopic polymerization, we could assign a probability $(P_r)_x$ that a monomer unit selected at random is a part of a polymer with degree of polymerization x. We expect this to be a function of p, which is the probability that any particular reactive site has been esterified. The probability that a monomer unit is still unreacted varies from 1 when $p = 0$ to 0 when $p = 1$. We surmise that

Population of five monomers

$p = 0$

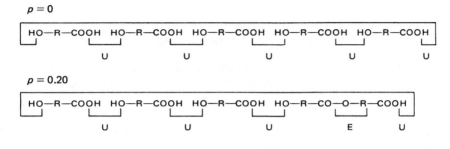

$p = 0.20$

Population of 100 monomers

$$\left[\begin{array}{l}\text{UEEEE EUEEU EUEEE UUEUE UUEUU UEUEU EEEUU EUUUU EUEUE EEEUU}\\\text{EUEEE UEEEE EEUUE EUEUU UUUEE EEUEU UUUUE UUUUU EUEEE UUUEU}\end{array}\right]$$

Summary:

	x	N_x	xN_x
Monomer (—UU—)	1	25	25
Dimer (—UEU—)	2	15	30
Trimer (—UEEU—)	3	2	6
Tetramer (—UEEEU—)	4	4	16
Pentamer (—UEEEEU—)	5	2	10
Hexamer (—UEEEEEU—)	6	1	6
Heptamer (—UEEEEEEU—)	7	1	7
		50	100

Figure 6.1 Generation of molecular weight distribution by random reaction

$$(P_r)_1 = 1 - p \tag{6.6}$$

The likelihood that a unit has become part of a dimer is p times $(P_r)_1$. Each additional unit is less likely, since the cumulative probability is represented by a product of the probabilities of each step. Thus we see that

$$(P_r)_x = (1 - p)p^{x-1} \tag{6.7}$$

Also,

$$\sum_{x=1}^{\infty} (P_r)x = \sum_{x=1}^{\infty} (1 - p)p^{x-1} = 1 \tag{6.8}$$

Now we can call $(P_r)_x$ the *mole fraction of x-mer*, since it is a normalized probability (i.e., the summation over all values of x is unity). According to Eq. (6.7), the degree of polymerization with the largest mole fraction *always* is 1, namely, monomer. Keep in mind that we are dealing with a particular polymerization scheme in which all species are at equilibrium. The number-average degree of polymerization x_n for a distribution represented by Eq. (6.7) is

$$x_n = \frac{\text{number of monomer units originally charged}}{\text{number of molecules in system}} = \frac{1}{1 - p} \quad (6.9)$$

Disregarding the contribution to molecular weight made by the end groups, the *molecular weight* of a polymer M is related to the *degree of polymerization* x by

$$M = M_0 x \quad (6.10)$$

where M_0 is the molecular weight of a monomer unit in the chain. We can write Eq. (6.7) for x and for x_n and combine the two to give

$$\ln \frac{(P_r)_x}{(P_r)_{x_n}} = (x - x_n) \ln p \quad (6.11)$$

But $-\ln p \cong 1 - p = 1/x_n$, so

$$\ln \frac{(P_r)_x}{(P_r)_{x_n}} \cong 1 - \frac{x}{x_n} \quad (6.12)$$

This "normalized" equation for mole fraction as a function of degree of polymerization is shown in Fig. 6.2.

A second kind of average degree of polymerization can be defined as

$$x_w = \frac{\sum W_x x}{\sum W_x} \qquad x = 1, 2, 3, \ldots \quad (6.13)$$

where W_x is proportional to the weight of x-mer and x_w is the *weight-average degree of polymerization*. The weight of x-mer from Eq. (6.7) is

$$W_x = x(1 - p)p^{x-1} M_0 \quad \text{and} \quad \sum W_x = \frac{M_0}{1 - p} \quad (6.14)$$

Thus we have

$$x_w = \frac{\sum (P_r)_x x^2}{\sum (P_r)_x x} = \frac{\sum x^2(1 - p)p^{x-1}}{\sum x(1 - p)p^{x-1}} = \frac{1 + p}{1 - p} \quad (6.15)$$

The weight fraction of x-mer w_x is given by

$$w_x = \frac{W_x}{\sum W_x} = x(1 - p)^2 p^{x-1} \quad (6.16)$$

As in Eq. (6.12), we can have a "normalized" distribution:

$$\ln \frac{w_x}{w_{x_n}} = \ln \frac{x}{x_n} + (x - x_n)\ln p \cong \ln \frac{x}{x_n} - \frac{x}{x_n} + 1 \quad (6.17)$$

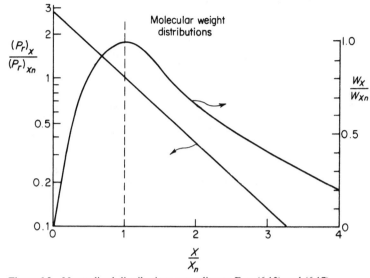

Figure 6.2 Normalized distributions according to Eqs. (6.12) and (6.17)

This also is plotted in Fig. 6.2. Higher averages of degree of polymerization are sometimes useful:

$$x_z = \frac{\Sigma(P_r)_x x^3}{\Sigma(P_r)_x x^2} \tag{6.18}$$

$$x_{z+1} = \frac{\Sigma(P_r)_x x^4}{\Sigma(P_r)_x x^3} \tag{6.19}$$

What we have been describing is sometimes called the *most probable distribution.* It gives $x_w/x_n = 1 + p$. The ratio of 2 when p approaches unity is found experimentally for many stepwise and also for many chain polymerizations, although the reasoning can be different. Under conditions of chain polymerization with all polymers initiated and growing *simultaneously,* the distribution can be much narrower. For example, in a butyl lithium-initiated polymerization to give living polymers, the simultaneous growth of polymer chains gives a *Poisson distribution* [2].

$$(P_r)_x(\text{mole fraction of } x\text{-mer}) = \frac{e^{1-x_n}(x_n - 1)^{x-1}}{(x - 1)!} \tag{6.20}$$

or

$$\ln (P_r)_x = 1 - x_n + (x - 1) \ln (x_n - 1) - \ln [(x - 1)!] \tag{6.21}$$

Table 6.2 Analysis of distribution from Fig. 6-1

Degree of polymerization x	Number of moles N_x	Mole fraction $P_r = N_x/\Sigma N_x$	Relative number of monomer units in x-mer $W_x/M_0 = x(P_r)_z$	Weight fraction $w_x = W_x/\Sigma W_x$	$x w_x$	$x^2 w_x$
1	25	0.50	0.50	0.25	0.25	0.25
2	15	0.30	0.60	0.30	0.60	1.20
3	2	0.04	0.12	0.06	0.18	0.54
4	4	0.08	0.32	0.16	0.64	2.56
5	2	0.04	0.20	0.10	0.50	2.50
6	1	0.02	0.12	0.06	0.36	2.16
7	1	0.02	0.14	0.07	0.49	3.43
	50	1.00	2.00	1.00	3.02	12.64

$$x_n = \frac{\Sigma\, x(P_r)_x}{\Sigma\,(P_r)_x} = \frac{2.00}{1.00} = 2.00$$

$$x_w = \frac{\Sigma\, x^2 (P_r)_x}{\Sigma\, x(P_r)_x} \quad \text{or} \quad \frac{\Sigma\, x w_x}{\Sigma\, w_x} = \frac{3.02}{1.00} = 3.02$$

$$x_z = \frac{\Sigma\, x^3 (P_r)_x}{\Sigma\, x^2 (P_r)_x} \quad \text{or} \quad \frac{\Sigma\, x^2 w_x}{\Sigma\, x w_x} = \frac{12.64}{3.02} = 4.19$$

$$w_x(\text{weight fraction of } x\text{-mer}) = \frac{x(P_r)_x}{x_n} \tag{6.22}$$

$$\frac{x_w}{x_n} = 1 + \frac{x_n - 1}{x_n^2} \tag{6.23}$$

It can be seen that even when x_n is small, say about 10, x_w/x_n is nearly unity. Polymers can be produced on a commercial scale with

$$\frac{M_w}{M_n} = \frac{x_w}{x_n} < 1.1$$

by using butyl lithium as the initiator. Most "living polymerization" systems, such as those using Ziegler-Natta, metallocene, group transfer, and some cationic initiation, can yield narrow distributions when impurities are carefully avoided.

We can further illustrate the differences in the various averages by examining the sample of Fig. 6.1 (Table 6.2). In this example, monomer ($x = 1$) has the highest mole fraction but not the highest relative weight (W_x). The average degrees of polymerization are also calculated according to the equations presented. As mentioned before, the usefulness of each of these numbers may depend on the particular property we are looking at. The reactivity of a urethane prepolymer may depend entirely on x_n. On the other hand, the viscosity of a polymer melt often varies more directly with x_w.

Example 6.2

Ten thousand elephants, while crossing the Alps in single file, are thrown into a panic by the sudden blast of a Punic trumpet. Each animal who panics entwines his trunk firmly about the tail of the preceding beast. In a quick but accurate survey, Hannibal finds that there are now as many elephants who are part of strung-together trios as there are singles (unattached elephants). How many of the ponderous pachyderms panicked?

Solution: We assume a most probable distribution for which Eq. (6.16) applies. According to Hannibal, the total weight of elephants in trios, w_3, is the same as the total weight in singles, w_1, assuming that all elephants weigh the same. Thus,

$$w_1 = 1(1 - p)p^0 = w_3 = 3(1 - p)p^2$$

and

$$p = \left(\frac{1}{3}\right)^{1/2} = 0.577$$

That is, 57.7% of the elephants panicked.

6.3 EMPIRICAL DISTRIBUTION MODELS

The distribution of molecular weights has been shown to be important in many diverse applications, including flow of melts and solutions, aging and weathering behavior, adhesion, and flocculation. Because of the difficulties involved in measuring a distribution in detail, one shortcut has been to postulate a reasonable mathematical form or model for the distribution and then evaluate the parameters from x_n and x_w [3]. In general, such a model would give w_x, the weight fraction of x-mer, as a function of x, x_w, x_n, and perhaps other parameters. The "most probable" distribution Eq. (6.7), has only one parameter p, which can be evaluated from either x_n [Eq. (6.9)] or x_w [Eq. (6.15)]. If the experimentally determined ratio of x_w/x_n is not nearly 2 for high-molecular-weight material, the model cannot be used. The *Schulz distribution* is a more general form of the most probable distribution:

$$w_x = \frac{a}{x_n\Gamma(a + 1)} \left(\frac{ax}{x_n}\right)^a \exp\left(-\frac{ax}{x_n}\right) \tag{6.24}$$

where $x_w/x_n = (a + 1)/a$ and $\Gamma(a + 1)$ is the gamma function of $a + 1$. When $a = 1$ and x is large, this becomes the same as the most probable distribution, since $x_n = 1/(1 - p)$ and $(x - 1) \ln p \cong -x/x_n$. However, any value of x_w/x_n can be

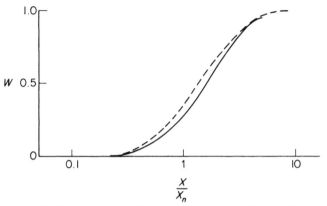

Figure 6.3 Cumulative weight fraction W as a function of ratio of degree of polymerization to number-average value for Schulz (solid line) and Wesslau (dashed line) models

inserted. Qualitatively, the curves will resemble Fig. 6.2 except that the overall breadth will reflect the influence of the polydispersity index x_w/x_n.

It will be seen in the next section that some experimental methods for measuring distribution result in a cumulative distribution function $W(x)$, from which the differential distribution $w_x = dW/dx$ can be derived. We are speaking now of distributions in the higher ranges of x_n, where the approximate $\Sigma w_x = \int w_x \, dx$ is applicable. It has been observed that a plot of $W(x)$ against x on logarithmic probability paper often gives a straight line. In terms of w_x this becomes the *Wesslau distribution:*

$$w_x = \frac{1}{\beta \pi^{1/2}} \frac{1}{x} \exp\left(-\frac{1}{\beta^2} \ln^2 \frac{x}{x_0}\right) \tag{6.25}$$

where $\beta^2 = \ln(w_w/x_n)^2$ and $x_0 = x_n \exp (\beta^2/4)$. Once again, specification of x_w and x_n defines the entire distribution. As with the Schulz model, only the central value of the distribution, reflected by x_n, and the breadth, measured by x_w/x_n, can be changed. On a cumulative weight $[W(x)]$ basis, the Schulz and Wesslau models give similar curves (Fig. 6.3). On a differential (w_x) basis, qualitative differences become apparent (Fig. 6.4).

Example 6.3

The molecular weight distribution for a certain polymer can be described as

$$w_x = \frac{kx}{10 + x^2}$$

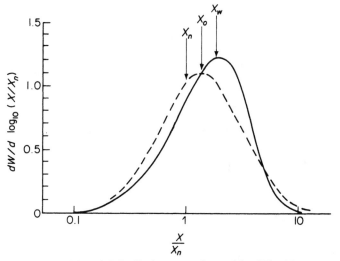

Figure 6.4   Differential distribution curves for models of Fig. 6.3

where $x = 1, 2$, and 3 only and w_x is the weight fraction of polymer with degree of polymerization x. What is the value of k? What is x_w? What is x_n?

Solution: The condition for k is that

$$\Sigma w_i = 1 = k\Sigma\left(\frac{x}{10 + x^2}\right)$$

$$k\left(\frac{1}{11} + \frac{2}{14} + \frac{3}{19}\right) = 0.392k = 1$$

so

$$k = \frac{1}{0.392} = 2.55$$

In analogy with Eqs. (6.3) and (6.2),

$$x_w = \Sigma w_i x_i = 2.55 \left(\frac{1}{11} + \frac{4}{14} + \frac{9}{19}\right) = 2.17$$

$$\frac{1}{x_n} = \Sigma\left(\frac{w_i}{x_i}\right) = 2.55\left(\frac{1}{11} + \frac{1}{14} + \frac{1}{19}\right) = 0.548$$

$$x_n = 1.82$$

When dealing with high molecular weights ($x > 100$), it becomes convenient to treat distributions as being continuous rather than finite. As an example, let us assume that the cumulative weight fraction $W(x)$ of a polymer with degree of polymerization less than x is given by

$$
\begin{aligned}
W(x) &= 0.01(x - 10) &\quad &\text{for } 10 < x < 110 \\
W(x) &= 0 &\quad &\text{for } x < 10 \\
W(x) &= 1 &\quad &\text{for } x > 110
\end{aligned}
\tag{6.26}
$$

The *differential* weight distribution w_x then is

$$
w_x = \frac{dW(x)}{dx} = 0.01
\tag{6.27}
$$

The degree of polymerization is obtained by integration and normalization.

$$
x_n = \frac{\displaystyle\int_1^\infty w_x \, dx}{\displaystyle\int_1^\infty w_x/x \, dx} = \frac{\displaystyle\int_{10}^{110} 0.01 \, dx}{\displaystyle\int_{10}^{110} 0.01/x \, dx} = 41.7
\tag{6.28}
$$

$$
x_w = \frac{\displaystyle\int_1^\infty x w_x \, dx}{\displaystyle\int_1^\infty w_x \, dx} = \frac{\displaystyle\int_{10}^{110} 0.01 \, x \, dx}{\displaystyle\int_{10}^{110} 0.01 \, dx} = 60.0
\tag{6.29}
$$

The continuous models are particularly useful when analyzing experimental data on distributions where a cumulative weight is obtained first and must be converted mathematically to a differential distribution.

6.4 MEASUREMENT OF DISTRIBUTION

Both x_n and x_w can be measured independently (Secs. 6.6 and 6.7) in order to get some idea of polydispersity. The ultracentrifuge, when applicable, gives a distribution directly. However, until the advent of gel permeation chromatography, fractionation of polymer from dilute solution was used. Usually there is a decrease in solubility with increasing molecular weight. Polymer can be put in dilute solution and progressively precipitated by addition of increment of a nonsolvent [4]. The nonsolvent is added to give a visible precipitate. Then the system is heated to a temperature at which it is homogeneous, after which it is slowly cooled over a period of hours to a standard temperature. A variation is to cool the solution to successively lower temperatures and separate a fraction at each temperature rather than at each increment of nonsolvent. The polymer-rich phase that separates, the so-called *coacervate,* usually is a small volume compared to the total volume of the system and is viscous or gelatinous. One drawback of this process of coacervation is that low-molecular-weight polymer is occluded in the coacervate even under

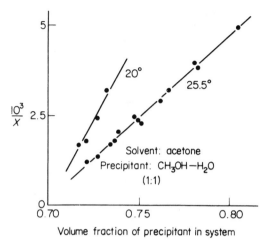

Figure 6.5 Decrease in solubility of poly(vinyl acetate) with added nonsolvent, increasing degree of polymerization x, and decreasing temperature [5]

conditions of slow cooling. Since this is more serious with higher molecular weights and with higher starting concentrations, a rule of thumb may be employed that the beginning concentration should be less than 1% polymer regardless of molecular weight and also that the concentration times the intrinsic viscosity (Sec. 7.4) should be less than 1. Another scheme is to redissolve each fraction and reprecipitate it under the same conditions of solvent-nonsolvent ratio and temperature. Figure 6.5 shows the relationship between the degree of polymerization of material remaining in solution and solvent composition when poly(vinyl acetate) is precipitated from an acetone solution by addition of a methanol-water mixture.

The procedure is tedious, because hours or days are required for clean separation of each fraction. Even then, each fraction is heterogeneous with $x_w/x_n = 1.05$ to 1.2 at best. Rather than precipitate the polymer from a dilute solution, it is possible to dissolve a solid polymer fractionally [6]. Diffusion through the solid is slow, so it is essential that the material be spread out in a very thin film. A typical technique for fractionating polyethylene uses a 2-in-diameter column about 2 ft long packed with ordinary sand (40 to 200 mesh). A hot solution of polyethylene in xylene is introduced and cooled so that the entire charge of polymer, a few grams, is precipitated on the surface of the sand, giving a thin film. Then a solvent or a solvent mixture is added to elutriate the low-molecular-weight portion of the polymer. When no more polymer is elutriated at the solvent composition or temperature, the solvent-to-nonsolvent ratio is increased or the temperature is raised so that a higher-molecular-weight fraction can be extracted. Columns of this type can be automated with a few solenoid-operated valves and a fraction collector in

Table 6.3 Analysis of typical fractionation data

	Raw experimental data			Calculated results	
Fraction no. i	Fraction of total weight recovered w_i	Degree of polymerization $x_i (\times 10^{-3})$	Σw_i	$\dfrac{\Sigma w_{i-1} + w_i/2}{W_i}$	$(dW/dx)_i$ $(\times 10^3)$
1	0.050	1.1	0.050	0.025	0.040
2	0.100	2.5	0.150	0.100	0.060
3	0.120	4.1	0.270	0.210	0.063
4	0.083	5.8	0.353	0.313	0.052
5	0.075	7.4	0.428	0.390	0.043
6	0.110	9.4	0.538	0.483	0.039
7	0.120	12.5	0.658	0.598	0.030
8	0.130	17.5	0.788	0.723	0.0165
9	0.130	29.0	0.918	0.853	0.0072
10	0.082	64.0	1.000	0.959	0.0013

order that fractionation of a sample into 10 to 20 fractions can be carried out overnight without immediate supervision. In this column method one does not have to wait for gravitational settling out of coacervate, and it is the speed with which fractionation can proceed that is its advantage.

In both precipitation and dissolving, a single equilibrium stage of separation is achieved. Multistaged separations are not usually convenient. In either case, the raw data consist of a series of i fractions each of weight w_i and an average molecular size $\bar{x}_i$. A cumulative curve can then be plotted in which the ordinate is $\Sigma w_{i-1} + w_i/2$ and the abscissa is $\bar{x}_i$. A smooth line through the points gives a curve that is a good approximation to $W(x)$, the cumulative molecular-weight distribution function. Differentiation of this curve then gives $w_x = dW(x)/dx$, the differential molecular-weight distribution function. The reason that the raw values of w_i and x_i cannot be plotted as a histogram is that the intervals of neither variable are uniform. An example (Table 6.3, Fig. 6.6) shows the long high-molecular-weight tail typical of most polymers. The uncertainty in the slope at the high end also is apparent. In Fig. 6.7 the advantage of a logarithmic plot in showing greater symmetry is seen.

A packed column technique is now popular under the names "gel filtration" and "gel permeation chromatography" (GPC). The first name was applied by Flodin and co-workers [7] to the separation obtained when water-soluble polymers are eluted from a column packed with swollen, cross-linked, hydrophilic dextran (Sephadex) beads. The second name was attached by Moore [8] to the separation of organic-soluble polymers in columns packed with swollen, cross-linked, lyophilic polystyrene beads. Both names are misleading, since the gel is not necessary, po-

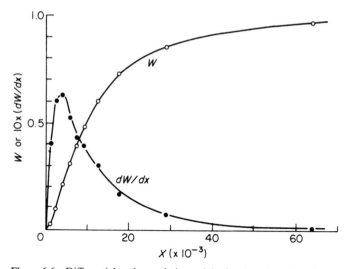

Figure 6.6 Differential and cumulative weight fraction plotted against degree of polymerization for distribution of Table 6.3

rous glass beads being equally applicable; filtration is not part of the process, small molecules being retarded and large ones passed rather than the other way around; and chromatography classically implies formation of colors, which seldom occurs in either technique. The term *size exclusion chromatography* (SEC) has been recommended.

The principle of the method is illustrated in Fig. 6.8, where a mixture of small and large molecules has been deposited at one end of a column packed with porous beads. Initially there is a concentration gradient causing diffusion of polymer into the bead. However, the large molecules cannot penetrate the beads. A continuous

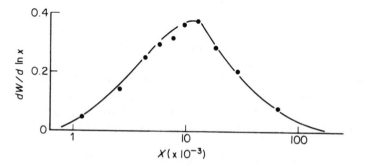

Figure 6.7 Data of Table 6.3 plotted as differential distribution on a logarithmic basis

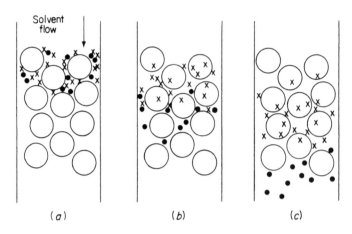

Solvent flow

(a)   (b)   (c)

x Small molecules

• Large molecules

Figure 6.8 Process of gel permeation (size exclusion) chromatography [9]: (a) sample injection; (b) elution; (c) continued elution

flow of solvent sweeps the large molecules along and reverses the concentration gradient for the small ones so that they now diffuse back out of the beads. This process is repeated again and again as the sample moves through the column. Eventually, when the sample is elutriated from the other end, large molecules will emerge first and the small ones, retarded by the diffusion in and out of the packing, will emerge last. This is just the opposite from what adsorption or filtration would have accomplished, as in gas chromatography. When porous glass beads (about 100 mesh) are used as packing, the pore sizes can be measured by mercury intrusion. Pore diameters of 10 to 250 nm are useful for polymers in the 10^3 to 10^7 molecular weight range. More common is packing comprised of swollen, cross-linked polymer beads. Such particles can be produced by suspension polymerization of a mixture of styrene, divinyl benzene, and a hydrocarbon diluent. Under conditions of constant temperature, flow rate, and concentration, for the same physical system, the retention volume V_r, the volume of solvent that must be elutriated from the system between the time the sample is introduced and the time it appears in the effluent, is a function only of the molecular size of the polymer.

The calibration procedure for size exclusion chromatography is illustrated in Figs. 6.9 and 6.10. Two runs are made with injections of mixtures of polystyrene (narrow) molecular weight standards into a series of two columns which are packed with cross-linked polystyrene particles about 10 μm in diameter. Each of the columns has an inside diameter of 7.5 mm and a length of 300 mm. Tetrahydrofuran flows at 1.0 ml/min with a total pressure drop over the columns of about 20 atm at

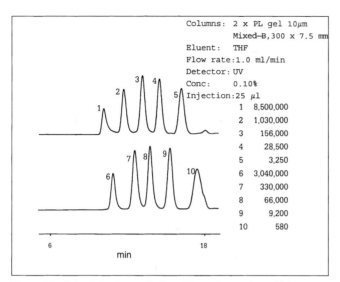

Columns:	2 x PL gel 10μm
	Mixed—B,300 x 7.5 mm
Eluent:	THF
Flow rate:	1.0 ml/min
Detector:	UV
Conc:	0.10%
Injection:	25 μl

1	8,500,000
2	1,030,000
3	156,000
4	28,500
5	3,250
6	3,040,000
7	330,000
8	66,000
9	9,200
10	580

Figure 6.9 Injection of two solutions (five polymers in each) in separate runs yields chromatograms in which ultraviolet (UV) absorption is used to measure the concentration of polymer in the effluent (courtesy of Polymer Laboratories Ltd., Amherst, Mass.)

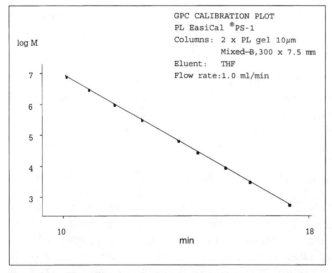

GPC CALIBRATION PLOT
PL EasiCal ®PS-1

Columns:	2 x PL gel 10μm
	Mixed—B,300 x 7.5 mm
Eluent:	THF
Flow rate:	1.0 ml/min

Figure 6.10 The calibration plot based on the 10 samples used in Fig. 6.9 is linear over a range of four decades (courtesy of Polymer Laboratories Ltd., Amherst, Mass.). The construction of the plot is carried out using a computer program (EasiCal is a registered trademark of Polymer Laboratories)

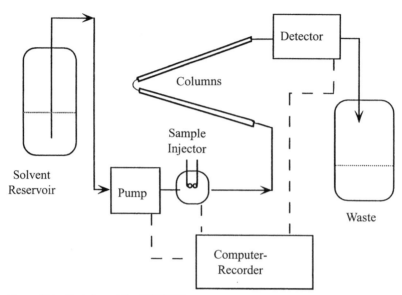

Figure 6.11 Basic layout for GPC/SEC

room temperature. When the times for elution at each of the 10 peaks are plotted against the logarithm of the corresponding molecular weights, a very linear calibration curve results covering four decades of "polystyrene equivalent" molecular weight. The entire calibration procedure takes less than 1 h. Of course, the pumping rate must be very reproducible in order for the elution times to be meaningful measures of molecular size.

The molecular weight distribution of a new sample can be estimated by the relative concentration of polymer at each effluent volume, which now corresponds to a known molecular weight. Some workers have found that calibration curves in which $M[\eta]$ is plotted against V_r coincide for polymers that differ in branching or in composition, thus giving a "universal" calibration. The quantity $[\eta]$, the intrinsic viscosity, is discussed in Sec. 7.3.

The elements of a basic system are shown in Fig. 6.11. The packed columns may each be 30 to 60 cm long. Often two or three will be placed in series. Even with swollen beads, there will be a void volume of around 30%. Solvent flow is maintained at a constant rate by a precision pump. A sample is injected after a flat baseline has been established. Concentration is measured by a detector, often a differential refractometer. Other devices are used also. Most commercial models for GPC/SEC use a laboratory computer to plot the concentration eluent volume data and also to calculate the appropriate molecular weight averages, the calibration data having been stored in the computer's memory. In addition to the differential refractometer, concentration can be detected by conductivity, radioactivity,

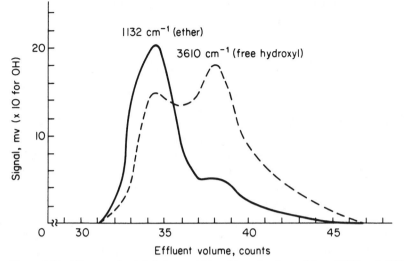

Figure 6.12 Mixture of poly(ethylene oxide)s with molecular weights of 300 and 4000 analyzed at ether and end group wave numbers (maximum signal for OH, 1.8 mV) [10]

light absorbance, or light scattering. Another detection device used in conjunction with a refractometer is a differential viscometer (See Sec. 7.3).

Infrared absorbance as a measure of concentration adds another dimension to the distribution analysis, because one can choose a band in the infrared region specific to one functional group and therefore to one component in a mixture or in a copolymer. For example, the ratio of hydroxyl to ether group in poly(ethylene oxide) varies with the molecular size.

$$HO—CH_2CH_2(—O—CH_2CH_2—)_nOH$$
Poly(ethylene oxide)

By measuring the concentration of hydroxyl and ether groups in the effluent from successive runs in the same system, the ratio of hydroxyl to ether, and therefore the molecular size, is obtained directly (Fig. 6.12). The distribution of a monomer within a copolymer distribution can also be measured by this technique. More than one analyzer can be used to monitor the output simultaneously. For example, a laser light scattering apparatus (see Sec. 6.7) can be used in tandem with a differential refractometer. The refractometer gives a measure of the concentration, and the light scattering (together with the concentration) gives a measure of the molecular weight of the polymer in absolute terms. This does away with the need for a separate calibration since the system now is yielding a distribution curve directly.

Column chromatography for molecular weight determination usually uses only one solvent at one temperature. In more general terms, high-performance liquid chromatography (HPLC) is used for the analysis of mixtures of drugs, natural products, or other complex materials. Often solvent temperature or composition is programmed to change over the analysis time.

6.5 MEASUREMENT OF MOLECULAR WEIGHT

We accept nowadays as the valid indication of molecular size the molecular weights of polymers measured by the same techniques that have been used for small molecules. Before the 1920s, many chemists were reluctant to believe that such large, covalently bonded structures were possible. The molecular weights were regarded as "apparent" values brought about by a physical association of smaller molecules in a "colloidal" state. Perhaps we should not dismiss all the reservations that those scientists had when we look at experimental results. No measurement is foolproof, and it is a brave person indeed who lists a molecular weight of a polymer in the hundred-thousand range to four or five figures. Even the so-called absolute methods we are about to discuss depend on extrapolations that are not completely objective, since some mathematical model or form invariably must be postulated.

Methods for determining molecular weight can be *relative* or *absolute.* Many properties of polymers that depend on molecular weight, such as solubility, elasticity, adsorption on solids, and tear strength, can be correlated with an average molecular weight. Once correlated, the property can be used as a measure of molecular weight. In practice, viscosity of melts and dilute solutions is the most often used *relative* method (see Sec. 7.3). Absolute methods can be categorized by the kind of average they measure.

6.6 NUMBER–AVERAGE MOLECULAR WEIGHT METHODS

The measurement of the concentration of end groups when one knows the exact number per molecule is a method of counting the number of molecules. In the hydroxy-acid ester example, a simple titration by standard sodium hydroxide would give the total number of free carboxyl groups. Since there is but one such group for each polymer molecule, we have a molecule-counting method. End group counting can be more sophisticated, as in the use of infrared or isotopic analysis. In any of these we derive a number-average molecular weight M_n from Eq. (6.9) together with Eq. (6.10):

$$M_n = x_n M_0 \tag{6.30}$$

where M_0 is the molecular weight of a repeating unit in the polymer.

The magnitude of the depression of the melting point or the elevation of the boiling point of a solvent by a nonionized solute depends on the number of solute molecules present but not on their chemical nature. These *colligative* properties are used to estimate the molecular weight of a solute. The melting point depression is widely used in organic chemistry for the characterization of small molecules. The lowering of the melting temperature, $T_f - T$, is related to the normal melting temperature of the solvent T_f, its heat of fusion ΔH_f, and molecular weight M_1, and is also related to the molecular weight of the solute M_2 and the relative weights of solvent and solute w_1 and w_2, respectively [11]:

$$T_f - T = \frac{RT_f^2}{\Delta H_f} \frac{M_1}{M_2} \frac{w_2}{w_1} \qquad (6.31)$$

where R is the gas constant. For a solution of 1 g of polyethylene with $M_2 = 500$ in 100 g of camphor ($T_f = 178.4°C$, $\Delta H_f = 2.58$ kcal/mol, $M_1 = 152$), $T_f - T$ is 0.48°C. As the molecular weight of the solute increases, measurement of the temperature lowering demands more sensitive equipment. In order to measure the number of molecules by colligative methods, it is necessary to use dilute solutions. In practice, we often extrapolate results to infinite dilution by some suitable linearizing techniques. Osmotic pressure is a useful measuring method. We place on opposite sides of a semipermeable membrane (typically cellophane) a polymer solution and pure solvent (Fig. 6.13). The membrane is selected to pass solvent but

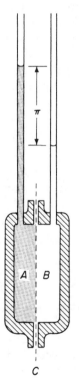

Figure 6.13 Osmotic pressure cell containing polymer solution A and solvent B separated by semipermeable membrane C. The osmotic pressure is π

not polymer, the diffusion rate of polymer above some low molecular weight being negligibly small. To establish equilibrium, the pressure on the solution side must be greater. Either by waiting for equilibrium or by measuring and compensating for pressures automatically, osmotic pressure π can be measured at several concentrations. According to thermodynamic theory:

$$\left(\frac{\pi}{c}\right)_{c=0} = \frac{RT}{M} \tag{6.32}$$

where π = osmotic pressure (g/cm²) = $h\rho$
h = difference in liquid levels at equilibrium, cm
ρ = density of solvent, g/cm³
c = concentration, g/cm³
T = absolute temperature, K
M = molecular weight, g/mol
R = gas constant, 8.48×10^4 g·cm/mol·K

The difference between osmotic pressure and vapor pressure lowering can be illustrated by considering a situation where we take pure solvent at total pressure $P(0)$, increase the vapor pressure by increasing the total pressure to $P(0) + \pi$, then decrease the vapor pressure by adding a solute with a mole fraction N_2 (Table 6.4). Assuming ideal solutions and perfect gases, the criterion for equilibrium of solvent between the first and last conditions reduces to having the vapor pressure of solvent p in pure solvent at $P(0)$ equal to that in solution at $P(0) + \pi$. The change in vapor pressure of solvent with total pressure $P(T)$ is given by the Poynting equation [12],

$$\frac{dp}{dP(T)} = \frac{V_1}{V_g} \tag{6.33}$$

where V_1 is the molar volume of liquid solvent and V_g is the molar volume of gaseous solvent. Assuming a perfect gas, we can replace V_g by RT/p and integrate:

$$RT \int_{p(0)}^{p(\pi)} d(\ln p) = V_1 \int_{P(0)}^{P(0) + \pi} dP(T) \tag{6.34}$$

$$RT \ln \frac{p(\pi)}{p(0)} = V_1 \pi \tag{6.35}$$

where the vapor pressure is increased as indicated by the limits when the total pressure is increased by the osmotic pressure π. The subsequent decrease in vapor pressure is accomplished by adding solute according to Raoult's law, $p(N_2, \pi) = N_1 p(0, \pi)$. Our final condition is that $P(N_2, \pi)$ be equal to $p(0)$. Also, when N_1 is nearly 1, $-\ln N_1$ can be approximated by $1 - N_1$, which is N_2, the mole fraction of

Table 6.4 Characterization of osmotic pressure*

Condition	A	B	C
Mole fraction of solvent	1.0	1.0	N_1
Mole fraction of solute (polymer)	0	0	N_2
Vapor pressure of solvent	$p(0, 0)$	$p(0, \pi)$	$p(N_2, \pi)$
Total pressure on system	$P(0)$	$P(0) + \pi$	$P(0) + \pi$

*Criterion for equilibrium between A and C: $p(0, 0) = p(N_2, \pi)$ when partial molar free energy of solvent is same in A and B. Raoult's law: $p(N_2, \pi) = N_1 p(0, \pi)$.

solute. If we replace N_2 by cV_1/M_2 with appropriate attention to units, we produce Eq. (6.36).

$$RT \ln \frac{p(\pi)}{p(0)} = RT \ln \frac{p(0, \pi)}{p(N_2, \pi)} = \qquad (6.36)$$
$$-RT \ln N_1 = RTN_2 = \frac{RTcV_1}{M_2} = V_1\pi$$

Thus it becomes clear that the osmotic pressure is large compared to the vapor pressure lowering because it corresponds to the change in total pressure that causes the lowering. Rearranging Eq. (6.33) and putting in finite changes in vapor pressure for a change in total pressure equal to the osmotic pressure, it is seen that the ratio is that of the volumes of equal weights of vapor and liquid, usually a factor of several hundredfold.

$$\frac{\pi}{\Delta p} = \frac{V_g}{V_1} \qquad (6.37)$$

Equation (6.32) holds only at infinite dilution (or in a Flory "theta" solvent, Sec. 7.3). At finite concentrations the osmotic pressure is best represented by an attenuated power series.

$$\frac{\pi}{c} \frac{M}{RT} = 1 + A_2 Mc + A_3 M^2 c^2 \qquad (6.38)$$

where A_2 and A_3 are the *second* and *third virial coefficients.* Often, A_3 can be taken as equal to $(A_2/2)^2$, so that Eq. (6.38) can be rewritten:

$$\left(\frac{\pi}{c}\right)^{1/2} = \left(\frac{RT}{M}\right)^{1/2} \left(1 + \frac{A_2 Mc}{2}\right) \qquad (6.39)$$

Plots of $(\pi/c)^{1/2}$ versus concentration often are linear (Fig. 6.14) and allow extrapolation to infinite dilution. The second virial coefficient is obtained from such plots by dividing the slope by the intercept and by $M/2$.

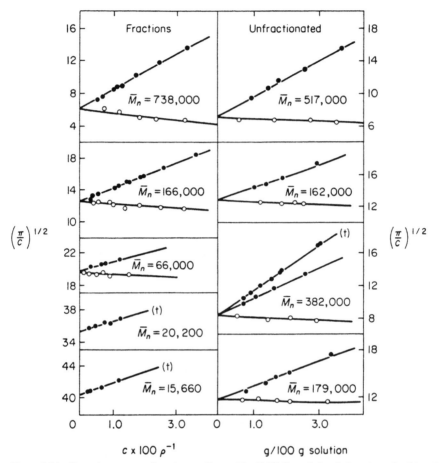

Figure 6.14 Osmotic pressure plotted according to Eq. (6.39) for poly(methyl methacrylate) in toluene [●(t)], acetone (●), or acetonitrile (○) [13]

Example 6.4

The following data are available for polymer A at 27°C. At a polymer concentration of 6.60 g/liter, the osmotic pressure, π, is 1.82 cm of solvent. At a concentration of 14.00 g/dl, the osmotic pressure, π, is 5.44 cm of solvent. Solvent density = 850 g/m³. Estimate M_n and the second virial coefficient.

Solution: There are various ways to handle the data. However, if R is used as J/mol·K, the other units should be used in a compatible way. It is useful to remember that the unit Pa is equivalent to J/m³.

Convert pressures to Pa:

$$\left(\frac{1.82}{100}\right)m \times 850 \text{ kg/m}^3 \times 9.81 \text{ N/kg} = 152 \text{ Pa}$$

and

$$c = 6.60 \times 10^3 \text{ g/m}^3$$

Thus,

$$\frac{\pi}{c} = 23.03 \times 10^{-3} \text{ J/g}$$

Similarly, 5.44 cm of solvent = 454 Pa and $\pi/c = 32.43 \times 10^{-3}$ J/g. Using Eq. (6.39) for each point and then simplifying:

$$\frac{RT}{M_n} = \frac{23.03 \times 10^{-3}}{(1 + A_2 M_n \, 6.60 \times 10^3/2)^2}$$

$$= \frac{32.43 \times 10^{-3}}{(1 + A_2 M_n \, 14.00 \times 10^3/2)^2}$$

$$(1.408)^{1/2}(1 + A_2 M_n \, 6.60 \times 10^3/2) = 1 + A_2 M_n \, 14.00 \times 10^3/2$$

$$1.187 - 1.000 = \left(\frac{A_2 M_n}{2}\right)(14.00 \times 10^3 - 1.187 \times 6.60 \times 10^3)$$

$$\frac{A_2 M_n}{2} = \frac{0.187}{6.17 \times 10^3} = 30.3 \times 10^{-6} \text{ m}^3/\text{g}$$

Going back to Eq. (6.39):

$$\frac{RT}{M_n} = \frac{23.03 \times 10^{-3}}{(1 + 30.3 \times 6.60 \times 10^{-3})^2} = 16.0 \times 10^{-3} \text{ J/g}$$

Here

$$R = 8.3145 \text{ J/mol·K} \qquad T = 300 \text{ K} \qquad M_n = 156 \times 10^3 \text{ g/mol}$$

so

$$A_2 = \frac{2 \times 30.3 \times 10^{-6}}{156 \times 10^3} = 3.88 \times 10^{-10} \text{ m}^3\text{mol/g}^2$$

Osmotic pressure probably is the most popular of the colligative methods. On the other hand, membrane preparation is tricky, equilibrium may be slow to be reached, and the effect observed, π, *decreases* as M increases. The procedure has been automated in various ways. In one modern automatic recording membrane osmometer, the osmotic pressure can be measured in 5 to 30 min. The upper cell (Fig. 6.15a) is flushed with several volumes (a few milliliters) of solution. Diffusion of solvent from the lower cell through the membrane causes the flexible stainless-

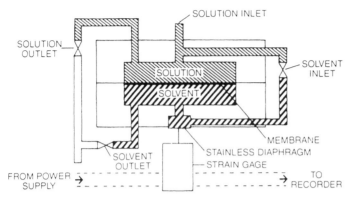

SOLUTION INLET

SOLUTION OUTLET

SOLVENT INLET

SOLUTION

SOLVENT

MEMBRANE

STAINLESS DIAPHRAGM

SOLVENT OUTLET

STRAIN GAGE

FROM POWER SUPPLY →

TO RECORDER

(a)

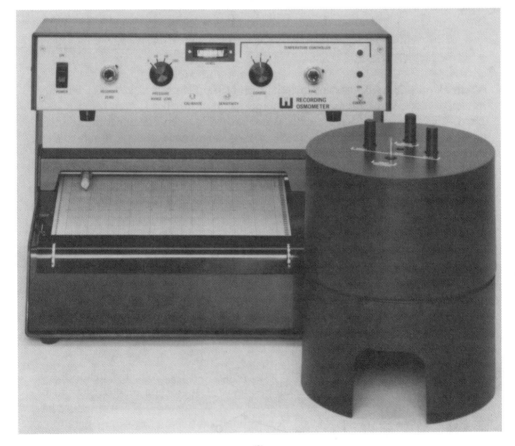

(b)

Figure 6.15 Recording automatic membrane osmometer [14]. (a) Cell volume for solution and solvent is about 0.5 ml each. (b) External view of compact unit (courtesy of Wescan Instruments, Inc., Santa Clara, Calif.)

steel diaphragm to move. This, in turn, decreases the pressure in the lower cell until equilibrium is reached. The strain gauge indicates the corresponding osmotic pressure on the recorder via a calibrated circuit. The entire instrument is very compact (Fig. 6.15b). With care, molecular weights of 5000 to 500,000 can be measured with about 1% accuracy. Below 5000, many polymers can penetrate the common membrane materials. When a low-molecular-weight polymer is placed in an osmometer, the rate of diffusion of the polymer through the membrane is discernible, since it results in a drift in the pressure necessary to maintain a null flow of solvent. Because the rate of diffusion decreases with increasing molecular size, the drift in osmotic pressure with time can be calibrated to yield the molecular weight of a specific polymer despite membrane penetration [15]. This is not an absolute method, of course.

For molecular weights less than about 20,000, another technique has been automated. In the so-called vapor pressure osmometer there is no membrane [16]. A drop of solution and a drop of pure solvent are placed on adjacent thermistors. The difference in solvent activity brings about a distillation of solvent from the solvent bead to the solution. The temperature change that results from the differential evaporation and condensation can be calibrated in terms of the number-average molecular weight of the solute. The method is rapid, although multiple concentrations and extrapolation to infinite dilution are still required.

6.7 WEIGHT–AVERAGE MOLECULAR WEIGHT METHODS

The scattering of light by small particles is a familiar phenomenon as, for example, the appearance of dust particles in a beam of sunlight. Similarly, if polymer molecules are dissolved in a solvent, the light scattered by the polymer far exceeds that scattered by the solvent and is an absolute measure of molecular weight (Fig. 6.16). The Debye relationship is [17]

$$\frac{H'c}{\kappa} = \frac{1}{MP(\theta)} + 2A_2c \tag{6.40}$$

where H' is a lumped constant including geometric factors and also the change in refractive index with polymer concentration for the particular system being investigated. This latter relationship usually is established in a separate experiment using a differential refractometer. The intensity of light is measured at angle θ and concentration c. The second virial coefficient A_2 and a complex function of molecular shape $P(\theta)$ usually are derived from the data. The light intensity factor κ is derived from the raw galvanometer reading I_g when the photocell of Fig. 6.16 is at angle θ. One must compensate for geometry and the scattering by solvent, I_{gs}.

$$\kappa = \frac{(I_g - I_{gs})\sin\theta}{1 + \cos^2\theta} \tag{6.41}$$

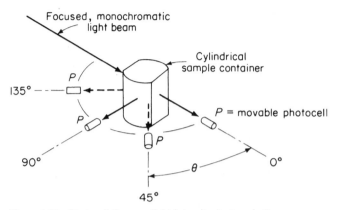

Figure 6.16   Photocell P senses light intensity I_g at angle θ

Since it is known that $P(\theta) = 1$ at $\theta = 0$, it is customary to extrapolate to $\theta = 0$ as well as $c = 0$. This can be done by plotting H' c/κ vs. concentration at constant values of θ and then plotting the intercept $1/MP(\theta)$ vs. $\sin^2 (\theta/2)$ to give the intercept $1/M$. Both can be done simultaneously in a *Zimm plot* (Fig. 6.17), where a network results. As in osmotic pressure, measurements in a poor solvent give a low value for A_2. Margerison and East [19] have worked out a complete example for polystyrene in benzene.

In one commercial instrument, 15 to 18 photodiode detectors are arranged in fixed positions to measure laser light scattered at angles between zero and 180°. For batch testing, a cell volume of 10 ml is used. The separate determination of change in refractive index with concentration, dn/dc, still is needed. In another model, a flow-through cell volume of only 67 μl is used to monitor light scattered as part of a size exclusion chromatograph (SEC). A differential refractometer (which can also be used for measuring dn/dc) is placed in series to measure concentration on the same stream. In either case, software is available which simplifies greatly the manipulation of experimental results. Using four concentrations in successive batch runs, all the intermediate parameters and a Zimm plot can be achieved in minutes when the apparatus output is fed directly to a personal computer. An example of the application of light scattering with SEC illustrates the increased scattering intensity at the highest molecular size (lowest retention volume) (Fig. 6.18). Volumes of 50 μl each of three samples were injected at concentrations of 0.1, 0.2, and 0.3 g/liter for molecular weights of 600, 200, and 30 × 10^3, respectively. The size of the signal seen by the differential refractive index detector reflects the concentrations. The size of the signals seen by the photodetectors is greatest for the highest molecular weight at all angles (which ranged from 19.0° to 132.5°).

The extrapolation to $\theta = 0$ can be eliminated if scattering can be measured at forward angles of less than 2° from the incident beam. This becomes possible when

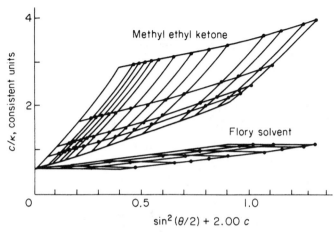

Figure 6.17 Zimm plots for scattering of light by poly(vinyl acetate) with $M_w = 3.0 \times 10^6$ in a good solvent, methyl ethyl ketone, and a poor one, methyl isopropyl ketone:n-heptane, 73.2:26.8. The ordinate has been adjusted to the same intercept with each solvent. (After [18].)

the light source is a small, single transverse-mode laser [20]. The optical diagram is relatively simple since no provision needs to be made for varying the angle of measurement. As in the case of the osmometer, the apparatus can be very compact. The weight-average molecular weight can be derived conveniently from this arrangement. However, when additional information such as the shape and size of the polymer molecules in solution is desired, the measurement of the angular dependence of scattered light intensity much be made. In *dynamic light scattering,* correlation of time-dependent scattering intensities at various angles is used to calculate diffusional properties of the polymer in solution.

To see what kind of average molecular weight is measured by light scattering, one can examine the contributions made by individual species to the factor κ. For a monodisperse polymer i at $\theta = 0$,

$$\frac{H'c_i}{\kappa_i} = \frac{1}{M_i} \tag{6.42}$$

For a polydisperse polymer,

$$\kappa = \Sigma\kappa_i = \Sigma H'c_i M_i \tag{6.43}$$

and

$$c = \Sigma c_i \tag{6.44}$$

The average molecular weight $\overline{M}$ is obtained from the overall scattering and concentration:

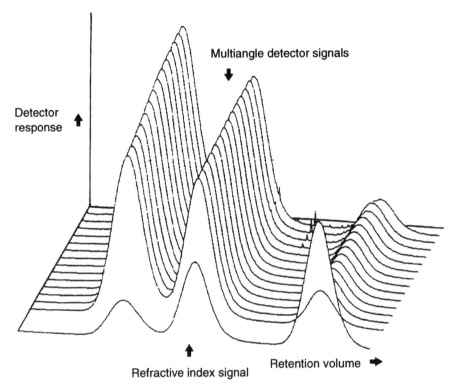

Figure 6.18 A three-dimensional plot of light scattering for three almost monodisperse polystyrenes (600,000, 200,000, and 30,000) measured by GPC/SEC in toluene. The differential refractive index signal is in the foreground. The signals from the DAWN photodetectors can be converted to a Zimm plot by a computer program, ASTRA [courtesy of Wyatt Technology Corp., Santa Barbara, Calif.; ASTRA software and DAWN light-scattering apparatus are registered trademarks of Wyatt Technology].

$$\overline{M} = \frac{\kappa}{H'c} = \frac{\sum H'c_i M_i}{H'c} = \frac{\sum c_i M_i}{c} \tag{6.45}$$

Thus it is seen that since c_i/c is the weight fraction of component i, w_i, the average molecular weight is M_w, the weight-average molecular weight.

A major practical problem in light scattering is the presence of dust particles or gelled polymer particles. A small concentration of these obscures the scattering by ordinary polymer particles and results in $\overline{M}_w = \infty$. Removal of dust by ultracentrifugation may be necessary, making this a rather expensive tool. Fortunately, solutions being eluted from an SEC usually have already been cleaned up to the point where dust particles have been eliminated. Kim and Cotts compared several light-scattering devices as adjuncts to a GPC/SEC system. They concluded that a low-angle device was convenient, but that a multiple-angle detector supplied more information [21].

The ultracentrifuge itself can be used to measure a molecular weight distribution. Under exceptionally high gravitational fields, polymers will settle out of a solution just as macroscopic beads fall through a viscous liquid according to Stokes' law. In an ultracentrifuge [22] the Brownian motion of the molecules is balanced against a centrifugal field and either the rate of settling or the equilibrium in the field is observed. A cell holding the solution is placed in a head that rotates on air bearings up to 70,000 rpm. The concentration gradient across the cell is estimated using refractive index as a measure. Because extrapolation to $c = 0$ is still necessary and the difference in refractive index between solution and solvent must be large, the ultracentrifuge has been very successful only with certain classes of polymers, for example, proteins. It has not been as widely used as osmometry or light scattering as an absolute measure of molecular weight. A definite advantage of the ultracentrifuge is that the entire distribution is obtained, not just one lumped average molecular weight. Not only is M_w measured but M_n, M_z, etc., are measured directly, without recourse to a mathematical model.

KEYWORDS FOR CHAPTER 6

Section

1. Weight-average molecular weight
   Number-average molecular weight
   Degree of polymerization
   Polydispersity index (PDI)
   Monodisperse
   Polydisperse
2. Most probable distribution
   Poisson distribution
3. Schulz distribution
   Wesslau distribution
   Differential weight distribution
4. Fractionation
   Coacervate
   Size exclusion chromatography (SEC)
   Gel permeation chromatography (GPC)
   Colligative property
6. Osmotic pressure
   Osmometer
   Second virial coefficient
   Theta (Flory) solvent
7. Light scattering
   Zimm plot

PROBLEMS

6.1 A polymer sample contains an equal number of moles N of species with degrees of polymerization $x = 1, 2, 3, 4, 5, 6, 7, 8, 9$ and 10. What is x_n? What is x_w? ($N_1 = N_2 = N_3$, etc.)

6.2 Three samples of polymer are mixed without reaction. Calculate M_n and M_w for the mixture:

Sample	M_n	M_w	Weight in mixture, g
A	1.2×10^5	4.5×10^5	200
B	5.6×10^5	8.9×10^5	200
C	10.0×10^5	10.0×10^5	100

6.3 The molecular weight distribution for a certain polymer can be described as

$$w_x = \frac{k(x)^{1/2}}{5 + x} \qquad x = 1, 4, 9, \text{ and } 16 \text{ only}$$

where w_x is the weight *fraction* of polymer with degree of polymerization x. What is the value of k? What is the weight-average degree of polymerization, x_w? What is x_n?

6.4 A polymer sample has a molecular weight distribution described by

$$w_x = kx^2 \qquad x = 1, 2, 3, \text{ and } 4 \text{ only}$$

where w_x is the weight fraction of polymer with degree of polymerization x. Calculate x_n and x_w.

6.5 One gram of polymer A ($x_n = 2000$, $x_w = 5000$) is mixed with 2 g of polymer B ($x_n = 6000$, $x_w = 10,000$). Calculate the degree of polymerization of the mixture that would be derived from osmotic pressure measurements at several concentrations.

6.6 A polymer blend is made up of four lots of monodisperse polymers, A, B, C, and D (with $k = 1, 2, 3,$ and 4, respectively), which can be described by:

$$x_w = 2800k^{1.5} \qquad x_n = 1400k$$

The blend contains $1000/k^2$ g of each lot. For the blend:

(a) What degree of polymerization can be expected from measurements made using a light scattering apparatus?

(b) What degree of polymerization can be expected from measurements of osmotic pressure at various concentrations?

6.7 A polymer sample is made up by evaporating together equal volumes (50.0 ml each) of three polymer solutions, A, B, and C. All three solutions are characterized by osmotic pressure measurements in a Flory "theta" solvent. What is the polydispersity index of the mixture?

Polymer ident.	Conc., g/dl	Osmotic pressure, cm of solvent	Distribution type
A	1.250	4.000	Monodisperse
B	0.730	3.250	Most probable*
C	0.308	3.600	Most probable*

*In each case, $x_n > 100$.

6.8 The molecular weight of a polymer is 10,000 according to an osmotic pressure experiment in a theta solvent. What osmotic pressure (atm) do you expect to encounter at a concentration of 1.18 g/dl? A light-scattering experiment in a good solvent gives a molecular weight of 34,500. Why is there a difference in measured molecular weight? Would the difference be smaller or greater if the osmotic pressure were measured in a good solvent? (R = 0.08206 liter · atm/mol · K, and T = 303 K.)

6.9 Phenyl isocyanate (MVW = 118) is expected to react quantitatively with primary alkyl hydroxyl groups to form urethanes and with water to form diphenyl urea.

$$\phi-N{=}C{=}O + HO-R \rightarrow \phi-NH-CO-OR$$
Isocyanate   Alcohol   Urethane

$$2\phi-N{=}C{=}O + H_2O \rightarrow \phi-NH-CO-NH-\phi + CO_2$$
Diphenyl urea

Analysis of 100 g of a poly(ethylene oxide) with two hydroxyls per molecule results in 5.923 g of isocyanate being consumed with the evolution of 210 cm³ (at 1 atm, 25°C) of CO_2. Estimate x_n and the weight fraction of water in the polymer.

6.10 The following data are available for polymers A and B in the same solvent at 27°C:

Conc. c_A, g/dl	Osmotic pressure π_A, cm of solvent	Conc. c_B, g/dl	Osmotic pressure π_B, cm of solvent
0.500	1.25		
0.800	2.40	0.400	1.60
1.100	3.91	0.900	4.44
1.300	4.94	1.400	8.95
1.500	6.30	1.800	13.01

Solvent density = 0.85 g/cm³; polymer density = 1.15 g/cm³.

(*a*) Plot $(\pi/c)^{1/2}$ versus c.
(*b*) Estimate M_n and the second virial coefficient for each polymer.
(*c*) Estimate M_n for a 25 : 75 mixture of A and B.

(d) If $M_w/M_n = 2.00$ for A and for B, what is M_w/M_n for the mixture of part c?

6.11 For a water-soluble polymer, what would be the upper limit on molecular weight measurable by freezing point depression?

6.12 Plot the Schulz distribution in cumulative form on logarithmic probability paper for $x_w/x_n = 2$ and 5. What parameter for the Wesslau model gives the best fit?

6.13 In a particular reactor, polymerization appears to occur by two simultaneous mechanisms which give rise to two separate molecular weight distributions (as seen by chromatography). A high-molecular-weight portion (62.5 wt% of the total batch) can be resolved by solvent fractionation. It has $M_w = 550,000$ and $M_n = 210,000$. The entire batch has a $M_w = 360,000$, but M_n cannot be measured unequivocally. Assuming that the lower-molecular-weight portion (37.5 wt% of the total batch) has a "most probable" distribution, find the M_n of the lower-molecular-weight portion and the M_n of the entire batch. It can be assumed that the lower-molecular-weight portion has M_n high enough so that p is essentially equal to 1.

6.14 A polymer has a continuous cumulative distribution given by

$$W(x) = \ln (ea^2x^2 + 1) - \ln (a^2x^2 + 1)$$

where e is 2.7183 and a is an adjustable parameter. Derive expressions for the differential weight distribution and for x_w and x_n in terms of a. What is x_w/x_n?

6.15 Compare the cumulative distribution of Prob. 6.12 with the Wesslau model on logarithmic probability paper using the product ax as the measure of molecular weight.

6.16 How could the model of Prob. 6.12 be modified to allow x_w/x_n to be an independent variable?

6.17 Show that the cumulative weight distribution for the Wesslau model, Eq. (6.25), is given by

$$W(x) = \int_0^x w_x\, dx = \frac{1 + \mathrm{erf}(Y/\beta)}{2}$$

where $Y = \ln(x/x_0)$ and erf is the normalized error function.

6.18 A polymer sample has a molecular weight distribution described by

$$w_x = k(x)^{1/2} \qquad x = 1, 2, 3, 4, 5 \text{ only}$$

where w_x is the weight fraction of x-mer. Calculate $\bar{x}_n$, $\bar{x}_w$, and the polydispersity index.

6.19 Polymers A and B are monodisperse polystyrenes. Polymer A has a molecular weight three times that of B. Polymer C, on the other hand, is a polydisperse polystyrene for which M_w is 2.0×10^5. Deduce the M_n for polymer C from the following measurements on a mixture of all three polymers: The mixture contains 25 g of A, 50 g of B, and 25 g of C. Light scattering gives a molecular weight of 112,500, while osmotic pressure gives a molecular weight of 60,000. $x_w/x_n = 2$

6.20 The molecular weight distributions of polymers A and B each can be represented by the "most probable" distribution. Also, $x_n > 100$ for each of them. When 200 g of A is mixed with 100 g of B, the polydispersity index $(x_w/x_n) = 4$ for the mixture. What is $(x_w)_A/(x_w)_B$?

6.21 By fractional precipitation, a polymer P with the "most probable" distribution and a number-average molecular weight $M_n = 150,000$ is cut into only two fractions, A and B, with $M_w = 250,000$ and $325,000$, respectively. If the polydispersity of A is the same as that of B, what is the weight of fraction A obtained from 200 g of initial polymer P? What is the polydispersity of A and B?

6.22 A monodisperse polyester with a degree of polymerization of 1.00×10^5 is partly hydrolyzed by reaction with water. A total of 10.0 g of water is reacted with 0.0100 mol of polymer. Random scission occurs.

(a) What is the weight-average degree of polymerization of the final, partly hydrolyzed polymer?

(b) What is the weight fraction of trimer in resulting polymer?

6.23 In a polyesterification, 10-hydroxydecanoic acid is heated and water is removed.

Run A: 4.5 mol of water are removed from 5 mol of monomer.
Run B: 4.9 mol of water are removed from 5 mol of monomer.

(a) What are x_w and x_n for the polymer of Run B?

(b) If the two resultant polymers are combined without reaction, what is the x_w and x_n of the mixture?

6.24 A molecular weight distribution is given by

$$W = \log x - 1 \qquad 10 < x < 100$$
$$W = 0 \qquad x < 10$$
$$W = 1 \qquad x > 100$$

where W is the cumulative weight fraction of polymer above degree of polymerization x. Calculate x_n and x_w.

6.25 A molecular weight distribution is represented by

$$\frac{dW}{dx} = Bxe^{-ax} \qquad 0 < x < \infty$$

(a) Express B in terms of x_n.

(b) Determine the ratio of x_w to x_n.

(c) Plot $dW/d(x/x_n)$ versus $\log(x/x_n)$ for the range $0.1 < (x/x_n) < 10$.

REFERENCES

1. Flory, P. J.: "Principles of Polymer Chemistry," chap. 8, Cornell Univ. Press, Ithaca, N.Y., 1953.
2. Swarc, M.: "Polymerization and Polycondensation Processes," p. 96, Adv. in Chem. Series, No. 34, ACS, 1962.
3. Tung, L. H., in M. J. Cantow (ed.): "Polymer Fractionation," Academic Press, New York, 1967.
4. Kotera, A., in M. J. Cantow (ed.): "Polymer Fractionation," Academic Press, New York, 1967.
5. Blease, R. A., and R. F. Tuckett: *Trans. Faraday Soc.,* 37:571 (1941).
6. Elliott, J. H., in M. J. Cantow (ed.): "Polymer Fractionation," Academic Press, New York, 1967.
7. Porath, J., and P. Flodin: *Nature,* 183:1657 (1959): P. Flodin, "Dextran Gels and Their Applications in Gel Filtration," Pharmacia, Uppsala, Sweden, 1962.
8. Moore, J. C.: *J. Polymer Sci.,* A2:835 (1964).
9. Rodriguez, F., R. A. Kulakowski and O. K. Clark: *Ind. Eng. Chem. Prod. Res. Develop.,* 5:121 (1966).
10. Terry, S. L., and F. Rodriguez: *J. Polymer Sci.,* part C(21):191 (1968).
11. Glasstone, S.: "Thermodynamics for Chemists," p. 340, Van Nostrand, Princeton, N.J., 1947.
12. Glasstone, S.: "Thermodynamics for Chemists," p. 236, Van Nostrand, Princeton, N.J., 1947.
13. Fox, T. G., J. B. Kinsinger, H. F. Mason, and E. M. Schuele: *Polymer,* 3:71 (1962).
14. "Recording Membrane Osmometers," Wescan Instruments, Inc., Santa Clara, Calif. 1987.
15. Hudson, B. E., Jr.: *ACS Div. Polymer Chem., Preprints,* 7:467 (1966).
16. Neumayer, J. J.: *Anal. Chim. Acta,* 20:519 (1959).
17. Hohenstein, W. P., and R. Ullman, in P. H. Groggins (ed.): "Unit Processes in Organic Synthesis," 5th ed., p. 919, McGraw-Hill, New York, 1958.
18. Schultz, A. R.: *J. ACS,* 76:3422 (1954).
19. Margerison, D., and G. C. East: "Introduction to Polymer Chemistry," p. 89, Pergamon, New York, 1967.
20. McConnell, M. L.: *American Laboratory,* 10:63 (May 1978).
21. Kim, S. H., and P. M. Cotts: *J. Appl. Polym. Sci.,* 42:217 (1991).
22. Flory, P. J.: "Principles of Polymer Chemistry," p. 303, Cornell Univ. Press, Ithaca, N.Y., 1953.

General References

Berne, B. J., and R. Pecora (eds.): "Dynamic Light Scattering," Plenum, New York, 1976.
Carroll, B. (ed.): "Physical Methods in Macromolecular Chemistry," Dekker, New York, vol. 1, 1969; vol. 2, 1972.
Cazes, J. (ed.): "Liquid Chromatography of Polymers and Related Materials," Dekker, New York, 1977.
Cazes, J., and X. Delamare (eds.): "Liquid Chromatography of Polymers and Related Materials, II," Dekker, New York, 1980.
Chu, B.: "Laser Light Scattering," Academic Press, New York, 1974.
Dawkins, J. V. (ed.): "Developments in Polymer Characterization—1," Applied Science, London, 1978.
Epton, R. (ed.): "Chromatography of Synthetic and Biological Polymers," 2 vols., Wiley, New York, 1978.
Ezrin, M. (ed.): "Polymer Molecular Weight Methods," ACS, Washington, D.C., 1973.
Glöckner, G.: "Polymer Characterization by Liquid Chromatography," Elsevier Applied Science, New York, 1987.
Hamilton, R. J., and P. A. Sewell: "Introduction to High Performance Liquid Chromatography," Wiley, New York, 1978.
Huglin, M. B. (ed.): "Light Scattering from Polymer Solutions," Academic Press, New York, 1972.
Hunt, B. J., and S. R. Holding (eds.): "Size Exclusion Chromatography," Chapman & Hall, New York, 1989.

Hunt, B. J., and M. I. James (eds.): "Polymer Characterization," Chapman & Hall, New York, 1993.

Janca, J. (ed.): "Steric Exclusion Liquid Chromotography of Polymers," Dekker, New York, 1983.

Kratochvil, P.: "Classical Light Scattering from Polymer Solutions," Elsevier Applied Science, New York, 1987.

Pecora, R. (ed.): "Dynamic Light Scattering. Applications of Photon Correlation Spectroscopy," Plenum, New York, 1985.

Peebles, L. H., Jr.: "Molecular Weight Distributions in Polymers," Wiley-Interscience, New York, 1971.

Provder, T. (ed.): "Size Exclusion Chromatography," ACS, Washington, D.C., 1984.

Provder, T. (ed.): "Detection and Data Analysis in Size Exclusion Chromatography," ACS, Washington, D.C., 1987.

Provder, T. (ed.): "Computer Applications in Applied Polymer Science II: Automation, Modeling, and Simulation," ACS, Washington, D.C., 1989.

Provder, T. (ed.): "Chromatography of Polymers—Characterization by SEC and FFF," ACS, Washington, D.C., 1993.

Rempp, P., and G. Weill (eds.): "Polymer Thermodynamics and Radiation Scattering," Hüthig & Wepf, Basel, Switzerland, 1992.

Roe, R. J. (ed.): "Computer Simulation of Polymers," Prentice Hall, Englewood Cliffs, N.J., 1991.

Slade, P. E. (ed.): "Techniques and Methods of Polymer Evaluation," vol. 4: "Polymer Molecular Weights," part II, Dekker, New York, 1975.

Smith, C. G., W. C. Buzanowski, J. D. Graham, and Z. Iskandarani (eds.): "Handbook of Chromatography: Polymers," vol. 2, CRC Press, Boca Raton, Fla., 1993.

Smith, C. G., N. E. Skelly, C. D. Chow, and R. A. Solomon: "Handbook of Chromatography, Polymers," CRC Press, Boca Raton, Fla., 1982.

Snyder, L. R., and J. J. Kirkland: "Introduction to Modern Liquid Chromatography," Wiley, New York, 1979.

Tung, L. H.: "Fractionation of Synethetic Polymers—Principles and Practices," Dekker, New York, 1977.

Wu, C.-S. (ed.): "Handbook of Size Exclusion Chromatography," Dekker, New York, 1995.

Yau, W., J. J. Kirland, and D. Bly: "Modern Size-Exclusion Liquid Chromatography," Wiley, New York, 1979.

SEVEN

VISCOUS FLOW

7.1 DEFINITIONS

Whenever matter is subjected to external forces, the response involves a flow or a deformation of some sort. *Rheology* is defined as the science of flow and deformation of matter. The breadth of the field is indicated by the motto of the Society of Rheology. The motto is a quotation from the Greek philosopher Heraclitus, who said "Everything flows." While many forms of matter are indeed studied by rheologists, polymers in the form of melts, solutions, and suspensions have always been of great interest since the founding of the society in 1929.

In this chapter, the emphasis is on the characterization of flow, considering polymers to be viscous fluids with or without time-dependent properties. In Chap. 8, the emphasis is on the elastic and inelastic responses of polymers, considering them primarily as solids and once again with or without time-dependent properties. In Chap. 9, examples of the ultimate (failure) behavior of polymers are described.

We have used the term "viscosity" up to now without a strict definition. In general, we know that polymer solutions flow more slowly through a tube than do solvents alone under the same pressure. Molten polymers flow much more slowly than do their solutions. Since many operations in production and fabrication involve flow in pipes, between rolls, and through slots and stirring in vessels, it is well for us to consider now some of the generalizations we can make regarding the effect of chemical and physical variables on flow properties.

Viscosity is a measure of the energy dissipated by a fluid in motion as it resists

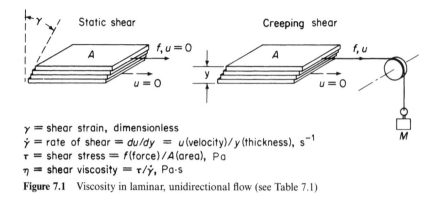

γ = shear strain, dimensionless
$\dot{\gamma}$ = rate of shear = du/dy = u(velocity)/y (thickness), s^{-1}
τ = shear stress = f(force)/A(area), Pa
η = shear viscosity = $\tau/\dot{\gamma}$, Pa·s

Figure 7.1 Viscosity in laminar, unidirectional flow (see Table 7.1)

an applied shearing force. The dissipation is a form of friction and, in an adiabatic system, results in raising the temperature of the system. In laminar, unidirectional flow, the terms shear stress τ and the rate of shear $\dot{\gamma}$ are used to indicate the applied force and the response of the fluid. In the lamellar picture (Fig. 7.1), a certain force per unit of area f/A is required to maintain a constant-velocity gradient, in this case u/y. The quantities of interest are

$$\tau = \text{shear stress} = \frac{f}{A} \tag{7.1}$$

$$\dot{\gamma} = \text{rate of shear} = \frac{u}{y} \tag{7.2}$$

$$\eta = \text{shear viscosity} = \frac{\tau}{\dot{\gamma}} \tag{7.3}$$

For many liquids of lower molecular weight, Newton's law applies; i.e., the viscosity is a constant independent of the magnitude of τ or $\dot{\gamma}$. For many polymer melts and solutions, the shear stress and shear rate are not proportional over all ranges, so that the non-Newtonian viscosity as defined by Eq. (7.3) is not a constant. Another important quantity is the kinematic viscosity, the ratio of the absolute viscosity η to the density of the fluid ρ. Since units are often confusing in working out problems, an example is worked out in conjunction with Fig. 7.1 (Table 7.1). The rate of energy dissipation per unit volume $\dot{Q}$ for a flowing liquid is simply stated:

$$\dot{Q} = \eta\dot{\gamma}^2 = \tau\dot{\gamma} = \frac{\tau^2}{\eta} \tag{7.4}$$

If stress is in dynes per square centimeter and rate of shear in seconds, $\dot{Q}$ has the dimension of ergs per cubic centimeter per second. If stress is in pascals, then $\dot{Q}$ has the dimension of watts per cubic meter. In the example of Fig. 7.1, $\dot{Q}$ is 23.95 MW/m^3 (or 5.72 cal/cm^3·s). Remember that 1 W is 1 J/s or 0.2389 cal/s.

Table 7.1 Viscosity in laminar, unidirectional flow

		SI units	Metric units	Engineering units
Given:	Mass M	0.454 kg	454 g	1.00 lb_m
	Velocity u	0.305 m/s	30.5 cm/s	1.00 ft/s
	Gap width y	61×10^{-6} m	6.1×10^{-3} cm	2.00×10^{-4} ft
	Area A	9.3×10^{-4} m^2	9.3 cm^2	0.0100 ft^2
	Constant g	9.81 m/s^2	981 cm/s^2	32.2 ft/s^2
Then:	Stress τ	4.79 kPa	4.79×10^4 dyn/cm^2	3.22×10^3 lb_m/ft·s
	Shear rate $\dot{\gamma}$	5.00×10^3 s^{-1}	5.00×10^3 s^{-1}	5.00×10^3 s^{-1}
	Viscosity η	0.958 Pa·s	9.58 poise	0.644 lb_m/ft·s

The viscous flow behavior of polymer melts and solutions is important as a relative measure of molecular weight. Also, since most forming operations involve flow, viscosity data often are gathered to predict processing characteristics. Some manufacturing methods with their typical associated shear rates are [1]:

Compression molding $\quad \dot{\gamma} = 1$ to 10 s^{-1}
Calendering $\quad\quad\quad\quad\ \ \dot{\gamma} = 10$ to 10^2 s^{-1}
Extrusion $\quad\quad\quad\quad\quad\ \dot{\gamma} = 10^2$ to 10^3 s^{-1}
Injection molding $\quad\quad\ \ \dot{\gamma} = 10^3$ to 10^4 s^{-1}

The most common instrument used for low-viscosity liquids is the capillary viscometer with liquid flowing under its own potential head (Fig. 7.2). The calibration equation for such a viscometer is [2]:

$$\frac{\eta}{\rho} = At - \frac{B}{t^2} \tag{7.5}$$

where B/t^2 is a correction factor for the kinetic energy losses at the entrance and exit to the capillary.

For flow when η is not constant but varies with τ or $\dot{\gamma}$, externally pressurized capillary and rotational viscometers are widely used (see Sec. 7.8).

7.2 POLYMER SHAPES IN SOLUTION [3, 4]

A treatment of amorphous polymers differs from that for low-molecular-weight materials because of the long covalent-bonded structure that must be accommodated. We shall treat here in outline some mathematical models for polymer shapes in solution that can be extended to polymer melts and cross-linked melts (rubbers).

A polymer molecule in solution is not a stationary piece of string, but instead is a constantly coiling and uncoiling chain whose conformation in space is ever

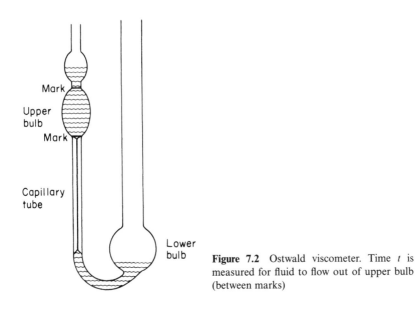

Mark

Upper
bulb

Mark

Capillary
tube

Lower
bulb

Figure 7.2 Ostwald viscometer. Time t is measured for fluid to flow out of upper bulb (between marks)

changing. In a very dilute solution the individual molecules can be considered as acting independently. Each molecule can be pictured as a string of beads with a tendency to coil in upon itself to form a spherical cloud of chain segments having radial symmetry.

In order to see how the average size of the molecule's cloud changes with molecular weight, several approaches varying in their realism have been used. A simple model is that of the *freely jointed chain* (Fig. 7.3). If the chain has n links of length a, this is a "random flight problem." Obviously, the average distance traveled $\bar{r}_0$ in flight of any number of links will be zero, since all starting directions are equally probable. However, the average square of the distance traveled $\bar{r}_0^2$ is not zero but increases linearly with the number of links:

$$\bar{r}_0^2 = na^2 \qquad (7.6)$$

This is a rigid mathematical consequence of the assumed model and not an approximation. For example, polyethylene is composed of methylene groups, CH_2, of molecular weight 14, separated by a bond distance of about 1.54 Å. Therefore, polyethylene of molecular weight 140,000 would have a root-mean-square (rms) end-to-end distance $(\bar{r}_0^2)^{1/2}$ of 15.4 nm. A more realistic approach includes the *constant* bond angle characteristic of any chemical combination. If θ' is the bond angle (109.5° for tetrahedral carbon) and $\theta = 180° - \theta'$,

$$\bar{r}_0^2 = na^2 \frac{1 + \cos \theta}{1 - \cos \theta} \qquad \text{or} \qquad 2na^2 \text{ for carbon} \qquad (7.7)$$

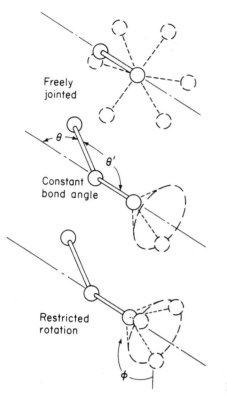

Freely
jointed

Constant
bond angle

Restricted
rotation

Figure 7.3 Polymer chain conformations

We are still assuming free rotation about each single bond. For the polyethylene of our previous example we would now get 21.8 nm.

However, we know that energy is required to rotate even one methylene group past another as in the paraffins. This is the phenomenon that has led to conformational analysis as a tool in the stereochemistry of organic compounds. Because there are energy barriers to rotation, the trans conformation is favored, although the probability of rotation increases with temperature. The gauche positions (Fig. 7.4) are more favorable than the cis position, or any other eclipsed conformation, but not as favorable as the trans position. In terms of the end-to-end distance, symmetrical, restricted *rotation* leads to

$$\overline{r_0^2} = na^2 \left(\frac{1 + \cos \theta}{1 - \cos \theta}\right)\left(\frac{1 + b}{1 - b}\right) \qquad (7.8)$$

where b = average value of cos ϕ. Also, $b = 0$ for free rotation. Naturally, bulkier groups or planar groups tend to decrease rotation and give stiffer polymers. For purposes of calculating chain dimensions, the energetically favored rotational

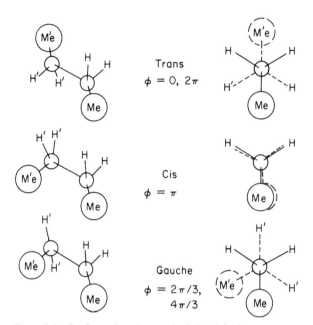

Figure 7.4 Conformation about a single bond for butane

states can be used to the exclusion of the intermediate conformations. In the case of carbon-chain polymers, the trans and 2 gauche positions are the obvious choices of *rotational isomeric states.* If all 3 positions are equally probable, b = 0 (in Eq. 7.8) just as it was for free rotation.

Of course, r_0 is not the diameter of the polymer molecule "particle" or cloud. The rms distance of the elements of a chain from its center of gravity $(\overline{s^2})^{1/2}$ (radius of gyration) is related for linear polymers by

$$\overline{s^2} = \frac{\overline{r^2}}{6} \quad \text{and} \quad \overline{s_0^2} = \frac{\overline{r_0^2}}{6} \tag{7.9}$$

Branching of the polymer chain has the effect of decreasing r and s.

Other factors must be considered before experimental data can be interpreted theoretically. For example, random flight, even with constant bond angles and restricted rotation, does not keep more than one segment of the polymer chain from occupying the same volume at the same time. Thus a certain amount of volume is excluded, even under conditions such that no other forces act on the molecule.

In Table 7.2 some experimental values of molecular dimensions are compared with those calculated from Eq. (7.7) assuming free rotation about fixed bond angles. The experimental values for $\overline{r_0^2}/M$ in Table 7.2 were obtained from measurements (in poor solvents) of the intrinsic viscosity (Sec. 7.3). The steric factor σ is

Table 7.2 Unperturbed molecular dimensions [5]

Polymer	Repeat unit	M/unit (n)	Temperature, °C	$K' \times 10^5$ $K' = [\eta]_\theta/M^{0.5}$	$(\overline{r_0^2}/M)^{0.5} \times 10^{11}$ cm exp.*	$(\overline{r_0^2}/M)^{0.5} \times 10^{11}$ cm calc.†	σ Steric factor*‡	$(\overline{r_0^2}/n)^{1/2}$ exp. Å §	$(\overline{r_0^2}/n)^{1/2}$ calc. (Å) ¶
Polyisobutylene	$-CH_2-C(CH_3)(CH_3)-$	56 (2)	24 / 95	106 / 91	795 / 757	412 / 412	1.93 / 1.84	4.21 / 4.01	2.18 / 2.18
Polystyrene	$-CH_2-CH(C_6H_5)-$	104 (2)	70	75	710	302	2.35	5.12	2.18
Poly(methyl methacrylate)	$-CH_2-C(CH_3)(C{=}O)(OCH_3)-$	100 (2)	30	65	680	310	2.20	4.81	2.18
cis-Polyisoprene	$CH_3, H / C{=}C / CH_2-, -CH_2$	68 (4)	0–60	119	830	485	1.71	3.42	2.00
trans-Polyisoprene	$CH_3, CH_2- / C{=}C / -CH_2, H$	68 (4)	60	232	1030	703	1.46	4.25	2.90
Poly(dimethyl siloxane)	$-O-Si(CH_3)(CH_3)-$	74 (2)	20	81	730	456	1.60	4.44	2.77
Polyethylene [7]	$-CH_2-CH_2-$	28 (2)	100	230	1030	582	1.77	3.85	2.18

* $(\overline{r_0^2}/M)^{0.5} = (K'/\phi)^{1/3}$, where $\phi = 2.1 \times 10^{21}$ dl/g·mol·cm³; 1 Å = 0.1 nm.
† From known angles and bond distances with free rotation.
‡ Ratio of exp. to calc. value of $(\overline{r_0^2})^{0.5}$.
§ From $(\overline{r_0^2}/M)_{\text{exp.}}$.
¶ From $(\overline{r_0^2}/M)_{\text{calc.}}$.

the ratio of the experimentally determined value of $(\overline{r_0^2})^{1/2}$ to that calculated from Eq. (7.7). It includes the effects of restricted rotation and excluded volume. Often, substitution on a chain has two compensating effects on $(\overline{r_0^2}/M)^{1/2}$. The phenyl group in polystyrene compared to the hydrogen in polyethylene makes the molecule more compact. That is, the molecular weight per chain atom is about four times as great.

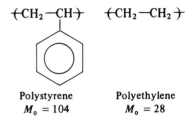

Polystyrene
$M_0 = 104$

Polyethylene
$M_0 = 28$

However, the bulky side group makes the chain stiffer, so that the polymer cannot coil up as readily, making the rms end-to-end distance per chain atom greater. The first effect outweighs the second, and it can be seen that polystyrene molecules occupy less volume per unit weight than polyethylene molecules of the same molecular weight at comparable temperatures. The ratio of volume is $(710/1030)^3$. On the other hand, the volume per chain atom (Table 7.2, second-last column) is greater for polystyrene. The ratio here is $(5.12/3.85)^3$.

A polymer molecule dissolved in a solvent will be more or less expanded depending on the degree to which solvent and polymer associate. If a polymer is in a "good" solvent, the segments of the polymer associate with the solvent molecules rather than with each other, expanding the total volume occupied by a single polymer cloud. We can define an expansion factor α by

$$\alpha = \left(\frac{\overline{r^2}}{\overline{r_0^2}}\right)^{1/2} \tag{7.10}$$

where $(\overline{r^2})^{1/2}$ is the actual rms end-to-end distance and $(\overline{r_0^2})^{1/2}$ is that when $\alpha = 1$ (unperturbed or unswollen dimension). Values of α have been determined experimentally and found to fit a theoretically derived relationship [6]

$$(\alpha^5 - \alpha^3) = C'M^{1/2}\left(1 - \frac{\theta_f}{T}\right) \tag{7.11}$$

where θ_f is the Flory temperature, C' is a constant for a given polymer-solvent combination, and M is the molecular weight. The Flory temperature θ_f is approximately the temperature at which polymer of infinite molecular weight precipitates from the solvent. This is logical, because this is the point where polymer segments associate more with each other than they do with the solvent, reducing α to unity. Like the interaction parameter, θ_f is characteristic of a given polymer-solvent combination. Kurata and Stockmayer [7] list a number of examples of θ_f.

Table 7.3 Dependence of α^3 on molecular weight [8]
Polyisobutylene in Cyclohexane, 30°C

Molecular weight $\times 10^{-3}$	Intrinsic viscosity, dl/g	$\alpha^3 = [\eta]/[\eta]_\theta$
9.5	0.145	1.39
50.2	0.47	1.96
558	2.48	3.10
2720	7.9	4.46

The experimental values of $(\overline{r_0^2}/M)^{1/2}$ in Table 7.2 were obtained in poor solvents at the point of precipitation. Under these conditions, the relationship between intrinsic viscosity and molecular weight, to be discussed in the next section, reduces to a form that gives $\overline{r_0^2}/M$ directly. Light scattering can also be used, but it is less convenient. The exact value of α also depends on the molecular weight. Some experimental values of α in a good solvent are shown in Table 7.3.

7.3 DILUTE SOLUTIONS AND INTRINSIC VISCOSITY

Manipulation of dilute solution viscosities yields an important parameter of a polymer in a given solvent, the intrinsic viscosity $[\eta]$. We can define $[\eta]$ as the ratio of specific viscosity η_{sp} to concentration c at infinite dilution.

$$[\eta] = \lim_{c \to 0} \frac{\eta - \eta_s}{c\eta_s} = \lim_{c \to 0} \frac{\eta_{sp}}{c} \tag{7.12}$$

where η and η_s are viscosities of solution and solvent, respectively. It should be pointed out that only η and η_s have the dimensions of viscosity. Specific viscosity η_{sp} and relative viscosity $\eta_r = \eta/\eta_s$ are dimensionless. Intrinsic viscosity, reduced viscosity η_{sp}/c, and inherent viscosity $(\ln \eta_r)/c$ all have the dimensions of inverse concentrations. By convention, c is usually expressed as grams of polymer per deciliter (100 ml) of solution. The somewhat confusing nomenclature is summarized in Table 7.4. If either η_{sp}/c or $\ln \eta_r/c$ is plotted against c, a linear plot (Fig. 7.5) corresponding to the following equations may result:

Huggins [9]:

$$\frac{\eta_{sp}}{c} = [\eta] + k'[\eta]^2 c \tag{7.13}$$

Kraemer [10]:

$$\frac{\ln \eta_r}{c} = [\eta] - k''[\eta]^2 c \tag{7.14}$$

Table 7.4 Viscometric terms

Symbol	Name	Common units
η	Solution viscosity	Poise or Pa·s
η_s	Solvent viscosity	Poise or Pa·s
$\eta_r = \eta/\eta_s$	Relative viscosity	Dimensionless
$\eta_{sp} = \eta_r - 1$	Specific viscosity	Dimensionless
$(\ln \eta_r)/c$	Inherent viscosity	dl/g
η_{sp}/c	Reduced viscosity	dl/g

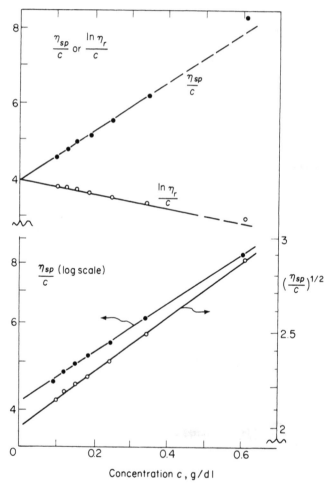

Figure 7.5 Correlations for viscosity of dilute solutions

Table 7.5 Parameters for Fig. 7.5

Equation	$[\eta]$	Slope factor
7-13	3.95	$k' = 0.41$
7-14	3.97	$k'' = 0.12$
7-15	3.99	$k''' = 0.36$
7-22b	4.14	$K'' = 0.12$

For many polymers in good solvents

$$k' = 0.4 \pm 0.1 \qquad \text{and} \qquad k'' = 0.05 \pm 0.05$$

The equations cited are applicable only in dilute solutions where η_r is less than about 2. In analogy to the treatment of osmotic data [Eq. (6.43)], we can linearize data to higher concentrations if we postulate an equation of the form [11]:

$$\left(\frac{\eta_{sp}}{c}\right)^{1/2} = [\eta]^{1/2} + \frac{k'''}{2}[\eta]^{3/2}c \tag{7.15}$$

This equation and another mentioned later [Eq. (7.25)] are compared with the Huggins and Kraemer equations in Fig. 7.5 and Table 7.5 for a set of experimental data.

Staudinger [12] proposed on empirical grounds that $[\eta]$ was proportional to molecular weight for a given polymer-solvent combination. The more general Mark-Houwink-Sakurada relationship [13] with two constants, K' and a, is

$$[\eta] = K'M^a \tag{7.16}$$

Typical values of K' and a are summarized in Table 7.6.

The practical importance of intrinsic viscosity cannot be overstated. It is the most common measure of molecular weight for high polymers. The average molecular weight measured by viscosity M_v lies between M_n and M_w, which are measured by osmotic pressure and light scattering, respectively (see Secs. 6.5 and 6.6). The experimental results give

$$[\eta] = \sum_{x=1}^{\infty} w(x)[\eta](x) \tag{7.17}$$

where $[\eta]$ is the intrinsic viscosity of the whole polymer made up of fractions weighing $w(x)$ each and possessing an intrinsic viscosity $[\eta](x)$ each. Combination of Eq. (7.16) with Eq. (7.17) yields

$$M_v = [\Sigma w(x)M(x)^a]^{1/a} \tag{7.18}$$

$$\frac{[\eta]_A}{[\eta]_B} = \frac{K'M_A^a}{K'M_B^a} = \left(\frac{M_A}{M_B}\right)^a$$

Table 7.6 Parameters for the Mark-Houwink-Sakurada equation [13]

Polymer	Solvent	Temperature, °C	$K' \times 10^5$	a
Cellulose triacetate	Acetone	25	8.97	0.90
SBR rubber	Benzene	25	54	0.66
Natural rubber	Benzene	30	18.5	0.74
	n-Propyl ketone	14.5	119	0.50
Polyacrylamide	Water	30	68	0.66
Polyacrylonitrile	Dimethyl formamide	25	23.3	0.75
Poly(dimethylsiloxane)	Toluene	20	20.0	0.66
Polyethylene	Decalin	135	62	0.70
Polyisobutylene	Benzene	24	107	0.50
	Benzene	40	43	0.60
	Cyclohexane	30	27.6	0.69
Poly(methyl methacrylate)	Toluene	25	7.1	0.73
Polystyrene				
Atactic	Toluene	30	11.0	0.725
Isotactic	Toluene	30	10.6	0.725
Poly(vinyl acetate)	Benzene	30	22	0.65
	Ethyl n-butyl ketone	29	92.9	0.50
Poly(vinyl chloride)	Tetrahydrofuran	20	3.63	0.92

or

$$x_v = [\Sigma w(x) x^a]^{1/a} \qquad (7.19)$$

For the example of Chap. 6 (Table 6.2) the student can verify that x_v is equal to 2.74 or 2.83 when the exponent of Eq. (7.19) is equal to 0.5 or 0.67, respectively. In either case, it should be noted, M_v is much closer to M_w than to M_n. If the exponent $a = 1$, then $M_v \equiv M_w$.

In comparison to light scattering, viscosity measurements involve far less time and equipment. In comparison to osmotic pressure, viscosity has the advantage that the observable quantity increases as molecular weight increases (Fig. 7.6). Of course, some absolute method such as light scattering or osmotic pressure measured on fractions where x_w/x_n is small (for example, $x_w/x_n < 1.2$) must be used to "calibrate" Eq. (7.16) for a given polymer and solvent.

In a semiautomatic Ubbelohde viscometer (Fig. 7.7), the time of flow is recorded by means of photocells, which sense the passage of the fluid meniscus. The Ubbelohde (see also Fig. 7.25) has the advantage over the Ostwald model (Fig. 7.2) that the initial charge of solution in the viscometer can be diluted by adding more

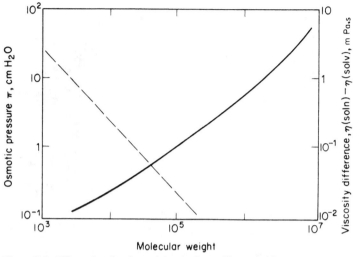

Figure 7.6 Effect of molecular weight on observable quantities

solvent without changing the calibration values [A and B in Eq. (7.5)]. Thus the viscosity-concentration data needed for extrapolation to infinite dilution can be measured from a single initial charge. Commercial devices also are available for carrying out automatically the entire procedure of filtering, charging, measuring flow time, and flushing out. The cost of these rather elaborate systems can be justified when a great many samples must be run, as in the quality control department of a large polymer-producing plant.

Another device which simplifies the measurement of dilute solution viscosity is the differential viscometer. One design [14] consists of two capillary tubes in series (Fig. 7.8). When fluid is pumped at a constant rate through the two tubes, the pressure drops are sensed by differential pressure transducers, DPT_1 and DPT_2. With solvent flowing, the ratio of the two pressures, $(P_2/P_1)_s$, is a constant for a given piece of apparatus. Now a dilute polymer solution is placed in the sample loop. By turning the injection valve, solvent continues to flow through the first tube, but polymer solution flows through the second tube, giving rise to a new ratio of pressures, $(P_2/P_1)_p$. The relative viscosity is simply

$$\eta_r = \left(\frac{P_1}{P_2}\right)_s \left(\frac{P_2}{P_1}\right)_p \qquad (7.20)$$

A more complicated version of the same apparatus has been used as a detection device for column chromatography [15]. In one example, an in-line differential viscometer was used together with a differential refractometer to measure relative viscosity and concentration, respectively. Because of the very low polymer concentrations, extrapolation to infinite dilution is not needed.

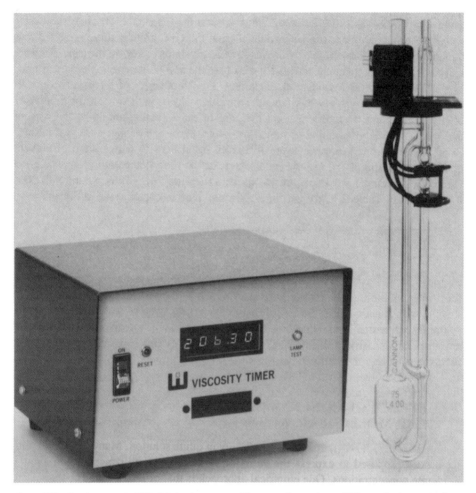

Figure 7.7   Semiautomatic Ubbelohde viscometer. (Photograph courtesy of Wescan Instruments, Inc., Santa Clara, Calif.)

The viscous (DV) and refractometric (DRI) signals for a polystyrene sample in 1,2,4-trichlorobenzene at 145°C shown in Fig. 7.9 have been corrected (normalized) for the volume delay between the detectors [16]. As one would expect, the viscosity signal peaks sooner than the concentration, since the higher molecular weights contribute more to the viscosity. Also, it can be noted that there will be an unavoidable uncertainty in the reduced viscosity, $(\ln \eta_r)/c$ at either end of the chromatogram, where the signal from one detector or the other merges with noise in the output.

The connection between Eq. (7.16) and Eqs. (7.6), (7.10), and (7.11) is made

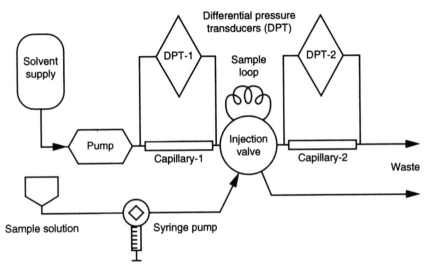

Figure 7.8 Schematic diagram of a differential viscometer (Viscotek Y-501) [14]

by assuming the polymer molecule to be a random, spherical coil of beads when the molecular weight is high enough. At infinite dilution, each molecule acts separately from other polymer molecules. In a shear field (velocity gradient du/dy, s^{-1}, or $\dot{\gamma}$), the sphere rotates with a frequency ω rad/s, where $\omega = \dot{\gamma}/2$.

The work necessary to rotate this complex sphere, part beads, part perturbed solvent, part immobilized solvent, is measured by the viscosity. Several theories [17] lead to the idea that

$$[\eta] = \Phi \frac{(\overline{r^2})^{3/2}}{M} \tag{7.21}$$

where Φ is a universal constant. When $\overline{r^2}$ (cm^2) and M (g/mol) are measured by light scattering and $[\eta]$ (dl/g) is measured under the same conditions, Φ can be evaluated. A number of such measurements gives $\Phi = 2.1 \pm 0.2 \times 10^{21}$ (dl/mol·cm^3) for unfractionated polymers. Values of 2.5 and 2.7 in place of 2.1 have been recommended for fractionated and "well-fractionated" polymers, respectively [13]. Since $(\overline{r_0^2}/M)^{1/2}$ is nearly constant (Table 7.2), it is convenient to combine Eqs. (7.10) and (7.21) to give

$$[\eta] = \Phi \left(\frac{\overline{r_0^2}}{M}\right)^{3/2} M^{1/2} \alpha^3 \tag{7.22}$$

This states that the intrinsic viscosity should vary with $M^{1/2}$ in any solvent at the Flory point ($T = \theta_f$) or that $a = 0.5$ in Eq. (7.16). At any higher temperature, α

Normalized signal

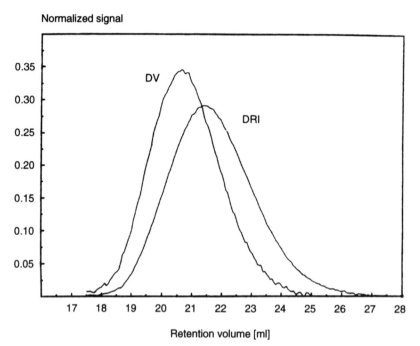

Retention volume [ml]

Figure 7.9 Normalized chromatographic signals from differential viscosmeter (DV) and differential refractive index monitor (DRI) for a wide-distribution polystyrene sample (M_n = 123,000, M_w/M_n = 2.24). [From R. Lew, D. Suwanda, S. T. Balke, and T. H. Moury, "Quantitative High-Temperature Size-Exclusion Chromatography of Polyolefins," *J. Appl. Polym. Sci.: Appl. Polym. Symp.*, 52:125 (1993); reprinted by permission of John Wiley & Sons, Inc.]

varies with M according to Eq. (7.11). If $\alpha \gg 1$, then $\alpha^5 \gg \alpha^3$ and $\alpha^3 \sim M^{0.3}$. In practice, Eq. (7.16) expresses Eq. (7.22) combined with Eq. (7.11) to a good approximation when $0.5 < a < 0.8$.

The quantity $[\eta]M$ is termed the *hydrodynamic volume* and is directly proportional to the volume occupied by each molecule according to Eq. (7.21). In Sec. 6.4 it was noted that this quantity is a useful parameter for the "universal" calibration of a chromatographic (GPC) molecular weight measurement. Molecular weight by itself does not necessarily correlate with the ability of a molecule to diffuse in and out of the porous packing. However, the size of the molecule as represented by the hydrodynamic volume is closely related to its diffusing ability. Even then, the "universal" calibration may not take into account specific interactions that may occur between the polymer being analyzed and the polymer constituting the packing. The intrinsic viscosity for purposes of calibration has to be measured in the same solvent and at the same temperature as the chromatographic separation. Thus, the absolute molecular weight of a new polymer can be estimated from the GPC elution curve provided that the apparatus is calibrated, say, with polystyrene standards of known intrinsic viscosity and molecular weight and the intrinsic viscosity of the new polymer sample is measured. A rather elegant combination consists of a differential viscometer, a light scattering instrument, and a differential refractometer, all monitoring the effluent of a GPC/SEC apparatus. A universal calibration curve, $[\eta]M$ vs. effluent volume, can be obtained directly from such an arrangement as long as the signals are appropriately calibrated and coordinated [18].

The value of α obviously characterizes the "goodness" of a solvent. Since Φ and $(\overline{r_0^2}/M)^{1/2}$ are relatively constant, the variation of $[\eta]$ with solvent for a polymer of given molecular weight is a good measure of the solvating ability of the solvent and of the volume of solution occupied per unit weight of polymer at high dilution.

Because $\overline{r_0^2}/M$ is known to be relatively insensitive to temperature or solvent character, the ratio of $[\eta]$ to $[\eta]_\theta$ (in a solvent at the Flory temperature) gives a good idea of α^3 no matter what solvent was used to measure $[\eta]_\theta$. Thus, taking values of K' and a from Table 7.6, we get for polyisobutylene in cyclohexane at 30°:

$$\alpha^3 = \frac{[\eta]}{[\eta]_\theta} = \frac{K'M^a \ (\text{cyclohexane, 30°C})}{K'M^a \ (\text{benzene, 24°C})} = 0.258M^{0.19} \qquad (7.23)$$

So for $M = 10^6$, $\alpha^3 = 3.56$, that is, the individual polymer molecules occupy almost four times the volume in cyclohexane than they do in any theta (Flory) solvent.

It was mentioned in Sec. 2.5 that polymers in contact with solvents of like solubility parameter δ should show the greatest solvation effects, i.e., swelling of networks or viscosity of solutions. Now we can identify this solvating ability with α, the expansion factor in Eq. (7.10).

Example 7.1

Polymers of monomer M are made in several molecular sizes and dissolved in solvent A or B. From the information given below, calculate the expansion factor and the hydrodynamic volume in solvent A for a polymer with a molecular weight of 10,000,000. Assume that the polymers are monodisperse.

	Intrinsic viscosity, dl/g	
Mol. wt.	In A	In B
1,00,000	4.000	2.000
200,000	1.100	—
50,000	—	0.447

Solution: Establish Mark-Houwink-Sakurada parameters.

$$\text{In A:} \quad \frac{4.000}{1.100} = \left(\frac{1,000,000}{200,000}\right)^a \qquad a = 0.802$$

$$\text{In B:} \quad \frac{2.000}{0.447} = \left(\frac{1,000,000}{50,000}\right)^b \qquad b = 0.500$$

Therefore, we surmise that B is a theta solvent. For a molecular weight of 10,000,000:

$$\text{In A: } \frac{[\eta]_A}{4.000} = \left(\frac{10,000,000}{1,000,000}\right)^{0.802}$$
$$[\eta]_A = 25.35 \text{ dl/g}$$
$$\text{In B: } \frac{[\eta]_B}{2.000} = \left(\frac{10,000,000}{1,000,000}\right)^{0.500}$$
$$[\eta]_B = 6.325 \text{ dl/g}$$

The expansion factor is:

$$\alpha = \left(\frac{[\eta]_A}{[\eta]_B}\right)^{1/3} = 1.59$$

The hydrodynamic volume is:

$$[\eta]_A M = 10,000,000 \times 25.35 = 253.5 \times 10^6 \text{ dl/mol}$$

Example 7.2

Polystyrene with a molecular weight of 1,000,000 in a theta solvent (ethylcyclohexane, 70°C) has an intrinsic viscosity of 0.75 dl/g. Calculate the steric factor.

Solution: Calculate the number of repeat units in a polymer with molecular weight 1,000,000:

$$n = \text{number of steps} = 2 \times \text{repeat units} = 2 \times \frac{1,000,000}{104}$$
$$= 19,230$$

The length of each step is 0.154 nm.

$$(\overline{r^2})^{1/2} = (2 \times 19,230)^{1/2} \times 0.154 = 30.2 \text{ nm}$$

Calculate the root-mean-square end-to-end distance from the intrinsic viscosity in a theta solvent:

$$0.75 \text{ dl/g} = 2.1 \times 10^{21} \times \frac{(\overline{r^2})^{3/2}}{1,000,000}$$
$$(\overline{r^2})^{3/2} = 3.57 \times 10^{-16} \text{ cm}^3$$
$$(\overline{r^2})^{1/2} = 7.09 \times 10^{-6} \text{ cm} = 70.9 \text{ nm}$$

The steric factor is:

$$\sigma = \frac{70.9}{30.2} = 2.35$$

7.4 EFFECT OF CONCENTRATION AND MOLECULAR WEIGHT

Equations (7.13), (7.14), and (7.15) hold only in dilute solution. Many other equations have been proposed to express the variation of viscosity with concentration even at moderate concentrations. One empirical approach is that of Lyons and Tobolsky [19], who proposed

$$\frac{\eta_{sp}}{c[\eta]} = \exp\left(\frac{k_L'[\eta]c}{1 - bc}\right) \tag{7.24}$$

In the limit of very dilute solutions, k_L' should be the same as k' in the Huggins equation [Eq. (7.13)]. Under some conditions this one equation can represent the low shear (Newtonian) viscosity of a single polymer from dilute solution to the solvent-free melt [20]. The constant b can be positive or negative. The equation resembles a theoretical relationship derived by Fujita and Kishimoto based on fractional free volume contributions by polymer and solvent [21, 22]. A simpler form, in which b is zero, was used by Martin [23] some years earlier:

$$\log \frac{\eta_{sp}}{c} = \log [\eta] + K''[\eta]c \tag{7.25}$$

This equation (see Fig. 7.5) has been very popular, because it involves only two fitted constants, yet it can correlate data to high concentrations. It, together with some other empirical equations, is treated more extensively by Ott and Spurlin [23]. As one approaches the pure molten polymer, that is, $c > 50$, the particular solvent used seems to have little bearing on the viscosity, and η(solution) = η(polymer)v_2, where v_2 is the volume fraction of polymer [24].

Combination of Eq. (7.16) with (7.13), (7.14), (7.15), or (7.25) gives the effect of molecular weight on η for solutions. For polymer melts, several approaches are suggested. Log-log plots of η versus M often can be resolved into two straight lines. Typical behavior of melts over a large range of molecular weight is shown in Fig. 7.10. The quantity plotted, $\bar{z}_w$, is equal to the weight-average degree of polymerization times the number of backbone, "chain," atoms per repeat unit. The upper section in Fig. 7.10 has a slope of 3.5 ± 0.5, and the lower section has a slope closer to 1 or 2. For each section we can write

$$\log \eta = A \log M_w + B \tag{7.26}$$

The break in the curve is thought to represent the molecular weight at which polymer entanglements become serious enough to contribute heavily to the viscosity. Single polymer molecules no longer diffuse separately, but drag along neighbors. An analogy would be the removal of a piece of string from a pile of strings. As the average length of strings increases, the difficulty of removing the string increases. Some values of this "entanglement molecular weight" are given in Table 7.7. Over a limited range of molecular weight, another relationship has some usefulness [24]:

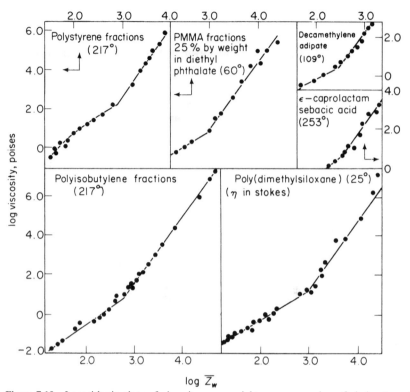

Figure 7.10 Logarithmic plots of viscosity η vs. weight-average number of chain atoms Z_w for six polymers [24]; 1 poise = 0.1 Pa·s

$$\log \eta = A + BM^{1/2} \qquad (7.27)$$

One view of polymer motion that helps to visualize the exponential dependence of viscosity on molecular size has been put forward by De Gennes [25]. A polymer in the melt state is surrounded by segments of all its neighboring polymer chains, which form a series of barriers to motion so that the polymer is confined in a kind of conduit or tube. The creeping flow of a polymer in the melt, termed *reptation* (moving like a reptile), is easier along the tube formed by these neighbors than it is sideways, where simultaneous cooperative motion of many segments is needed for effective movement. Under these conditions, the diffusion of the long, wormlike chain of length L through the tube of length L can be shown to require a maximum relaxation time proportional to L^3. The viscosity of the melt, being proportional to the maximum relaxation time, also will *scale with* (depend on) molecular length to the third power. Since the experimental value is 3.4, the reptation theory has been accepted by many as a reasonable view.

Table 7.7 Applicability of the empirical law log $\eta = 3.4 \log Z_w + K$ to long polymer chains [24]

Polymer solvent pair	[Polymer], wt fraction	$v_2 Z_c$ at break[*]	Highest Z studied	Temperature range studied, °C	Polymers and mol wt detm'n[†]
		Range of applicability			
	Linear nonpolar polymers				
Polyisobutylene	1.0	610	53,000	−9 to 217	$f:(\eta)$
Polyisobutylene-xylene	0.05 to 0.20	~1400	140,000	25	$W \& f:(\eta)$
Polyisobutylene-decalin	0.05 to 0.20	–	140,000	25	$W \& f:(\eta)$
Polystyrene	1.0	730	7,300	130 to 217	$f:(\eta)$
Polystyrene-diethyl benzene	0.14 to 0.44	~1000	22,500	30 to 100	$f:(\eta)$
Poly(dimethylsiloxane)	1.0	~950	34,000	25	$W:\pi$, L.S.
	Linear polar polymers				
Poly(methyl methacrylate)-diethyl phthalate	0.25	208	24,000	60	$f:(\eta)$
Decamethylene sebacate	1.0	290	960	80 to 200	$W:E$
Decamethylene adipate	1.0	280	1,320	80 to 200	$W:E$
Decamethylene succinate	1.0	290	990	80 to 200	$W:E$
Diethylene adipate	1.0	290	1,190	0 to 200	$W:E$
ω-Hydroxy undecanoate	1.0	<326	1,443	90	$W:E$
	Effect of branching				
Poly(ϵ-caprolactam):					
Linear	1.0	324	2,300	253	$W:E$
Tetrachain	1.0	390	1,560	253	$W:E$
Octachain	1.0	550	2,000	253	$W:E$

[*]Z_c is the lowest chain length for which equation is valid for the solution wherein v_2 is the volume fraction of polymer.

[†]f = fraction, W = whole polymer. Chain lengths determined from η = intrinsic viscosity, π = osmotic pressure, L.S. = light scattering data, E = end group analysis.

Example 7.3

A monodisperse polymer with a molecular weight of 133,000 has an intrinsic viscosity in a theta solvent at 35°C of 2.10 dl/g and a melt viscosity at 200°C of 250 kPa·s. What will be the value of the intrinsic viscosity (same solvent and temperature as before) for a new sample made from the same monomer but with a melt viscosity (at 200°C) of 835 kPa·s?

Solution: Assume Eq. (7.26) with a slope of 3.5 and molecular weight M_2 for the second sample:

$$\log\left(\frac{825}{250}\right) = 3.5 \log\left(\frac{M_2}{133,000}\right)$$
$$M_2 = 187,000$$

In a theta solvent:

$$\frac{[\eta]_2}{[\eta]_1} = \left(\frac{187}{133}\right)^{1/2}$$
$$[\eta]_2 = 2.49 \text{ dl/g}$$

Melt flow is often used to characterize molecular size for polyolefins and for many rubbers. Melts are easier to handle than the very hot solutions needed for polyolefins. In the case of some rubbers, the presence of nonrubber ingredients or gels makes solutions difficult or impossible to prepare.

The molecular size of polyolefins conventionally is reported in terms of the *melt index,* referring to the grams of polymer extruded in 10 min through a die that is 2.1 mm in diameter and 8.0 mm in length. The temperature and pressure must be specified [ASTM D1238]. Instead of weighing the extrudate, the rate of movement of the plunger forcing the polymer through the die can be sensed and indicated (Fig. 7.11). Typically, a force of 298 kPa is used at a temperature of 190°C for polyethylene and 230°C for polypropylene. The actual test time can vary from 15 s to several minutes, but the result is expressed as the flow that would take place in a 10-min period. Melt indexes of 0.1 to 50 are common. The test has been criticized because the low length-to-diameter ratio in the die allows elastic effects to mask differences in melt viscosity (see Sec. 7.8), but it still is widely used in commercial practice.

The molecular size of rubber, both natural and synthetic, often is reported in terms of a viscosity measured in a particular *rotational* viscometer. *Mooney viscosity* refers to the torque in arbitrary units on a disk rotating in a confined melt. The temperature and apparatus usually are specified when the test is reported (ASTM D1646).

7.5 EFFECT OF TEMPERATURE AND PRESSURE

Viscous flow can be pictured as taking place by the movement of molecules or segments of molecules in jumps from one place in a lattice to a vacant hole. The total "hole concentration" can be regarded as space free of polymer, or "free volume." Doolittle proposed [26] that the viscosity should vary with the free volume in the following way:

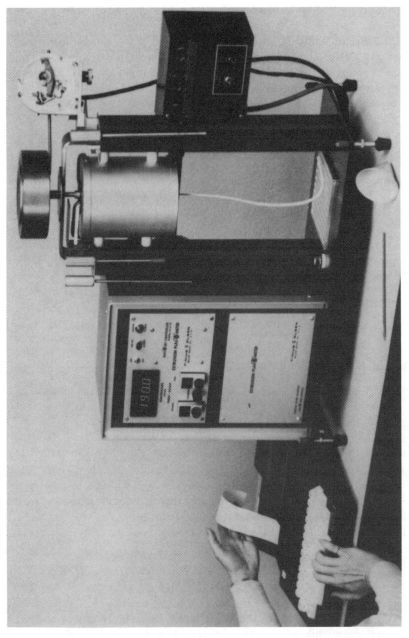

Figure 7.11 Melt Indexer. (Photograph courtesy of Tinius Olson Testing Machine Co., Willow Grove, Pa.)

247

$$\log \eta = A + \frac{BV_0}{V_f} \tag{7.28}$$

where A and B are constants, V_f is the *free volume* equal to $V - V_0$, V is the specific volume (per unit weight), and V_0 is the specific volume extrapolated from the melt to 0 K without change of phase. The fractional free volume $f' = V_f/V$ is expected to vary with temperature as follows:

$$f' - f'_g = \alpha \Delta T \tag{7.29}$$

where α is the difference between the expansion coefficients for melt and glass, $\Delta T = T - T_g$, and f'_g is the fractional free volume at T_g. For many polymers α is about $4.8 \times 10^{-4} \text{K}^{-1}$. Also, Williams, Landel, and Ferry [27] found empirically that

$$\log\left(\frac{\eta}{\eta_g}\right) = \frac{-17.44(T - T_g)}{51.6 + (T - T_g)} = \frac{-17.44\Delta T}{51.6 + \Delta T} \tag{7.30}$$

where η_g is the viscosity at T_g. We can combine Eqs. (7.28), (7.29), and (7.30) if we replace V_0 in Eq. (7.28) by V. This is not too great a change, since, unlike V_f, V does not change rapidly with temperature. First we write Eq. (7.28) as the difference of two conditions to eliminate the constant A.

$$\log \frac{\eta}{\eta_g} = -\frac{B(f' - f'_g)}{f_g f'} \tag{7.31}$$

Equation (7.29) is used to eliminate f':

$$\log \frac{\eta}{\eta_g} = \frac{(-B\alpha\Delta T)}{f'_g(f'_g + \alpha\Delta T)} = \frac{(-B/f'_g)\Delta T}{(f'_g/\alpha) + \Delta T} \tag{7.32}$$

Identification of Eq. (7.32) with Eq. (7.30) shows that $f'_g = 51.6\alpha$ or 0.025. The conclusion is drawn that the glass transition occurs for all materials (polymers and inorganic glasses, low-molecular-weight compounds) at a temperature where the fractional free volume is 2.5% of the total, since the empirical relationships used have been found to apply for all these materials.

When $T - T_g > 100°C$, Eq. (7.30) does not work as well as the more general one [24]:

$$\log \frac{\eta}{\eta_R} = B'\left(\frac{1}{T^a} - \frac{1}{T_R^a}\right) \exp\left(-\frac{\beta}{M}\right) \tag{7.33}$$

where η_R is the viscosity at a reference temperature T_R for molecular weight M and a usually is an integer, while B' and β are fitted constants. Some typical results are seen in Fig. 7.12.

One would expect the compression of a melt to decrease in the free volume

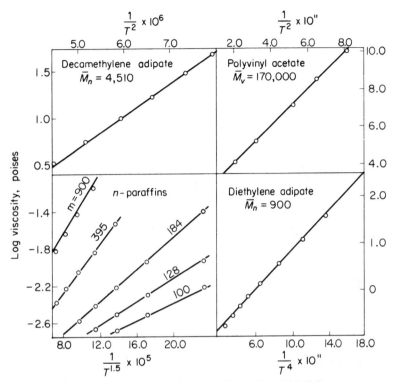

Figure 7.12 Viscosity-temperature curves according to Eq. (7.33) [24]

and consequently increase the viscosity. It can be seen that the viscosity of polystyrene is affected more by pressure than that of polyethylene (Fig. 7.13). Because increasing pressure will increase the glass transition temperature, it is possible to use Eq. (7.30) with a pressure-dependent T_g to correlate viscosity if one knows the form of the dependence [29, 30]. For several polymers, $d(T_g)/dP$ is reported to be in the range of 0.16 to 0.43°C/MPa. Polystyrene, for example, has a value of 0.3°C/MPa [29]. It is harder to generalize about the temperature and pressure dependence of polymer solutions. Also, logarithmic plots of viscosity vs. concentration are not linear, especially in concentrated solutions (Fig. 7.14).

7.6 MODELS FOR NON-NEWTONIAN FLOW

It has been assumed until now that the viscosity is a constant for any polymer with a given molecular weight, solvent, concentration, temperature, and pressure. Two

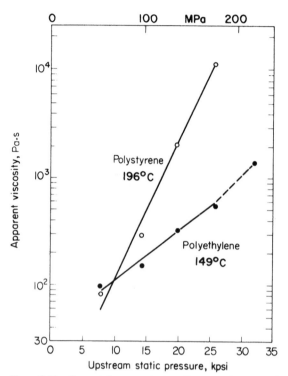

Figure 7.13 Corrected apparent viscosity vs. upstream static pressure for polyethylene and polystyrene [28]

more variables are the rate at which deformation takes place and the length of time at a given rate of shear. For even if we hold the rate of deformation constant, the viscosity may increase or decrease with time. If the effect is temporary and viscosity goes back to its original value (after the system has relaxed long enough with no deforming forces imposed), we call the behavior *rheopexy* (viscosity increases with time of deformation) or *thixotropy* (viscosity decreases with time of deformation) (Fig. 7.15). One expects this behavior when viscosity is partially due to intermolecular structures that take some time to be established or destroyed. Ionic polymers in water tend to be thixotropic, because the polymers probably aggregate when relaxed but disperse and act individually when sheared.

The variation of viscosity with rate of deformation can be shown by a combination of any two variables τ, $\dot{\gamma}$, or η, since η can be defined as a variable by Eq. (7.3). Behavior can be shear-thickening (viscosity increases with $\dot{\gamma}$ or τ) or shear-thinning (viscosity decreases with $\dot{\gamma}$ or τ) (Fig. 7.16). Most examples of shear-thickening are particulate dispersions such as latexes, slurries, and concentrated

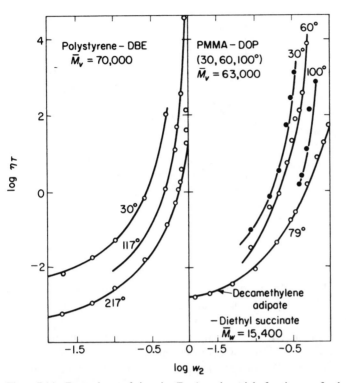

Figure 7.14 Dependence of viscosity (Pa·s) on the weight fraction w_2 of polymer in solution at various temperatures. DBE is dibenzyl ether, PMMA-DEP is poly(methyl methacrylate) in diethyl phthalate [24]

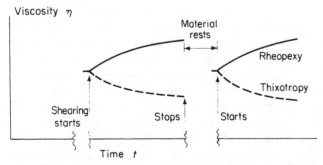

Figure 7.15 Schematic behavior of time-dependent viscosity, which increases (rheopectic behavior) or decreases (thixotropic behavior) with time during shearing at a steady rate. Resting the material allows the pattern to be observed again

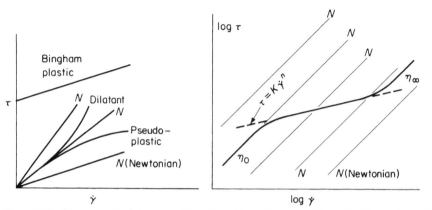

Figure 7.16 Stress vs. rate of shear on arithmetic and logarithmic coordinates for Newtonian and non-Newtonian fluids

suspensions. On the other hand, in some slurries and concentrated suspensions there may be almost no flow, $\dot{\gamma} = 0$, until τ reaches some "yield value" (Fig. 7.16). Ordinary toothpaste approaches this "Bingham plastic" behavior.

The typical behavior of polymer solutions and melts is shear-thinning. Often the older terms *dilatant* (for shear-thickening) and *pseudoplastic* (for shear-thinning) will be encountered. The development of a single mathematical model to express the behavior of all such materials has not been successful. One of the simplest approaches [31] is to approximate sections of log τ–log $\dot{\gamma}$ plots by straight lines—the "power-law" model (Fig. 7.16):

$$\tau = K\dot{\gamma}^n \qquad \eta = K\dot{\gamma}^{n-1} \tag{7.34}$$

It can be seen from flow curves for a polymer melt (Fig. 7.17) that Eq. (7.34) would serve well over several decades of shear rate but would fail as the melt approaches Newtonian behavior. Even in polymer solutions (Figs. 7.18 and 7.19) the power law is adequate over a decade or so in shear rate. In the case of the solutions, the lines are drawn according to a different mathematical model [32].

The power-law model is attractive in its simplicity. However, as stress or rate of shear becomes small, the viscosities of almost all polymer melts, solutions, latexes, and some other dispersions appear to approach a constant value η_0, the zero-shear-rate viscosity (Fig. 7.16). In the case of the power law, as τ or $\dot{\gamma}$ goes to zero, viscosity becomes infinite when n is less than 1. The Ellis model is a mathematical combination of Newtonian behavior at low values of τ, and power-law behavior at high values of τ [34].

$$\frac{\eta_0}{\eta} - 1 = k\tau^a \qquad \text{("Ellis")} \tag{7.35}$$

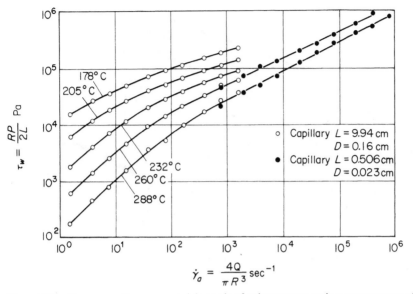

Figure 7.17 Flow curves for a commercial sample of polystyrene at various temperatures obtained in a ram-extrusion type of capillary viscometer [1]

or

$$\eta_0 \dot{\gamma} = \tau + k\tau^{1+a} \tag{7.36}$$

For polymer solutions, there is evidence that the viscosity approaches a constant value less than η_0 but greater than that of the solvent as the rate of shear becomes very large. In most two-parameter models, this viscosity at infinite shear rate η_∞ is zero (Fig. 7.17).

By going to three or more parameters, there is a better possibility of including η_∞ as a finite number. The Ellis model is an exception, since η_∞ must be zero. However, the Williamson [35], Krieger-Dougherty [36], Eyring-Powell [37], and Reiner-Philippoff [38] models can be written with finite values for η_0 and η_∞.

Some years ago, Denny and Brodkey proposed an equation based on kinetic theories which included both η_0 and η_∞ [39]. The equation contains four other parameters. However, when two of them are equal to each other, the equation reduces to

$$\frac{\eta_0 - \eta}{\eta - \eta_\infty} = (\theta_t \dot{\gamma})^a \tag{7.37}$$

A few years later, Cross used this equation to correlate the flow properties of silicones and polyurethanes in both steady and oscillatory flows [40]. The constant θ_t has the units of time, and Cross demonstrated that it can be regarded as a *charac-*

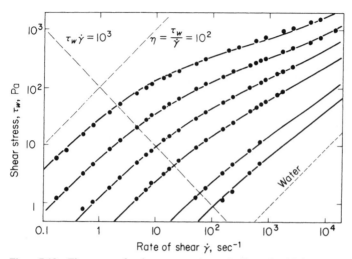

Figure 7.18 Flow curves for six concentrations of polyacrylamide in water. Lines are drawn according to a model [32]

teristic relaxation time for a non-Newtonian fluid [41]. In practice, η_∞ often is negligible, so only three constants need to be fitted to experimental data.

$$\frac{\eta_0}{\eta} = 1 + (\theta_t \dot{\gamma})^a \tag{7.38}$$

Equation (7.38) appears on logarithmic coordinates (Fig. 7.20) as two straight lines corresponding to the Newtonian regime and the power-law regime joined by a transition region. The reduced shear rate $(\theta_t \dot{\gamma})$ is unity when $\eta = \eta_0/2$. The limiting line at very high shear rates is a power law [Eq. (7.34)] in which $n = (1 - a)$. A very similar diagram results when the Ellis model is plotted with stress replacing the rate of shear on the abscissa.

Example 7.4

Assume that the flow data for the most concentrated polyacrylamide solution in Fig. 7.18 can be approximated by the Cross model [Eq. (7.38)] with a zero shear viscosity of 25 Pa·s. What is the "characteristic relaxation time," θ_t?

Solution: The stresses at shear rates of 10 and 10^4 s^{-1} are roughly 130 and 1700 Pa, respectively. The corresponding apparent viscosities are 13 and 0.17 Pa·s. In Eq. (7.38):

$$\frac{25}{13} = 1 + (10\theta_t)^a \qquad \text{and} \qquad \frac{25}{0.17} = 1 + (10^4\theta_t)^a$$

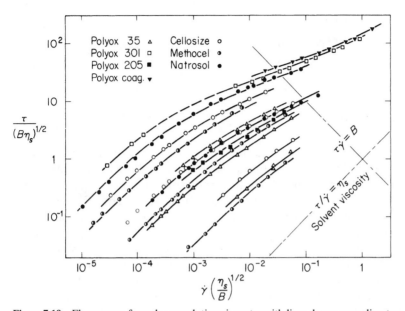

Figure 7.19   Flow curves for polymer solutions in water with lines drawn according to a model [33]

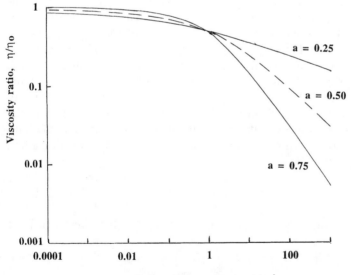

Figure 7.20   Equation (7.38) plotted on logarithmic coordinates for three values of the exponent a

Eliminating θ_t by taking ratios yields:

$$\left(\frac{10^4}{10}\right)^a = \frac{146}{0.92} \qquad a = 0.73 \qquad \frac{1}{a} = 1.37$$

Then

$$(10\theta_t) = \left(\frac{25}{13} - 1\right)^{1.37} \qquad \text{and} \qquad \theta_t = 0.090 \text{ s}$$

Often only a narrow range of shear rates or stresses is important. In molecular weight characterization, η_0 is used. In capillary flow at high shear rates, the shear stress falls off rapidly away from the wall. Whether the viscosity at the centerline is finite or infinite may have little effect on any calculations. Thus the power law can be quite adequate in calculating such things as true shear rate at the wall for a non-Newtonian fluid.

In most cases, η_0 can be estimated with more confidence than η_∞. Mechanical degradation [42], static pressure effects [28], melt fracture [43], kinetic energy effects [2], or mechanical heating [44] may obscure high-shear-rate data. On the other hand, a reliable measure of η_0 usually is limited only by the patience of the investigator and the long-term stability of the system he is measuring. It follows that η_0 should precede η_∞ as a parameter in most models. Of course, many values of η_0 in the literature were not measured directly but are extrapolated from other data according to an assumed model.

The various models can be manipulated to provide methods for deriving η_0 or $(\eta_r)_0$ from data at finite shear rates. A plot of $1/\eta_r$ vs. τ^a is suggested by the Ellis model, with a increasing from about 0.6 at $\log (\eta_r)_0 = 1$ to $a = 1.4$ at $\log(\eta_r)_0 = 5$. The intercept is $1/(\eta_r)_0$. For determination of $(\eta_r)_\infty$, a plot of $1/[(\eta_r)_0 - (\gamma_r)]$ vs. τ^{-2} or $1/\dot\gamma$ may be helpful. The intercept, of course, is $1/[(\eta_r)_0 - (\eta_r)_\infty]$ in either case.

To the engineer, a mathematical model for viscosity is a means to an end. Among other quantities, bulk flow rates in pipe and other conduits are usually desirable. The simple problem of deriving an equation for the velocity profile in laminar flow becomes a formidable task with some models. The popularity of the power law is due in good measure to its tractability in mathematical manipulations.

A radial velocity distribution in round pipes can be obtained by integration, making use of the linear stress distribution:

$$\tau = -\frac{r}{2} \frac{\Delta P}{L} \tag{7.39}$$

where $\Delta P/L$ is the pressure drop per unit length of pipe (usually a constant). Then

$$\int_0^u du = -\int_{\tau_w}^\tau \dot\gamma \, dr = \frac{2}{\Delta P/L} \int_{\tau_w}^\tau \dot\gamma \, d\tau \tag{7.40}$$

It can be seen that, in the absence of an explicit equation, the velocity distribution can be obtained by graphical or numerical integration.

All these manipulations assume that data corresponding to $\dot{\gamma}$ and τ are available. With data from most rotational viscometers this is nearly the case if $\dot{\gamma}$ and τ are evaluated at the geometric mean radius (see Sec. 7.7).

In a capillary viscometer the "true" rate of shear at the wall $\dot{\gamma}_w$ is related to that for a Newtonian fluid $(\dot{\gamma}_w)_a$ [45, 46]:

$$\dot{\gamma}_w = \frac{3 + z}{4}(\dot{\gamma}_w)_a \tag{7.41}$$

where z is $d(\log \dot{\gamma})/d(\log \tau)$ at the point in question, $(\dot{\gamma}_w)_a = (4/\pi)(Q/r^3)$, and Q is the volume rate of flow.

Many authors have fitted capillary data to empirical equations using the "apparent" rate of shear $(\dot{\gamma}_w)_a$. Of course, this method is perfectly adequate as long as all operations are to be carried out in geometrically similar conduits. However, the use of such correlations in different geometries demands a rather laborious conversion of information to "true" rate of shear.

7.7 MEASUREMENT OF VISCOSITY

The parallel-plate geometry of Fig. 7.1 is seldom used in the routine measurement of viscosity. However, the behavior of capillary and rotational viscometers can be analyzed by using the same model despite the apparent differences in operation. In Fig. 7.21, three typical viscometer geometries are pictured as moving liquid layers, so that the analogy with our definition is apparent. Only where inertial effects can be ignored is such a simplification possible. In the case of the rotational viscometers we measure a rate of rotation Ω in radians per second and a torque $\mathcal{T}$ in Newton·meters (1 N·m = 10^7 dyn·cm). The average volumetric flow dV/dt and the pressure drop per unit length $\Delta P/L$ are measured in the capillary or pipe. The torque or pressure drop is proportional to the shear stress, and the rate of rotational or flow rate is proportional to the rate of shear. In calculating the stress, only geometric parameters enter into the proportionality constant. For the rate of shear, the model that relates stress to rate of shear may enter in. The expressions for stress and rate of shear will now be derived by considering, for each geometry, (1) a force balance involving shear stress at radius r; (2) an expression for the shear γ_r, which can then be differentiated to give the rate of shear $\dot{\gamma}_r$, also at a radius r; and (3) insertion of a general expression relating $\dot{\gamma}$ to τ, the power-law model:

$$\dot{\gamma} = m\tau^z \tag{7.42}$$

Cup-and-Bob Viscometer (Figs. 7.21 and 7.22)

Bob radius = R_1 cm   Bob length = L cm
Cup radius = R_2 cm   Rotational velocity of cup = Ω rad/s
Measured torque = $\mathcal{T}$, N·m (independent of radius)

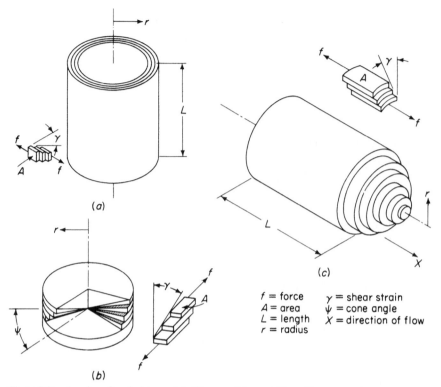

Figure 7.21   Laminar flow in (a) coaxial, (b) cone-plate, and (c) capillary geometry

(a) *Stress at radius r.* Consider a cylinder with radius $R_1 < r < R_2$. The force f dragging this cylinder of area $A = 2\pi rL$ is equal to the shear stress times the area. The torque is equal to rf.

$$f = 2\pi r L \tau_r = \frac{\mathcal{T}}{r} \tag{7.43}$$

$$\tau_r = \frac{\mathcal{T}}{2\pi r^2 L} \tag{7.44}$$

(b) *Shear and shear rate at radius r.* Regarded as a parallel-plate viscometer, the shear γ_r at radius r for a differential angular rotation of $d\theta$ is

$$\gamma_r = r\frac{d\theta}{dr} \tag{7.45}$$

The rate of shear becomes

$$\dot{\gamma}_r = \frac{d\gamma_r}{dt} = r\frac{d(d\theta/dr)}{dt} = r\frac{d^2\theta}{dr\,dt} = r\frac{d(d\theta/dt)}{dr} \tag{7.46}$$

$$\frac{V_\phi}{r} = \frac{\mathcal{T}}{4\pi\eta L}\left(\frac{1}{r^2} - \frac{1}{R_2}\right)$$

$$\text{at } r = R_1$$

$$V_\phi = R_1\Omega$$

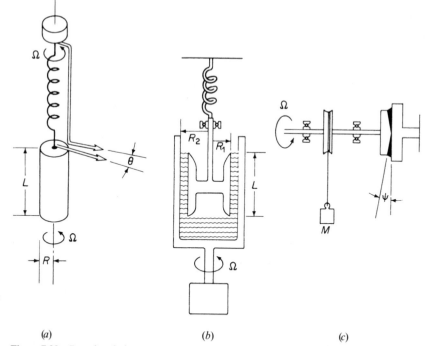

(a)   (b)   (c)

Figure 7.22 Rotational viscometers: (a) extremely long cylinder in infinite volume of fluid; (b) coaxial cylinder (cup and bob); (c) cone-plate. Rate of rotation Ω is in radians per second

For $d\theta/dt$ we can write Ω, the rotational velocity in radians per second, so that

$$\dot{\gamma}_r = r \frac{d\Omega}{dr} \tag{7.47}$$

(c) *Insertion of model.* With the cup rotating at Ω_c, we combine Eqs. (7.42), (7.44), and (7.47):

$$\dot{\gamma}_r = r \frac{d\Omega}{dr} = m\tau_r^z = m\left(\frac{\mathcal{T}}{2\pi r^2 L}\right)^z = C_z r^{-2z} \tag{7.48}$$

where

$$C_z = m\left(\frac{\mathcal{T}}{2\pi L}\right)^z \tag{7.49}$$

Assuming that m and z are independent of r and Ω, we can integrate from R_1 to r:

$$\int_0^{\Omega_r} d\Omega = C_z \int_{R_1}^r r^{-(2z+1)} \, dr \tag{7.50}$$

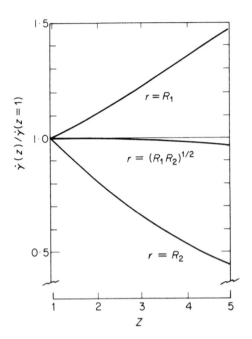

Figure 7.23 Relative change in rate of shear $\dot{\gamma}$ with increasing shear-thinning behavior [increasing $z = d(\ln \dot{\gamma})/d(\ln \tau)$] for $R_1/R_2 = 0.9$

$$\Omega_r = \frac{C_z(r^{-2z} - R_1^{-2z})}{-2z} \tag{7.51}$$

At $\Omega_r = \Omega_c$,

$$\Omega_c = \frac{C_z(R_1^{-2z} - R_2^{-2z})}{2z} \tag{7.52}$$

and eliminating C_z between Eqs. (7.48) and (7.52) we get

$$\dot{\gamma}_r = \frac{2z\Omega_c r^{-2z}}{R_1^{-2z} - R_2^{-2z}} \tag{7.53}$$

Although Eqs. (7.44) and (7.53) can be applied at any value of r, the advantage of using a geometric-mean radius $r = (R_1R_2)^{1/2}$ is apparent in Fig. 7.23. Equations (7.44) and (7.53) become

$$\tau_{gm} = \frac{\mathcal{T}}{2\pi R_1 R_2 L} \tag{7.54}$$

$$\dot{\gamma}_{gm} = \frac{2z\Omega_c}{(R_2/R_1)^z - (R_1/R_2)^z} \tag{7.55}$$

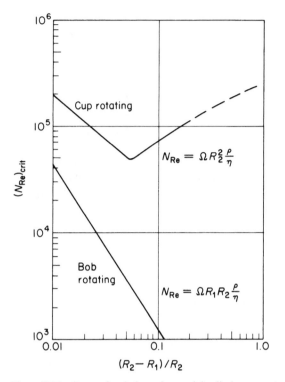

Figure 7.24 Onset of turbulence in coaxial cylinder geometry [47]

Repetition of the derivation with the bob rotating at Ω_b and the cup stationary will give the same expressions with Ω_b replacing Ω_c. The two cases are not equivalent, however, because turbulence is induced in the system at lower values of Ω for the bob rotating than for the cup rotating. In the former case, the material on the surface of the bob is rotating faster than any other in the system. The centrifugal force eventually throws material outward and causes *Taylor turbulence*, in which a helical path is followed by fluid particles and a series of toroids are formed in the annular space. The limiting values of stability for Newtonian fluids are correlated by using a Reynolds number as indicated in the diagram (Fig. 7.24).

A useful extension of the cup-and-bob viscometer is the bob rotating in an infinite amount of material (Fig. 7.22). Setting $R_2 = \infty$ and solving Eqs. (7.44) and (7.53) at R_1, we have:

$$\tau_{R_1} = \frac{\mathcal{T}}{2\pi R_1^2 L} \tag{7.56}$$

$$\dot{\gamma}_{R_1} = 2z\Omega_b \tag{7.57}$$

For a very long bob rotating in a Newtonian fluid ($z = 1$), the rate of shear depends only on the rate of rotation and is independent of viscosity, length, and radius.

Example 7.5

A coaxial cylinder viscometer is available with radii $R = 1.50$ cm and 1.60 cm. It is calibrated by measuring the torque needed to rotate the inner cylinder at 200 rpm in a Newtonian standard oil of viscosity $= 12.5$ Pa·s. The torque is read as 92.5% of full scale on a meter.

(a) What are the power-law parameters for a polymer solution which gives readings of 40.0 and 72.0% of full scale at rotational rates of 50 and 200 rpm?

(b) What is the rate of shear at the inner surface when the polymer solution is measured at 200 rpm? How does this compare with the geometric mean value?

Solution:

(a) The rate of shear at 200 rpm for geometric mean radius when $z = 1$ is:

$$\Omega_c = 200 \times \frac{2\pi}{60} = 20.9 \text{ s}^{-1}$$

$$\dot{\gamma}_{gm} = 2 \times 20.9 \left(\frac{1.60}{1.50} - \frac{1.50}{1.60} \right) = 324 \text{ s}^{-1}$$

For oil:

$$\tau_{gm} = 12.5 \times 324 = 4050 \text{ Pa}$$

so

$$\text{Full scale} = \frac{4050}{0.925} = 4380 \text{ Pa}$$

For the solution, ignoring the correction for $z > 1$:

$$\tau_{gm} = 0.40 \times 4380 = 1750 \text{ Pa} \qquad \text{at } \dot{\gamma}_{gm} = 81 \text{ s}^{-1}$$
$$\tau_{gm} = 0.72 \times 4380 = 3160 \text{ Pa} \qquad \text{at } \dot{\gamma}_{gm} = 324 \text{ s}^{-1}$$
$$\frac{3,160}{1,750} = \left(\frac{324}{81} \right)^n \qquad n = 0.425 \qquad \text{so} \qquad z = \frac{1}{n} = 2.35$$

$$K = \frac{1750}{(81)^{0.425}} = 270 \qquad \text{(where Pa and s}^{-1} \text{ are used)}$$

(b) Using Eq. (7.53):

$$\dot{\gamma}_1 = 2 \times 2.35 \times 20.9 \left[1 - \left(\frac{1.50}{1.60} \right)^{4.70} \right] = 375 \text{ s}^{-1}$$

This is 1.16 times the geometric mean value. The geometric mean value is not changed much by including z:

$$\dot{\gamma}_{gm} = 2 \times 2.35 \times 20.9 \left[\left(\frac{1.60}{1.50} \right)^{2.35} - \left(\frac{1.50}{1.60} \right)^{2.35} \right] = 323 \text{ s}^{-1}$$

Cone-and-Plate Viscometer (Figs. 7.21 and 7.22)

Radius = R cm
Cone angle = ψ rad (preferably very small, say less than 0.1 rad; 1 rad = 57.296°)
Measured torque = $\mathcal{T}$ N·m
Cone rotates at Ω rad/s

(a) *Stress at radius r.* Consider a ring of material of radius r and width dr. The force df dragging this ring past the ring below or above it is the shear stress times the area. The torque is equal to $r\, df$. It will be shown shortly that τ_r is independent of r, so that we can integrate from 0 to R to get the total torque.

$$df = 2\pi r \tau_r\, dr = \frac{d\mathcal{T}}{r} \tag{7.58}$$

$$\int_0^{\mathcal{T}} d\mathcal{T} = \int_0^R 2\pi r^2 \tau_r\, dr \tag{7.59}$$

$$\tau_r = \frac{3\mathcal{T}}{2\pi R^3} \tag{7.60}$$

(b) *Shear and rate of shear at radius r.* Regarded as a parallel plate viscometer, the shear $d\gamma_r$ at radius r for a differential angular rotation of $d\theta$ is

$$d\gamma_r = \frac{r\, d\theta}{\text{thickness}} = \frac{r\, d\theta}{r\psi} = \frac{d\theta}{\psi} \tag{7.61}$$

The rate of shear also is independent of r:

$$\dot{\gamma}_r = \frac{d\gamma_r}{dt} = (\psi)^{-1} \frac{d\theta}{dt} = \frac{\Omega}{\psi} \tag{7.62}$$

This vindicates the statement made previously that τ_r should be independent of r.

Insertion of a model is not necessary, since Eqs. (7.60) and (7.62) represent the desired solutions. The popularity of the cone-and-plate viscometer is due in no small measure to the simplicity of analysis that results from stress and rate of shear being independent of r.

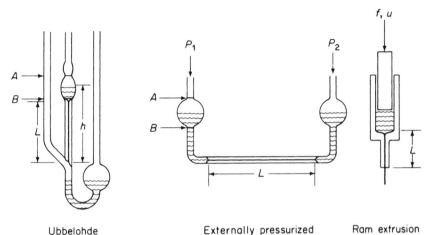

Figure 7.25 Capillary viscometers. Stress is proportional to h, $P_1 - P_2$, or ram force f. Rate of shear is proportional to $1/$(time of flow between marks) or to ram velocity u

Capillary Viscometer (Figs. 7.21 and 7.25)

Radius $= R$ cm
Pressure drop $= \Delta P/L$ Pa/cm $= (P_1 - P_2)/(x_2 - x_1)$ by convention
Flow rate $= Q$ cm^3/s

$$Q = \frac{\pi R^4 \Delta P}{8 \eta L}$$

(a) *Stress at radius r.* Consider a cylinder of material of radius r. The force f dragging this cylinder of area $A = 2\pi r L$ is equal to the shear stress times the area. The force also is equal to the pressure on the plug of radius r times the normal area of the plug πr^2.

$$f = 2\pi r L \tau_r = \Delta P \pi r^2 \qquad (7.63)$$

$$\tau_r = \frac{\Delta P r}{2L} \qquad (7.64)$$

(b) *Shear and rate of shear at radius r.* Regarded as a parallel-plate viscometer, the shear γ_r at radius r for a differential translation dx in the x direction (parallel to the axis of the tube) is

$$\gamma_r = -\frac{dx}{dr} \qquad (7.65)$$

The rate of shear becomes the velocity gradient, since the velocity is $u_r = (dx/dt)_r$:

$$\dot{\gamma}_r = \frac{d\gamma_r}{dt} = -\frac{d^2x}{dr\,dt} = -\left(\frac{du}{dr}\right), \tag{7.66}$$

(c) *Insertion of model.* We combine Eqs. (7.42), (7.64), and (7.66):

$$\dot{\gamma}_r = \left(-\frac{du}{dt}\right)_r = m\left(\frac{\Delta P}{2L}\right)^z r^z = \psi' r^z \tag{7.67}$$

where $\psi' = m(\Delta P/2L)^z$. Assuming m and z to be independent of r (not always a good assumption) and assuming $u = 0$ at $r = R$, we can integrate

$$\int_0^{u_r} -du = \psi' \int_R^r r^z\,dr \tag{7.68}$$

$$u_r = \frac{\psi'(R^{z+1} - r^{z+1})}{z + 1} \tag{7.69}$$

At $r = 0$, $u = u_m$:

$$u_m = \frac{\psi'(R^{z+1})}{z + 1} \tag{7.70}$$

Eliminating ψ' between Eqs. (7.69) and (7.70), we get

$$u_r = u_m\left[1 - \left(\frac{r}{R}\right)^{z+1}\right] \tag{7.71}$$

The average velocity $\bar{u}$ is given by the flow rate per unit cross-sectional area and also by integrating u_r for each ring of area $2\pi r\,dr$ and dividing by the total cross-sectional area:

$$\bar{u} = \frac{Q}{\pi R^2} = \int_0^R u_r \frac{2\pi r\,dr}{\pi R^2} \tag{7.72}$$

Using Eq. (7.71) for u_r, and integrating, we get $\bar{u}$ in terms of u_m:

$$\bar{u} = u_m \frac{z + 1}{z + 3} \tag{7.73}$$

Eliminating ψ' between Eqs. (7.67) and (7.70) yields:

$$\dot{\gamma}_r = u_m \frac{(z + 1)r^z}{R^{z+1}} \tag{7.74}$$

Substituting for u_m from Eqs. (7.72) and (7.73) gives the shear rate at any radius r in terms of the flow rate, the radius, and z:

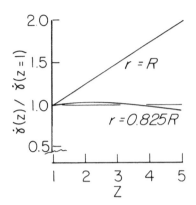

Figure 7.26 Change in true rate of shear $\dot\gamma$ compared to that for a Newtonian fluid ($z = 1$), $\dot\gamma_{app}$, at the wall and at an intermediate radius for tube flows

$$\dot\gamma_r = \frac{(z + 3)(Q)(r/R)^z}{\pi R^3} \tag{7.75}$$

At the wall, rate of shear and stress are at a maximum so that

$$\dot\gamma_w = \frac{(z + 3)Q}{\pi R^3} \tag{7.76}$$

$$\tau_w = \frac{\Delta P R}{2L} \tag{7.77}$$

It can be seen that the true rate of shear at the wall differs from that calculated for a Newtonian fluid with the same flow rate Q by the factor $(z + 3)/4$, often called the Rabinowitsch correction [48]. Actually, it is not necessary to invoke the power-law model as long as z is defined as $d(\ln Q)/d[\ln (\Delta P/L)]$ [49]. Cram and Whitwell [45] pointed out that by referring the stress and rate of shear to a radius other than the wall, the correction factor could be minimized, but not so successfully as in the concentric cylinder geometry. At a radius that is 82.5% of the radius at the wall, stress and rate of shear are given by Eqs. (7.64) and (7.65) as:

$$\tau^+ = \frac{0.825(R\,\Delta P)}{2L} \tag{7.78}$$

$$\dot\gamma^+ = \frac{(z + 3)Q(0.825)^z}{\pi R^3} \tag{7.79}$$

where reference radius is $R^+ = 0.825R$. The apparent rate of shear is:

$$\dot\gamma_{app}^+ = \frac{4Q(0.825)}{\pi R^3} \tag{7.80}$$

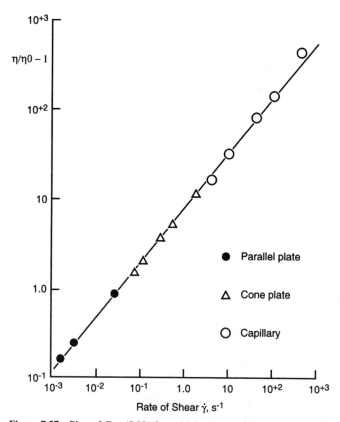

Figure 7.27 Plot of Eq. (7.38) for a high-molecular-weight poly(dimethylsiloxane) measured using parallel-plate, cone-plate, and capillary viscometers at 23°C. The fitted parameters are: $\theta_t = 34.5$ s, $a = 0.604$, and $\eta_0 = 98.0$ kPa·s

Out to a value of $z = 4$ (Fig. 7.26), there is a maximum difference of only 3.2% between the apparent rate of shear (which requires no knowledge of z) and the corrected rate of shear. Unfortunately, most practicing rheologists ignore the advantage of this method and continue to refer measurements to wall conditions rather than to an intermediate radius.

In order to characterize a polymer melt over a very large range of shear rates, several measuring instruments may be combined. In the example of Fig. 7.27, an uncross-linked silicone rubber was measured using three separate instruments which covered a wide range of applicability. The data are shown as fitted to the Cross model [Eq. (7.38)].

Example 7.6

A power-law fluid is characterized in a cone-plate viscometer (cone angle = 0.500°). The measured viscosity is 0.2000 Pa·s at 6 rpm and 0.0800 Pa·s at 60 rpm. What $\Delta P/L$ is required to cause a flow of 1.00 liter/s in a horizontal pipe with an inner diameter of 2.00 cm?

Solution:

$$\dot{\gamma}(6 \text{ rpm}) = \frac{6 \times 2\pi/60}{0.00873} = 72.0 \text{ s}^{-1}$$
$$\tau = 0.2000 \times 72.0 = 14.4 \text{ Pa}$$
$$\dot{\gamma}(60 \text{ rpm}) = 720 \text{ s}^{-1}$$
$$\tau = 0.0800 \times 720 = 57.6 \text{ Pa}$$

Using the ratio of Eq. (7.34) for the two conditions:

$$\frac{57.6}{14.4} = \left(\frac{720}{72.0}\right)^n \quad \text{and} \quad n = 0.602 = \frac{1}{z}$$

Also,

$$57.6 = K(720)^{0.602} \quad \text{and} \quad K = 1.10$$

At the wall of the pipe, using Eqs. (7.76) and (7.77):

$$\dot{\gamma}_w = \frac{4.66 \times 1000 \text{ cm}^3/\text{s}}{\pi \times 1.00 \text{ cm}^3} = 1480 \text{ s}^{-1}$$
$$\tau_w = (\Delta P \times 1 \text{ cm}/2L) = 1.10(1480)^{0.602}$$
$$\Delta P/L = 178 \text{ Pa/cm} \quad (= 0.18 \text{ atm/m})$$

7.8 ELASTIC LIQUIDS (NORMAL STRESSES AND MELT FRACTURE)

This extended treatment of one-dimensional, time-independent, non-Newtonian flow is an example of the complexity added by any new flow anomaly. When polymers are sheared, we have been assuming that the materials are purely viscous and have no rigidity, i.e., no shear modulus. This is far from the truth for polymer melts and even for some polymer solutions. The general treatment of combined viscous and elastic effects can be approached from the use of a time-dependent modulus (viscoelasticity, Chap. 8).

Some aspects of elastic behavior accompany and complicate attempts at steady-state measurements and will be mentioned here [43, 50–52].

If pressures normal to the direction of flow are measured in a cone-and-plate viscometer, the situations illustrated in Fig. 7.28 may arise. Generally speaking, such normal stresses increase with the rate of shear, the concentration of polymer

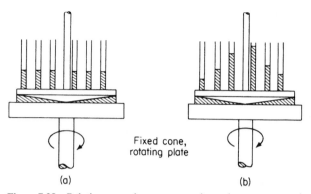

(a)   (b)

Fixed cone,
rotating plate

Figure 7.28 Relative normal pressures as shown by manometers inserted in stationary cone of cone-plate viscometer: (*a*) purely viscous behavior; (*b*) viscoelastic behavior

in solution, the molecular weight of the polymer, and the amount of branching in the polymer. These stresses are a source of instability in rotational instruments, pushing the cone away from the plate and thus lowering the rate of shear at a given rotational speed. At some critical speed, a material exhibiting high normal stresses will relieve the situation by suddenly pulling out of the shearing zone, literally jumping out of a cup-and-bob instrument.

Relaxation of this elastic energy at the exit of the pipe results in an increase in diameter over that of the pipe or die (die swell) (Fig. 7.29). As the amount of elastic energy put into the polymer entering the pipe increases, the elastic limit is reached and a pulsating irregularity in flow appears (Fig. 7.29). Such "melt fracture" is a major limitation on extrusion processes. Changing the geometry of the pipe entrance to a more gradual constriction can raise the maximum flow rate at which

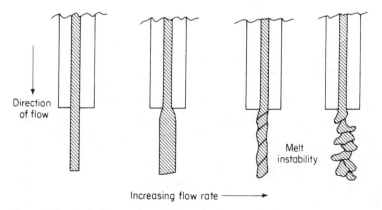

Direction
of flow

Melt
instability

Increasing flow rate ⟶

Figure 7.29 Typical stages in extrusion of polymer melt through capillary of diameter D_s. The die swell β is the ratio of the extrudate diameter to D_s

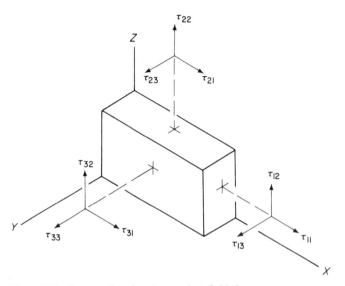

Figure 7.30 Stresses (force/area) exerted on fluid element

these irregularities appear. However, slip-stick flow at the pipe wall and distortion at the exit are harder to cure. We can illustrate some features of elastic liquids by considering the flow process in a more general way than we did before. Consider the stresses that prevail on a particle of fluid in motion (Fig. 7.30). On each face we have three stresses (force/area). The nine stresses, each labeled τ_{ij}, form a stress tensor variously represented as τ_{ij} or

$$\tau_{ij} = \begin{pmatrix} \tau_{11} & \tau_{12} & \tau_{13} \\ \tau_{21} & \tau_{22} & \tau_{23} \\ \tau_{31} & \tau_{32} & \tau_{33} \end{pmatrix} \tag{7.81}$$

A torque balance about each axis can be used to show that it is necessary that $\tau_{21} = \tau_{12}$, $\tau_{13} = \tau_{31}$, and $\tau_{32} = \tau_{23}$. In the case of simple shear (Sec. 7.1) only τ_{21} and τ_{12} are involved. The normal stresses τ_{11}, τ_{22}, and τ_{33} will be recognized as the hydrostatic pressure for any fluid at rest or for a Newtonian fluid in motion.

The rate-of-strain tensor $\dot{\varepsilon}_{ij}$ has nine components each in the same relative position as the stress with the same subscript. As before, $\dot{\varepsilon}_{12} = \dot{\varepsilon}_{21}$, $\dot{\varepsilon}_{23} = \dot{\varepsilon}_{32}$, and $\dot{\varepsilon}_{13} = \dot{\varepsilon}_{31}$. In a simple tensile test, for example, $\dot{\varepsilon}_{11}$ would represent the rate of elongation responding along the axis where stress τ_{11} is acting. The rate of shear defined earlier is $2\dot{\varepsilon}_{12}$. The most common viscous flow experiment is simple shearing flow. Experimentally, it is found that the following relationships are important (η and P_n are *not* constants) [53]:

$$\tau_{12} = \eta\dot{\gamma}_{12} \qquad \text{viscosity} \tag{7.82}$$

$$\tau_{22} - \tau_{11} = P_n(\dot{\gamma}_{12})^2 \qquad \text{first normal stress difference} \qquad (7.83)$$

$$\tau_{33} - \tau_{22} \cong 0 \qquad \text{second normal stress difference} \qquad (7.84)$$

In a cone-plate viscometer, $\tau_{22} - \tau_{11}$ represents a force tending to push apart the cone and the plate. If the net force f on the cone of radius R is measured (above the force due to atmospheric pressure), the first normal stress difference can be calculated [53]:

$$\tau_{22} - \tau_{11} = \frac{2f}{\pi R^2} \qquad (7.85)$$

Experimentally, the force f is affected by edge conditions, so that this formula is only approximate. The continuum mechanics of fluid dynamics can best be described using vector and tensor notation. Among the general references at the end of this chapter, many use this approach and apply it to polymer fabrication techniques such as extrusion and molding.

Bagley proposed a method of taking into account the elastic contribution to energy used in capillary flow [50]. The extrusion of the polymer melt or solution requires a pressure in excess of that predicted from Eq. (7.77). This can be visualized as being due to a fictitious additional length corresponding to the *recoverable shear strain*, γ_R, stored elastically in the melt while it is in the capillary. Thus, the "true" stress at the wall, $(\tau_w)_t$, is proportional to the slope of the ΔP, L/R plot (Fig. 7.31) according to

$$(\tau_w)_t = \frac{\Delta P}{2(L/R + e)} \qquad (7.86)$$

where e is the dimensionless *end correction*. Also,

$$e = C_t + \gamma_R/2 \qquad (7.87)$$

where C_t is the *Couette* part of the correction term arising from the establishment of a velocity profile in the capillary. With various grades of polyethylene, Bagley found that e invariably reached a plateau value at the stress corresponding to melt fracture (Fig. 7.31). The behavior shown in Fig. 7.31 is not always observed with other polymers, but the end correction method is widely used to reconcile melt flow behavior when results from various pieces of capillary-flow apparatus must be compared.

7.9 TURBULENT FLOW

A rather spectacular phenomenon is that of friction reduction in highly turbulent flow of dilute polymer solutions. Fabula has shown that small amounts of poly(ethylene oxide) in water reduce the pressure drop for a given flow rate to as little as

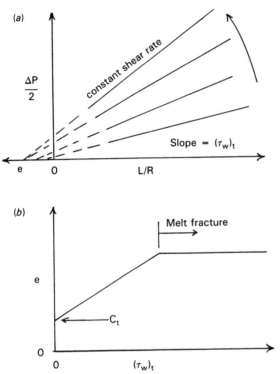

Figure 7.31 Extrapolation of measurements at constant apparent shear rate, $4Q/\pi R^3$, in a set of dies with a range of values for L/R (a) yields values of the end correction e and wall stress $(\tau_w)_t$, which can be correlated (b) in terms of a stress-strain diagram

one-fourth that for water alone [54]. In the experiments summarized in Fig. 7.32, it was found that the pressure drop, which is proportional to $(\tau_{12})_w$, steadily decreased as a high-molecular-weight ($M_v = 5.8 \times 10^6$) poly(ethylene oxide) was added. The Reynolds number N_{Re} is given by

$$N_{Re} = \frac{D\bar{u}\rho}{\eta} \tag{7.88}$$

where D = pipe diameter, cm
$\bar{u}$ = average velocity, cm/s
ρ = density, g/cm^3
η = viscosity, poise (g/s·cm; 1 poise = 0.1 Pa·s)

For water in a 1-cm-diameter pipe, $D\rho/\eta \cong 100$, so the range of $\bar{u}$ represented in Fig. 7.32 corresponds to N_{Re} between 40,000 and 200,000. At the concentrations generally used for drag reduction, ordinary shear viscosity is relatively unaffected.

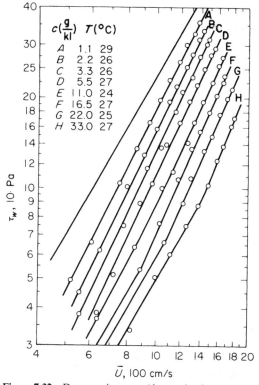

Figure 7.32 Decrease in stress (drag reduction) at constant average velocity $\bar{u}$ by addition of small amounts of high-molecular-weight poly(ethylene oxide) [54]

But even at very low concentrations, *elongational viscosity* is increased dramatically. This can be demonstrated by the long, pituitous strings that can be drawn from solutions by dipping in an object and then withdrawing it rapidly. Other manifestations of the same phenomenon are the elastic recoil of stirred solutions when stopped suddenly and the inhibition of vortex formation in a draining vessel.

It is the high elongational viscosity that may change the rate of energy dissipation. Turbulent flow dissipates energy because of the exchange of momentum between fluid domains. These domains, or *eddies,* can be characterized by measuring velocity fluctuations in turbulent flow. In any system there will be a spectrum of eddy sizes. Turbulence is sometimes pictured as "eddies within eddies within eddies." The withdrawal of material from the viscous sublayer along the wall is decreased by high elongational viscosity. Also, the small eddies throughout the conduit are decreased in intensity as the result of viscous damping.

Drag reduction has been suggested as a means of extending the distance that water from fire hoses can be discharged. For a fixed pressure drop, a smaller diame-

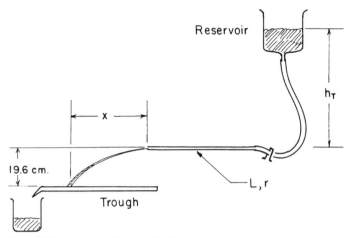

Figure 7.33 Apparatus for turbulent flow measurements

ter hose will transmit the same flow of water enabling firemen to drag less weight per unit length when pulling hoses into buildings. Storm sewer capacity during temporary overloads can be increased by injection of polymers. A ship's speed can be increased by injection of a polymer solution at the bow. Oil-soluble polymers have been used to assist the flow of crude oil in the trans-Alaska pipeline as well as in pipes carrying oil from offshore platforms to facilities on the shore. Friction reduction of 30 to 50% is achieved using as little as 10 ppm polymer [55].

For a quantitative demonstration [56], a simple apparatus (Fig. 7.33) consists of a reservoir that holds about 2000 cm³ of liquid and a flexible tube 200 cm long with an inside diameter of about 1 cm. This in turn is connected to a horizontal precision-bore tube. The efflux from the horizontal tube is allowed to fall freely a

Table 7.8 Turbulent flow measurement

Apparatus:	A 2000-cm³ beaker + a precision-bore tube (Fig. 7.33)
Typical data (water, 25°C):	Tube length, L = 91.3 cm Jet distance, x = 30.0 cm Tube radius, r = 0.15 cm Jet height, y = 19.6 cm Head, h_T = 150 cm η/ρ = 0.0090 cm²/s (25°C)
Equations:	Velocity, u = 5.00x Reynolds number, Re = $2ru/(\eta/\rho)$ Friction factor, f = $[(g/50)(h_T/x^2) - 0.35](2r/L)$ (second term is kinetic energy correction)
Calculated result (water):	u = 150 cm/s Re = 5000 f = 0.0096 (compared to expected value of 0.0094)

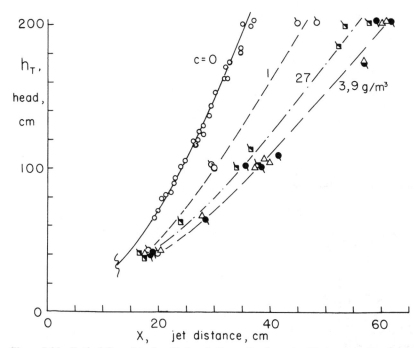

Figure 7.34 Typical flow data for dilute solutions of polyacrylamide in water. Results from several student groups are shown

vertical distance of 19.6 cm into a trough and the horizontal jet distance x is measured. From simple mechanics, the average velocity in the tube u is related to x and to the time t for liquid to fall the 19.6 cm under the force due to gravity.

$$u = \frac{x}{t} = x\left(\frac{981}{39.2}\right)^{1/2} = 5.00x \tag{7.89}$$

If the total head is 150 cm, x will be around 30 cm at room temperature for water flowing through a tube with the dimensions listed in Table 7.8. When water is replaced by a dilute polymer solution, much larger values of x result. Typical results are shown in Fig. 7.34. In engineering calculations the Reynolds number and the friction factor are of interest (Table 7.8). The reduction in friction by a polyacrylamide solution is seen in Fig. 7.35. Some polymers that are effective are poly-(ethylene oxide) (Polyox FRA, Union Carbide Corp.) and polyacrylamide (Separan AP30, Dow Chemical Co.).

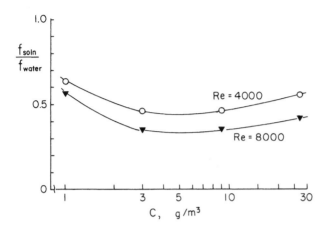

Figure 7.35 Data of Fig. 7.34 presented as ratio of friction factors at constant Reynolds number

KEYWORDS FOR CHAPTER 7

Section

1. Rheology
   Viscosity
   Shear stress
   Rate of shear
   Newtonian, Non-Newtonian
   Laminar flow
   Ostwald viscometer
2. Freely jointed chain
   Free rotation
   Rotational isomeric states
   Restricted rotation
   Steric factor
   Excluded volume
   Root-mean-square end-to-end distance
   Expansion factor
   Flory temperature
3. Intrinsic viscosity
   Inherent viscosity
   Reduced viscosity
   Relative viscosity
   Mark-Houwink-Sakurada equation

Huggins equation
Kraemer equation
Martin equation
Ubbelhode viscometer
Hydrodynamic volume
Universal calibration
4. Lyon and Tobolsky equation
Reptation
Melt index
5. Free volume
WLF equation
6. Power law
Ellis model
Cross model
Characteristic relaxation time
Shear thickening
Shear thinning
Thixotropy
Rheopexy
Zero-shear viscosity
7. Parallel-plate viscometer
Coaxial cylinder viscometer
Cone-plate viscometer
Capillary viscometer
Rabinowitsch correction
8. Melt (flow) fracture
Normal stress
Die swell
Recoverable shear strain
Bagley end correction
9. Turbulent flow
Drag (friction) reduction
Elongational viscosity

PROBLEMS

7.1 The following data are for a narrow molecular weight fraction of poly-(methyl methacrylate) in acetone at 30°C (density = 0.780 g/cm^3):

Concentration c, g/100 ml	Osmotic pressure π (h), cm solvent	Relative viscosity η_r
0.275	0.457	1.170
0.338	0.592	
0.344	0.609	1.215
0.486	0.867	
0.896	1.756	1.629
1.006	2.098	
1.199	2.710	1.892
1.536	3.725	
1.604	3.978	2.330
2.108	5.919	2.995
2.878	9.713	

Plot appropriately and estimate $[\eta]$, k' (Huggins), $\overline{M}$, and the second virial coefficient. Include dimensions for each. Knowing that $[\eta]_\theta = 6.5 \times 10^{-4} M^{0.5}$ for this polymer in a theta solvent at 30°C, calculate from these data $(\overline{r_0^2})^{1/2}$, $(\overline{r^2})^{1/2}$, and α, the expansion factor.

7.2 Two samples with narrow molecular weight distributions are prepared of a new polymer. Some measurements are made in acetone and in hexane.

Medium	Parameter	Sample A	Sample B
Acetone, 25°C	M_n (osmotic pressure)	8.50×10^4	Not run
	Second virial coeff.	Zero	
	Intrinsic viscosity	0.87 dl/g	1.32 dl/g
Hexane, 25°C	Intrinsic viscosity	1.25 dl/g	2.05 dl/g

(a) What is the M_n of sample B?
(b) What are the Mark-Houwink-Sakurada parameters (K' and a) in acetone and in hexane?

7.3 The system polyisobutylene-benzene has a theta temperature of 24°C. Describe how each of the quantities mentioned below would change for a given sample as the temperature is increased to 40°C from 24°C (increase, decrease, or no change). Explain why.

(a) Molecular weight measured by osmotic pressure.
(b) Osmotic pressure of a 1-g/dl solution.
(c) Intrinsic viscosity.
(d) Ratio of M_w (light scattering) to M_n (osmotic pressure).

7.4 A polymer based on isobutylene (FW = 56.0) has a steric factor of 1.93 at 24.0°C. What is the molecular weight of a sample which has an intrinsic viscosity of 0.780 dl/g in a good solvent at 24°C for which the expansion factor is 1.22?

7.5 A monodisperse polymer P is characterized in solvent S with the following results (all at 27°C):

Concentration, g/dl	Osmotic pressure, atm	Viscosity, Pa·s
0.000	—	0.00700
0.600	0.00500	0.00995
1.000	0.01125	0.01225

In theta solvent T, P has an intrinsic viscosity of 0.350 dl/g at 27°C.

(a) What osmotic pressure (atm) will be exhibited by a 0.25 g/dl solution of P in T at 27°C?

(b) What is the *expansion factor* in S for a new polymer sample P_{new} from the same monomer that exhibits a viscosity of 0.01225 Pa·s at a concentration of 0.600 g/dl at 27°C? The molecular weight of P_{new} is 100,000 (monodisperse).

7.6 For a certain monodisperse polymer in solution, the original concentration c_0 is unknown. At a series of dilutions, the osmotic pressure and viscosity are measured. Notice that the osmotic pressure is directly proportional to the concentration. Plot the data properly to obtain the absolute value of the hydrodynamic volume $[\eta]M$ in units of dl/mol.

Concentration	Osmotic pressure, atm, 27°C	Viscosity, centipoise, 27°C
c_0 divided by 10.0	2.00×10^{-3}	4.60
12.5	1.60×10^{-3}	3.62
15.0	1.33×10^{-3}	3.04
20.0	1.00×10^{-3}	2.40
Solvent alone	0.00	1.00

7.7 For polymer X, the hydrodynamic volume in a theta solvent at 40°C is 6.0 times that calculated for a chain of constant tetrahedral bond angle ($\theta = 70.53°$). If the repeat unit of four carbon chain atoms has a formula (molecular) weight of 115, what is the molecular weight of the monodisperse sample described by the following data (obtained in a theta solvent at 40°C)?

Concentration, g/dl	Viscosity, mPa·s
0.000	3.00
0.050	3.52
0.100	4.20
0.175	5.35
0.250	6.94

7.8 What is the root-mean-square end-to-end distance for a monodisperse polymer sample which is characterized by osmotic pressure, π, measurements, and by viscosity measurements, η.

Concentration c, g/dl	Osmotic pressure π, cm of solvent	Relative viscosity η_r
0.137	0.228	1.085
0.243	0.433	—
0.448	0.878	1.315
0.599	1.355	1.446

Solvent density = 0.780 g/cm³, T = 27.0°C.

7.9 The following data are obtained for solutions of two samples of polymer P in two solvents, A and B, all at 30°C. Solvent B is a theta solvent. The repeat unit of the carbon chain making up the polymer has a formula weight of 152 atomic mass units per two chain atoms. The carbon-carbon bond has a length of 1.54 Å and the usual tetrahedral bond angle.

Concentration, g/dl		Viscosity, Pa·s	
0.000	1.200	1.200	0.930
0.100	1.757	1.338	1.136
0.200	2.467	1.487	1.384
Solvent	A	A	B
Mol wt of polymer	1.00×10^6	2.00×10^5	1.00×10^6

Calculate:

(a) The root-mean-square end-to-end distance for the *unperturbed* polymer chain of mol wt = 1,000,000.
(b) The value of the steric factor.
(c) The value of the expansion factor in solvent A for the mol wt of 1,000,000.

7.10 A sample of polyacrylamide forms solutions in water that can be fitted by a Huggins equation plot over the range of $1.0 < \eta_r < 1.5$ with a slope of 0.700 (g/dl)² and an intercept of 1.35 dl/g. If the data from the Huggins equation at relative viscosities of 1.0 and 1.5 are used to fit the Martin equation, what specific viscosity can be predicted at a concentration of 5.0 g/dl?

7.11 The ratio of the intrinsic viscosity (dl/g) to the square root of molecular weight for polystyrene in solvent K at 25°C is 0.00360. In solvent J at the same temperature, the ratio of intrinsic viscosity to the square root of molecular weight for poly(vinyl acetate) is proportional to the 1/5 power of molecular weight. A polystyrene of molecular weight 1.85×10^5 has an intrinsic viscosity in K exactly 2.5 times that of a poly(vinyl acetate) sample of the same molecular weight in J. What is the expected *relative viscosity* for a poly-(vinyl acetate) sample with molecular weight 4.45×10^5 in solvent J at a concentration of 0.275 g/dl? The Huggins constant can be assumed to have a value of 0.400.

7.12 Three monodisperse samples of polystyrene are dissolved in the same solvent at 27°C. Relative viscosities are measured. Given the following information, calculate the root-mean-square end-to-end distance and the hydrodynamic volume, $[\eta]M$, for polymer sample C.

| Polymer sample | Relative viscosity at $c =$ | | Mol wt |
	1.0 g/dl	0.5 g/dl	
A	—	5.572	710,000
B	3.086	1.872	500,000
C	—	1.250	—

7.13 The following data are obtained for a series of polymers all made from the same monomer ($CH_2{=}CHX$, formula weight $= 92.0$, $T_g = 25°C$):

Sample	$[\eta]_\theta$, dl/g	Melt viscosity, poise (175°C)
A	0.21	3.6×10^2
B	0.25	5.0×10^2
C	0.32	8.1×10^2
D	0.50	33×10^2
E	0.85	0.14×10^6
F	1.13	1.0×10^6
G	1.37	3.9×10^6

The molecular weight of sample E measured by osmometry is 150,000. All samples are monodisperse.

(a) What is the "chain entanglement point" in terms of the number of chain atoms?

(b) At what intrinsic viscosity for sample E (in a good solvent) would the root-mean-square end-to-end distance be double the unperturbed length?

7.14 Three fractions are cut from a homopolymer. Each fraction has a very narrow molecular weight distribution. The following data are obtained. Fill in the missing data.

Fraction no.	Intrinsic viscosity at 25°C, dl/g		Melt viscosity at 200°C, Pa·s	Molecular weight, g/mol
	Solvent A	Solvent B		
1	1.2	2.0	?	1.3×10^5
2	1.8	3.6	4.0×10^5	?
3	2.5	?	?	5.6×10^5

What is the actual radius of gyration (in nm) of fraction 1 in solvent B at 25°C?

7.15 Assuming that the rms end-to-end distance is an approximation to the diameter of the spherical, coiled polymer in dilute solution, compare the *volume* occupied by one molecule of polyisobutylene (mol wt $= 10^6$) (*a*) as a solid at 30°C ($\rho = 0.92$ g/cm³), (*b*) in a theta solvent at 24°C (see Table 7.6), (*c*) in cyclohexane at 30°C (see Table 7.6).

7.16 From the information given below, calculate the ratio of the root-mean-square end-to-end distance in solvent A to that in solvent B for a polymer with a molecular weight of 1.00×10^7.

Mol wt	$[\eta]$ in A, dl/g	$[\eta]$ in B, dl/g
1.0×10^6	4.0	2.0
0.2×10^6	1.1	0.9
5.0×10^4	0.36	0.44

7.17 A typical laboratory capillary viscometer has a flow time t of water at 30°C of 200.0 s. The flow volume V is 5.0 cm³, capillary length L is 15 cm, and the average head $\bar{h}$ is $1.10L$. Estimate the diameter of the capillary disregarding end effects.

7.18 For poly(amyl zolate), assume that the hydrodynamic volume in a theta solvent at 30°C is 7.5 times that calculated for a chain of constant tetrahedral bond angle. If the repeat unit of four chain atoms has a molecular weight of 90, what is the molecular weight of the sample described by these data?

Theta solvent, 30°C	
c, g/dl	Viscosity, mPa·s
0.00	2.00
0.10	2.35
0.20	2.78
0.50	4.63
1.00	9.50

7.19 As a lecture demonstration it is desired to drop a plastic sphere (radius R, density $\rho = 1.200$ g/cm³) into a graduated cylinder full of a Newtonian solution (viscosity η, density $\rho_0 = 1.050$ g/cm³). According to Stokes' law, the terminal velocity of the sphere $\bar{u}$ is given by:

$$\frac{2}{9}R^2 g(\rho - \rho_0) = \eta\bar{u}$$

What values of R and η are reasonable for an effective experiment? The equation is valid only when the Reynolds number $N_{Re,p}$ is less than 0.1:

$$N_{Re,p} = \frac{2\bar{u}\rho_0 R}{\eta}$$

7.20 An empirical equation relating stress τ to rate of shear $\dot{\gamma}$ contains two constants, A and B:

$$\tau = \frac{A\dot{\gamma}}{1 + \sqrt{B\dot{\gamma}}}$$

Given that the zero-shear viscosity is 10 Pa·s for a certain sample and that the viscosity is 1.0 Pa·s at a stress of 10^3 Pa, calculate the viscosity at a rate of shear of 4×10^3 s⁻¹.

7.21 Three Ubbelohde viscometers are constructed and their parameters are identified by subscripts 1, 2, and 3, respectively. Construction is such that $h_1 = h_2 = h_3$, $L_2 = L_3 = 2L_1$, and $D_1 = D_2 = 0.5D_3$. The volume of a power-law liquid that flows through each in 100 s is measured (volumes V_1, V_2, and V_3). If $V_2/V_1 = \frac{1}{3}$, what is V_3/V_2?

7.22 For some polymer melts, viscosity plotted against shear stress on semilog paper gives a straight line.

(a) Write the corresponding equation for the rate of shear of a pseudoplastic melt in terms of the stress, the zero-shear viscosity, and the minimum number of remaining terms.

(b) Derive an equation for z as a function of stress, where

$$z = \frac{d(\ln \dot{\gamma})}{d(\ln \tau)}$$

7.23 The Ellis model for a non-Newtonian fluid can be written as

$$\frac{\eta_0}{\eta} - 1 = \left(\frac{\tau}{\tau_c}\right)^m$$

where η is the apparent viscosity at shear stress τ. Assuming that the shear rate can be given by $\dot{\gamma}_a = 4Q/\pi r^3$ and the stress by $\tau = \Delta Pr/2L$, calculate the flow rate in m³/h through a long, vertical pipe draining under its own head:

$\eta_0 = 10$ Pa·s, $\tau_c = 312$ Pa, $m = 0.75$, $\rho = 1.0$ g/cm³, $r = 1.0$ cm. Calculate a Reynolds number using the apparent viscosity.

7.24 A certain polymer solution can be characterized as a power-law fluid. Solution issues from a cylindrical storage tank through a horizontal capillary of length $L = 10.0$ cm and $r = 0.025$ cm. The time is measured for the level in the cylinder to fall from 981.0 to 968.0 cm (40.0 s) and from 742.0 to 721.5 cm (90.0 s). Predict the time it will take for the level to fall from 550.0 to 540.0 cm.

7.25 Solve Prob. 7.24 assuming the reservoir has a rectangular cross-section with two rectangular sides and two triangular sides. The apex of the triangles is at the entrance to the pipe.

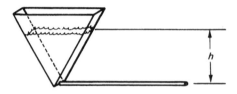

7.26 The following mathematical model is proposed for non-Newtonian flow:

$$A\eta = \frac{1 + B\tau^z}{1 + C\tau^z}$$

(a) Rewrite the equation in terms of $\dot{\gamma}$, τ, η_0, η_∞, C, and z.
(b) If $C > B$, is the material dilatant or pseudoplastic? Why?

7.27 A polymer solution can be represented by the equation relating stress τ to rate of shear $\dot{\gamma}$:

$$\tau = 2.70 \times 10^2 \ (\text{Pa})(s^{0.635})(\dot{\gamma}^{0.635})$$

What is the viscosity in mPa·s at a stress of 1000 Pa?

7.28 A coaxial viscometer with $R_2/R_1 = 1.10$, $R_1 = 2$ cm, $L = 5$ cm, is used to measure the viscosity of a fluid which follows the equation

$$A\dot{\gamma} = \tau\left[1 + \left(\frac{\tau}{B}\right)^{1.5}\right]$$

where $A = 100$ Pa·s and $B = 1000$ Pa. When $\tau = 800$ Pa at the geometric-mean radius, what is (a) the steady-state torque $\mathcal{T}$ in N·m, (b) the rotational speed of the cup in rpm, and (c) the ratio of rate of shear at the bob to that at the cup?

7.29 A power-law polymer solution, S, is characterized in a Brookfield viscometer, which consists of a vertical cylinder rotating in an "infinite" amount of liquid. Data are obtained as follows:

Liquid being tested	Rate of rotation, rpm	Scale reading
Newtonian Std., $\eta = 0.200$ Pa·s	6	175
S	6	200
S	60	400

Now S is pumped first through cylindrical pipe A and then splits into two flows through pipes B and C. Pipe A has a diameter twice that of B or C and a length that is 10 times that of B or C. What is the expected ratio of pressure drop over pipe A to that over pipe B? Neglect end effects.

7.30 A fluid is forced by a pressure drop $P_2 - P_1$ to flow through two filters F_1 and F_2. The filters are the same thickness and have the same total open area, but the pores of F_2 are half the diameter of those of F_1. If the fluid is Newtonian, what is the ratio of the flows $Q_1:Q_2$ through F_1 and F_2? What is the ratio if the fluid follows a power law?

If the fluid is non-Newtonian, is there a possibility of the ratio $Q_1:Q_2$ being less than 1? Assume that each filter is made up of a number of tubes of uniform length with no end effects. Assume all tubes of F_1 have the same diameter, etc.

7.31 Given the equation

$$\dot{\gamma} = A\tau + B\tau^2 + C\tau^3 \qquad \begin{aligned} A &= 1.0 \times 10^{-3} \\ B &= 2.0 \times 10^{-6} \\ C &= 3.0 \times 10^{-9} \end{aligned}$$

(a) What is the zero-shear viscosity in Pa·s? ($\dot{\gamma}$, s^{-1}, τ, Pa)
(b) Calculate the *ratio* of flow rate (cubic centimeters per second) in a capillary viscometer at a stress of $\tau = 1000$ Pa compared to that at $\tau = 10$ Pa. For the capillary: $r = 0.05$ cm, $L = 10.0$ cm.

7.32 If 0.500 cm^3 of polymer described by the following model is placed in a parallel-plate viscometer with a plate separation of 0.0100 cm and a weight of 0.500 g is applied to the movable plate by appropriate pulleys, how far will the plate move in 10 s? τ is in Pa.

$$\text{Model:} \qquad \dot{\gamma} = 2.00 \times 10^{-2}\, \tau^2$$

7.33 A power-law fluid flows through a slot of width W and height $2Y$ in the y direction under a pressure gradient $\Delta P/L$. Derive the following:

(a) Stress τ_y as a function of y and $\Delta P/L$
(b) Velocity u_y as a function of u_{max}, y, Y, and z
(c) Maximum rate of shear $\dot{\gamma}_w$ as a function of Q (flow rate, cm^3/s), W, Y, and z

$$\dot{\gamma}_y = m\tau_y^z \qquad W \gg Y$$

7.34 For a sample of poly(vinyl alcohol) in water at 21°C, the following data are obtained:

Concentration, g/dl	0	2.0	3.0	4.0	6.0	7.0	10.0
Viscosity, mPa·s	1.00	10.0	28.0	58.0	240	550	4000

From a Martin's plot, estimate $[\eta]$.

7.35 In order to simplify end corrections in a cup-and-bob viscometer, the ends can be conical. If each end of the bob is the cone of a cone-plate viscometer, what is the ratio of torque due to the ends of the bob to the torque due to flow in the annulus in the cylindrical part? Assume a bob of length L and radius KR, a cup of radius R, and a cone angle of $1 - K$. Assume a Newtonian fluid and refer torque in the annulus to the geometric-mean radius.

7.36 For a Bingham plastic, qualitatively, what unusual feature must a velocity profile in pipe flow exhibit? In a cup-and-bob viscometer, what discontinuity might be expected in the flow field for the same model? In a cone-plate viscometer, would any discontinuities be expected?

7.37 Estimate the parameters for Eqs. (7.34) and (7.35) for the flow curve at 260°C (polystyrene) in Fig. 7.17.

7.38 What is the maximum shear rate that one could use in a cone-plate viscometer with a Newtonian liquid of 100 Pa·s if the adiabatic heating rate is to be kept below 0.01°C/min? Assume a density of 0.95 g/cm² and a specific heat of 2.5 J/g·°C.

REFERENCES

1. Pezzin, G.: "Materie Plastiche ed Elastomerie," August, 1962, English translation, Instron Corporation, Canton, Mass.
2. Cannon, M. R., R. E. Manning, and J. D. Bell: *Anal. Chem.*, 32:355 (1960).
3. Flory, P. J.: "Principles of Polymer Chemistry," chaps. 10 and 14, Cornell Univ. Press, Ithaca, N.Y., 1953.
4. Morawetz, H. (ed.): "Macromolecules in Solution," 2d ed., Wiley-Interscience, New York, 1975.
5. Flory, P. J.: "Principles of Polymer Chemistry," p. 618, Cornell Univ. Press, Ithaca, N.Y., 1953.
6. Flory, P. J.: "Principles of Polymer Chemistry," p. 600, Cornell Univ. Press, Ithaca, N.Y., 1953.
7. Kurata, M., and W. H. Stockmayer: *Fortschr, Hochpolymer.-Forsch.*, 3(2):196 (1963).
8. Flory, P. J.: "Principles of Polymer Chemistry," p. 621, Cornell Univ. Press, Ithaca, N.Y., 1953.
9. Huggins, M. L.: *J. ACS,* 64:2716 (1942).
10. Kraemer, E. O.: *Ind. Eng. Chem.*, 30:1200 (1938).
11. Rodriguez, F.: Ph.D. thesis, Cornell Univ., 1958.
12. Staudinger, H.: "Die Hochmolekularen Organischer Verbindungen," Springer, Berlin, 1932.
13. Kurata, M., and Y. Tsunashima: in "Polymer Handbook," 3d ed., sec. VII, Wiley, New York, 1989.
14. Hitchcock, C. D., H. K. Hammons, and W. W. Yau: *Am. Lab.*, 26(11):26 (1994).
15. Haney, M. A.: *J. Appl. Polym. Sci.*, 30:3023 and 3037 (1985).

16. Lew, R., D. Suwanda, S. T. Balke, and T. H. Moury: *J. Appl. Polym. Sci: Appl. Polym. Symp.,* 52:125 (1993).
17. Flory, P. J.: "Principles of Polymer Chemistry," pp. 606.611, Cornell Univ. Press, Ithaca, N.Y., 1953.
18. Haney, M. A., D. Gillespie, and W. W. Yau: *Today's Chemist at Work,* 3 (11):39 (1994).
19. Lyons, P. F., and A. V. Tobolsky: *Polym. Eng. Sci.,* 10:1 (1970).
20. Rink, M., A. Pavan, and S. Roccasalvo: *Polym. Eng. Sci.,* 18:755 (1978).
21. Fujita, H., and A. Kishimoto: *J. Chem. Phys.,* 34:393 (1961).
22. Rodriguez, F.: *Polymer Letters,* 10:455 (1972).
23. Ott, E., H. Spurlin, and M. W. Grafflin (eds.): "Cellulose and Cellulose Derivatives," part III, p. 1216, Wiley-Interscience, New York, 1955.
24. Fox, T. G., S. Gratch, and S. Loshaek, in F. R. Eirich (ed.): "Rheology," vol. 1, chap. 12, Academic Press, New York, 1956.
25. De Gennes, P. G.: "Scaling Concepts in Polymer Physics," Cornell Univ. Press, Ithaca, N.Y., 1979.
26. Doolittle, A. K.: *J. Appl. Phys.,* 22:1471 (1951).
27. Williams, M. L., R. F. Landel, and J. D. Ferry: *J. ACS,* 77:3701 (1955).
28. Carley, J. E.: *Mod. Plastics,* 39:123 (December 1961).
29. Goldblatt, P. H., and R. S. Porter: *J. Appl. Polym. Sci.,* 20:1199 (1976).
30. Penwell, R. C., R. S. Porter, and S. Middleman: *J. Polym. Sci. A-2,* 9:463 (1970).
31. DeWaele, A.: *J. Oil Colloid Chem. Assoc.,* 6:33 (1925); Ostwald, W.: *Kolloid Z.,* 36:99 (1925).
32. Rodriguez, F., and L. A. Goettler: *Trans. Soc. Rheol.,* 8:3 (1964).
33. Rodriguez, F.: *Trans. Soc. Rheol.,* 10:169 (1966).
34. Ellis, S. B.: thesis, Lafayette College, Easton, Pa., quoted by M. Reiner, *J. Rheol.,* 1:14 (1929); also R. E. Gee and J. B. Lyon, *Ind. Eng. Chem.,* 49:956 (1957).
35. Williamson, R. V.: *Ind. Eng. Chem.,* 21:1108 (1929).
36. Krieger, I. M., and J. T. Dougherty: *Trans. Soc. Rheol.,* 3:137 (1959).
37. Ree, F. H., T. Ree, and H. Eyring: *Ind. Eng. Chem.,* 50:1036 (1958).
38. Bird, R. B., W. E. Stewart, and E. N. Lightfoot: "Transport Phenomena," pp. 11.15, Wiley, New York, 1960.
39. Denny, D. A., and R. S. Brodkey: *J. Appl. Phys.,* 33:2269 (1962).
40. Cross, M. M.: *J. Coll. Sci.,* 20:417 (1965).
41. Cross, M. M.: *Rheol. Acta,* 18:609 (1979).
42. Rodriguez, F., and C. C. Winding: *Ind. Eng. Chem.,* 51:1281 (1959).
43. Tordella, J.: *Trans. Soc. Rheol.,* 1:203 (1957); *Rheol. Acta,* 1:216 (1958).
44. Kearsley, E. A.: *Trans. Soc. Rheol.,* 6:253 (1962).
45. Cram, K. H., and J. C. Whitwell, *J. Appl. Phys.,* 26:613 (1955).
46. Krieger, I. M., and S. H. Maron: *J. Appl. Phys.,* 23:147, 1412 (1952).
47. Bird, R. B., W. E. Stewart, and E. N. Lightfoot: "Transport Phenomena," p. 96, Wiley, New York, 1960.
48. Rabinowitsch, B.: *Z. Physik. Chem.,* A145:1 (1929).
49. Schowalter, W. R.: *Chem. Eng. Educ.,* 6:14 (1972).
50. Bagley, E. B.: *J. Appl. Phys.,* 28:624 (1957); 31:1126 (1960).
51. Benbow, J. J., and P. Lamb: *SPE Trans.,* 3:7 (1963).
52. Lodge, A. S.: "Elastic Liquids," Academic Press, New York, 1964.
53. Markovitz, H., in E. H. Lee and A. L. Copley (eds.): "Proc. Fourth Intern. Congr. Rheol.," part 1, p. 189, Wiley-Interscience, New York, 1965.
54. Fabula, A. G., in E. H. Lee (ed.): "Proc. Fourth Intern. Congr. Rheology," part 3, p. 455ff., Wiley, New York, 1965.
55. Hoyt, J. W.: "Drag Reduction," in *Enc. Polym. Sci. Eng., 2d ed.,* 5:129 (1986).
56. Rodriguez, F.: *Eng. Educ.,* 65:245 (1974).

General References

Agassant, J. F., P. Avenas, J. Sergent, and P. J. Carreau: "Polymer Processing: Principles and Modeling," Hanser-Gardner, Cincinnati, Ohio, 1991.

Aklonis, J. J., and W. J. MacKnight: "Introduction to Polymer Viscoelasticity," 2d ed., Wiley, New York, 1983.

Barnes, H. A., J. F. Hutton, and K. Walters: "An Introduction to Rheology," Elsevier, New York, 1989.

Bekturov, E. A., and Z. Kh. Bakauova: "Synthetic Water-Soluble Polymers in Solution," 2d ed., Hüthig & Wepf, Basel, Switzerland, 1993.

Bikales, N. M. (ed.): "Water-Soluble Polymers," Plenum, New York, 1973.

Bird, R. B., R. C. Armstrong, and O. Hassager: "Dynamics of Polymeric Liquids: Vol. 1: Fluid Mechanics," 2d ed., Wiley, New York, 1987.

Bird, R. B., O. Hassager, R. C. Armstrong, and C. F. Curtiss: "Dynamics of Polymeric Liquids: Vol. 2: Kinetic Theory," 2d ed., Wiley, New York, 1987.

Brodkey, R. S.: "The Phenomena of Fluid Motions," Addison-Wesley, Reading, Mass., 1967.

Brydson, J. A.: "Flow Properties of Polymer Melts," Technomic, Lancaster, Penn., 1981.

Cogswell, F. N.: "Polymer Melt Rheology," Halsted-Wiley, New York, 1981.

Collyer, A. A., and D. W. Clegg (eds.): "Rheological Measurement," Elsevier, New York, 1988.

Collyer, A. A., and L. A. Utracki (eds.): "Polymer Rheology and Processing," Elsevier, New York, 1990.

Crochet, M. J., A. R. Davies, and K. Walters: "Numerical Simulation of Non-Newtonian Flow," Elsevier Applied Science, New York, 1984.

Cussler, E. L.: "Diffusion," Cambridge Univ. Press, Cambridge, UK, 1984.

Darby, R.: "Viscoelastic Fluids," Dekker, New York, 1976.

Dautzenberg, H., W. Jaeger, J. Kötz, B. Philipp, C. Seidel, and D. Stscherbina: "Polyelectrolytes," Hanser-Gardner, Cincinnati, Ohio, 1994.

De Gennes, P. G.: "Scaling Concepts in Polymer Physics," Cornell Univ. Press, Ithaca, N.Y., 1979.

Doi, M., and S. F. Edwards: "The Theory of Polymer Dynamics," Oxford, New York, 1986.

Eirich, F. R.: "Rheology," vols. 1 to 4, Academic Press, New York, 1956.1967.

Eisenberg, A., and F. E. Bailey (eds.): "Coulombic Interactions in Macromolecular Systems," ACS, Washington, D.C., 1986.

Ferry, J. D.: "Viscoelastic Properties of Polymers," 3d ed., Wiley-Interscience, New York, 1980.

Flory, P. J.: "Principles of Polymer Chemistry," Cornell Univ. Press, Ithaca, N.Y., 1953.

Flory, P. J.: "Statistical Mechanics of Chain Molecules (Revised)," Hanser-Gardner, Cincinnati, Ohio, 1991.

Forsman, W. C. (ed.): "Polymers in Solution," Plenum, New York, 1986.

Fredrickson, A. G.: "Principles and Applications of Rheology," Prentice-Hall, Englewood Cliffs, N.J., 1964.

Freed, K. F.: "Renormalization Group Theory of Macromolecules," Wiley, New York, 1987.

Fujita, H.: "Polymer Solutions," Elsevier, Amsterdam, 1990.

Gordon, G. V., and M. T. Shaw: "Computer Programs for Rheologists," Hanser-Gardner, Cincinnati, Ohio, 1994.

Han, C. D.: "Rheology in Polymer Processing," Academic Press, New York, 1976.

Isayev, A. I. (ed.): "Modeling of Polymer Processing," Hanser-Gardner, Cincinnati, Ohio, 1991.

Kurata, M.: "Thermodynamics of Polymer Solutions," Gordon & Breach, New York, 1982.

Kurata, M., and W. H. Stockmayer: *Fortschr. Hochpolymer.-Forsch.,* 3(2):196 (1963).

Larson, R.: "Constitutive Equations for Polymer Melts and Solutions," Butterworth, Stoneham, Mass., 1988.

Lenk, R. S.: "Polymer Rheology," Applied Science Publishers, London, 1978.

Macosko, C. W.: "Rheology: Principles, Measurements and Applications," VCH, New York, 1994.

Malkin, A. Ya.: "Rheology Fundamentals," ChemTec, Toronto, 1995.

Mashelkar, R. A., A. S. Mujumdar, and R. Kamal (eds.): "Transport Phenomena in Polymeric Systems," Prentice Hall, Englewood Cliffs, N.J., 1989.

Mena, B., A. Garcia-Rejon, and C. Rangel-Nafaile (eds.): "Advances in Rheology," 4 vols., Elsevier Applied Science, New York, 1984.

Middleman, S.: "The Flow of High Polymers," Wiley-Interscience, New York, 1968.

Middleman, S.: "Fundamentals of Polymer Processing," McGraw-Hill, New York, 1977.

Morawetz, H., (ed.): "Macromolecules in Solution," 2d ed, Wiley-Interscience, New York, 1975.

Patton, T. C.: "Paint Flow and Pigment Dispersion," Wiley, New York, 1979.

Schowalter, W. R.: "Mechanics of Non-Newtonian Fluids," Pergamon, New York, 1978.

Schulz, D. N., and J. E. Glass (eds.): "Polymers as Rheology Modifiers," ACS, Washington, D.C., 1991.

Tadmor, Z., and C. G. Gogos: "Principles of Polymer Processing," Wiley, New York, 1978.

Tanner, R. I.: "Engineering Rheology," Oxford Univ. Press, New York, 1986.

Tsvetkov, V. N.: "Rigid-Chain Polymers: Hydrodynamic and Optical Properties in Solution." Plenum, New York, 1989.

VanWazer, J. R., J. W. Lyons, K. Y. Kim, and R. E. Colwell: "Viscosity and Flow Measurement," Wiley-Interscience, New York, 1963.

Walters, K.: "Rheometry: Industrial Applications," Wiley, New York, 1980.

Whorlow, R. W.: "Rheological Techniques," Halsted (Wiley), New York, 1980.

Yamakawa, H.: "Modern Theory of Polymer Solutions," Harper & Row, New York, 1971.

EIGHT

MECHANICAL PROPERTIES
AT SMALL DEFORMATIONS

The study of mechanical properties of amorphous polymers at small deformations has been one of the most fruitful areas for theoreticians. The equation for equilibrium elasticity of cross-linked amorphous networks is one of the triumphs of statistical mechanics. Similarly, linear viscoelasticity has yielded to generalizations that are mainly empirical but have a theoretical explanation. On the other hand, partly crystalline and heterogeneous, reinforced materials are not so well correlated, even at small deformations. The situation regarding ultimate properties of all materials (tensile strength, fatigue strength, impact strength, burst strength) is even more nebulous, although there is a vast store of empirical knowledge.

8.1 ELASTICITY IN VARIOUS GEOMETRIES

In Chap. 7 we looked at polymer melts and solutions as liquids, and only in the later sections did we introduce the idea of elasticity. While it is useful to think of these as elastic liquids, rubber, glass, and materials of low flow under stress are more profitably generalized as elastic *solids* whose characteristic parameters are somewhat time-dependent. Some familiar mechanical properties are summarized in Table 8.1. Other quantities are used also. For example, "true" stress is sometimes used as force/(actual area) rather than the more conventional definition of force/(original area). In ordinary calculations for steel, wood, and other materials of

Table 8.1 Mechanical deformation—terminology and formulas

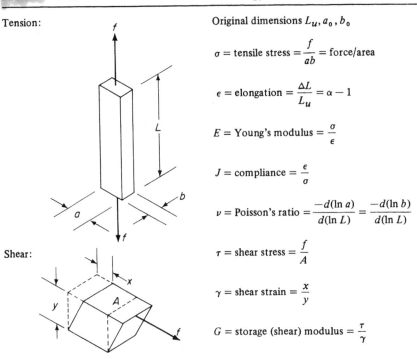

Tension:

Original dimensions L_u, a_0, b_0

$$\sigma = \text{tensile stress} = \frac{f}{ab} = \text{force/area}$$

$$\epsilon = \text{elongation} = \frac{\Delta L}{L_u} = \alpha - 1$$

$$E = \text{Young's modulus} = \frac{\sigma}{\epsilon}$$

$$J = \text{compliance} = \frac{\epsilon}{\sigma}$$

$$\nu = \text{Poisson's ratio} = \frac{-d(\ln a)}{d(\ln L)} = \frac{-d(\ln b)}{d(\ln L)}$$

Shear:

$$\tau = \text{shear stress} = \frac{f}{A}$$

$$\gamma = \text{shear strain} = \frac{x}{y}$$

$$G = \text{storage (shear) modulus} = \frac{\tau}{\gamma}$$

Bulk:

$P = $ hydrostatic pressure, $V = $ volume

$$B = \text{bulk modulus} = \frac{\Delta P}{\Delta V/V_0} = \text{compressibility}^{-1}$$

Isotropic materials:

$$E = 2G(1 + \nu) = 3B(1 - 2\nu) \tag{8.1}$$

$\nu = 0.5$ for incompressible liquids, most rubbers

Cantilever beam, end-loaded ($b = $ width):

$$Y = \frac{4fL^3}{ba^3 E} \tag{8.2}$$

Simple beam, center-loaded ($b = $ width):

$$Y = \frac{fL^3}{4ba^3 E} \tag{8.3}$$

Table 8.1 (*Continued*)

Beam in torsion (torque $\mathcal{T}$; angle of twist, ψ' rad):

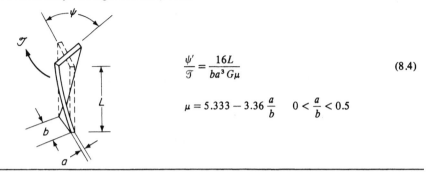

$$\frac{\psi'}{\mathcal{T}} = \frac{16L}{ba^3 G\mu} \tag{8.4}$$

$$\mu = 5.333 - 3.36\,\frac{a}{b} \qquad 0 < \frac{a}{b} < 0.5$$

construction, the materials are regarded as linearly elastic and a constant value is assumed for the modulus. We shall use some of these formulas where the moduli are functions of time of stress application or rate of stress application.

From the definition of Poisson's ratio, in Table 8.1, it is easily seen that when the volume of a material ($V = abL$) does not change on stretching, $v = 0.5$.

$$V = abL \tag{8.5}$$

$$\frac{d(\ln V)}{d(\ln L)} = \frac{d(\ln a)}{d(\ln L)} + \frac{d(\ln b)}{d(\ln L)} + 1 \tag{8.6}$$

If the changes in a and b with L are proportionately the same and V does not change with L,

$$2\frac{d(\ln a)}{d(\ln L)} + 1 = 0 \qquad v = 0.5 \tag{8.7}$$

Hooke's law, the direct proportionality between stress and strain in tension or shear, is often assumed. It will be shown that for cross-linked, amorphous polymers above T_g, a nonlinear relationship can be derived theoretically. For such materials $v = 0.5$. When v is not 0.5, it is an indication that voids are forming in the sample or that crystallization is taking place. In either case, neither the theoretical equation nor Hooke's law generally applies.

8.2 RUBBER ELASTICITY [1]

The first law of thermodynamics states that the change in internal energy of an isolated system ΔE is equal to the heat added to the system Q less work done by the system W:

$$\Delta E = Q - W \tag{8.8}$$

From the second law, we know that an increment of entropy dS is equal to an increment of heat dQ added reversibly at temperature T:

$$dS = \frac{dQ_{rev}}{T} \tag{8.9}$$

If we stretch a piece of rubber with force f a distance, dL, we also do work involving pressure and volume:

$$dW = P\,dV - f\,dL \tag{8.10}$$

Combining all these, we have [differentiating Eq. (8.8)]

$$dE = T\,dS - P\,dV + f\,dL \tag{8.11}$$

With the introduction of several assumptions, the chief one being that V does not change greatly on stretching (a good assumption for rubber), the "thermodynamic equation of state" for rubber is

$$f = \left(\frac{\partial E}{\partial L}\right)_{T,V} + T\left(\frac{\partial f}{\partial T}\right)_{P,\alpha} \tag{8.12}$$

and

$$\left(\frac{\partial f}{\partial T}\right)_{P,\alpha} = -\left(\frac{\partial S}{\partial L}\right)_{T,V} \tag{8.13}$$

where $\alpha = L/L_u$, the ratio of length to unstretched length at T.

Changes in ΔE for a system are identified with changes in temperature and with energy stored by bond bending and stretching. Changes in ΔS are changes due to differences in conformation. In this respect the concept of entropy as a measure of probability is useful. If a string is stretched tightly between two points, there is only one arrangement for the string in space to satisfy the condition. If the two points are moved together, the rest of the string can accommodate the change by numerous conformations in space. In other words, if all conformations in space are equally likely, the stretched condition is the least probable (only one way of being achieved, lowest entropy), and the two ends together is the most probable (many ways of being achieved, highest entropy).

The "ideal" rubber should respond to an external stress *only* by uncoiling, that is, $\partial E/\partial L = 0$. Experiments in which f, L, and T are varied confirm that for many amorphous, cross-linked materials above T_g, $\partial E/\partial L$ is very small (Fig. 8.1). Below T_g, we have said that polymer segments cannot deform in the time scale in which T_g was measured, so $\partial S/\partial L = 0$. In a glass we expect external stresses to be countered by bond bending and stretching.

Now, for an ideal rubber we have

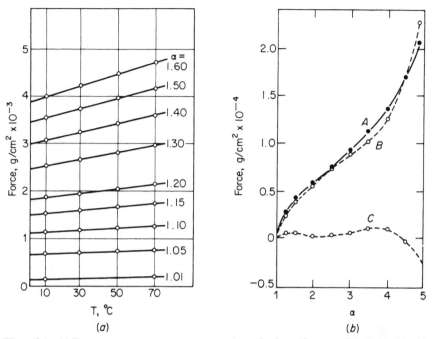

Figure 8.1 (a) Force vs. temperature at constant values of α for sulfur-cross-linked natural rubber [2]. (b) Force of retraction $A;\ -T(\partial S/\partial L)_{T,V},\ B;$ and $(\partial E/\partial L)_{T,V},\ C,$ obtained by using Eqs. (8.12) and (8.13) [2]. (Reproduced by permission from R. L. Anthony et al., *J. Phys. Chem.*, 46:826, © 1942, the Williams & Wilkins Company, Baltimore, Md.)

$$f = -T\left(\frac{\partial S}{\partial L}\right)_{T,V} \tag{8.14}$$

According to Boltzmann, the entropy of a system in a given state with respect to a reference state entropy is related to the ratio of the probability of that stage Ω_2 to that in the reference state Ω_1 by the equation

$$\Delta S = k \ln \frac{\Omega_2}{\Omega_1} \tag{8.15}$$

where k is the Boltzmann constant, the gas constant per molecule. What we need to know, then, is how does the probability of a polymer segment quantitatively decrease when it goes from a more probable state to a less probable state? The picture we have is a polymer segment that is part of a loose network. When we stretch the whole piece of rubber, we move the ends of the segment to new positions in the same proportion as we do the whole piece (an "affine" deformation). Because these chains are part of a network, the chain ends cannot diffuse to more probable positions. The rubber cannot "relax." Only removal of the external stress

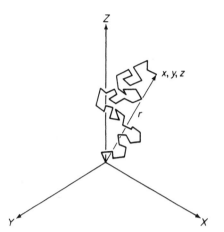

Figure 8.2 Random flight in three dimensions with resultant distance traveled equal to r

will allow the chain ends to return to their original distribution. The derivation of a distribution of chain lengths follows the same pattern as the random-flight problem. Let us now look at the probability $\Omega(x, y, z)$ that a polymer segment has one end fixed at the origin of a coordinate system (Fig. 8.2) and the other end at (x, y, z). A simple model for this probability is one in which Ω decays exponentially with the square of the distance from the origin.

$$\Omega(x, y, z) = \left(\frac{\beta^2}{\pi}\right)^{3/2} \exp\left[-\beta^2(x^2 + y^2 + z^2)\right] \qquad (8.16)$$

This is a Gaussian distribution. Here $\beta = (3/2)^{1/2}n^{-1/2}a^{-1}$ and n is the number of bonds each of length a. Of course, this will apply only at small deformations. After all, the polymer can be elongated only to its length of n bonds, whereas this model predicts a small but real probability of chain lengths increasing without limit. From Eq. (7.6), $\overline{r_0^2} = na^2$, and from geometry $r^2 = x^2 + y^2 + z^2$. Equation (8.16) is consistent with the idea that $\overline{r_0^2}$ is the most probable value of r^2, even though the most probable value for r, as well as for x, y, and z, is zero.

If we change the overall dimensions of the sample so that now $x_2 = \alpha_x x$, $y_2 = \alpha_y y$, and $z_2 = \alpha_z z$, the ratio of the probability in perturbed state 2 to that in unperturbed state 1 can be expressed as

$$\ln\frac{\Omega_2}{\Omega_1} = -\beta^2\left[(\alpha_x^2 - 1)x^2 + (\alpha_y^2 - 1)y^2 + (\alpha_z^2 - 1)z^2\right] \qquad (8.17)$$

But initially in our sample, chain ends were randomly oriented in all directions, so we can replace x^2, y^2, and z^2 each by their average value, $\overline{r_0^2}/3$, giving for each chain [Eqs. (8.15) and (8.17)]:

$$\Delta S = S_2 - S_1 = k \ln \frac{\Omega_2}{\Omega_2} = -k \frac{3}{2na^2} \frac{\overline{r_0^2}}{3} (\alpha_x^2 + \alpha_y^2 + \alpha_z^2 - 3) \qquad (8.18)$$

or

$$\Delta S = -\frac{k}{2} (\alpha_x^2 + \alpha_y^2 + \alpha_z^2 - 3)$$

Let us say that we have ζ effective chains in our system, or better, segments with ends that cannot move freely because they are tied in junctions (cross-links). Then, for the entire system,

$$\Delta S = -\frac{k\zeta}{2} (\alpha_x^2 + \alpha_y^2 + \alpha_z^2 - 3) \qquad (8.19)$$

Also,

$$f = -T\left(\frac{\partial \Delta S}{\partial L}\right)_{T,V} = -\frac{T}{L_u}\left(\frac{\partial \Delta S}{\partial \alpha_x}\right)_{T,V} \qquad (8.20)$$

If rubber is stretched in the x direction and the total volume remains constant,

$$\alpha_x \alpha_y \alpha_z = 1 \quad \text{and} \quad \alpha_y^2 = \alpha_z^2 = \frac{1}{\alpha_x} \qquad (8.21)$$

Then

$$f = \frac{kT\zeta}{L_u}\left[\alpha_x - \left(\frac{1}{\alpha_x}\right)^2\right] \qquad (8.22)$$

If retractive force per unit area $\sigma = f/A_u$, where A_u is the *original* cross-sectional area, and *original* volume $V_u = L_u A_u$, then

$$\sigma = kT\frac{\zeta}{V_u}\left[\alpha_x - \left(\frac{1}{\alpha_x}\right)^2\right] = RTN\left(\alpha - \frac{1}{\alpha^2}\right) \qquad (8.23)$$

where R is the gas constant per mole and N is the moles of polymer chains per unit volume.

Example 8.1

A cross-linked rubber with a density of 1.00 g/cm³ is composed of segments with a molecular weight of 10,000. What is the tensile stress at 100% elongation ($\alpha = 2.00$) at 25°C?

Solution:

$N = (1.00 \text{ g/cm}^3)(\text{mol}/10,000 \text{ g}) = 1.00 \times 10^{-4} \text{ mol/cm}^3$

$R = 8.3145 \text{ J/mol·K}$

$\sigma = 8.3145 \times 298 \times 1.00 \times 10^{-4} (2.00 - 0.25) = 0.434 \text{ J/cm}^3$

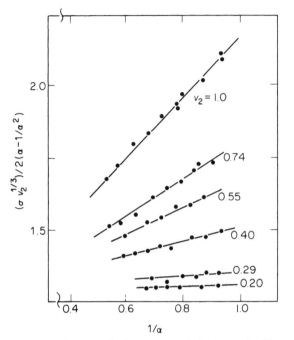

Figure 8.3 Stress-strain data for cross-linked natural rubber swollen to various degrees in benzene plotted according to Eq. (8.24) [4]

Since Pa = J/m³,

$$\sigma = 434 \text{ kPa} \quad \text{(or 62.8 psi)}$$

Equation (8.23) has been confirmed for rubbery materials when viscous effects can be made small. One way to decrease viscous effects is to swell the rubber to a new volume V_s by immersion in a low-molecular-weight solvent. Polymer molecules now are "lubricated," so viscous drag is minimized. Two other phenomena should be taken into account. In the first place, polymer segments at chain ends do not contribute to the modulus, because they are not restrained at one end. It is necessary to subtract from the chain density N the quantity $2\rho/M_n$, representing the density of chain ends per unit volume. The original number-average molecular weight before cross-linking M_n and the density ρ are required. In the second place, the deviation from Gaussian statistics at large deformations can be taken into account by using an additional constant C_2 to give the "Mooney-Rivlin" form of the equation. Assembled into one equation, we then get [3]:

$$\sigma = \left[\left(N - \frac{2\rho}{M_n}\right)RT + \frac{2C_2}{\alpha}\right]\left(\alpha - \frac{1}{\alpha^2}\right)\left(\frac{1}{v_2}\right)^{1/3} \tag{8.24}$$

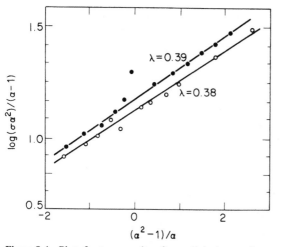

Figure 8.4 Plots for two samples of cross-linked natural according to Eq. (8.25) [6]. Stress is in kg/cm²

where σ is based on the original, *unswollen* cross section, α is the ratio of length to *swollen*, unstretched length, and $v_2 = V_u/V_s$. An example of natural rubber swollen in benzene (Fig. 8.3) gives a value for N that is relatively constant and values of C_2 that decrease as v_2 decreases. Many unswollen, homogeneous rubbers (gum or unfilled materials) are better represented by the empirical equation of Martin, Roth, and Stiehler [5]:

$$\ln \frac{\sigma \alpha^2}{\alpha - 1} = \ln \Lambda + \lambda \frac{\alpha^2 - 1}{\alpha} \qquad (8.25)$$

The constant λ normally has a value of 0.38 ± 0.02. The results for two samples of rubber (Fig. 8.4) show that this equation applies in compression as well as tension.

The differences between Eqs. (8.23), (8.24), and (8.25) (compared in Fig. 8.5) are quantitative rather than qualitative.

Example 8.2

The area under a reversible stress-strain diagram represents the energy stored per unit volume. How much work is done on an ideal rubber band that is slowly and reversibly stretched to $\alpha = 2.00$? The initial slope of the stress-strain curve is known to be 2 MPa, and the volume of the rubber band is 4.0 cm³.

Solution:

$$\text{Work} = \int f\, dL \qquad \sigma = \frac{f}{A_0} \qquad \alpha = \frac{L}{L_0}$$

$$\frac{\text{Work}}{\text{Volume}} = \frac{\text{work}}{A_0 L_0} = \int \sigma\, d\alpha = NRT \int (\alpha - \alpha^{-2})\, d\alpha$$

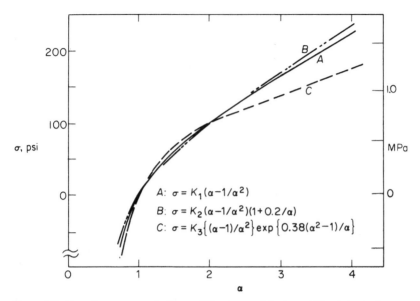

Figure 8.5 Equations compared with condition imposed that $\sigma = 100$ psi at $\alpha = 2$.

Integration from $\alpha = 1$ gives $NRT(\alpha^2/2 + \alpha^{-1} - 3/2)$. The initial slope (at $\alpha = 1$) $= (d\sigma/d\alpha) = 3NRT$ from differentiation of Eq. (8.23) and substituting $\alpha = 1$]. Then

$$3NRT = 2 \text{ MPa}$$

Thus, at $\alpha = 2$,

$$\frac{\text{Work}}{\text{Volume}} = \left(\frac{2}{3}\right) \text{MPa} \left(\frac{2^2}{2} + \frac{1}{2} - \frac{3}{2}\right) = (0.667) \text{ MJ/m}^3$$
$$\text{Volume} = 4.0 \times 10^{-6} \text{ m}^3$$
$$\text{Work} = (0.667) \times 4.0 = 2.67 \text{ J}$$

8.3 VISCOELASTICITY—MAXWELL MODEL

Although we have shown that stress σ is not directly proportional to strain, $\varepsilon = \alpha - 1$, for the ideal rubber, it is convenient in dealing with very small deformations of materials with both elastic and viscous nature to combine a linear, Hookean stress-strain relationship with a linear, Newtonian viscous relationship in order to illustrate the behavior of real materials. First, we introduce a simple mathematical model, the Maxwell element. Even though this model is inadequate for quantitative correlation of polymer properties, it illustrates the qualitative nature of real behavior. Furthermore, it can be generalized by the concept of a distribution of relaxation times so that it becomes adequate for quantitative evaluation. Maxwell's element is a simple one combining one viscous parameter and one elastic parameter. Mechanically, it can be visualized as the Hookean spring and a Newtonian dashpot in series:

σ, ϵ

Spring $\sigma_1 = E\epsilon_1$

Dashpot $\sigma_2 = \eta \dfrac{d\epsilon_2}{dt}$

σ, ϵ

where σ is the stress, ε is the elongation, $d\varepsilon/dt$ is rate of elongation with time, E is the Hooke's law constant or Young's modulus, and η is a viscosity coefficient. From the diagram it is apparent that

$$\sigma = \sigma_1 = \sigma_2 \qquad \varepsilon = \varepsilon_1 + \varepsilon_2$$

and

$$\frac{d\varepsilon}{dt} = \frac{d\varepsilon_1}{dt} + \frac{d\varepsilon_2}{dt} = \frac{1}{E}\frac{d\sigma}{dt} + \frac{\sigma}{\eta} \tag{8.26}$$

The behavior of this model in four experiments is considered now.

Creep

A fixed stress σ_0 is applied to the sample at time $t = 0$ (Fig. 8.6a). We desire the elongation as a function of time $\varepsilon(t)$. With stress constant in the spring, ε_1 is constant and $d\varepsilon_1/dt = 0$.

$$\frac{d\varepsilon}{dt} = \frac{d\varepsilon_1}{dt} + \frac{d\varepsilon_2}{dt} = \frac{\sigma_0}{\eta} \tag{8.27}$$

$$\int_{\varepsilon_0}^{\varepsilon(t)} d\varepsilon = \frac{\sigma_0}{\eta} \int_0^t dt \tag{8.28}$$

$$\varepsilon(t) = \varepsilon_0 + \frac{\sigma_0 t}{\eta} = \sigma_0\left(E^{-1} + \frac{t}{\eta}\right) \tag{8.29}$$

Often, it is useful to express the results as a time-dependent *compliance* $J(t)$, where

$$J(t) = \frac{\varepsilon(t)}{\sigma_0} = E^{-1} + \frac{t}{\eta} \tag{8.30}$$

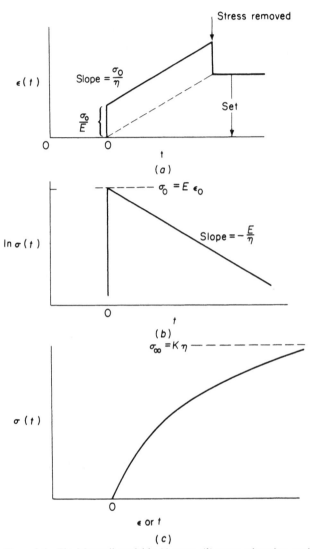

Figure 8.6 The Maxwell model in (a) creep, (b) stress relaxation, and (c) constant rate of extension.

The *linear viscoelastic region* is the region where $J(t)$ is independent of σ_0 or, more generally, where the dependence is expressible by an exact relationship like Eq. (8.23). It can be seen from the diagram that when the stress is removed, only the deformation of the spring is recovered. The flow of the dashpot is retained as a permanent "set." A real, un-cross-linked polymer often gives this type of behavior. Another popular method of presenting experimental creep data is to plot the elongation vs. time on log-log paper. A straight line on this kind of plot gives the Nutting equation [7],

$$\log \varepsilon(t) = \log A + s \log t \tag{8.31}$$

where $\log A$ and s are the intercept and slope of the plot, respectively.

Stress Relaxation

A fixed elongation ε_0 is applied at time $t = 0$ and held (Fig. 8.6b). We desire stress as a function of time $\sigma(t)$.

$$\frac{d\varepsilon}{dt} = \frac{d\varepsilon_1}{dt} + \frac{d\varepsilon_2}{dt} = 0 \tag{8.32}$$

$$E^{-1}\frac{d\sigma}{dt} + \frac{\sigma}{\eta} = 0 \tag{8.33}$$

$$\int_{\sigma_0}^{\sigma(t)} \frac{d\sigma}{\sigma} = -\frac{E}{\eta}\int_0^t dt \tag{8.34}$$

$$\ln \frac{\sigma(t)}{\sigma_0} = -\frac{Et}{\eta} \tag{8.35}$$

A time-dependent modulus $E(t)$ can be defined as

$$E(t) = \frac{\sigma(t)}{\varepsilon_0} = E \exp\left(-\frac{Et}{\eta}\right) \tag{8.36}$$

The term η/E, the *relaxation time*, is a measure of the rate at which stress decays. The linear viscoelastic region, of course, corresponds to $E(t)$ being independent of ε_0.

Constant Rate of Strain

Many commercial testing machines operate under conditions that approximate $d\varepsilon/dt = K_0$. We want stress as a function of time or of elongation, since $\varepsilon = K_0 t$ (Fig. 8.6c).

$$\frac{d\varepsilon_1}{dt} + \frac{d\varepsilon_2}{dt} = K_0 \tag{8.37}$$

$$E^{-1}\frac{d\sigma}{dt} + \frac{\sigma}{\eta} = K_0 \tag{8.38}$$

$$\int_0^{\sigma(t)} \frac{d\sigma}{K_0\eta - \sigma} = \frac{E}{\eta}\int_0^t dt \tag{8.39}$$

$$\sigma(t) = K_0\eta\left[1 - \exp\left(-\frac{Et}{\eta}\right)\right] = K_0\eta\left[1 - \exp\left(-\frac{E\varepsilon}{K_0\eta}\right)\right] \qquad (8.40)$$

Example 8.3

When a 10.0-cm-long polymer sample is subjected to a constant tensile load of 15.0 kPa, it stretches to a total length of 12.5 cm in 1200 s. Upon removal of the load, the sample recovers only to 10.85 cm. Representing the polymer as a Maxwell element, estimate the time it would take for an identical (untested) sample to relax to 0.100 its original stress when a constant elongation of 85% is applied.

Solution: In creep, the set (permanent deformation) represents the deformation of the dashpot only, since the spring returns to its original length upon removal of the load.

$$\text{Set} = \frac{10.85 - 10.00}{10.00} = 0.085 \qquad \text{(dimensionless)}$$

The recoverable strain is due to deformation of the spring.

$$\text{Spring deformation} = \frac{12.50 - 10.85}{10.0} = 0.165$$

At time = 0 (after loading),

$$\varepsilon(0) = 15.0 \text{ kPa } E^{-1} = 0.165 \qquad \text{and} \qquad E = 90.9 \times 10^3 \text{ Pa}$$

After 1200 s,

$$\varepsilon(1200) = \frac{12.50 - 10.00}{10.0} = 0.250$$

$$15.0 \times 10^3 \left(E^{-1} + \frac{1200}{\eta}\right) = 0.250$$

$$\eta = 212 \times 10^6 \text{ Pa·s}$$

$$\text{Relaxation time} = \frac{\eta}{E} = \theta = 2.33 \times 10^3 \text{ s}$$

In stress relaxation,

$$\ln\left[\frac{\sigma(t)}{\sigma_0}\right] = \ln 0.85 = \frac{-t}{421}$$

$$t = 68 \text{ s}$$

Example 8.4

If the sample of the preceding example is stretched at a constant rate of 0.60 cm/min, what will be the stress at (a) 1 min and (b) 100 min?

Solution:

$$K_0 = \frac{0.60}{10 \times 60} = 0.0010 \ \text{s}^{-1}$$

(a) $\sigma(60 \ \text{s}) = 0.0010 \times 212 \times 10^6 \left[1 - \exp\left(\frac{-60}{233 \times 10^3}\right)\right] = 5.39 \times 10^3 \ \text{Pa}$

(b) $\sigma(6000 \ \text{s}) = 212 \times 10^3 \left[1 - \exp\left(\frac{-6000}{2.33 \times 10^3}\right)\right] = 973 \ \text{kPa}$

Harmonic Motion

If a sinusoidal force acts on a Maxwell element, the resulting strain will be sinusoidal at the same frequency, but out of phase. The same holds true if the strain is the input and the stress the output. For example, let the strain be a sinusoidal function of time with frequency ω (rad/s),

$$\varepsilon = \varepsilon_m \sin \omega t \tag{8.41}$$

The motion of the Maxwell element (with modulus E and relaxation time $\theta = \eta/E$) is given by

$$\frac{d\varepsilon}{dt} = \frac{1}{E}\frac{d\sigma}{dt} + \frac{\sigma}{\theta E} \tag{8.42}$$

The resulting stress (see App. 2) will be directly proportional to E. Also, the magnitude will be affected by the product $\omega\theta$ and will *lead* the strain by an angle $\delta = \cot^{-1}\omega\theta$:

$$\sigma = \varepsilon_m E \frac{\omega\theta}{(1 + \omega^2\theta^2)^{1/2}} \sin(\omega t + \delta) \tag{8.43}$$

The situation is shown in Fig. 8.7. A more common notation for sinusoidal tests is the complex dynamic modulus E^*, which is made up of a dynamic modulus E' and a *loss* modulus E'':

$$E^* = E' + iE'' = \frac{\sigma^*}{\varepsilon^*} \tag{8.44}$$

Let the strain be a complex oscillating function of time with maximum amplitude ε_m and frequency ω:

$$\varepsilon^* = \varepsilon_m \exp(i\omega t) \tag{8.45}$$

The real strain is the real part of the complex strain ε^*. The resulting ratio of stress to strain can be written (App. 2) as

$$\frac{\sigma^*}{\varepsilon^*} = \frac{E\omega^2\theta^2}{1 + \omega^2\theta^2} + i\frac{\omega\theta E}{1 + \omega^2\theta^2} \tag{8.46}$$

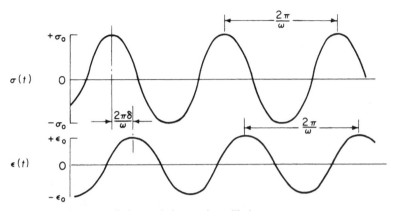

Figure 8.7 The Maxwell element in harmonic oscillation

or

$$\sigma^* = \sigma_m \exp(i\omega t + \delta)$$

that is, σ^* leads ε^* by a loss angle δ. Once again, the actual stress is the real part of the complex stress σ^*. Identification with Eq. (8.44) leads to

$$E' = \frac{E\omega^2\theta^2}{1 + \omega^2\theta^2} \qquad E'' = \frac{E\omega\theta}{1 + \omega^2\theta^2} \tag{8.47}$$

$$\tan\delta = \frac{1}{\omega\theta} = \frac{E''}{E'} \tag{8.48}$$

Another term sometimes used is the dynamic tensile viscosity η_T':

$$\eta_T' = \frac{E''}{\omega} \tag{8.49}$$

The variations of E', E'', and $\tan\delta$ with θ are shown in Fig. 8.8. E'' also is a measure of energy lost per cycle per unit volume (u_e), since

$$u_e = \pi E'' \varepsilon_{max}^2 \tag{8.50}$$

where ε_{max} is the maximum amplitude of strain. The dissipation of energy in a cyclic process is sometimes known more familiarly as *hysteresis*.

It is noteworthy that in an oscillatory experiment we can detect a peak in loss angle or loss modulus at a transition point such as T_g or T_m (where $\omega = 1/\theta$), whereas creep, stress relaxation, and constant rate of strain give only changes in level of modulus.

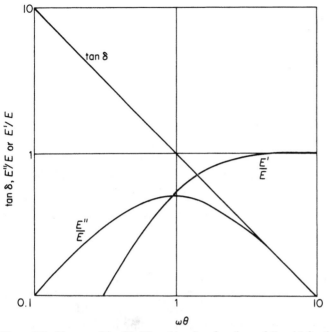

Figure 8.8 The quantities E', E'', and tan δ as functions of E and θ for the Maxwell element

8.4 DYNAMIC MEASUREMENTS

A number of commercial instruments are made that apply a harmonic stress or strain to a sample and measure the response [8, 9]. In a dynamic mechanical (thermal) analyzer, often abbreviated as DMA or DMTA, modulus and loss can be measured as a function of controlled frequency, amplitude, and temperature. Commercial machines may include provisions for testing materials in tension, bending, or compression [10–12]. Such devices have the advantage that the cycles are repeated many times and the sample can come to a steady state. Analysis of in-phase and out-of-phase responses follows along the lines of the Maxwell model discussed in the previous section. Variations in geometry and frequency can permit the examination of many materials. A disadvantage of the technique is that the energy dissipated within the sample may cause the temperature to increase. One way to minimize the effect is to store only a small amount of energy in the sample initially and then to measure the rate of dissipation without any additional inputs of energy. This is done in stress relaxation. A dynamic mechanical analog is the pendulum. In the common pendulum a mass on the end of a string is given an initial input of energy (by raising it above its resting place). Potential energy is converted to kinetic energy as the mass returns to its rest position and again into potential energy as it rises. The frequency of oscillation of the mass is independent of amplitude.

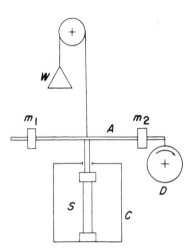

Figure 8.9 Torsion pendulum. Sample S in conditioning chamber C acts as the restoring element. The moment arm A with masses m_1 and m_2 is counterbalanced by weight W. Amplitude of oscillation is recorded on drum D using thermal or light-sensitive paper

A torsion pendulum is a particular useful device for examining the modulus and internal friction of polymers. When the sample geometry is in the form of a cone-and-plate or concentric cylinders, the pendulum can be used for very weak gels, which are hard to characterize in conventional testing machines. Recording models are available which give a complete record of amplitude vs. time. However, even manual estimation of frequency of oscillation and the number of cycles to decrease the amplitude by some factor can be quite useful.

In any torsion pendulum the sample is the energy storing and dissipating element (S in Fig. 8.9), and the external masses (m) can be used to regulate the frequency of oscillation ω. The stiffness of the sample (torque per unit twist $\mathcal{T}/\psi'$) depends on the shear modulus G and a geometric factor K''. Formulas for K'' are tabulated (Table 8.2). The overall equation is

$$\frac{\mathcal{T}}{\psi'} = GK'' = \omega^2 I \left[1 + \left(\frac{\Delta}{2\pi} \right)^2 \right] \tag{8.51}$$

The last term contains the logarithmic decrement Δ, which is the logarithm of the ratio of amplitudes A for successive cycles n:

$$\Delta = \ln\left(\frac{A_n}{A_{n+1}} \right) \tag{8.52}$$

It represents the fraction of stored energy lost per cycle. The moment of inertia of the moving part of the system usually is due mainly to external masses and can be calculated from

$$I_r = \frac{m'L^2}{12} \tag{8.53}$$

for a transverse rod of total mass m' and length L rotating about its center, and

$$I_m = mr^2 \tag{8.54}$$

for a mass m at a distance r from the center of rotation.

Example 8.5

The restoring element for a torsion pendulum is a piece of plasticized vinyl labora-tory tubing, and the formula for a thin-walled tube can be used (Table 8.2). The appartus outlined in Fig. 8.9 is used with $I = 7.4 \times 10^3$ g·cm², $L = 10$ cm, $t = 0.15$ cm, $D = 0.85$ cm. Using the period of 1.67 s and $\Delta = 0.23$ (Fig. 8.10), calculate G and the loss modulus, G'' where G'' is the product of G and Δ. It should be noted that, with the units employed above, G will have the units of g/cm·s², which is equivalent to dyne/cm².

Solution:

$$K'' \text{ [Eq. (8.51)]} = \frac{0.15 \times \pi \times 0.85^3}{4 \times 10} = 7.2 \times 10^{-3} \text{ cm}^3$$

$$\omega = \frac{2\pi}{\text{period}} = \frac{2\pi}{1.67} = 3.76 \text{ rad/s}$$

Using Eq. (8.51):

$$G(7.2 \times 10^{-3}) = (3.76)^2 (7.4 \times 10^3) \left(\frac{1 + 0.23}{2\pi} \right)$$

$$G = 14 \times 10^6 \text{ dyne/cm}^2 = 1.4 \text{ MPa}$$

$$G'' = 0.23 \times 1.4 \text{ MPa} = 0.32 \text{ MPa}$$

Table 8.2 Geometric factor K'' for torque per unit twist

Sample geometry	Geometric factor, K''	
1. Beam in torsion	See Table 8.1	
2. Thin-wall tube: D = average diameter t = wall thickness L = length	$\left(\dfrac{t\pi D^3}{4L} \right)$	(8.55)
3. Sample between concentric cylinders: R_1, R_2 = radii of inner and outer cylinders L = length of sample	$\left(\dfrac{4\pi L}{R_1^{-2} - R_2^{-2}} \right)$	(8.56)
4. Sample between a cone and a plate: R = radius of cone ψ = angle between cone and plate	$\left(\dfrac{2\pi R^3}{3\psi} \right)$	(8.57)

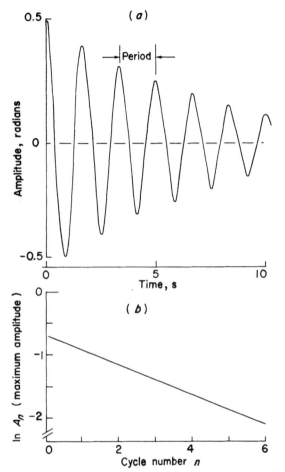

Figure 8.10 Typical analysis of torsion pendulum. (*a*) Oscillating trace made by tip of moment arm gives a period of 1.67 s corresponding to a frequency of 0.600 Hz or 1.20 π rad/s. (*b*) Plot according to Eq. (8.52) gives a logarithmic decrement of 0.23

When oscillations are slow, as in this example, manual measurement of successive amplitudes is as simple as making marks with a pencil at the extreme position of the moment arm for each cycle. Faster oscillations require some mechanical recording device, especially when the amplitudes are very small. Figure 8.19 (Sec. 8.7) illustrates the use of the torsion pendulum for multiple transitions.

In one modification the restoring element is not simply the polymer sample, but some semirigid substrate in or on which the polymer is mounted. If blotting paper is impregnated with polymer, the restoring element does not change in rigidity very much at transitions, but the loss peaks are quite easily detected [13]. In the "torsional braid" pendulum, a glass braid is impregnated with polymer. Because of the inert nature of the glass, the torsion pendulum will respond with changes in damping and stiffness, not only for transitions but also for chemical reactions such as oxidation, polymerization, and condensation [14, 15].

8.5 REAL POLYMERS—FIVE REGIONS OF VISCOELASTICITY

The time-dependent modulus for a typical amorphous polymer, poly(methyl methacrylate), measured by stress relaxation at various temperatures is shown in Fig. 8.11. One way to define the five regions of viscoelasticity [16] is to take a cross section of the modulus-time curve at some fixed relaxation time—say 100 h, 1 h, or 0.001 h (Fig. 8.12). In each of these there are two reasonably well-defined plateaus. The first, at $E = 10^9$ or 10^{10} Pa, is the region of glassy behavior; the second is the region of rubbery behavior. Between the two plateaus we have the main *glass transition.* Figure 8.12 illustrates the rate dependence of this important temperature. Finally, at high temperatures, *rubbery flow* marks the transition to the *liquid flow* region.

Horizontal shifting of stress-relaxation curves from Fig. 8.11 to correspond with that at some reference temperature T_R gives the composite curve in Fig. 8.13. This would be an academic exercise except that this time-temperature superposition is real and has been confirmed by experiments covering extremes of time scales at the same temperature. The horizontal shifts [log $t(T)$ − log $t(T_R)$] along the time axis in the vicinity of T_g are correlated by the WLF equation (7.30).

$$\log a_T = \log\frac{t(T)}{t(T_g)} = \frac{-17.44(T - T_g)}{51.6 + (T - T_g)} \tag{8.58}$$

Obviously, the curves could be shifted to any temperature other than T_g also by the same equation ($T_R = 115°C$ in Fig. 8.13, while $T_g = 105°C$).

The significance of this generalization cannot be overemphasized. Again and again, we find in the literature methods of superposing time and temperature for mechanical and other properties in amorphous and partially amorphous materials. Whatever modifications are introduced usually reduce the behavior back in the direction of Eq. (8.58).

8.6 GENERALIZED MAXWELL MODEL [18]

In Fig. 8.14 we compare stress-relaxation curves for a single Maxwell element with the same glassy modulus as the "master curve" but with various values of η. Obviously, this simple model cannot approach the behavior of the real system. But suppose we used a model made up of Maxwell elements in parallel held at fixed elongation ε_T which is the same for all elements:

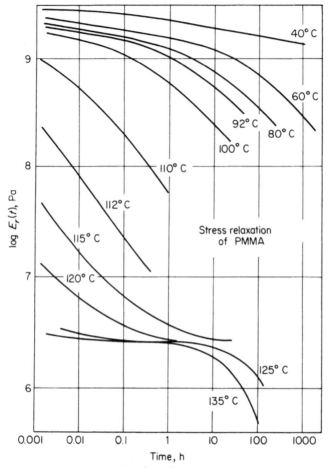

Figure 8.11 Log $E_r(t)$ vs. log t for unfractionated poly(methyl mechacrylate) of $M_v = 3.6 \times 10^6$ [17]

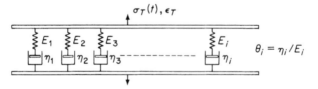

The total time-dependent force $\sigma_T(t)$ is the sum of the forces acting on the individual elements σ_i.

$$\sigma_T(t) = \sum \sigma_i = \sum \varepsilon_T E_i \exp\left(-\frac{t}{\theta_i}\right) \tag{8.59}$$

The overall time-dependent modulus $E_T(t)$ is also definable:

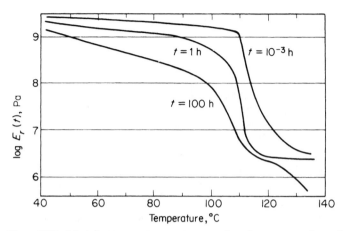

Figure 8.12 Modulus-temperature master curve based on cross sections of Fig. 8.11

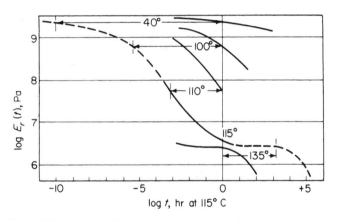

Figure 8.13 Modulus-time master curve based on time-temperature superposition of data in Fig. 8.11. Times referred to temperatures of 115°C

$$E_T(t) = \frac{\sigma_T(t)}{\varepsilon_T} = \sum E_i \exp\left(-\frac{t}{\theta_i}\right) \qquad (8.60)$$

The synthesis of $E_T(t)$ from known values of E_i and θ_i is simplified by the use of semilog paper.

For example, one can derive $E_T(t)$ for $0 < t < 200$ s when

i	E_0, dyn/cm$^2 \times 10^{-8}$	E_0, MPa	θ, s
1	1.000	10.0	100
2	0.667	6.67	50
3	0.333	3.33	25

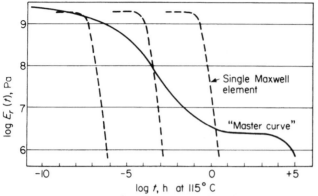

Figure 8.14 Master curve for poly(methyl methacrylate) compared with single Maxwell elements

For each element, $E_i(t)$ is given by a straight line on semilog paper with intercept $(E_i)_0$ and a negative slope of $(\theta_i)^{-1}$ (each divided by 2.303, since we are using paper based on logarithms to the base 10) (Fig. 8.15). Adding the curves arithmetically gives $E_T(t)$ directly:

$$E_i(t) = (E_i)_0 \exp\left(-\frac{t}{\theta_i}\right) \tag{8.61}$$

$$E_T(t) = \sum_{i=1}^{3} E_i(t) \tag{8.62}$$

Note that $E_T(t)$ is *not* a straight line in Fig. 8.15.

Example 8.6

A cross-linked polymer can be represented by three Maxwell elements in parallel with constants

$$E_1 = E_2 = E_3 = 100 \text{ kPa}$$
$$\theta_1 = 10 \text{ s} \qquad \theta_2 = 100 \text{ s} \qquad \theta_3 = \text{infinity}$$

(a) What stress is required for a sudden elongation to three times the original length?
(b) What will be the stress after being held at three times the original length for 100 s?
(c) What will be the stress after being held at three times the original length for 100,000 s?

Solution:

(a) At $t = 0$ and $\varepsilon_T = 2.00$ [Eq. (8.60)]:

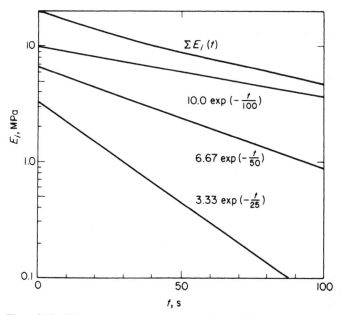

Figure 8.15 Time-dependent modulus for individual Maxwell elements and for the sum of three elements in parallel $\Sigma E_i(t)$

$$\sigma(0) = 2.00(3 \times 100 \text{ kPa}) = 600 \text{ kPa}$$

(b) $\sigma(100 \text{ s}) = 200 \text{ kPa} [\exp(-10) + \exp(-1) + \exp(0)] = 274 \text{ kPa}$

(c) $\sigma(100{,}000 \text{ s}) = 200 \text{ kPa}(0 + 0 + 1) = 200 \text{ kPa}$

For an infinite number of elements, the continuous analog is

$$E_T(t) = \int_0^\infty E(\theta) \exp\left(-\frac{t}{\theta}\right) d\theta \tag{8.63}$$

or

$$E_T(t) = \int_{-\infty}^{+\infty} \overline{H}(\log \theta) \exp\left(-\frac{t}{\theta}\right) d(\log \theta) \tag{8.64}$$

where $\overline{H}(\log \theta) = 2.303 \, \theta E(\theta)$. Both $E(\theta)$ and $\overline{H}(\log \theta)$ are called the distribution of relaxation times. A number of mathematical models have been proposed for the relationship between $\overline{H}$ and θ. When these are inserted in Eq. (8.64), the infinite

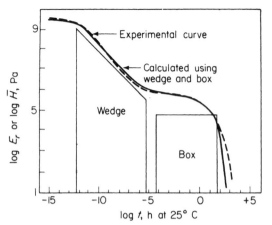

Figure 8.16 Master curve for polyisobutylene as determined by superposition of experimental stress-relaxation data (dashed line) and from the insertion of the wedge and box models in Eq. (8.64) [18]

array is reduced to something represented by a few parameters. Two elementary models are the box and the wedge:

Box:

$$\overline{H} = E_m \; \theta_3 < \theta < \theta_m \qquad \text{and} \qquad \overline{H} = 0 \qquad \text{for } \theta < \theta_3 \text{ or } \theta > \theta_m \quad (8.65)$$

Wedge:

$$\overline{H} = \frac{M_0}{(\theta)^{1/2}} \; \theta_1 < \theta < \theta_2 \qquad \text{and} \qquad \overline{H} = 0 \qquad \text{for } \theta < \theta_1 \text{ or } \theta > \theta_2 \quad (8.66)$$

The functions that result when these distributions are inserted into Eq. (8.64) are the incomplete gamma function (for the wedge) and the exponential integral function (for the box), each evaluated between limits of θ (Fig. 8.16). A comparison of $E_T(t)$ for polyisobutylene with the fit based on a box and a wedge shows the degree of approximation of the wedge to the glass-transition region and the box to the rubbery flow region (Fig. 8.16).

8.7 EFFECT OF MOLECULAR WEIGHT

As long as the molecular weight exceeds that of a polymer segment, glassy behavior and T_g should be relatively insensitive to further increases in molecular weight. For our purposes here, this can be taken as 50 to 100 chain atoms, although there is no clear-cut definition of a segment for all situations. However, rubbery behavior, according to Eq. (8.23), depends on the chain density N in moles per unit volume. In fact, as α approaches unity,

$$E = \frac{d\sigma}{d\varepsilon} = \frac{d\sigma}{d\alpha} = 3RTN \qquad (8.67)$$

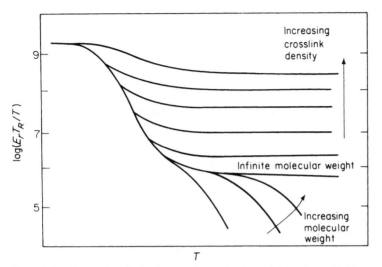

Figure 8.17 Qualitative effects of increasing molecular weight and cross-linking on the master curve

Therefore, we expect E to increase linearly with absolute temperature and with cross-link density for network polymers. In Fig. 8.17, lines corresponding to various cross-link densities are shown. The effect of T on E is corrected in the ordinate (T_R is any reference temperature). Very high cross-link densities preserve pseudo-glassy behavior to high temperatures. A thermoset such as a phenol-formaldehyde resin remains glassy with no measurable mechanical transition in the absence of chemical change. But a network polymer has an infinite molecular weight. For a finite molecular weight, the cross-links are temporary ones and are due to entanglements or to the time needed for segments to diffuse by one another. The effect of molecular weight on the curves also is shown (Fig. 8.17). At a low enough molecular weight, there is no rubbery plateau and the glass goes directly to a melt.

The actual level of the rubbery plateau modulus has been shown to be a function of polymer density and unperturbed chain dimensions [19]. The starting point is the realization that the plateau modulus is due to polymer entanglements which increase with molecular weight. The rubbery plateau shear modulus, G_N^0, is related to the molecular weight between entanglements, M_e, by

$$G_N^0 = \frac{\frac{4}{5}\rho k N_A T}{M_e} \tag{8.68}$$

where ρ is the density, k is the Boltzmann constant, and N_A is Avogadro's number. For most incompressible materials including most rubbery melts, Poisson's ratio is $1/2$, so $E = 3G$ (Table 8.1).

A packing length, p_c, is introduced, which is related to the unperturbed dimension (Sec. 7.2) by

$$\frac{1}{p_c} = \left(\frac{\overline{r_0^2}}{M}\right) \rho N_A \tag{8.69}$$

where M is the molecular weight. It will be recalled that $\left(\dfrac{\overline{r_0^2}}{M}\right)$ is nearly constant for each homopolymer at a given temperature. It can be shown that the molecular weight between entanglements can be related to the packing length by

$$M_e = B^{-2} p_c^3 (\rho N_A) \tag{8.70}$$

where B is a dimensionless factor equal to about 0.051 at 140°C. Thus, by combining equations, the modulus can be estimated from a measurement of the unperturbed dimension.

$$G_N^0 = \frac{\frac{4}{5} B^2 k T}{(p_c)^3} \tag{8.71}$$

As an example, the values for poly(methyl methacrylate) used by Fetters et al. [19] are $\left(\dfrac{\overline{r_0^2}}{M}\right) = 0.425$ Å^2·mol/g and $\rho = 1.13$ g/cm^3 at 140°C.

The calculated values are:

$$p_c = 3.46 \text{ Å} \qquad \text{[from Eq. (8.69)]}$$
$$M_e = 10.5 \times 10^3 \text{ g/mol}$$
$$G_N^0 = 0.29 \text{ MPa} \qquad \text{[from Eq. (8.71)]}$$

This is very close to an experimentally measured value of $G_N^0 = 0.31$ MPa.

Measured values of the plateau modulus range from 0.18 MPa for poly(dimethyl siloxane) to 2.50 MPa for polyethylene [19]. However, it can be seen that the level of the rubbery modulus is not affected by molecular weight, since the packing length depends on the ratio of unperturbed end-to-end distance to molecular weight, a relatively constant number for any one homopolymer. On the other hand, the breadth of the plateau in terms of time or temperature span is increased as the molecular weight is increased (Fig. 8.17).

8.8 EFFECT OF CRYSTALLINITY

The main response of a partially cyrstalline material to an external stress is borne by the amorphous fraction. In materials of low crystallinity, say, less than 30% crystallinity, the crystallites function as temperature-dependent cross-links. We expect the glassy and rubbery plateaus to exist as before, but we expect the rubbery region to undergo a time-independent, fairly sharp transition at the melting tem-

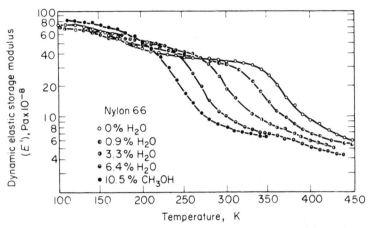

Figure 8.18 Dynamic modulus vs. temperature for nylon 66 containing various amounts of water or methanol [20]

perature T_m. Of course, smaller crystallites and those formed at lower temperatures have lower T_m's than others. This blurs the transition.

With highly crystalline materials, the glass transition may become less pronounced. The behavior of nylon 66 is illustrative (Fig. 8.18). The glass transition is quite sensitive to moisture but involves a modulus change of only about 10-fold compared with 1000-fold for a completely amorphous material (Fig. 8.13). However, both above and below T_g there is a continuous loss in stiffness due to decrease in crystalline content.

When an oscillating stress is imposed on the material, the fraction of energy dissipated per cycle (logarithmic decrement) is a sensitive measure of transitions. The data for partially crystalline tetrafluoroethylene (Fig. 8.19) show four transitions below the melting temperature, which is about 600 K. At 176 K the transition is thought to be the hindered rotation of small segments of the polymer chains. The transition of 292 K marks changes in crystalline morphology from triclinic to disordered hexagonal and a change in the helical repeating unit of 13 to 15 chain atoms. At 303 K there is further disordering of the hexagonal lattice to an irregular repeating unit. At 400 K we have the main glass transition T_g.

8.9 EFFECT OF FILLERS

In the region of small deformations, fillers have much the same role as crystallites, especially if the filler surface is wetted by the polymer. The behavior of carbon black in a cross-linked natural rubber is typified by Fig. 8.20 showing an increase

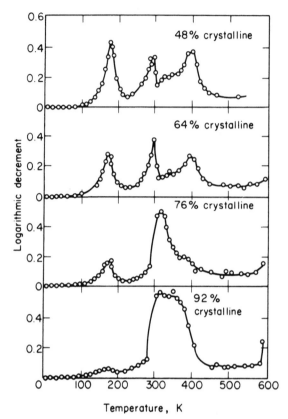

Figure 8.19 Variation with temperature of the logarithmic decrement for samples of poly(tetrafluor-ethylene) of various degrees of crystallinity [21]

in stiffness (spring constant) with loading and a moderate increase in damping, i.e., fractional energy loss per cycle. Note that the stiffness increases with tempera-ture in accordance with Eq. (8.23) when no filler is added. The situation is more complex when filler is present. The modulus of a glassy polymer is increased by fillers, and T_g may be increased somewhat also (Fig. 8.21).

8.10 OTHER TRANSITIONS

There are other transitions in molecular conformations that give rise to loss peaks in oscillatory experiments such as the torsion pendulum. Often the modulus changes by such a small amount at a minor transition that performance properties are not noticeably affected. It has been pointed out that an additional amorphous

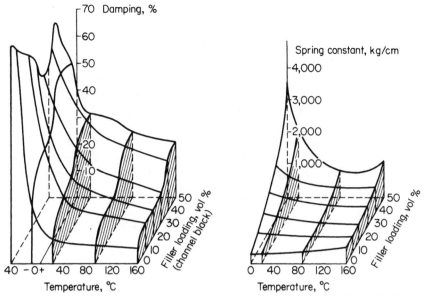

Figure 8.20 Contour representation of dependence of dynamic properties of natural rubber on filler loading and temperature (after Ecker [22])

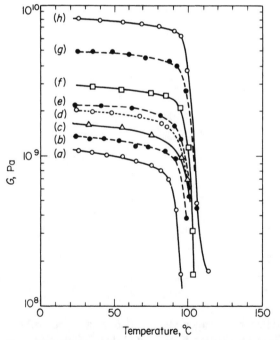

Figure 8.21 Shear modulus vs. temperature for (a) polystyrene and for polystyrene containing (b) 20% calcium carbonate, (c) 20% asbestos, (d) 20% mica, (e) 40% asbestos, (f) 60% asbestos, (g) 40% mica, and (h) 60% mica [23]

change occurs well below the main glass transition temperature of poly(tetrafluor-ethylene) (Fig. 8.19). In some literature the main T_g is designated as the α transition, and transitions at successively lower temperatures are termed β and γ. For nylon 66, loss peaks identified with the amorphous region are seen at 200 K (β transition) and at 135 K (γ transition) [24]. The storage modulus (Fig. 8.18) scarcely changes at those temperatures.

Much more controversial is the observation of a loss peak above T_g in the amorphous melt. This $T_{1,1}$ transition temperature represents a change from one liquid state to another [25, 26]. For polystyrene $T_{1,1}$ is about 70°C higher than T_g; and for polyisobutylene it is about 105°C higher than the T_g measured by the same method.

KEYWORDS FOR CHAPTER 8

Section

1. Poisson's ratio
   Storage modulus
   Young's modulus
   Tensile strain
   Tensile stress
   Compliance
2. Ideal rubber
   Rubber elasticity
3. Maxwell model
   Creep
   Stress relaxation
   Harmonic motion
4. Torsion pendulum
   Log decrement
5. Viscoelasticity (five regions)
   Glassy plateau
   Rubbery plateau
6. Generalized Maxwell model
10. β Transition

PROBLEMS

8.1 An unstressed rubber band becomes longer when heated. A stretched rubber band with constant load becomes shorter when heated. Is this behavior consistent with Eq. (8.23)? Explain.

8.2 An ideal rubber is stretched to 1.10 times its unstretched length at 27°C (300

K) supporting a stress of 150 psi. After aging 10 days in air at 127°C, the stress at 1.10 times its unstretched length is only 100 psi (at 127°C). If the original polymer contained 0.015 mol of cross-links per unit volume originally, how many moles of cross-links per unit volume were lost or gained on aging?

8.3 The following data are obtained on a sample of natural rubber. Calculate the stress to be expected at 100% elongation for the dry rubber (in Pa) at 27°C.

Swelling in benzene at 27°C:

$$\chi = 0.44$$
$$\upsilon_2 = 0.20$$
$$V_1 = 100 \text{ cm}^3/\text{mol}$$
$$R = 8.31 \text{ J/mol·K}$$

8.4 An ideal rubber has a nonlinear stress-strain curve. If a 1.00-cm cube of such a rubber is compressed to 0.98 cm by a mass of 3 kg, what stress is going to be required to stretch a 5-cm strip of the same rubber to a length of 7.5 cm?

8.5 An ideal rubber band is stretched to a length of 15.0 cm from its original length of 6.00 cm. It is found that the stress at this length increases by an increment of 1.5×10^5 Pa when the temperature is raised 5°C (from 27° up to 32°). What modulus ($E = \sigma/\varepsilon$) should we expect to measure at 1% elongation at 27°C? Neglect any changes in volume with temperature.

8.6 The area under a reversible stress-strain curve represents the energy stored per unit volume. For an ideal rubber strip, 40.0 J of work is expended in extending it to an elongation of 100% at 27.0°C. If the rubber strip is 30.0 cm long, 1.25 cm thick, and 2.50 cm wide, what is the chain density in mol/cm³?

8.7 Blocks A and B are connected by three ideal, flexible rubber cables. The cables are as follows:

Cable	Diameter, cm	Length, ft
1 (neoprene)	0.250	2.00
2 (SBR)	0.500	2.50
3 (silicone)	0.350	3.00

The force needed to separate block A from B is as follows:

Distance between blocks, ft	Force, newtons
2.00	0.0
2.35	100.0
3.55	485.0
4.00	625.0

What force is required to separate the blocks a distance of 5.00 ft if cable 3 is cut so that only cables 1 and 2 connect A with B?

8.8 An ideal rubber band is stretched to 1.20 times its unstretched length (20% elongation) at room temperature (27°C). The force required is 2.00 N. The sample is allowed to relax to its original length. Now the temperature of the relaxed sample is raised to 127°C and it is aged there for one week. At the end of the week, the force required to stretch the sample to 1.30 times its unstretched length (measured at 127°C) is 2.25 N. Assume that the overall dimensions of the rubber band are not changed significantly by temperature or aging. If the original sample contained 0.020 mol of cross-links per unit volume, how many moles of cross-links per unit volume were lost or gained upon aging?

8.9 A polymer can be represented by two Maxwell elements connected in parallel, each with the same spring modulus. In a stress relaxation experiment the stress decreases to 23.7% of its initial value after 10 min. If the relaxation time of the first element is 10 min, what is the relaxation time of the second element?

8.10 A certain polymer sample can be represented by a single Maxwell element. When a tensile stress of 10^3 Pa is applied for 10 s, the maximum length attained by the sample is 1.15 times the original length. After removal of the stress (at 10 s), the length is only 1.10 times the original length. What is the relaxation time of the element?

8.11 A Maxwell element is placed in parallel with an ideal rubber band. In a stress relaxation experiment, the initial stress supported at a fixed elongation (stretched length = 1.5 × unstretched length) is 145,000 Pa. After 250 s, the stress decays to 120,000 Pa. However, if the fixed elongation corresponds to 2.5 × unstretched length, the initial stress is 367,000 Pa. Going back to the first experiment (with fixed elongation = 1.5 × unstretched length), what *stress* should be expected after 500 s?

8.12 A Maxwell-type element is constructed of the usual viscous dashpot but with an "ideal rubber" band in place of the usual Hookean spring. In a creep experiment with a constant load of 1000 Pa, and at a temperature of 27°C, the total deformation in tensile strain units after 3 h is $\varepsilon = 2.00$. When the load is removed (at 3 h), the resulting permanent set is 0.750 units. Now the same element (with no starting deformation) is made to undergo stress relaxation at 127°C with a fixed elongation of 1.5 strain units. What is the stress at time zero in the stress relaxation experiment?

8.13 Maxwell element A in creep at a constant stress of 10,000 Pa exhibits a total elongation of 0.65 strain units after 100 h and a permanent set of 1.00 unit on removal of the load after a total of 250 h. Now a generalized Maxwell model is constructed made up of elements A, B, and C. B has twice the spring modulus and one-half the relaxation time of A, while C has one-half the spring modulus and twice the relaxation time of A. If a sudden elongation of 1.80 units is applied and held constant on the generalized Maxwell model, what stress will be exerted by it after 100 h?

8.14 Assume that a polymer can be approximated by three Maxwell elements in parallel each having the same modulus but with viscosities for elements 1, 2, and 3 having the relationship

$$\eta_1 = 4\eta_2 = 9\eta_3$$

Plot log (σ/σ_0) vs. t/θ_1 and vs. log (t/θ_1) for the overall assembly between $t/\theta_1 = 0.01$ and $t/\theta_1 = 2$, where θ_1 is the relaxation time for element 1. What is the relaxation time [t for ln $(\sigma/\sigma_0) = -1$] for the assembly?

8.15 Two Maxwell elements are tested separately in creep experiments in which the stress is equal to σ_0. The time to reach an elongation ε of 150% is measured. At that exact time, the stress is removed and the permanent set is measured. Next, the two elements are combined in parallel (a generalized Maxwell model) and subjected to a stress relaxation experiment with a total elongation of ε_0. What is the ratio of the stress at 250.0 s compared to that at 50.0 s?

	Creep data	
	Time to reach 150% elong.	Permanent set
Element 1	120 s	30.0%
Element 2	186 s	55.0%

8.16 At times such that $t \ll \theta_1$, a polymer has a glassy modulus of E_g. At times greater than $t = \theta_2$, it exhibits liquid flow. Between θ_1 and θ_2, the transition can be described by a continuous distribution of relaxation times, $E(\theta) = Y/\theta^2$. Derive an equation for the time-dependent modulus $E(t)$ as a function of t, Y, θ_1, and θ_2. Also, express E_g in terms of Y, θ_1, and θ_2. Plot the curve of log $E(t)$ vs. log t from $t = 0.01$ to $t = 100$ s for the following set of constants: $E_g = 10^9$ Pa; $\theta_1 = 0.1$ s; $\theta_2 = 10$ s. How does the result compare with the relaxation of a single Maxwell element?

8.17 A "generalized" Maxwell model has a "distribution of relaxation times," E_i (θ) given by

$$E_i(\theta) = k_1\theta^{-2} + k_2 \quad \text{and} \quad \theta = 1, 2, \& 3 \text{ min only}$$

In a stress relaxation experiment, the initial stress of 2.50×10^8 dyn/cm² decays to 1.00×10^8 dyn/cm² after 1.50 min. What is the expected value of stress 3.75 min after the initial stress is applied?

8.18 Samples of polymers X and Y behave like perfect Maxwell elements with the same spring modulus and the same initial length (20.0 cm). When the two samples are connected in series with a constant force of 2.00 kPa, the assembly stretches to a total length of 90.0 cm in 120 s. At that time the force is removed and the assembly contracts to a length of 60.0 cm. Sample X now is 25.0 cm long and sample Y is 35.0 cm long.

If two new samples of X and Y (each 20.0 cm long) are connected in parallel and subjected to an instantaneous stretching to 30.0 cm, what fraction of the initial stress will remain after 5.00 min? (The assembly is held at a total length of 30.0 cm throughout this period.)

8.19 A minimum of two Maxwell elements in parallel is required to represent the "master curve" of (log) modulus vs. (log) time for a real polymer. In a hypothetical case the glassy plateau modulus (extending to time = 0) is 1.20 $\times$ 10^9 Pa and the rubbery plateau modulus (extending to time = ∞) is 0.300 $\times$ 10^7 Pa. If the time corresponding to a modulus of 2.00 $\times$ 10^8 Pa is 150 s, what are the appropriate values of the spring constants and relaxation times for the two elements?

8.20 The Voigt (or Kelvin) element resembles the Maxwell element except that the spring and dashpot are in parallel rather than in series. Show that the creep compliance of such an element is

$$J(t) = (E)^{-1}\left[1 - \exp\left(-\frac{tE}{\eta}\right)\right]$$

8.21 For 4 Voigt elements (see Prob. 8.20) in series for a creep experiment, plot compliance vs. time on log-log paper for each element and add them to get the total compliance. What constants do you obtain for a Nutting equation fitted between the times $t = 10$ and $t = 1000$ s?

	Element number			
	1	2	3	4
E_i, Pa	0	1.0×10^4	1.333×10^4	2.0×10^4
η_i, Pa·s	2.0×10^6	5.0×10^5	1.333×10^5	4.0×10^4

8.22 An amorphous polymer in a creep experiment behaves like a Maxwell element joined in series with a Voigt element. A constant load of 10^4 Pa is applied to a tensile specimen at time $t = 0$. After 10 h the strain is 0.05 cm/cm. The load is removed. The strain can be described *thereafter* by

$$\text{Strain} = \frac{3 + e^{-t'}}{100} \quad \text{where } t' = t - 10 \text{ (h)}$$

Identify and evaluate the four model parameters. Use proper units.

8.23 An ideal rubber has a density of 0.93 g/cm³ and an average molecular weight between cross-links of 1250 g/mol. It is placed in a torsion pendulum (rectangular specimen) and cooled to 100 K, which is 80°C below the glass transition temperature. In the pendulum it exhibits a frequency of 55 cycles/s. What frequency do we expect to observe at 27°C? Assume that changes in dimensions of the sample are negligible between the extremes of temperature. In the glassy state it resembles room-temperature polystyrene, that is, a shear modulus of 3.0 GPa.

8.24 Samples of a given amorphous polymer are strained 1% at a variety of temperatures. Draw a typical diagram for stress (measured after 10 s) vs. temperature for the polymer. Identify on the diagram (a) T_g and (b) the rubbery plateau. How would the diagram be changed if:

(c) The material were degraded to 10% of the original molecular weight?
(d) The material were loosely cross-linked?
(e) The diagram were plotted using stress measured after 10 h?

8.25 In a certain test a constant stress is instantly applied to a dumbbell sample (in tension). Three materials are tested. A rubber reacts by deforming rather rapidly to start and then more and more slowly until it reaches an equilibrium extension. A thermoplast reacts by deforming almost instantly to some substantial extension and then deforming further linearly with time. A thermoset reacts by deforming almost instantly to an equilibrium extension.

(a) Describe the behavior of each material with an appropriate combination of dashpots and springs.
(b) Describe what happens to each model when the stress is instantaneously removed.

8.26 How much of a shift in T_d (temperature at peak in mechanical loss) do you expect from the WLF equation when the time scale of the test is decreased by a factor of 10?

8.27 Creep data for a sample of plasticized poly(vinyl chloride) at six temperatures can be represented by the Nutting equation with constants as follows: $J(t) = A't^n$ and $10 \text{ s} < t < 1000 \text{ s}$.

Temperature T, °C	A' s^n/Pa	n
37	9.5×10^{-10}	0.08
44	1.3×10^{-9}	0.19
50	1.3×10^{-9}	0.68
53	2.2×10^{-8}	0.37
63	2.2×10^{-7}	0.18
72	5.2×10^{-7}	0.10

Assuming $T_g = 50°C$, construct a plot of $J(t)$ vs. log t at $50°C$ over the range of t from 10^{-3} to 10^7 s by choosing shift factors according to the WLF equation. If a sample has a strain of 0.0005 units after 5 s of creep at constant stress at $50°C$, how long will it take (in seconds) to grow to a strain of 0.050 unit?

REFERENCES

1. Flory, P. J.: "Principles of Polymer Chemistry," chap. 11, Cornell Univ. Press, Ithaca, N.Y., 1953.
2. Anthony, R. L., R. H. Caston, and E. Guth: *J. Phys. Chem.*, 46:826 (1942).
3. Mullins, L., and A. G. Thomas, in L. Bateman (ed.): "The Chemistry and Physics of Rubber-like Substances," chap. 7, Wiley, New York, 1963.
4. Gumbrell, S., L. Mullins, and R. S. Rivlin: *Trans. Faraday Soc.*, 49:1945 (1953).
5. Martin, G. M., F. L. Roth, and R. D. Stiehler: *Trans. Inst. Rubber Ind.*, 32:189 (1956); *Rubber Chem. Technol.*, 30:876 (1957).
6. Wood, L. A.: *Rubber Chem. Technol.*, 32:1 (1959).
7. Nielsen, L. E.: "The Mechanical Properties of Polymers," p. 55, Reinhold, New York, 1962.
8. Nielsen, L. E.: "Mechanical Properties of Polymers and Composites," vol. 1, chap. 4, Dekker, New York, 1974.
9. Whorlow, R. W.: "Rheological Techniques," chap. 5, Halsted (Wiley), New York, 1980.
10. Wetton, R. E., in J. Dawkins (ed.): "Developments in Polymer Characterization," chap. 5, Applied Science, New York, 1984.
11. Xue, G., J. Dong, and J. Ding, in N. P. Cheremisinoff and P. N. Cheremisinoff (eds.): "Handbook of Advanced Materials Testing," chap. 40, Dekker, New York, 1995.
12. Grubb, D. T., in R. W. Cahn, P. Haasen, and E. J. Kramer (eds.): "Materials Science and Technology: Structure and Properties of Polymers," chap. 7, VCH, New York, 1993.
13. Koleske, J. V., and J. A. Faucher: *Polym. Eng. Sci.*, 19:716 (1979).
14. Kiran, E., and J. K. Gillham: *Polym. Eng. Sci.*, 19:699 (1979).
15. Gillham, J. K.: *Polym. Eng. Sci.*, 19:676 (1979).
16. Tobolsky, A. V.: "Properties and Structure of Polymers," chap. 4, Wiley, New York, 1960.
17. McLoughlin, J. R., and A. V. Tobolsky: *J. Colloid Sci.*, 7:555 (1952).
18. Tobolsky, A. V.: "Properties and Structure of Polymers," chap. 3, Wiley, New York, 1960.
19. Fetters, L. J., D. J. Lohse, D. Richter, T. A. Witten, and A. Zirkel: *Macromolecules*, 27:4639 (1994).
20. Sauer, J. A.: *SPE Trans.*, 2:57 (1962).
21. Sinnot, K. M.: *SPE Trans.*, 2:65 (1962).
22. Gehman, S. D.: *Rubber Chem. Technol.*, 30:1202 (1957).
23. Nielsen, L. E., R. A. Wall, and P. G. Richmond: *SPE J.*, 11:22 (1955).
24. Murayama, T.: "Dynamic Mechanical Analysis of Polymeric Material," p. 65, Elsevier, New York, 1978.
25. Boyer, R. F.: *Polym. Eng. Sci.*, 19:732 (1979).
26. Gillham, J. K.: *Polym. Eng. Sci.*, 19:749 (1979).

General References

Aklonis, J. J., and W. J. MacKnight: "Introduction to Polymer Viscoelasticity," 2d ed., Wiley, New York, 1983.
Arridge, R. G. C.: "An Introduction to Polymer Mechanics," Taylor & Francis, Philadelphia, 1985.

Bateman, L. (ed.): "The Chemistry and Physics of Rubber-like Substances," Wiley, New York, 1963.

Blokland, R.: "Elasticity and Structure of Polyurethane Networks," Gordon & Breach, New York, 1969.

Boyer, R. F. (ed.): "Technological Aspects of the Mechanical Behavior of Polymers," Wiley-Interscience, New York, 1975.

Boyer, R. F., and S. Keinath (eds.): "Molecular Motion in Polymers by ESR," Harwood Academic, New York, 1980.

Brown, R. P. (ed.): "Physical Testing of Rubber," 2d ed., Elsevier, New York, 1986.

Brown, R. P., and B. E. Read (eds.): "Measurement Techniques for Polymeric Solids," Elsevier Applied Science, New York, 1984.

Clegg, D. W., and A. A. Collyer (eds.): "Mechanical Properties of Reinforced Thermoplastics," Elsevier Applied Science, New York, 1986.

Collyer, A. A., and D. W. Clegg (eds.): "Rheological Measurement," Elsevier, New York, 1988.

Collyer, A. A., and L. A. Utracki (eds.): "Polymer Rheology and Processing," Elsevier, New York, 1990.

Craver, C. D., and T. Provder (eds.): "Polymer Characterization: Physical Property, Spectroscopic, and Chromatographic Methods," ACS, Washington, D.C., 1990.

Deanin, R. D.: "Polymer Structure—Properties—Applications," Cahners, Boston, 1971.

Dunn, A. S. (ed.): "Rubber and Rubber Elasticity," Wiley-Interscience, New York, 1975.

Eirich, F. R. (ed.): "Rheology," vols. 1–4, Academic Press, New York, 1956–1964.

Enikoloyan, N. S. (ed.): "Filled Polymers I. Science and Technology," Springer, Secaucus, N.J., 1990.

Ferry, J. D.: "Viscoelastic Properties of Polymers," 3d ed., Wiley, New York, 1980.

Flory, P. J.: "Principles of Polymer Chemistry," Cornell Univ. Press, Ithaca, N.Y., 1953.

Flory, P. J.: "Statistical Mechanics of Chain Molecules (Revised)," Hanser-Gardner, Cincinnati, Ohio, 1991.

Goldman, A. Y.: "Prediction of the Deformation Properties of Polymeric and Composite Materials," ACS, Washington, D.C., 1993.

Gordon, G. V., and M. T. Shaw: "Computer Programs for Rheologists," Hanser-Gardner, Cincinnati, Ohio, 1994.

Haward, R. N. (ed.): "The Physics of Glassy Polymers," Halsted (Wiley), New York, 1973.

Hwang, C. R., C. C. Lin, and G. Matis: "Computer Aided Analysis of the Stress/Strain Response of High Polymers," Technomic, Lancaster, Pa., 1989.

Krishnamachari, S. I., and L. J. Broutman: "Applied Stress Analysis of Plastics," Chapman & Hall, New York, 1993.

Lal, J., and J. E. Mark: "Advances in Elastomers and Rubber Elasticity," Plenum, New York, 1986.

Lenz, R. W., and R. S. Stein (eds.): "Structure and Properties of Polymer Films," Plenum, New York, 1973.

Macosko, C. W.: "Rheology: Principles, Measurements and Applications," VCH, New York, 1994.

Mark, J. E., A. Eisenberg, W. W. Graessley, L. Mandelkern, E. T. Samulski, J. L. Koenig, and G. D. Wignall: "Physical Properties of Polymers," 2d ed., ACS, Washington, D.C., 1993.

Matsuoka, S.: "Relaxation Phenomena in Polymers," Hanser-Gardner, Cincinnati, Ohio, 1992.

McCauley, J. W., and V. Weiss (eds.): "Materials Characterization for Systems Performance and Reliability," Plenum, New York, 1986.

McCrum, N. G., C. P. Buckley, and C. B. Bucknall: "Principles of Polymer Engineering," Oxford, New York, 1988.

Meier, D. J. (ed.): "Molecular Basis of Transitions and Relaxations," Gordon & Breach, New York, 1978.

Mena, B., A. Garcia-Rejon, and C. Rangel-Nafaile (eds.): "Advances in Rheology," 4 vols., Elsevier Applied Science, New York, 1984.

Miller, M. L.: "The Structure of Polymers," Reinhold, New York, 1966.

Mitchell, J., Jr. (ed.): "Applied Polymer Analysis and Characterization," Hanser-Gardner, Cincinnati, Ohio, 1992.

Murayama, T.: "Dynamic Mechanical Analysis of Polymer Material," 2d ed., Elsevier, Amsterdam, 1982.

Nielsen, L. E., and R. F. Landel: "Mechanical Properties of Polymers and Composites," 2d ed., Dekker, New York, 1993.

Ogorkiewicz, R. M. (ed.): "The Engineering Properties of Plastics," Oxford, New York, 1970.

Ogorkiewicz, R. M. (ed.): "Thermoplastics: Properties and Design," Wiley-Interscience, New York, 1974.

Peterlin, A. (ed.): "Plastic Deformation of Polymers," Dekker, New York, 1971.

Proc. Roy. Soc. London Ser. A, No. 1666 (19 Nov. 1976). This issue has five papers on rubber elasticity, pp. 297–406.

Read, B. E., and G. D. Dean: "The Determination of Dynamic Properties of Polymers and Composites," Wiley, New York, 1978.

Ritchie, P. D. (ed.): "Physics of Plastics," Van Nostrand-Reinhold, Cincinnati, 1965.

Samuels, R. J.: "Structured Polymer Properties," Wiley-Interscience, New York, 1974.

Schultz, J., and S. Fakirov: "Solid State Behavior of Linear Polyesters and Polyamides," Prentice Hall, Englewood Cliffs, N.J., 1990.

Serafini, T. T., and J. L. Koenig: "Cryogenic Properties of Polymers," Dekker, New York, 1968.

Shah, V.: "Handbook of Plastics Testing Technology," Wiley, New York, 1984.

Skrzypek, J. J., and R. B. Hetnarski: "Plasticity and Creep," CRC Press, Boca Raton, Fla., 1993.

Spells, S. J. (ed.): "Characterization of Solid Polymers: New Techniques and Developments," Chapman & Hall, New York, 1994.

Struik, L. C. E.: "Internal Stresses, Dimensional Instabilities and Molecular Orientations in Plastics," Wiley, New York, 1990.

Thomas, E. L. (ed.): "Materials Science and Technology. Structure and Properties of Polymers," VCH, New York, 1993.

Tobolsky, A. V.: "Properties and Structure of Polymers," Wiley, New York, 1960.

Treloar, L. R. G.: *Reports Prog. Phys.,* 36:755 (1973); also reprinted as *Rubber Chem. Tech.,* 47:625 (1974).

Treloar, L. R. G.: "The Physics of Rubber Elasticity," 3d ed., Oxford Univ. Press, New York, 1975.

Turner, S.: "Mechanical Testing of Plastics," Wiley, New York, 1986.

Van Krevelen, D. W., and P. J. Hoftyzer: "Properties of Polymers," 2d ed., Elsevier, New York, 1976.

Ward, I. M., and D. W. Hadley: "An Introduction to the Mechanical Properties of Solid Polymers," Wiley, New York, 1993.

Williams, J. G.: "Stress Analysis of Polymers," 2d ed., Halsted-Wiley, New York, 1980.

Wunderlich, B.: "Macromolecular Physics," 3 vols., Academic Press, New York, 1976–1980.

Zachariades, A. E., and R. S. Porter (eds.): "The Strength and Stiffness of Polymers," Dekker, New York, 1983.

Zachariades, A. E., and R. S. Porter (eds.): "High Modulus Polymers," Dekker, New York, 1987.

NINE

ULTIMATE PROPERTIES

9.1 FAILURE TESTS

Most industrial tests of polymer systems are carried to failure with some degree of simulation between test and end use. A manufacturer of plastic cups is more interested in a fast breaking test that simulates dropping a cup than in slow, steady deformation. The compromise that is reached is between a specific test (dropping a cup, wearing out a tire on the road, parachuting greater weights until the cords on the chute break) and a more general one (impact strength, sandpaper abrasion, high-speed tensile strength).

Usually, the *ultimate strength* of a material is the stress at or near failure. For most materials, failure is catastrophic with a complete break. However, some materials, especially spherulitic crystalline polymers, reach a point where a large inelastic deformation starts (*yielding*) but continue to deform and absorb energy long beyond that point. Even when strength is defined unequivocally and measured with care and statistical confidence, it can be a misleading criterion for the unwary designer. Although it is one of the most widely specified parameters, tensile strength is almost never the limiting factor in a practical application. Seldom, if ever, is a polymer subjected in an end use to a single, steady deformation in the absence of aggressive environments or stress concentrators. In garments, rugs, instrument cases, tires, book covers, and upholstery, polymers fail from repeated stresses, from impact, from penetration by sharp objects, or by propagation of a tear. It is only insofar as the tensile strength correlates with toughness, tear resistance, and fatigue strength that it has real merit.

Toughness can be defined as the energy absorbed at failure. The area under the stress-strain curve has the units of joules per cubic meter when stress is in pascals and strain is in meters per meter. If stress is in pounds per square inch and strain is expressed as inches per inch, the area under the curve has the units of inch-pounds per cubic inch, which also represents energy per unit volume. The energy may be stored elastically as in unfilled rubber, or it may be dissipated as heat as in the permanent deformation of a crystalline material. A rod made from a glass with a strength of 100 MPa can be used to suspend 10 times the weight that a rubber rod with a strength of 10 MPa can withstand. But because the rubber may have an elongation at break that is a thousand times larger than that of the glass, the energy that the rubber rod can absorb before breaking may be many times that of the glass rod.

There is an important qualitative difference between the mechanical properties at small deformations and those at large. The modulus, for example, truly is a *material property*. On the other hand, most ultimate properties are *sample properties*. Failure invariably takes place at a defect or stress concentrator. Every value reported must be regarded as an average that represents a distribution. A typical frequency function found useful for rubber testing is the double exponential [1], which can be represented in integral form by

$$\sigma_b(n) = \overline{\sigma_b} + s \ln \left(- \ln \frac{n - 0.5}{N} \right) \tag{9.1}$$

where $\sigma_b(n)$ is the strength of the nth sample when all N samples of a population are ranked in order of decreasing strength. The mode of the distribution is $\overline{\sigma_b}$, and the standard deviation is s. An analysis of 24 tests shows values for tensile strength that vary from 25 to 29 MPa (3500 to 4100 psi) (Fig. 9.1). All points are equally valid, since all are part of the expected distribution. If the twenty-fourth point had a tensile strength of 20 MPa (3000 psi), one would be justified in discarding it as being atypical of the distribution. In making plots for this purpose, one arranges the samples in order of decreasing tensile strength.

In presenting the strength data for fibers, a very similar distribution first proposed by Weibull (originally in 1939) has become popular [2]. When the stress at break covers a wide range, the logarithm scale for the stress may be used in place of the arithmetic scale of Fig. 9.1. Also, the length of the fiber tested may be taken into account. A long fiber is more likely than a short fiber to contain a flaw that limits stress. In a typical application, both the average strength and the slope of the probability curve were found to vary with strain rate and sample length for ultrahigh-strength polyethylene fibers [3]. In testing plastics and rubber, the use of dumbbell-shaped specimens is standard, although a longer straight length in the narrow section of the specimen could have the same effect as the length of a fiber.

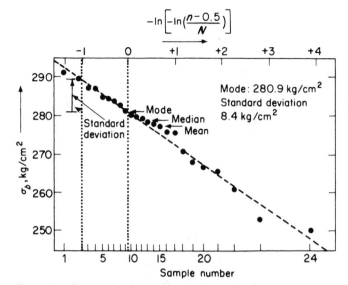

Figure 9.1 Stress at break plotted according to Eq. (9.1). All points are typical [1]

9.2 CONSTANT-RATE-OF-STRAIN TESTS

A common tensile test involves elongating a dumbbell-shaped sample held in jaws that separate at a constant rate (Fig. 9.2). The stress is measured as a function of time. Because the section of uniform cross section of the sample does not elongate at a constant rate, an independent measure of elongation is required if a stress-strain curve is to be constructed from the experiment. The true rate of extension in the center portion can be estimated by measuring with a ruler the separation of two *benchmarks* that were originally a convenient distance apart, say 1 in. At each half-inch or inch interval, a pip may be placed on the record of stress to indicate increments of strain. An *extensometer* (extension-measuring device) for materials that elongate between 2 to 10 times before breaking can be a pair of clamps that are connected by a tape. As the clamps separate, the tape moves with one and is pulled through the other. If the tape has a pattern that generates a signal at equal intervals, pips can be put on the stress record automatically. When the extension at break is very small, as with glassy materials, an elaborate means of monitoring extension is to follow and record the position of several luminescent bench marks by means of photocells and a servomotor. Despite the difficulty of measuring strain accurately, the dumbbell samples have the great advantage that ultimate failure will take place in the center of the sample and will not be affected by stress concentration at the jaws.

Molded, die-cut, or lathe-cut rings can be used for rubbery samples. The rate of strain is uniform, so that the stress-time record is also a stress-strain record.

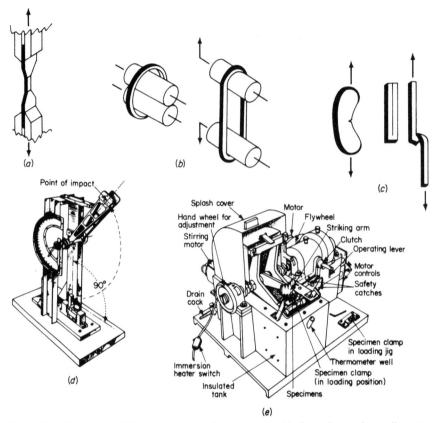

Figure 9.2 Mechanical failure tests: (a) tensile test on dumbbell specimen; (b) tensile test on ring specimen; (c) notched crescent and trousers tear specimens; (d) Izod impact test machine [2]; (e) motor-driven brittleness temperature tester [4]

With care the strength measured on ring samples is the same as that on dumbbells. However, stress concentration at the holder is difficult to avoid even when the holder is lubricated or rotated. Also, the initial portion of the stress-strain curve may be distorted because the stress needed to deform the ring into an oval may mask the stress due to uniform elongation.

Tests with fibers offer another complication. With the uniform cross section, any local yielding or necking is likely to start at the holders and the effective length of the sample may be increased by "jaw penetration." Long samples or multiple samples wound around spools at either end will minimize the uncertainty in length.

Anyone who has tried to open a cellophane-wrapped article has a real appreciation for the difference between the stress needed to rupture a smooth-edged film and the stress needed to propagate a tear. An ordinary piece of cellophane tape

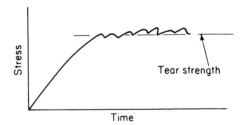

Figure 9.3 Typical trousers tear test diagram

with a cross section of a thousandth of a square inch will support a load of many pounds. The slightest notch with a razor on one edge will reduce the supportable load to several ounces. Although there are a number of tear-testing geometries to choose from, the *trousers tear test* has the merit that the stress to propagate the tear often stays at a constant value, making it easier to estimate than some others where the tear propagates so rapidly that only a peak stress can be read (Figs. 9.2 and 9.3).

We have pointed out that the extension at break for a glassy sample often is so small that it becomes hard to measure. In a flexural test, the elongation occurs on one side of the sample and compression on the other. A small elongation at break corresponds to a large flexural deflection, which is easily measured. The simple beam with sliding supports is used most often.

For an adhesive bond, the peel strength (normal to the bonded surface) or the shear strength (in the plane of the bonded surface) may be measured. For another application, the stress supportable in compression before yielding may be the most important parameter. This would be true for a rigid foam, for example.

9.3 BREAKING ENERGY

If more than enough energy is applied to rupture the sample, we can measure directly the total *energy to cause rupture,* or *toughness,* which is the area under the stress-strain curve. One such simple device is the impact tester, where a hammer has a certain potential energy before and after breaking the sample in flexure or tension. The energy used in breaking the sample is proportional to the difference in height before and after.

$$\text{Impact strength} = \text{energy to break sample per unit thickness} \qquad (9.2)$$

$$= (h_1 - h_2)\frac{W}{d} \text{ ft·lb/in}$$

Although most impact test machines in use today are calibrated in the older units of ft·lb/in, it is possible to express the results in the SI units of joules for energy and meters for width. One ft·lb/in equals 53.4 J/m.

The Izod test (Figs. 9.2 and 9.4) uses a cantilever beam, while the Charpy test

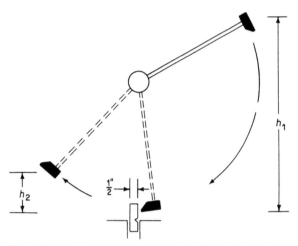

Figure 9.4   Impact testing (notched Izod). Original and final heights h_1 and h_2 of hammer determine strength of sample of thickness d held in vise

uses a simple beam. Since the notch and the width of the sample are specified, the energy is usually given per unit of notched thickness. In practice, to account for friction in the apparatus, h_0 is used rather than h_1; h_0 is the final height with no sample. Also, some energy goes into accelerating the broken piece of sample. This can be subtracted from the impact strength, but usually it is not. The impact strength of a bar sample is measured only on glassy or crystalline materials, since most rubbers will not fracture in this test.

A pendulum arm is used to break rubber and soft plastics in the "brittleness temperature" test (Fig. 9.2). The energy to break is not measured, but the temperature is progressively lowered until the samples do fracture. The brittleness temperature so established often corresponds to T_g or T_m.

9.4 CREEP FAILURE

A single *creep test* in tension or flexure gives information on long-term dimensional stability of a load-bearing element. When combined with a variable temperature, the test measures the *deflection temperature,* where creep rate changes rapidly. Once again, this is close to T_g. The load at which deflection took place also must be specified. A more specific test holds the sample in pipe form under fixed pressure until rupture occurs. Conducted in an aggressive environment such as soapy water or organic solvents, we get a type of *environmental-stress cracking.* Although such tests are useful in predicting the service life of a product, they are even more useful when run at different temperatures and then combined by time-temperature superposition (Sec. 9.6) in order to estimate service life at some constant temperature.

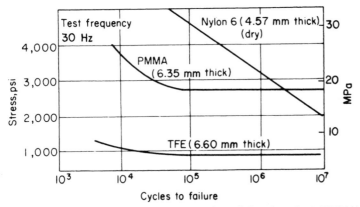

Figure 9.5 Fatigue life curves for nylon 6, poly(methyl methacrylate) (PMMA), and poly(tetrafluoro-ethylene) (TFE) [5]

9.5 FATIGUE

Most materials subjected to a stress repeatedly will fail even when that stress is well below the ultimate stress measured in a single deformation to the point of failure. When a stress is applied in an alternating fashion, as in a flexing test, the sample usually fails after a certain number of cycles. The lower the applied stress, the more cycles needed to cause failure. For some materials, the stress reaches a limit below which failure does not occur in a measurable number of cycles. Poly-(tetrafluoroethylene) and poly(methyl methacrylate) in Fig. 9.5 exhibit such an *en-durance limit* or *fatigue strength,* whereas nylon 6 does not. Rigid plastics such as the crystalline and glassy thermoplastics fail catastrophically, often with a sharp increase in temperature at the time of failing. A rubber being *flex-tested* may fail more gradually by the development of cracks or by the growth of a cut artificially introduced before testing. Also, in a rubber, the temperature rise due to mechanical hysteresis may be great enough to accelerate oxidative degradation.

9.6 REDUCED-VARIABLE FAILURE CORRELATIONS

The relationship between stress and strain at large values of strain follows the same pattern discussed in the preceding chapter. We expect glasses and highly cross-linked networks to be brittle, rubbers to have high elongations before breaking, and crystalline materials to be strong and tough if oriented and to "neck" and yield if not oriented. However, all of these types of behavior depend on the time scale of the test, the temperature at which it is run, the presence of fillers or other polymers, and even sample geometry and history. In general [6]:

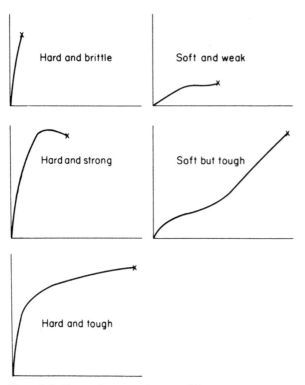

Figure 9.6 Types of stress-strain curves [6]

A soft and weak material will have a low modulus, a low yield stress, and a moderate to high elongation at the break point (Fig. 9.6).

A hard and brittle material will have a high modulus and a low elongation at break. It may not yield before breaking.

A soft but tough material has a low modulus but a high elongation and high stress at break. It may have a low yield stress.

The hard, strong material has high modulus, high yield stress, and high breaking stress and perhaps a moderate elongation at break.

The hard, tough material has a high modulus, high elongation, and high stress at break.

Any amorphous polymer that is stretched undergoes some orientation of polymer segments. In this oriented state, crystallization may occur, which will increase the effective number of cross-links (see Sec. 3.4). If crystallization does not occur, the behavior of the sample to rupture may be described by the same equation as small deformations. Figure 9.7 shows that the stress-strain curves for some elastic fibers correlate well (give a linear plot) with Eq. (8.25) up to the rupture point

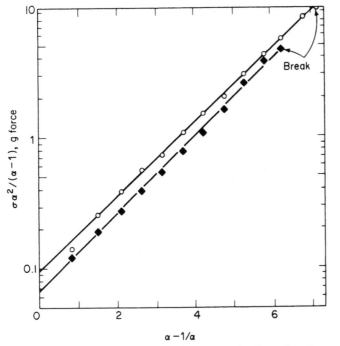

Figure 9.7 Stress-strain data to break for two spandex fibers plotted according to Eq. (8.25) [7]

(about 600% elongation). The most often used mechanical test is one that measures stress at constant strain rate. The ultimate tensile stress and strain at rupture varies with temperature and rate of strain. For a cross-linked unfilled rubber, which does not crystallize on stretching, the shift factors of the WLF equation (9.3) (see below) allow a reasonable superposition over a wide range of conditions. In Fig. 9.8, the stress at break σ_b is multiplied by the ratio of a reference temperature ($T_s = 263$ K) to the test temperature in accordance with the theory of rubber elasticity (Sec. 8.2). The shift factors a_T used to multiply the rate of extension R (meters per meter per second) are given approximately by the WLF equation in modified form, since T_s is not T_g (Fig. 9.9):

$$\log a_T = - \frac{8.86(T - T_s)}{101.6 + T - T_s} \tag{9.3}$$

A similar reduced curve can be produced for the elongation at break ε_b (Fig. 9.10). The same shift factors have been used here. The curves of stress and strain at break taken together can be used to construct a *failure envelope* in which we have eliminated the parameters of time and temperature. This general relationship is shown schematically in Fig. 9.11. Lines *OA*, *OB*, and *OC* are typical for constant-

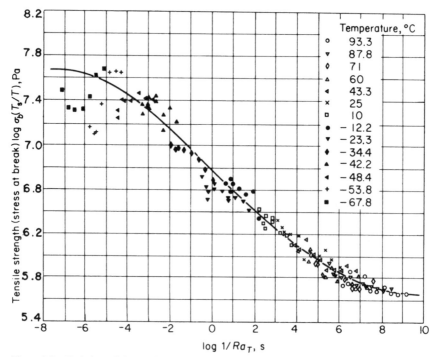

Figure 9.8 Variation of the tensile strength for a cross-linked styrene-butadiene rubber with reduced strain rate Ra_T. Reference temperature is 263 K [8]

rate-of-strain experiments at decreasing temperatures or increasing strain rates, respectively. In the case of strain rates at constant temperature, we can superpose a time grid on the diagram, since each strain on each line corresponds to a value of elapsed time.

This is illustrated in Fig. 9.12, where the dashed lines indicate test times (isochronal lines) that increase from a to g. The lowest line, OA, is the *equilibrium* elastic line for a cross-linked material. Its position is subject to the cross-link density and temperature but independent of time. All the constant-rate-of-strain lines exhibit higher stresses at the same strain, because part of the stress is used to dissipate viscous energy in the sample. Several other cases are illustrated in Fig. 9.11. For example, let us stretch a sample from O to point D and then hold it at constant stress. This becomes a creep experiment. By following the isochronal lines, we can depict the time-dependent strain that culminates when the point F on the equilibrium curve is reached. If at point D we held the strain constant, we would have a stress-relaxation experiment culminating in point E. On the other hand, if we extended the sample to point G and held it at constant stress or strain, the diagram predicts eventual failure. Experimental curves showing the stress relaxation culminating in failure for a synthetic rubber are shown in Fig. 9.13.

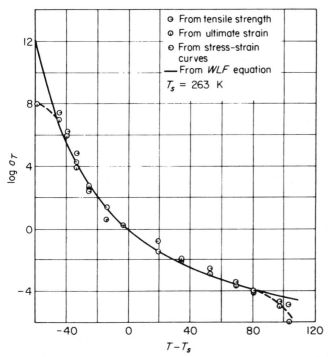

Figure 9.9 Experimental values of shift factor a_T from experiment and from WLF-type equation in which the reference temperature is 263 K, not T_g [8]

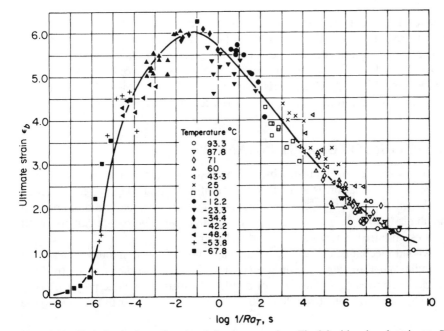

Figure 9.10 Variation in the strain at break for same sample as Fig. 9.8 with reduced strain rate Ra_T [8]

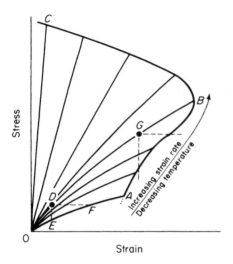

Figure 9.11 Schematic representation of failure envelope enclosing curves at constant strain rate. Dashed lines illustrate stress relaxation and creep under different conditions [8]

Actual failure envelopes for five cross-linked rubbery polymers are summarized in Fig. 9.14. The ordinate is the product of stress σ_b and strain ε_b at break, while the abscissa is the strain multiplied by the equilibrium modulus E_e. This allows superposition for samples varying in cross-link density. At low strains, the slope is unity, which corresponds to

$$\sigma_b = E_e\left(\frac{\varepsilon_b}{\varepsilon_b + 1}\right) \tag{9.4}$$

In general, the behavior of Eq. (9.4) is closer to that of Eq. (8.25) than that of Eq. (8.23). Curves for all the polymers in Fig. 9.14 are similar in shape. Natural rubber is not completely described, because at high elongations and lower temperatures it does crystallize. In Fig. 9.15 we show schematically the behavior of such crystallizable polymers. There is a characteristic increase in slope at high elongations because of the self-reinforcing nature of the oriented crystallites.

9.7 FRACTURE OF GLASSY POLYMERS

There are two major mechanisms by which glassy polymers yield in response to stress. Shear deformation is a localized phenomenon that may appear as shear bands. When a polystyrene cube is uniaxially compressed, the bands appear at a large angle (about 38°) to the compression direction. The bands can be about 1 μm thick and are made up of oriented polymer.

A second mechanism is crazing. Crazes are typically seen in a tensile test on a thin sample. What appear to be fine cracks at right angles to the applied stress are not cracks but crazes, "narrow zones of highly deformed and voided polymer" [10]. The typical craze contains from 20 to 90% voids and the rest fibrils oriented in the

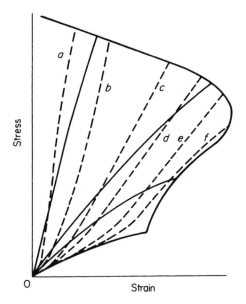

Figure 9.12 Isochronal lines in a hypotheti-cal failure envelope (time increases from *a* through *f*)

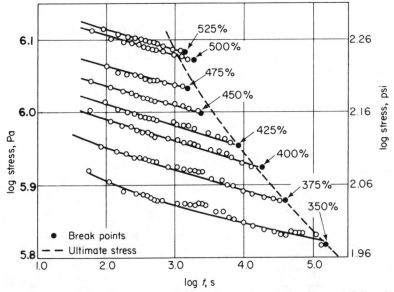

Figure 9.13 Stress relaxation of a cross-linked styrene-butadiene rubber at 1.7°C at elongations from 350 to 525%. Solid points indicate rupture [9]

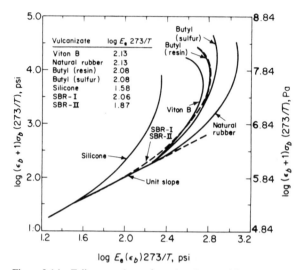

Figure 9.14 Failure envelopes for vulcanizates of five different polymers [9]

direction of the applied stress. The shear bands, in contrast, have very little void volume. Unlike actual cracks, both crazes and shear bands are capable of supporting stresses because of the oriented polymer involved. The failure of a glassy polymer in tension may involve both mechanisms [10–14].

When a crack propagates, a plastically deformed zone usually precedes the crack (Fig. 9.16). To quote Kambour [12], "Optimists concentrate on plastic deformation in crazes as a source of toughness or stress relief on polymers, while pessimists focus on crazing as the beginning of brittle fracture." The surfaces left on the crack retain a layer of craze material. The dimensions of a craze that has not yet developed into a crack are shown in the profile obtained for a craze in a thin film

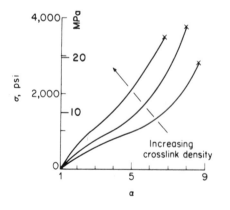

Figure 9.15 Typical stress-strain curves for a rubber that crystallizes at high elongations

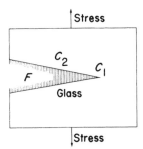

Figure 9.16 In response to a stress the unoriented glassy polymer is pulled into fibrils at the craze tip C_1. At the crack tip C_2 the fibrils rupture leaving a layer (typically about 500 nm thick) of craze material on the fracture surface F.

of polystyrene (Fig. 9.17). The total length of this craze was 189 μm. The individual fibrils connecting the two sides were about 15 nm in diameter [15]. Fibrils are an essential feature of crazes. The energy absorbed in their orientation process contributes to polymer toughness.

If crack propagation is carried out in the presence of a liquid that can act as a plasticizer, the stress needed to cause crazing may be decreased by the greater ability of the polymer to be obtained as fibrils [11, 15, 16]. *Environmental stress cracking* of polymers is studied as a practical test of performance for pipes and other load-bearing applications where the polymer will be exposed to an aggressive environment. The effects are seen not only in glassy polymers, but also in semicrystalline materials such as polyethylene.

To quantify the processes occurring when a crack propagates, a model system is an elliptical hole of length $2c$ in a large sheet of material (Fig. 9.18). The stress at fracture σ_f should be related to the modulus of the material E (since E represents energy storage capability) and to the energy γ to create the new surface area in the crack. Another approach relates the strain energy released per unit area of crack

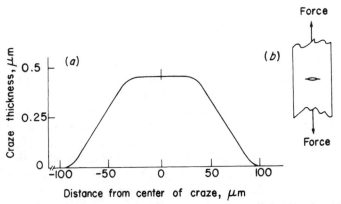

Figure 9.17 (a) Thickness profile for an isolated air craze in a thin polystyrene film [15]. (b) Thickness is measured in the direction of the applied force and is a measure of fibril length in the craze

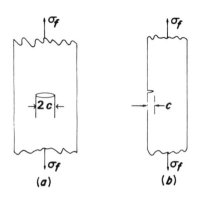

Figure 9.18 (a) A model system for crack propagation is an elliptical hole of length 2c in a large sheet of material subjected to a tensile stress σ_f. (b) It is often convenient to introduce the crack as a notch in a single edge of a sheet

growth. When the *strain energy release rate* $\mathfrak{g}$ exceeds 2γ, fracture will result. For the model system, the latter approach gives

$$\mathfrak{g}_{Ic}^2 = \frac{\sigma_f^2 \pi c}{E} \tag{9.5}$$

Unfortunately, various groups who study brittle fracture of polymers throughout the world choose to express their results in different ways. A related parameter used by some to describe the dependence of stress at fracture crack size is the *critical stress intensity factor* K_{Ic}:

$$K_{Ic}^2 = 2E\gamma = \mathfrak{g}_{Ic} E \tag{9.6}$$

Some authors refer to K_{Ic} as the *fracture toughness*.

In an experiment corresponding to Fig. 9.18b, the specific geometry is accommodated in a constant Y, so that

$$\sigma_f^2 Y^2 = \frac{K_{Ic}^2}{c} \tag{9.7}$$

Values of Y are tabulated for many geometries [13, 17]. For the sample of polycarbonate (Fig. 9.19), the value of K_{Ic} is 4.0 MPa·m$^{1/2}$, with little dependence on the manner in which the notch was introduced or on the rate of crack propagation (within limits). For many glassy materials, including polystyrene and poly(methyl methacrylate), K_{Ic} is relatively independent of thickness. The failure process in polycarbonate is somewhat more complex. In a thin sheet the plate surfaces can contract laterally and the plastic zone extends from one surface to the other. In a thick sheet, lateral contraction is constrained and part of the crack tip is put in triaxial tension, greatly reducing the overall stress needed for failure. Sheets of polycarbonate less than 3 mm thick may have a K_{Ic} of about 5.5 MPa·m$^{1/2}$. The

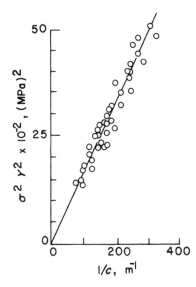

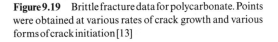

Figure 9.19 Brittle fracture data for polycarbonate. Points were obtained at various rates of crack growth and various forms of crack initiation[13]

sample of Fig. 9.18 was about 5 mm thick. Very thick sheets approach a K_{Ic} of about 2.2 MPa·m$^{1/2}$. The two extreme values reflect the two different failure mechanisms that predominate at the extremes in thickness.

From the standpoint of convenience, the impact "strength" (Sec. 9.3) is the most often reported measure of toughness for brittle materials. The equipment is inexpensive, the test is fast, and the results for notched samples of specified dimensions are satisfactorily reproducible. On the other hand, even in a notched sample, much energy is used in initiating the actual crack, and the propagation energy is not clearly distinguishable.

Fracture toughness (and impact strength) is increased by various inclusions. Rigid fibers or particles that adhere well to the glassy polymer matrix can spread the applied force over a larger zone, thus decreasing local stress concentration. Inclusion of amorphous particles above their T_g increases energy dissipation by viscous deformation. Cross-linking a glassy polymer inhibits the orientation processes and usually decreases impact strength. In a typical thermoset glass such as a phenolic resin, cross-linking contributes high temperature stability. Fiber or particulate reinforcement is used to keep up the impact strength. Heterophase systems (see next section) are commonly employed when toughness is a major criterion.

Despite what has been said about the obvious differences in fracture mechanisms for glasses and rubbers, some generalizations remain. For example, the stress at failure is linear in the logarithm of time to fail (Fig. 9.20) for virtually all materials when one ignores minor variations [18].

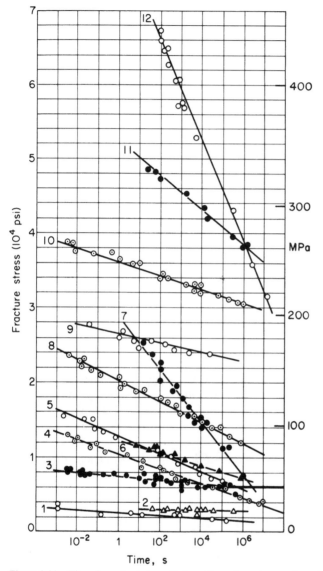

Figure 9.20 Time-dependent strength for polymeric and nonpolymeric materials [18]: 1, silver chloride; 2, polyvinyl chloride; 3, aluminum; 4, poly(methyl methacrylate); 5, zinc; 6, celluloid; 7, rubber; 8, nitrocelluloid; 9, platinum; 10, silver; 11, phosphoric bronze; 12, nylon 6 (oriented)

9.8 HETEROPHASE SYSTEMS

Partly crystalline polymers will vary in stress-strain behavior depending on rate and temperature. Some three-dimensional representations for commercial polyethylene (Figs. 9.21 and 9.22) show the behavior of yield point and elongation with test conditions. Yielding often coincides with the onset of *necking* (Sec. 3.4) and marks the limit of rubbery behavior in the amorphous portion and bond bending and stretching in the crystalline portion that can be borne before spherulites become unstable (Fig. 9.23). Once a crystalline polymer is oriented, as in typical drawn fibers, the ultimate elongations are much lower. The ultimate tensile strengths are higher mainly because they are now based on the cross section of the

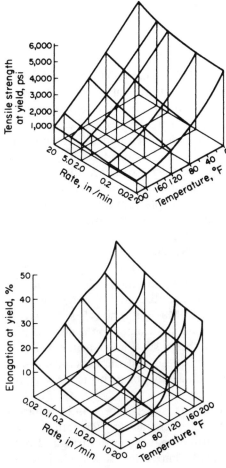

Figure 9.21 Effect of temperature and rate of extension on tensile stress at yield for an ethylene-butene copolymer [19]

Figure 9.22 Effect of temperature and rate of extension on elongation at yield for the ethylene-butene copolymer of Fig. 9.21 [19]

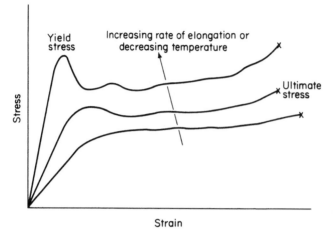

Figure 9.23 Typical stress-strain curves for polyethylene

drawn fiber rather than on an originally spherulitic sample. Undrawn nylon 66 might typically have an ultimate tensile strength of 50 MPa at a breaking elongation of 200%. Based on the cross section of a drawn fiber, the tensile strength (tenacity) may be over 500 MPa. When a spherulitic material is stressed below T_g, the rearrangement to oriented crystallites becomes very difficult. Reding and Brown [20] found that in poly(chlorotrifluorethylene) cracks tend to propagate along the radii of spherulites and sometimes between two spherulites. For this reason, large, well-formed spherulites usually are avoided. Long-term creep tests leading to stress cracking in polyethylene that is above its T_g may involve spherulite growth also. These spherulites eventually reach the size of macroscopic flaws and can act as stress concentrators.

Since fibers consist primarily of oriented crystallites, perhaps it is unfair to classify them as heterophase. However, the generalizations regarding time-temperature superposition that work so well with amorphous polymers do not apply to fibers. Fibers do exhibit viscoelasticity qualitatively like the amorphous polymers. It comes as a surprise to some that J. C. Maxwell, who is best known for his work in electricity and magnetism, should have contributed to the mathematics of viscoelasticity. It was when he was using a fiber, a silk thread, as the restoring element in a charge-measuring device that he noticed that the material was not perfectly elastic and showed time effects. The model that bears his name was propounded to correlate the real behavior of a fiber.

The high melting temperature desirable in a crystalline fiber can be achieved by using a polar structure. Rayon, nylon, the polyesters, the acrylics, cotton, wool, and silk all contain ester, amide, or hydroxyl groups that can form strong hydrogen bonds. A corollary is that moisture and heat will almost always have a large effect on the physical properties of fibers that is magnified by the large exposed surface area. The terms used in the fiber industry for strength and diameter differ from those in other polymer industries. When one considers a typical fiber or yarn, the actual cross section may be very difficult to describe (See Fig. 12.33). However, it is a very simple exercise to obtain a measure of *linear density* by weighing a strand of measured length. Also, when a given strand is stretched to the breaking point in tension, the actual force, rather than the force per unit area, is easily measured. A measure of linear density is the *tex* of a fiber, defined as the weight per unit length, in grams per kilometer. The strength is expressed as the *tenacity* (the stress per unit of linear density), defined as stress per unit of linear density, usually New-tons per tex (N/tex). The older terms still often encountered are *denier* (grams per 9000 m) and grams (force) per denier. It can be seen that 1 g/denier = 0.0883 N/tex. Some authors favor the usage of dN/tex (= 1.13 g/denier). The relationship between density, linear density, tenacity, and strength for a round fiber of diameter D can be illustrated. An ordinary nylon fishing line of density 1.14 g/cm³ with a diameter of 0.850 mm has a rating of 60.0 lb (force) at break.

$$\text{Cross-sectional area} = A = \frac{\pi D^2}{4} = \frac{\pi 0.0850^2 \text{cm}^2}{4} = 0.567 \times 10^{-3} \text{cm}^2$$

$$\text{Linear density} = \rho A 10^5 = (1.14 \text{ g/cm}^3)(0.567 \times 10^{-3} \text{cm}^2)10^5 \text{ cm}$$
$$= 647 \text{ tex} \qquad (\text{or } 5822 \text{ denier})$$

$$\text{Stress at break} = 60.0 \text{ lb(force)} = 27.2 \text{ kg} = 267 \text{ N}$$

$$\text{Tenacity} = 267/647 = 0.413 \text{ N/tex} = 4.13 \text{ dN/tex} = 4.67 \text{ g/denier}$$

In conventional engineering units the strength is also given by

$$\text{Strength} = \frac{\text{force}}{\text{area}} = \frac{60.0}{87.9} \times 10^{-6} \text{in}^2 = 68.3 \times 10^3 \text{ psi} = 471 \text{ MPa}$$

Nylon and silk show a uniform decrease in tenacity with increasing humidity, which is understandable in terms of a weaker crystal lattice with as much as 8% sorbed water competing for the hydrogen bonds (Fig. 9.24). Cotton, conversely, increases in tenacity as it sorbs up to 25% moisture, presumably because its cellular structure is then able to distort and allow more molecular chains to support stress along the major axis. Rayon, which is the same as cotton chemically but differs crystallographically, shows the more normal decrease with moisture at high humid-ities. The strengths indicated are for high-tenacity nylon and rayon. The overall balance of properties does not always make the strongest fiber the best for a partic-ular application. The effect of temperature on rayon and nylon also reflects the relative stability of the hydrogen-bonded structure in each (Fig. 9.25).

Almost all applications of rubber make use of particulate fillers. Because auto-motive tires take about 70% of all the rubber consumed in the United States and because all tires are reinforced with carbon black, the economic importance of this one class of fillers becomes apparent. The term "reinforcement" implies an increase in the magnitude of some ultimate property. However, tensile strength is not the

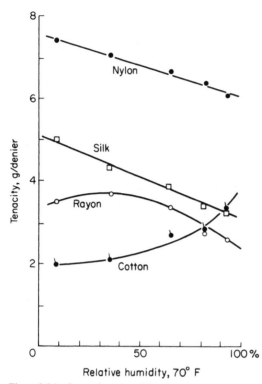

Figure 9.24 Strength vs. humidity for common fibers [21]

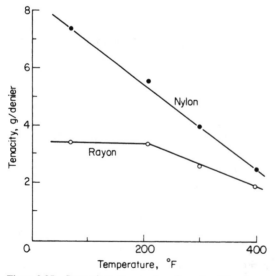

Figure 9.25 Strength vs. temperature for nylon 66 and rayon [21]

only property that may be reinforced. An improvement in tear strength, abrasion resistance, or fatigue may also justify the use of the term. The typical effect of carbon blacks on the ultimate tensile strength of natural rubber and a styrene-butadiene copolymer illustrates a basic difference in behavior of two classes of rubber (Figs. 9.26 and 9.27). Natural rubber crystallizes on stretching even though it may be well above its normal melting temperature (see Sec. 3.4). This is a reversible process in that the crystallites melt on release of the stress. The crystallites act as massive cross-links that spread the concentration of stress that would otherwise be located in a few bonds over a much wider area. Such a material is "self-reinforcing" in that it is stronger at a highly stressed point than elsewhere. At ordinary temperatures polychloroprene, some polyurethanes, polyisobutylene, and poly(propylene oxide) rubbers also are self-reinforcing, as evidenced by the fact that unfilled, cross-linked polymers of each can be made with tensile strengths of

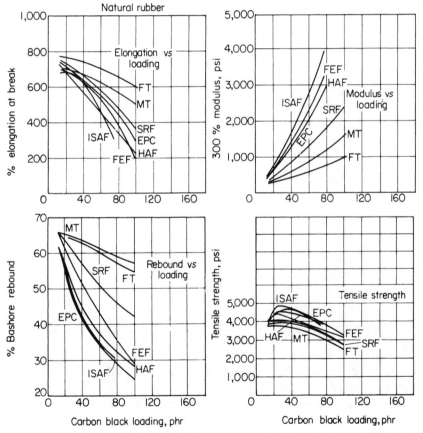

Figure 9.26 Properties of carbon-black-reinforced natural rubber vulcanizates [22]. Letters refer to specific black types; 1000 psi = 6.895 MPa

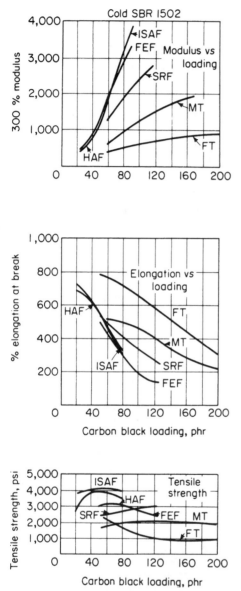

Figure 9.27 Properties of carbon-black-reinforced styrene-butadiene rubber (SBR 1502, cold rubber) vulcanizates [22]. Letters refer to specific black types; 1000 psi = 6.895 MPa

over 20 MPa (2900 psi) with high elongations. Addition of fillers such as carbon black increases tensile strength only slightly for this class of polymers, although tear strength may be improved greatly, and abrasion resistance may be very much improved.

Rubbers that do not crystallize on stretching must be compounded with fillers to attain a tensile strength of more than a few MPa. The styrene-butadiene rubber (Fig. 9.27) typifies this class, which includes poly(dimethyl siloxane), ethylene-propylene copolymers, and butadiene-acrylonitrile copolymers. It must be emphasized that we are speaking of typical behavior and relative changes in properties. Tensile strength and other properties are affected by cross-link density, temperature, and rate of testing, and other compounding ingredients (Sec. 10–2).

Ultimate properties are sample properties, and so the skill of the person making and testing the sample enters in also. Despite the many standardized procedures promulgated by the ASTM, among others, it is an incontrovertible fact that an experienced laboratory worker can always produce samples that outperform those put together by a neophyte using the same raw materials and equipment. The importance of the laboratory technician in providing reliable and reproducible characterizations should never be underestimated.

An unsophisticated rationalization of the fact that fillers (or crystallites) often increase tensile strength has been given by Bueche [23]. In a simple network (Fig. 9.28), the breaking of highly oriented strand A shifts the entire stress formerly borne by that strand to its few immediate neighbors. If they are near to the stress to rupture themselves already, they may break under the new added load and the

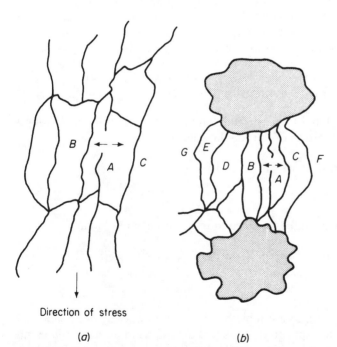

Direction of stress

(a)   (b)

Figure 9.28 Stress-sharing mechanism for filler or crystallite reinforcement: (a) polymer under stress without filler particles; (b) with filler particles or crystallites about 10 to 50 nm in diameter

flaw will propagate catastrophically. On the other hand, with a particle anchoring many strands, the load formerly borne by A can be shared among many strands rather than just the immediate neighbors (Fig. 9.28). The filler particle, in addition to acting as a cross-link, has the virtue of extremely high functionality, binding many chains to a rigid surface.

It is important that the filler be bonded to the polymer for this theory to apply. A test case is where polymer and filler are not similar chemically and thus have poor adhesion. Ethylene-propylene terpolymer and finely divided silica make such a system. Under the same conditions of mixing and cross-linking, a gum (unfilled) compound has a tensile strength of less than 1 MPa. A loading of 100 g of silica for 100 g of polymer raises this to 3.6 MPa. However, when the system is stressed, the polymer pulls away from the filler surface, forming vacuoles. Such vacuoles have been observed in carbon black-filled systems under the electron microscope. An indirect effect that they have is to cause opacity, a whitening at the stressed point, due to the light scattered at the additional interfaced formed. The net volume of the system has to increase also as the result of vacuole formation. By placing a rubber band of the material in a dilatometer filled with water and stretching the band by means of a wire that issues through the capillary, one can measure the stress-strain behavior and the volume change as seen in the capillary section simultaneously (Fig. 9.29). When the polymer-silica band is extended, the volume does increase by a substantial amount (Fig. 9.30). However, if we incorporate a *coupling agent* that can react with the filler surface and with the polymer, we may be able to prevent the vacuole formation. Such a compound is a mercaptosilane:

$$HS-CH_2CH_2CH_2-Si(-O-CH_3)_3 \quad \text{or} \quad HS-R(-OMe)_3$$
3-Mercaptopropyltrimethoxysilane

We expect the filler, which is SiO_2 in the bulk but has a surface of Si—OH groups, to react with the alkoxy portion of the coupling agent:

$$HS-R(-OMe)_3 + HO-Si\text{\textlbrackdbl} \rightarrow HS-RO-Si\text{\textlbrackdbl} + MeOH$$
$$(OMe)_2 \quad \text{Methanol}$$

The mercapto portion can be expected to react during cross-linking:

$$PR\cdot + HS-R-O-Si\text{\textlbrackdbl} \rightarrow \cdot SR-O-Si\text{\textlbrackdbl} + PRH$$

| Peroxide radical | Mercapto surface | Radical on surface | Peroxide residue |

The surface radical may add to a double bond of the polymer as in initiation of polymerization. If the mercapto functionality is absent, as in an amyl silane, no reaction with polymer should occur. It is obvious (Fig. 9.30) that the mercaptosilane has increased tensile strength and prevented the filler and polymer from part-

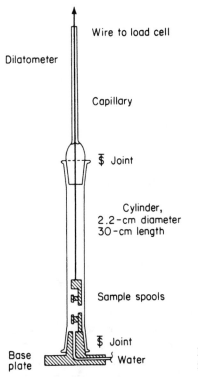

Dilatometer

Wire to load cell

Capillary

⚡ Joint

Cylinder,
2.2-cm diameter
30-cm length

Sample spools

⚡ Joint

Base
plate

Water

Figure 9.29 Dilatometer for measurement of volume during stretching of rubber [24]

ing at substantial energy inputs (area under the stress-strain curve). The amyl silane has no such effect.

In the carbon black reinforcement of rubber, the chemistry of the surface is important, as the previous work would indicate. The amount and functionality of oxygen, nitrogen, and hydrogen on the surface varies with the preparative method employed. Particle size itself is important, the generalization being that the finer the particle size the greater the reinforcement. Matters of ease of incorporation and fabrication as well as cost and other properties enter in to complicate the choice for any one application.

Particulate fillers are used also in adhesives and protective coatings. The function in the latter generally is pigmentation rather than reinforcement. Thermoset resins may be filled also, but the most common additive would be fibrous. The term *composites* has been applied to heterophase materials when the dimensions involved approach the macroscopic. Plywood and corrugated paper are two homely examples. Fibrous glass is used in several forms with polymers. As in the case of silica-reinforced rubber discussed above, coupling agents are useful to promote adhesion between the glass, which is about 50% SiO_2 and 50% oxides of aluminum, calcium, and other cations. The surface contains hydroxyl groups that

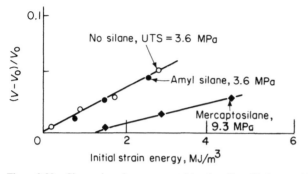

Figure 9.30 Change in volume on stretching for silica-filled, peroxide-cross-linked ethylene-propylene terpolymer [24]

can react with alkoxy silanes. The organofunctional alkyl group on the silane is tailored to react with the particular polymer.

Chopped strands 1 to 10 mm in length can be incorporated in thermoset or thermoplastic materials about as easily as the particulate fillers. Each strand may be made up of 204 individual filaments whose diameter is 5 to 20 μm. Since the modulus and tensile strength of the glass (Table 9.1) both are much higher than that of the typical plastic, it is not surprising that the stiffness and strength of most plastics can be increased by compounding with glass. In looking at the effect of glass on tensile and flexural strength of the nylons (Table 9.2), the sacrifice in elongation should not be ignored. The total energy that can be absorbed by the system before failing is much smaller for nylon 66 with glass, for example, than without, as indicated by the product of stress and elongation at break. Even at very short test times, the impact strength is not changed. On the other hand, the designer of a nylon gear or wheel has no need for high elongation, so that the dimensional stability (higher modulus and higher heat distortion temperature) and added strength are obtained at the loss of a property he or she did not particularly cherish from the beginning.

Chopped strands of glass several inches long can be loosely bound as a mat

Table 9.1 Properties of some reinforcing fibers [25]

	Density, g/cm³	Tensile strength, GPa	Tensile modulus, GPa
Glass (E-glass)	2.5	3.4–4.5	70–85
Carbon from:			
Polyacrylonitrile	1.7–1.9	2.3–7.1	230–490
Pitch	1.6–2.2	0.8–2.3	38–820
Rayon	1.4–1.5	0.7–1.2	34–55
Aramid (Kevlar)	1.4	2.4–2.8	60–200

Table 9.2 Mechanical properties of glass-reinforced thermoplastics [26]

Material	Percent glass loading	Tensile strength, 73°F, psi ×1000 (D 638)*	Elongation, %, 73°F (D 638)	Flexural strength, 73°F, psi ×1000 (D 790)	Impact strength, notched Izod, 73°F ft·lb/in (D 256)	Heat distortion temp, °C at 264 psi (D 648)	Rockwell hardness (D 785)	Specific gravity (D 792)
Nylon 6/10 raw		8.5	85 to 300		1.2	–	R111	1.09
Short fiber	30	17 to 19	3.0	22 to 26	1.4 to 2.2	204	E35-45, R118	1.30
Long fiber	30	19.0	1.9	23	3.4	216	E70-75	1.30
Nylon 66 raw		9.0	60 to 300	12.5	1.0 to 2.0	66 to 86	R108-R118	1.13 to 1.15
Short fiber	30	18.5 to 23	3.0	26.5 to 32	1.2 to 2.0	204 to 243	E50-55-R120	1.37
Long fiber	30	20	1.5	28	2.5	243	E60-70	1.37
Nylon 6 raw		7.0	25 to 320	8	1.0 to 3.6	67 to 70	R103-R118	1.12 to 1.14
Short fiber	30	17 to 24	3	22.5 to 32	1.3 to 2.0	204 to 216	E45-50, M90	1.38
Long fiber	30	21.0	2.0	27	3.0	216	E55-60	1.37
Polycarbonate raw		9.5	60 to 110	13.5	2.5	129 to 138	M70-R118	1.2
Short fiber	20	12 to 18.5	2.5 to 3	17 to 25	1.5 to 2.5	141 to 146	M92-R118	1.35
Long fiber	20	14 to 18.5	2.2 to 5	18.5	2.5 to 3.0	146	H80-90	1.35
Polypropylene raw		4.3	200 to 700	6	1.0	57 to 63	R85-110	0.90 to 0.91
Short fiber	20	6.0	3.0	7.5		110	M40	1.05
Long fiber	20	8	2.2	10	3.5	139	M50	1.05
Polyacetal raw		10.0	15	14	1.4	124	M94-R120	1.425
Short fiber	20	10 to 13.5	2 to 3	14 to 15	0.8 to 1.4	157 to 163	M70-75-95	1.55
Long fiber	20	10.5	2.3	15	2.2	163	M75-80	1.55
Polyethylene raw		1.2	50 to 600	4.8	0.5 to 16	32 to 41	(Shore) D50-60	0.92 to 0.94
Short fiber	20	6	3.0	7	1.1	107	R60	1.10
Long fiber	20	6.5	3.0	8	2.1	127	R60	1.10
Polysulfone raw		10.2	50 to 100		1.3	174	M69-R120	1.24
Short fiber	30	16	2	21	1.8	182	–	1.41
Long fiber	30	18.5	2.0	24	2.5	167	E45-55	1.37

* ASTM test method. Note: Most favorable figures for short-fiber performance are based upon results with nominal ¼-in fibers. Not included in the table are glass-reinforced styrene, SAN, ABS, polyurethane, and PPO, for which comparable data between short and long fibers are not available. 10^3 psi = 6.895 MPa.

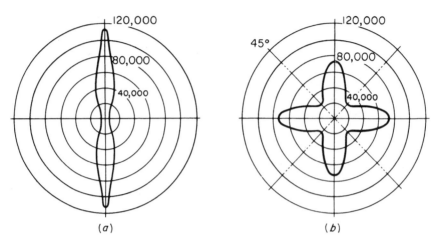

Figure 9.31 Tensile strength as a function of direction for fiber-reinforced plastics [27]: (*a*) unidirectional fibers; (*b*) cross-plied fibers; 1000 psi = 6.895 MPa

which is porous and in which the strands are randomly oriented in two dimensions. This form is suitable for impregnation by a liquid prepolymer. After polymerization or cross-linking (curing) under pressure, the composite will comprise a network-polymer matrix in which are embedded the individual strands. A woven glass cloth might be used in place of the mat or in combination with it. In this case there will be a variation in strength with the angle between the axis of the fibers and the direction of stress (Fig. 9.31). The filament-wound structure represents the ultimate in directional strength (Fig. 9.32). With careful fabrication and with the aid of coupling agents, the tensile strength of the glass is approached. In one glass-reinforced polyester system, the ultimate flexural strength varied with treatment. The resin itself had a strength of only 115 MPa (16.7 × 10³ psi) compared to 924 MPa (134 × 10³ psi) for an untreated glass-resin filament-wound ring. With proper sizing and with application of a vinyl silane coupling agent, the strength rose to 1.17 GPa (170 × 10³ psi). In this last condition, the system was 87% glass by weight (75% by volume). In some applications where weight is important, the ratio of the

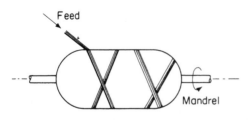

Figure 9.32 A cylindrical pressure vessel being fabricated by winding glass filaments over a mandrel

strength to the density of a material becomes a dominant factor. The ratio of strength divided by density, the *specific strength,* is often expressed in inches when the engineering system of units is used. For the composite just described, a strength of 170×10^3 lb_f/in^2 divided by a density of 0.077 lb/in^3 yields a specific strength of 2.2×10^6 in. The "cancellation" of pounds mass by pounds force is a delightful peculiarity of the older system of units. In SI units, the specific strength for the same material is 550 kJ/kg. A high-strength steel alloy would have a specific strength about one-third that of the composite [28].

The properties of several other reinforcing fibers are shown in Table 9.1. Carbon fibers are made by pyrolysis of precursor fibers, and their properties differ according to the source. Glass, carbon, and aramid fibers have also been used in composite for applications ranging from golf clubs to automobile transmission shafts. Probably the most often used matrix is some form of epoxy resin.

Another kind of composite is the polymer "alloy." Two polymers of marginal compatibility and widely differing moduli will act somewhat like the glass and polymer, having a rigid, load-bearing element and at the same time a softer, energy-absorbing element. Styrene-butadiene copolymer ($T_g = -50°C$) is combined with polystyrene ($T_g = +100°C$) to give a "rubber-reinforced" plastic. In long-time-scale tests such as ordinary tensile testing, the rigidity and strength of pure polystyrene are approached. However, in a rapid test, such as the Izod impact test, the rubbery portion will deform, absorbing much more energy than the brittle polystyrene would have. The toughness of alloys together with their dimensional stability makes them quite versatile. Some other examples are butadiene-acrylonitrile copolymer in poly(vinyl chloride) and acrylonitrile-butadiene-styrene (ABS) terpolymer with polycarbonates. Applications include refrigerator door liners, radio and TV cabinets, pipe and fittings, and women's shoe heels.

KEYWORDS FOR CHAPTER 9

Section

1. Ultimate strength
   Toughness
   Yield strength
   Weibull distribution
2. Trousers tear
3. Izod impact strength
5. Fatigue strength
7. Crazing
   Environmental stress cracking
   Fracture toughness
8. Tenacity
   Denier

Tex
Reinforcement
Coupling agent
Specific strength
Alloy

PROBLEMS

9.1 Three different rubber compositions give the following strengths in a standard tensile test:

Sample	Strength for compositions, psi		
	A	B	C
1	3300	2310	1660
2	3250	2310	1600
3	3125	2200	1600
4	3000	2150	1510
5	2850	1990	1325

Are any samples atypical? If a fourth composition were run and gave a value of 2000 psi for a single sample, how confident would you be that this was the average strength?

9.2 For a very large population of samples ($N = 10^4$), what fraction of the population will have a tensile strength greater than $\overline{\sigma}_b + s$? What fraction will have a tensile strength greater than $\overline{\sigma}_b - s$ and $\overline{\sigma}_b - 2s$?

9.3 For the average performance indicated by the solid lines in Figs. 9.8 and 9.10, plot breaking stress vs. time to break in the manner suggested by Fig. 9.20. How does the slope compare with the slopes for other polymers? How important is the elongation at break in this kind of plot?

9.4 Why does the addition of glass fibers increase the heat-distortion temperature of crystalline polymers, such as polyethylene and polypropylene, but not of glassy ones, such as polycarbonate and polysulfone (Table 9.2)?

9.5 What is the "true" stress at break based on actual cross section for SBR rubber loaded with FT black (Fig. 9.27) as a function of black loading? What is the corresponding strength in grams per denier assuming a density that is the weighted average of that for rubber (0.94 g/cm³) and that for carbon black (1.8 g/cm³)?

9.6 An amorphous polymer can be approximated by a Maxwell element in which a spring flies apart when it reaches some maximum elongation ε_{sb}. Show how this model leads to a typical failure envelope. Why is this model inadequate for a cross-linked SBR rubber?

REFERENCES

1. May, W.: *Trans. Inst. Rubber Ind.,* 40:T109 (1964); *Rubber Chem. Technol.,* 37:826 (1964).
2. Hallinan, A. J., Jr.: *J. Qual. Tech.,* 25(85):1993
3. Schwartz, P., A. Netravali, and S. Sembach: *Text. Res. J.,* 56:502 (1986).
4. "Book of Standards," pp. 286 and 356, ASTM, Baltimore, 1958.
5. Riddell, M. N., G. P. Koo, and J. L. O'Toole: *Polymer Sci. Eng.,* 6:363 (1966).
6. "Symposium on Plastics," ASTM, Philadelphia, 1944.
7. Higgins, T. D.: Union Carbide Corp., S. Charleston, private communication, 1969.
8. Smith, T. L.: *J. Polymer Sci.,* 32:99 (1958); part A,1:3597 (1963).
9. Smith, T. L., and P. J. Stedry: *J. Appl. Phys.,* 31:1892 (1960). Also Smith, T. L.: *J. Appl. Phys.,* 35:27 (1964).
10. Andrews, E. H., in R. N. Haward (ed.): "The Physics of Glassy Polymers," chap. 7, Wiley, New York, 1973.
11. Kramer, E. J., in E. H. Andrews (ed.): "Developments in Polymer Fracture—1," Applied Science Publishers, London, 1979.
12. Kambour, R. P.: *J. Polymer Sci.:D,* 7:1 (1973).
13. Williams, J. G.: *Polymer Eng. and Sci.,* 17:144 (1977).
14. Sauer, J. A., and K. D. Pae, in H. S. Kaufman and J. J. Falcetta (eds.): "Introduction to Polymer Sciences and Technology," Wiley-Interscience, New York, 1977.
15. Lauterwasser, B. D., and E. J. Kramer: *Phil. Mag. A,* 39:469 (1979).
16. Graham, I. D., J. G. Williams, and E. L. Zichy: *Polymer,* 17:439 (1976).
17. Brown, H. F., and J. E. Srawley: *ASTM, STP* 410 (1966).
18. Hsaio, C. C.: *Phys. Today,* 19:49 (March 1966).
19. Pritchard, J. E., R. L. McGlamery, and P. J. Boeke: *Mod. Plastics,* 37(2):132 (October 1959).
20. Reding, F. P., and A. Brown: *Ind. Eng. Chem.,* 46:1962 (1954).
21. "Tensile Stress-Strain Properties of Fibers," Reprint T-1, Instron Corp., Canton, Mass., 1958.
22. Studebaker, M. L., in G. Kraus (ed.): "Reinforcement of Elastomers," pp. 356 to 357, 362, Wiley-Interscience, New York, 1965.
23. Bueche, F.: *Rubber Chem. Technol.,* 32:1269 (1959).
24. Schwaber, D. M., and F. Rodriguez: *Rubber Plastics Age (London),* 48:1081 (1967).
25. Matsuda, H. S.: *Chem. Tech.,* 18(5):310 (1988).
26. *Mod. Plastics,* 43:102 (March 1966).
27. "Scotchply Reference Manual," 3M Company, St. Paul, Minn., 1968.
28. Throckmorton, P. E., H. M. Hickman, and M. F. Browne: *Mod. Plastics,* 41:140 (November 1963).

General References

Andrews, E. H. (ed.): "Developments in Polymer Fracture—I," Applied Science Publishers, London, 1979.

Beaumont, P. W. R., J. M. Schultz, and K. Friedrich: "Failure Analysis of Composite Materials," Technomic, Lancaster, Pa., 1990.

Brostow, W., and R. Corneliussen (eds.): "Failure of Plastics," Macmillan, New York, 1986.

Bucknall, C. B.: "Toughened Plastics," Applied Science Publishers, London, 1977.

Bucknall, C. B., in D. R. Paul and S. Newman (eds.): "Polymer Blends," vol. 2, chap. 14, Academic Press, New York, 1978.

Cheremisinoff, N. P. (ed.): "Product Design and Testing of Polymeric Materials," Dekker, New York, 1990.

Clegg, D. W., and A. A. Collyer (eds.): "Mechanical Properties of Reinforced Thermoplastics," Elsevier Applied Science, New York, 1986.

Deanin, R. D., and A. M. Crugnola (eds.): "Toughness and Brittleness of Plastics," ACS, Washington, D.C., 1976.

Enikoloyan, N. S. (ed.): "Filled Polymers I. Science and Technology," Springer, Secaucus, N.J., 1990.

Harper, C. A. (ed.): "Handbook of Plastics, Elastomers, and Composites," 2d ed., McGraw-Hill, New York, 1992.

Haward, R. N. (ed.): "The Physics of Glassy Polymers," chap. 7. Halsted (Wiley), New York, 1973.

Hertzberg, R. W., and J. A. Manson: "Fatigue of Engineering Plastics," Academic Press, New York, 1980.

Kausch, H. H.: "Polymer Fracture," Springer-Verlag, New York, 1979.

Kausch, H.-H. (ed.): "Crazing in Polymers," vol. 2, Springer, Secaucus, N.J., 1990.

Kessler, S. L. (ed.): "Instrumented Impact Testing of Plastics and Composite Materials," American Society for Testing and Materials, Philadelphia, 1987.

Kinloch, A. J., and R. J. Young: "Fracture Behaviour of Polymers," Elsevier Applied Science, New York, 1983.

Lipatov, Y.: "Polymer Reinforcement," ChemTec, Toronto, 1995.

Mark, J. E., A. Eisenberg, W. W. Graessley, L. Mandelkern, E. T. Samulski, J. L. Koenig, and G. D. Wignall: "Physical Properties of Polymers," 2d ed., ACS, Washington, D.C., 1993.

McCauley, J. W., and V. Weiss (eds.): "Materials Characterization for Systems Performance and Reliability," Plenum, New York, 1986.

Milewski, J. V., and H. S. Katz (eds.): "Handbook of Reinforcements for Plastics," Van Nostrand-Reinhold, New York, 1987.

Nielsen, L. E., and R. F. Landel: "Mechanical Properties of Polymers and Composites," 2d ed., Dekker, New York, 1993.

Ogorkiewicz, R. M. (ed.): "Thermoplastics: Properties and Design," Wiley-Interscience, New York, 1974.

Ogorkiewicz, R. M. (ed.): "The Engineering Properties of Plastics," Oxford, New York, 1977.

Plueddemann, E. P.: "Silane Coupling Agents," 2d ed., Plenum, New York, 1990.

Richardson, M. O. W. (ed.): "Polymer Engineering Composites," Applied Science Publishers, London, 1977.

Ritchie, P. D. (ed.): "Physics of Plastics," Van Nostrand, Princeton, N.J., 1965.

Roulin-Moloney, A. C.: "Fractography and Failure Mechanisms of Polymers and Composites," Elsevier, New York, 1989.

Schultz, J., and S. Fakirov: "Solid State Behavior of Linear Polyesters and Polyamides," Prentice Hall, Englewood Cliffs, N.J., 1990.

Shah, V.: "Handbook of Plastics Testing Technology," Wiley, New York, 1984.

Spells, S. J. (ed.): "Characterization of Solid Polymers: New Techniques and Developments," Chapman & Hall, New York, 1994.

Turner, S.: "Mechanical Testing of Plastics," Wiley, New York, 1986.

Ward, I. M., and D. W. Hadley: "An Introduction to the Mechanical Properties of Solid Polymers," Wiley, New York, 1993.

Williams, J. G.: "Stress Analysis of Polymers," 2d ed., Halsted-Wiley, New York, 1980.

Williams, J. G.: "Fracture Mechanics of Polymers," Wiley, New York, 1984.

Zachariades, A. E., and R. S. Porter (eds.): "The Strength and Stiffness of Polymers," Dekker, New York, 1983.

CHAPTER TEN

SOME GENERAL PROPERTIES
OF POLYMER SYSTEMS

10.1 DESIGN CRITERIA

The final judgment on the suitability for a polymer for a given application usually involves a complex combination of properties. Some of these properties are inherent in the physical state of the polymer. A glassy material will not stretch, nor will an un-cross-linked, amorphous polymer support a stress indefinitely when it is above its T_g. Some properties are inherent in the chemical structure of the polymer. Hydrolysis, thermal dissociation, and toxicity may be fixed by the reactivity of certain groups within the polymer structure.

In any application, cost is an important factor. The price of the polymer in the form of a powder, latex, solution, or bale is just part of the economic picture. The cost of other ingredients, equipment for fabrication, labor, power, and all the indirect costs, may well overshadow the polymer's share. A first-line tire containing the best rubber sells for 10 to 20 times the cost of the raw rubber it contains. Even in a much less complicated structure, such as plasticized poly(vinyl chloride) laboratory tubing, the raw material cost may be less than one-tenth of the selling price.

Some properties that might be considered for a single application are [1]:

Appearance	Flexural strength
Hardness	Shear strength
Density	Impact resistance and toughness
Mechanical properties:	Rigidity
Tensile strength	Creep and cold flow
Compressive strength	Fatigue

Dimensional stability
Durability
Thermal properties:
  Coefficient of expansion
  Thermal conductivity
  Specific heat
  Heat distortion temperature
  Heat resistance
  Flammability

Electrical properties:
  Resistivity
  Dielectric strength
  Dielectric constant
  Power factor
  Arc resistance
Chemical resistance to:
  Acids, bases, solvents, oils,
  and fats

10.2 COMPOUNDING

The polymer characteristics we have discussed so far are useful guideposts in predicting the behavior of polymers in real applications. However, a lacquer is not an "infinitely dilute" solution, nor is a tire composed of "ideal" rubber. In many useful systems the polymer is one of several constituents, and perhaps not the most important one.

The largest volume synthetic rubber today, styrene-butadiene rubber (SBR), is a prime example of the complexities of polymer *compounding* for real life. By compounding we mean the mixing of polymer with other ingredients. In SBR these can be:

1. *Reinforcing fillers.* These improve tensile strength or tear strength. Carbon black and silica are typical.
2. *Inert fillers and pigments.* These may not change final properties in a desirable direction, but they may make the polymer easier to mold or extrude and also lower in cost. Clay, talc, and calcium carbonate are used.
3. *Plasticizers and lubricants.* Petroleum-based oils, fatty acids, and esters are most often used.
4. *Antioxidants.* These generally act as free-radical "sinks" and thus stop the chain reactions in oxidation.
5. *Curatives.* These are essential to form the network of cross-links that guarantee elasticity rather than flow. Sulfur for unsaturated and peroxides for saturated polymers are used with auxiliaries to control reaction rate.

One may ask why all this is necessary, since SBR (cross-linked) is almost an ideal rubber as defined earlier. The trouble with this "ideal" material is that it does not extrude smoothly, it degrades rapidly on exposure to warm air, and it has a tensile strength of about 500 psi (3.5 MPa). Proper compounding changes SBR to a smooth-processing, heat-stable rubber with a tensile strength of over 3000 psi (20 MPa)! However, theoretical equations no longer describe its mechanical behavior. The *compounder* then must be part scientist, part artist, and part statistician if he or she is to develop optimum properties in a material. It is virtually impossible to optimize all properties at once, so compounders have developed specialized "recipes" for girdles, tires, fan belts, tarpaulins, electrical insulation, etc. Over 9000 such recipes were *published* between 1948 and 1957 alone!

Figure 10.1 illustrates the means by which compounders have attempted to shorten the work of *formulation*. A grid of data points is determined experimentally and a "contour map" is prepared. A mathematical counterpart can be obtained with the help of a computer [2]. While such techniques are helpful, the experience of the compounder and his or her willingness to innovate usually determine the effectiveness.

Mixing machines for polymers vary from ball-and-pebble mills that disperse pigments in low-viscosity paints over periods of hours and days to heavy "internal mixers" that may use 1500 hp (about 1100 kW) to disperse 100 kg of carbon black in 200 kg of rubber in about 10 min.

One might divide polymer applications into those that involve extensive compounding and those that do not. Of course, there are many exceptions.

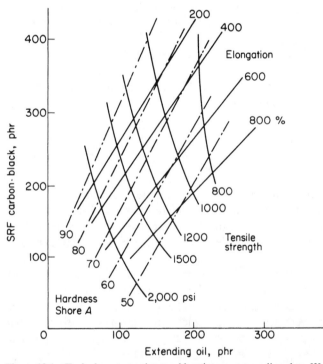

Figure 10.1 Typical contour plot to abbreviate compounding data [3]. Note that a compound with 200 *parts* each of oil and filler *per hundred* parts of rubber by weight (phr) still has a strength of about 1000 psi (7 MPa), quite acceptable for many mechanical applications. The compound, based on a high-molecular-weight ethylene-propylene terpolymer, also contains 5 parts zinc oxide, 1 part stearic acid, 1.5 parts sulfur, 1.5 parts tetramethyl thiuram disulfide, and 0.5 parts of benzthiazyl disulfide. Cross-linking takes place at 160°C for 20 min

Generally compounded	*Generally not compounded*
Rubber	Fibers
Thermosets	Thermoplasts (PVC is a wild exception)
Adhesives	
Protective and decorative coatings	

In the case of fibers, the art of compounding is of less importance than the art of *aftertreatment*. Dyeing, texturing, and fabrication of textiles are complex processes that can obscure the ideal properties of polymers in the final product.

10.3 HARDNESS

The most common measure of hardness is the distance into the material that a steel ball will penetrate under a specified load. A spring-loaded, ball-tipped indenter may be used so that the stress is not a linear function of penetration. Since the measurement is basically a compressive modulus, one expects stiff materials to be hard and flexible materials to be soft. In choosing a material for a gasket, hardness often is the only specification listed.

Surface hardness is also measured by scratch and abrasion resistance. As a rule, polymer systems cannot approach the surface hardness of silica glass, which is itself a highly cross-linked structure. Some highly cross-linked laminates do have scratch resistance nearly like that of glass.

10.4 DENSITY

Polymers and other ingredients of polymer systems are sold on a weight basis. However, in most applications they compete on a volume or strength basis. On a strength-volume basis, a reinforced plastic sheet may prove much more economical for an aircraft seatshell than a metal sheet that sells for one one-hundredth as much on a pound basis. Several factors affect density including pressure, temperature, and crystallinity.

It has been pointed out that specific volume increases with temperature for polymers, especially where melting occurs. Since crystallinity involves a closer packing of molecules than glassiness, density can be used to measure crystallinity. The specific gravity of polyethylene, for example, varies from 0.86 (extrapolated from the melt) when amorphous to 0.98 when highly crystalline at 25°C [4]. The effect of pressure is important in molding processes, where the high molding pressures alter the dimensions of a piece from what they will be at 1 atm. High pressures that increase density also increase viscosity, because less "free volume" is available for segmental motion (see Sec. 7.5). In a molding operation just above the normal melting temperature, a high pressure may induce crystallization, which stops flow altogether. In Fig. 10.2 the compressibility of an amorphous polymer, polystyrene, is shown together with the compressibility of branched and linear

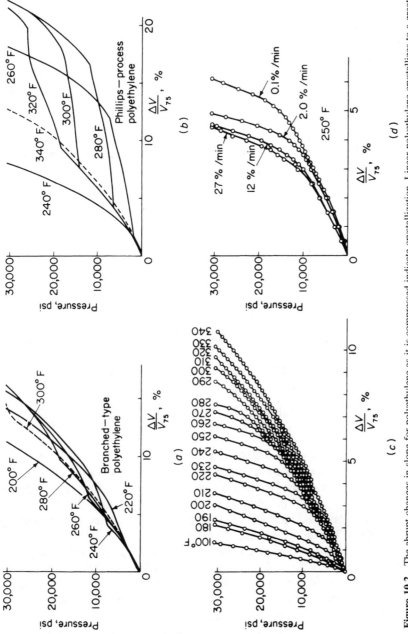

Figure 10.2 The abrupt changes in slope for polyethylene as it is compressed indicate crystallization. Linear polyethylene crystallizes to a greater extent than the branched polymer. The compressibility also is rate-dependent, as seen in (*d*). $\Delta V/V_{75}$, %, is the change in volume from 1 atm to the pressure indicated times 100/(volume at 75°F; 1 atm) [5]; 1000 psi = 6.895 MPa

369

polyethylene. Compressibility is expressed here as the change in volume $\Delta V/V_R$, where V_R is a reference volume at 1 atm and 75°F (23.9°C). The sudden change in volume with pressure for polyethylene is due to crystallization. The change is greater with linear than with branched polymer, because the degree of crystallinity induced is greater. Also illustrated in Fig. 10.2 is the fact that compression of a polymer is time-dependent too. When polystyrene is injection-molded at a pressure of 10,000 psi (69 MPa), there can be a significant difference in final dimensions of the piece depending on the length of time the melt is under pressure before cooling below T_g.

The atomic makeup of polymers affects density in much the same way it does low-molecular-weight compounds (Table 10.1). It should be remembered that a typical linear polymer is 15 to 30% more dense than the corresponding monomer.

The density of final compounded products usually is the weighted average of the ingredients (provided none are volatilized during fabrication). Some examples are given in Table 10.2.

10.5 THERMAL PROPERTIES

Many polymers have a coefficient of linear thermal expansion α_e in the range of 2 to 20×10^{-5} cm/cm·°C, compared to steel at about 1×10^{-5}. This complicates the design of molds for precision parts and the design of metal inserts in polymer parts. Of course, α_e varies with the state of the polymer, as indicated earlier in comments on the variations of specific volume at T_g and T_m (Sec. 3.4). Replacement of polymer by less expansile fillers lowers the overall expansion.

Thermal conductivity k_c of polymers is uniformly low. Values of $k_c = 0.05$ to 0.20 Btu/ft·h·°F are common.

$$\frac{242 \text{ Btu}}{\text{ft·h·°F}} = \frac{1 \text{ cal}}{\text{cm·s·°C}} = \frac{419 \text{ watt}}{\text{m·°C}}$$

Conductivity is not easily increased. A high concentration of a metal in powder or fiber form can raise it perhaps 10-fold. The thermal conductivity of the base resins in Table 10.3 can be increased by aluminum or copper metal. These also increase electrical conductivity. If low electrical conductivity, of the order of 10^{-16} (ohm·cm)$^{-1}$, must be combined with high thermal conductivity, the mixture of aluminas (Table 10.3) will increase the former by a factor of only 1.5 over that of the base epoxy resin, while the latter is increased by a factor of 12. Foaming with air or some other gas is used to decrease thermal conductivity. A foamed polystyrene with a density of 15kg/m³ and a $k_c = 0.040$ W/m°C is useful as insulation for a variety of applications, from picnic baskets to boxcars. Some values of k_c for typical foams are shown in Table 12.12 (Sec. 12.6).

A *specific heat* of 0.4 ± 0.1 cal/g·°C is typical for unfilled polymers. Composites generally have the average specific heat of the components. This is another property that varies with the physical state of the polymer in relation to T_g and T_m.

Table 10.1 Specific gravity of polymers

Chemical composition of polymer	Typical specific gravity of pure polymer near room temperature
Aliphatic hydrocarbons (polyethylene, polyisoprene)	0.8–1.0
Aromatic hydrocarbons and silicones (polystyrene)	1.0–1.1
Oxygen and nitrogen-containing polymers (cellulosics, polyesters, polyamides)	1.1–1.4
Chlorinated polymers	1.2–1.8
Fluorinated polymers	1.8–2.2

Table 10.2 Specific gravity of filled polymers

Parts by weight	Polymer	Specific gravity	Parts by weight	Filler	Specific gravity	Final specific gravity
100	Natural rubber	0.93	50	Carbon black	1.8	1.1
100	Natural rubber	0.93	100	Calcined clay	2.6	1.4
100	Epoxy resin	1.2	200	Glass fibers	2.5	1.8
100	Phenolic resin	1.3	100	Wood flour	0.9	1.1
100	Polyurethane	1.2	900 (pts by vol)	Nitrogen		0.12

Table 10.3 Thermal conductivity of various filled epoxy resins [6]

Filler	Volume percent of filler in compound	Thermal conductivity cal/cm·s × 10^4		
		Filler	Base resin	Filled compound
Aluminum, 30 mesh	63	4970	4.7	60.4
Sand, coarse grain	64	28	4.7	23.6
Mica, 325 mesh	24	16	4.7	12.2
Alumina, tabular	53	723	4.7	24.5
Alumina, 325 mesh	53	723	5.4	34.0
Copper powder	60	9180	5.4	39.0
Silica, 325 mesh	39	28	5.4	18.3
Mixture of: Alumina, tabular (20 to 30 mesh)	45	723	4.7	58.8
Alumina, 325 mesh	20			

The yielding of a polymer under load or under its own weight usually occurs at a *deflection temperature* that is in the vicinity of T_g or T_m. The older term, *heat distortion temperature,* is still used in many places.

Flammability is a function of physical form and chemical composition. An air-foamed material or a thin film presents extensive surface for burning compared to a heavy solid section. Chemical composition has the same effect that it has with lower-molecular-weight compounds. In general, we can rank pure polymers as follows:

Most flammable: Nitrated polymers
       Oxygen-containing polymers
       Hydrocarbon polymers
       Polyamides
Least flammable: Halogenated polymers

Flammability of polymers can be measured objectively by a variety of tests that present varying degrees of simulation of real-life situations. For many plastics the oxygen index method is used, since it requires rather small amounts of material (ASTM D2863). The minimum concentration of oxygen in an oxygen/nitrogen mixture that will just support combustion after the sample is ignited is termed the *oxygen index,* often called the *limiting oxygen index* (LOI) (Fig. 10.3). Some typical LOI values for pure polymers are 17.4% for polyethylene, 21.5% for nylon 66, 29.4% for polycarbonate, 47.0% for poly(vinyl chloride), and 95.0% for poly(tetra-fluoroethylene) [7]. Certain plasticizers (phosphate ester and halogenated waxes) and fillers (antimony trioxide combined with chlorinated hydrocarbons) can contribute flame resistance. On the other hand, nitroglycerine is an ideal plasticizer for nitrocellulose when the objective is to maintain the flammability of a rocket propellant [8].

10.6 ELECTRICAL PROPERTIES

Resistance is a familiar electrical property. The *volume resistivity* ρ_r is the resistance in ohms of a material 1 cm thick, t, and 1 cm^2 in area, A (Table 10.4). The resistance R of any other configuration is given by

$$R = \frac{\rho_r t}{A} \tag{10.1}$$

In the case of an insulated cable with inside radius R_1 and outside radius R_2 and length L,

$$R = \frac{\rho_r}{2\pi L} \ln \frac{R_2}{R_1} \tag{10.2}$$

High values of resistivity are common for organic polymers, 10^{12} to 10^{18} ohm·cm being typical. The actual value of resistivity depends on frequency and

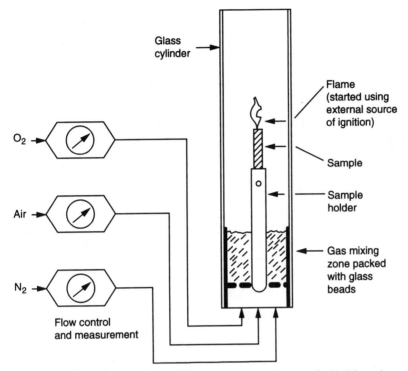

Figure 10.3 To measure the flammability, a sample about the size of a birthday cake candle is ignited by an external source such as a propane torch. The percentage of oxygen in a mixture of air with either oxygen or nitrogen which is barely able to sustain burning is called the oxygen index (ASTM D2863). A sample size of 6 × 3 × 100 mm is typical

voltage. It decreases with increasing temperature (Fig. 10.4). In any polymer system, fillers or absorbed water may provide conductive paths.

Electrical resistance can be lowered markedly by the addition of conductive fillers. A novel application involves the use of a conductive, elastic silicone rubber, as a metal-free ignition wire in automobiles. Certain carbon blacks impart conductivity:

Carbon black (Vulcan XC-72, Cabot Corp.), parts by wt/100 parts of rubber	Volume resistivity, ohm·cm
0	10^{14}
15	10^4
60	10

For getting good thermal and electrical contact between a computer chip and its ceramic support plate, a "die-attach adhesive" can be used. Silver- or gold-filled

Table 10.4 Electrical properties [9]

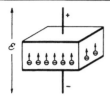

Volume resistivity is the ohmic resistance of the bulk dielectric material measured as though the material were a conductor. The resistance is expressed as the resistance of a cube 1 cm on a side measured between faces. This value is measured according to ASTM D 257. It is dependent on temperature, frequency, and voltage and will vary with the conditioning of the material.

Surface resistivity is defined as the resistance between two electrodes on the surface of an insulating material. It obviously can be measured only on materials having high intrinsic volume resistivity as a reduction of resistance. The units are ohms per square centimeter, and the test method is ASTM D 257. The value is dependent on temperature, frequency, and voltage, but is most affected by the humidity- and moisture-conditioning cycles to which it has been subjected, all of which must be given with the value.

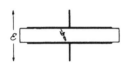

Dielectric strength is defined as the potential per unit of thickness that will cause catastrophic failure of a dielectric material. This value is measured according to ASTM D 149. The value is dependent on the method of application of potential, the nature of the potential, dc or frequency of ac, and the temperature, and it varies with the conditioning of the specimen, all of which must be specified with the value.

Arc resistance is the time in seconds that an arc may play across the surface of a material without rendering it conductive. The property is measured according to ASTM D 495 with the low-current, high-voltage arc. The failure may occur by carbonization, heating, and other means, and it is dependent on temperature, frequency, and conditioning, which, as well as the type of failure, must be specified.

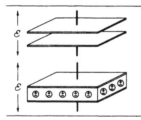

Dielectric constant is defined as the ratio of the capacity of a condenser made with a particular dielectric to the capacity of the same condenser with air as the dielectric. The test method for plastics is ASTM D 150. The value is frequency- and temperature-dependent and varies with conditioning, all of which must be specified with the value.

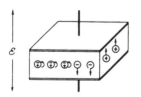

Dissipation factor is the ratio of the real power (in-phase power) to the reactive power (power 90° out of phase). It is related to the power factor, which is the ratio of the real power to the voltage-ampere product by the relation

$$\text{Dissipation factor} = \frac{\text{power factor}}{\sqrt{1 - \text{power factor}^2}}$$

Loss factor is the product of the dielectric constant and the power factor and is a measure of signal absorption. The dissipation factor is a measure of the conversion of the reactive power to real power, showing as heat. The mode of heating can be by electron or ion flow and by dipole rotation. It is variable with frequency, temperature, conditioning, and potential. The test method is ASTM D 150, and the conditions of the test and frequency must be specified.

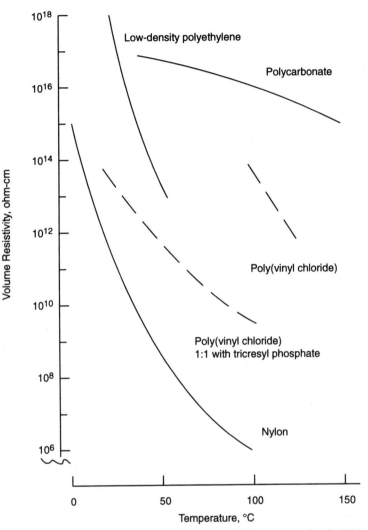

Figure 10.4   Most polymers decrease in resistivity as the temperature is raised [10, 11]

epoxy resins are available with resistivities of the order of 10^{-4} ohm·cm. A low resistivity often is desirable to prevent the accumulation of static charges, which can be merely annoying in clothing or carpeting but very dangerous in the presence of a combustible gas. Plastic tile floors for special rooms are often compounded with conductive fillers. The reciprocal of resistivity is the conductivity. The older units for conductivity $(ohm·cm)^{-1}$ have been replaced in much of the literature by the equivalent units siemen/cm, usually abbreviated as S/cm.

An important use for conducting polymers is as shielding against *electromagnetic interference* (EMI). Almost all devices that use electricity generate some electromagnetic radiation. This includes motors, TV sets, radios, computers, and light switches. Additionally, some sources of EMI occur naturally as in lightning, static buildup, and discharge from belts, carpets, and clothing. Conductive housings around computers have become necessary because of their particular vulnerability to chance signals that can disrupt stored information. *EMI shielding* can be achieved by coating an appliance housing with a conductive layer of metal or by making the entire housing from a plastic that has had its conductivity increased by a metallic filler.

Several classes of polymers span the range from insulator to conductor [12]. Polyacetylene (see also chap. 13) by itself has a conductivity of about 10^{-8} S/cm, which makes it, classically speaking, an insulator. However, doping (complexing) with large amounts of AsF_5 can increase the conductivity to as much as 1000 S/cm. This is still about 1000 times less than the conductivity of copper, silver, or gold. Since polyacetylene is not very resistant to oxidation, aromatic polymers with somewhat lesser conducting possibilities can become attractive because of their chemical stability. Almost all the common polymers being investigated (Table 10.5) are characterized by delocalized electron structures [13].

One conductive polymer which has received much attention is polyaniline [15]. The popularity of the material can be judged from the fact that, in the four-year period 1986–1989, almost 1000 papers and patents were published dealing with polyanilines. The average oxidation state of polyaniline is characterized by the term $(1 - y)$.

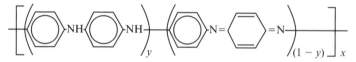

(Polyaniline with reduced y portion and oxidized $(1 - y)$ portion.)

There are three neutral states for polyaniline:

Leucoemeraldine, which is completely reduced, $(1 - y) = 0$
Emeraldine, with $(1 - y) = 0.5$
Pernigraniline, which is completely oxidized, $(1 - y) = 1$

When a compressed powder pellet of the emeraldine form ($M_w = 211{,}000$) is protonated using 1 M aqueous HCl, a conductivity of 15 S/cm is achieved. This represents an increase in conductivity of about 10^{10} times over the unprotonated polymer. Oxidation of aniline using ammonium persufate in aqueous HCl gives directly the crystalline, black-green precipitate appropriately called "emeraldine." The protonated polymer is believed to have a polysemiquinone radical cation structure.

Table 10.5 Some typical organic polymers that can be doped to become electrical conductors [13, 14]

Polymer repeat (monomer) unit		Dopant		Conductivity at 25°C, s/cm
Name	Structure	Formula	Conc. equivalents per double bond	
Acetylene (cis)		None	—	1.7×10^{-9}
		AsF$_5$	0.28	560
		Na	0.42	25
		Iodine	0.50	360
Phenylene		AsF$_5$	0.13	500
		Na	0.19	3000
Phenylene sulfide		AsF$_5$	0.33	1
Thiophene		Iodine	0.15	0.1

Source: Adapted with permission from Cowan, D.O., and F. M. Wiygul: *Chem. Eng. News*, 64(29):28 (1986). Copyright © 1986 American Chemical Society.

Aniline   Protonated emeraldine $[(1-y) = 0.5]$

Both fibers and films of the conductive polymer have been made. A self-protonated form results from the reaction of fuming sulfuric acid on emeraldine. Unlike "externally" protonated forms, the conductivity of this form in pellet form remains constant at about 0.1 S/cm over the pH range 0 to 7.

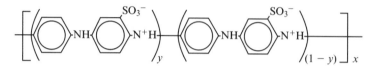

Self-protonated emeraldine

Also, this latter polymer dissolves in aqueous sodium hydroxide without decomposition.

Conducting polymers which can be molded by conventional melt processes have been made by attaching side groups to the conjugated polymers of Table 10.5. For example, poly(3-octylthiophene) in the undoped state can be molded by compression or injection just like other thermoplastics [16]. When films of this polymer are doped with $FeCl_3$, conductivities of over 10 S/cm are measured. Even a blend of 20 parts of the polymer with 80 parts of an ethylene-vinyl acetate copolymer can be doped with iodine to a conductivity of about 1 S/cm.

Some polymers increase in conductivity on exposure to light (*photoconductive* polymers). Poly(vinyl carbazole) that has been dyed (chemically reacted) with methylene blue is an example [14]. Others, like poly(vinylidene fluoride), change conductivity in response to pressure (*piezoelectric* polymers). Both of these polymers have obvious applications as sensors in various devices. Other applications for conductive polymers include electrical wiring, switches, and lightweight batteries. A most intriguing possibility is that of molecule-sized components in an integrated circuit. The idea of replacing silicon or gallium arsenide in computer chips by organic semiconductors conjures up the image of a synthetic brain. While a great deal of progress has been made in recent years, much more must be learned about controlling conductivity in polymers before such *molecular electronics* become reality [17].

Surface resistivity (Table 10.4) is especially important when the fabricated article may be subjected to a high humidity, which can alter the surface physically or chemically to give a lower resistance than the bulk of the material.

Dielectric strength (Table 10.4) is analogous to tensile strength. The voltage on a thin slab of material held between two electrodes is increased until catastrophic failure occurs, usually burning a hole completely through the slab. The shape of the electrodes, the rate of increase of voltage, and the thickness of the slab all affect the dielectric strength. Values of several hundred volts for a thickness of 25 μm are not uncommon.

Arc resistance (Table 10.4) of a polymer that decomposes to volatile products in the presence of an arc usually is better than for a polymer that is reduced by the arc to conductive carbon. For example, poly(methyl methacrylate) is "nontracking" because it depolymerizes to monomer (see Sec. 11.3).

The *dielectric constant* (K_c) has its mechanical analog in the stiffness modulus. Most often it is measured as the ratio of the capacitance of a parallel-plate capacitor with the material as the dielectric compared to the capacitance of the same capacitor in a vacuum—or air, which is about the same. The capacitance C_f in farads for a given plate spacing d in centimeters and plate area A in square centimeters is

$$C_f = \frac{K_c A}{d(3.6\pi 10^{-12})} \tag{10.3}$$

The capacitance is expressed in nano- (10^{-9}) or pico (10^{-12}) farads. Often the dielectric constant is measured in an alternating field. The reason more electrons can be stored on one plate of the capacitor at a given voltage with a solid dielectric is that electron fields in the material are distorted and counteract the impressed stress. Even a nonpolar polymer such as polyethylene can have valence electrons displaced to give an electronic displacement with a corresponding component of dielectric constant of about 2.0 to 2.5. A polar polymer such as poly(vinyl chloride), $+CH_2-CHCl+$, also can respond to an imposed field by actual orientation of molecular segments provided it is above T_g.

The *dissipation factor*, tan δ, is a measure of the hysteresis in charging and discharging a dielectric. One method of measuring it is in a Schering bridge circuit (Fig. 10.5) [18]. When resistance R_3 and capacitance C_{f_4} are adjusted to give a null at the galvanometer, the equivalent series capacitance of the specimen C_{f_1} is

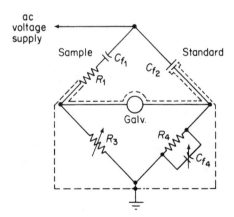

ac voltage supply

Sample   Standard

Galv.

Figure 10.5   Simplified Schering bridge [18]

$$C_{f_1} = C_{f_2} \frac{R_4}{R_3} \tag{10.4}$$

and

$$\tan \delta = 2\pi\omega_c C_{f_4} R_4 \tag{10.5}$$

where ω_c is the frequency, C_{f_4} is in farads, and R_4 is in ohms. Since C_{f_2} and R_4 are usually fixed, adjustment of two dials gives a null and the dial for R_3 can be calibrated directly in capacitance, while that for C_{f_4} can be calibrated directly in dissipation factor. The power factor, $\sin \delta$, is sometimes specified, as is the loss factor, $K_c \tan \delta$. The power factor is the ratio of energy loss in a dielectric to the volt-ampere input.

Since the energy loss in a dielectric appears as heat, placing the material in a high-frequency field represents a method of heating a thick section uniformly. This is especially useful, because heat conduction is so poor that bringing a thick piece to a high temperature in an oven might take hours. The relationship between material properties, sample dimensions, and heating source parameters is [19]

$$U_p = 1.77 \times 10^{-13}\pi\omega_c K_c \varepsilon^2 A(\tan \delta)d^{-1} = 10^3 Mc_p \frac{dT}{dt} \tag{10.6}$$

where U_p = power, watts
ω_c = frequency, Hz (cycles/s)
A = area, cm^2
K_c = dielectric constant
ε = voltage, rms volts
$\tan \delta$ = dissipation factor
d = sample thickness, cm
M = sample mass, kg
c_p = specific heat, J/g·°C
dT/dt = rate of heating, °C/s

Example 10.1

Given the following information for a slab of rubber that is to be heated, find dT/dt and U_p.

Source parameters: 10 kV at 1 MHz
Sample properties: K_c = 3.0, $\tan \delta$ = 0.035, c_p = 1.67 J/g·°C, density = 0.96 g/cm^3
Sample dimensions: A = 1610 cm^2, thickness = 2.54 cm
Mass = 1610 × 2.54 × 0.96 = 3926 g = 3.93 kg

Solution:

$$U_p = \frac{1.77 \times 10^{-13}\pi \times 10^6 \times 3.0(10^4)^2(1610)0.035}{2.54}$$

$$= 3.7 \times 10^3 \text{ J/s} = 3.7 \text{ kW}$$

$$dT/dt = \frac{3.7 \times 10^3}{10^3 \times 3.93 \times 1.67} = 0.56 \text{ °C/s}$$

(The actual expected dT/dt would be more like 0.40°C after accounting for heat losses at the surfaces.)

Dielectric heating is used for preheating plastics before molding, for welding and sealing plastic film and sheeting, and in woodworking for heating joints. In general, industrial dielectric heating uses frequencies in the range of 1 to 100 MHz [20]. The common microwave kitchen oven uses frequencies in the range of 915 to 2450 MHz.

Radio-frequency energy has been used to heat the adhesive between two polymers. In order to fasten a PVC skin to a polypropylene panel for an automobile interior, a water-based adhesive was used that had to be heated to 90°C to be activated [21]. Five seconds of exposure to a 10- to 100-MHz source at 0.03 kW/cm was sufficient to heat the adhesive, which has a high dielectric loss compared to the skin and panel materials. Neither the PVC nor the polypropylene was heated measurably. The radio-frequency heating was 10 times faster than conductive heating, which would have distorted the plastic skin and panel.

When a signal in the radio-frequency range is to be transmitted over long distances in coaxial cables, the dielectric loss becomes particularly objectionable.

The maxima in dielectric loss often parallel the maxima in mechanical loss for the same material. We expect maximum losses at transition temperatures (Fig. 10.6) for both electrical and mechanical deformations. In Fig. 10.7 we can see the increase in apparent main T_g (about 120°C) with increased frequency in the behavior of tan δ for a partly crystalline material.

Just as mechanical fatigue after many alternating cycles causes a polymer to fail at a lower stress than in a rapid single test, so insulation can fatigue under electrical stresses. *Treeing* is a poorly understood phenomenon in which conductive paths extend themselves even at rather low overall voltages because the stress at the needlelike end of the tree is so high (Fig. 10.8).

10.7 OPTICAL PROPERTIES

The transparency of most unfilled plastics is obvious in thin-film applications such as packaging. However, the number of polymers that are useful in thick sections (greater than 0.5 cm thick) is limited. Applications such as eyeglass lenses, contact

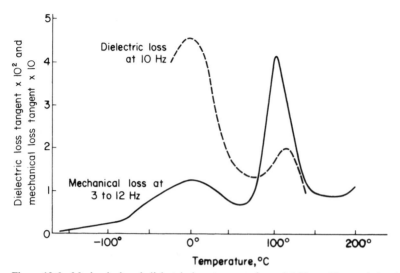

Figure 10.6 Mechanical and dielectric loss tangents for poly(chlorotrifluoroethylene). (After Saito *et al.* [22].)

lenses, and precision optics all place special demands on the mechanical and optical properties of the materials.

Transparency can be measured as the light intensity transmitted through a given thickness of material using light of a specified wavelength. Some relative values for useful optical polymers are shown in Table 10.6 along with other properties. The index of refraction of a material, n, is equal to the ratio of the velocity of light of a specific wavelength in a vacuum to the velocity of the same light in the material. The common wavelengths used are 486 nm (F line), 589 nm (D "sodium" line), and 651 nm (C line). The change in n with wavelength is summarized by the dispersion, D_s, or the Abbé value, defined as

$$D_s = \frac{n_F - n_C}{n_D - 1} = \frac{100}{\text{Abbé value}} \tag{10.7}$$

The first three materials listed are amorphous thermoplastics. CR-39 is a crosslinked, amorphous network (see Sec. 15.1). Most highly crystalline polymers are hazy because the crystals and the amorphous phases do not have the same index of refraction and light is scattered at the interfaces. Poly(4-methyl-1-pentene) is unusual in that the two phases have nearly the same index of refraction. The comparison of densities of the various plastics with crown glass shows why eyeglass lenses of plastic are popular despite their generally lower scratch resistance. Contact lenses that are hydrophilic may be made from copolymers of hydroxyethylmethacrylate, HEMA. Transparency is high since the lenses are essentially crosslinked gels containing 40 to 70% water.

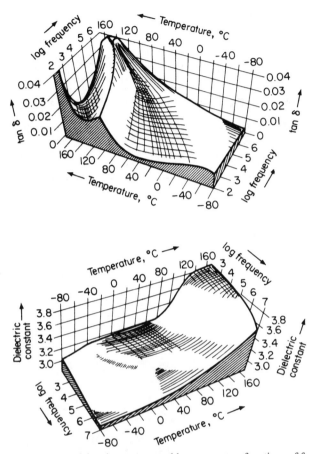

Figure 10.7 Dielectric constant and loss tangent as functions of frequency and temperature for poly-(ethylene terephthalate) [23]

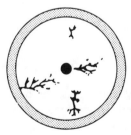

Figure 10.8 Treeing in high-voltage cable [24].

Table 10.6 Selected optical properties [25, 26]

| Material | Index of refraction | | | Abbé value | Density, g/cm³ | Trans., % | Linear exp. coeff., (°C)⁻¹ |
	n_F	n_D	n_c				
Poly(methylmethacrylate)	1.497	1.491	1.489	57.2	1.19	92	65×10^{-6}
Polystyrene	1.604	1.590	1.584	30.8	1.10	88	63×10^{-6}
Polycarbonate	1.593	1.586	1.576	34.0	1.20	89	68×10^{-6}
Poly(4-methyl-1-pentene)	—	1.466	—	56.4	0.83	—	117×10^{-6}
Allyl diglycol carbonate, CR-39™	—	1.499	—	57.8	1.32	—	120×10^{-6}
Crown glass	—	1.525	—	—	2.53	—	—

KEYWORDS FOR CHAPTER 10

Section

2. Compounding
   Antioxidant
   Curative
   Formulation
3. Hardness
5. Heat distortion temperature
   Oxygen Index
6. Volume resistivity
   Polyaniline
   Emeraldine
   Surface resistivity
   Dielectric strength
   Arc resistance
   Dielectric constant
   Dissipation factor
   Electromagnetic interference shielding (EMI)
   Dielectric heating
   Treeing
7. Index of refraction

PROBLEMS

10.1 A major concern of civic organizations and local governments has been the disposal of plastic containers. Bleach bottles and other blow-molded containers from polyolefins are lighter than water and litter the shores of our lakes and streams. What are some ways in which the problem might be attacked?

10.2 Polystyrene melt at 250°F is injected into a cavity in 0.4 min (compression rate of 10%/min) at 25,000 psi. What is the pressure after 10 min at constant volume and temperature? (Refer to Fig. 10.2.)

10.3 A silicone rubber cable is made with a conductive carbon-loaded core and an unfilled layer of insulation. If the resistivities are 10 and 10^{14} ohm·cm, respectively, and outer and inner diameters of the insulation are 0.3 and 0.1 cm, respectively, how fast will the cable heat up when a potential of 100 V dc is impressed on a cable 10 m long? Assume the heat capacity of both filled and unfilled rubber to be 0.5 cal/cm^3·°C and the cable to behave adiabatically.

10.4 If carbon black costs \$0.30/lb, oil \$0.15/lb, and EPDM rubber \$0.95/lb, what is the most economical composition (based on Fig. 10.1) that will yield a tensile strength of 1500 psi with a hardness of at least 50 (Shore A)? The carbon black has a density of 1.80 g/cm^3, the oil of 0.92 g/cm^3, and the rubber of 0.86 g/cm^3.

10.5 A wheel is cast from low-molecular-weight polystyrene in a stainless-steel mold at 300°F. If the mold is 5 in in diameter at 75°F, what is the diameter of the plastic wheel at 75°F? If the polymer had been molded under pressure to pack more material in the mold (see Fig. 10.2c), what pressure would have been needed to make the plastic wheel the same diameter as the mold at 75°F? Linear coefficients of thermal expansion are 1×10^{-5} (°F)$^{-1}$ for steel and 1.5×10^{-4} (°F)$^{-1}$ for the polymer.

10.6 A cubic container with a volume of 1.00 m^3 is to be insulated so that a charge of 100 kg of ice will take at least 20 h to melt completely. What thickness of insulation is required? Urethane foam insulation, $k = 0.050$ W/m·°C; heat of fusion for water is 79.7 cal/g (334 kJ/kg); average outside temperature is 40°C.

10.7 The thermal resistance of pipe insulation can be calculated from Eq. (10.2) if the reciprocal of the thermal conductivity of the insulation is taken as the equivalent of the electrical resistance. The rate of energy loss from the pipe then is directly proportional to the temperature difference and inversely proportional to the thermal "resistance." If a 6.0-cm diameter pipe carries chilled water at an average velocity of 3.0 m/s over a distance of 1.50 km, what thickness of insulation (same k as in last problem) is needed to keep the temperature from rising more than 2.0°C when the inside temperature is 10°C and the outside is 65°C? Neglect all resistance except that of the insulation.

REFERENCES

1. Winding, C. C., and G. D. Hiatt: "Polymeric Materials," p. 62, McGraw-Hill, New York, 1961.
2. Bertsch, P. F.: "Graphical Analysis of Compounding Data," BL-355, Elastomer Chemical Dept., DuPont, June 1959.
3. "Vistalon 3708/Ethylene-Propylene Terpolymer," Enjay Chemical Company, 1968.
4. Quinn, F. A., Jr., and L. Mandelkern: *J. ACS,* 80:3178 (1958).
5. Matsuoka, S., and B. Maxwell: *J. Polymer Sci.,* 32:131 (1958).
6. Colleti, W., and L. Rebori: *Insulation,* 11:27 (January 1965).
7. Khanna, Y. P., and E. M. Pearce: "Flammability of Polymers," in R. W. Tess and G. W. Poehlein (eds.): "Applied Polymer Science," 2d ed., ACS, Washington, D.C., 1985.
8. Ball, A. M.: *Chem. Eng. Progr.,* 57(9):80 (1961).
9. Levy, S.: *Mod. Plastics Encycl.,* 39(1A):28 (1961).
10. Mathes, K. N.: in E. Baer (ed.), "Engineering Design for Plastics," chap. 7, Reinhold, New York, 1964.
11. Davies, J., R. Miller, and W. Busse: *J. ACS,* 63:361 (1941).
12. Seanor, D. A. (ed.): "Electrical Properties of Polymers," Academic, New York, 1982.
13. Cowan, D. O., and F. M. Wiygul: *Chem. Eng. News,* 64(29):28 (1986).
14. Gibson, H. W.: *Polymer,* 25:3 (1984).
15. MacDiarmid, A. G., and A. J. Epstein: in "Science and Applications of Conducting Polymers," p. 117, A. Hilger, Philadelphia, 1991.
16. Osterholm, J. E., J. Laakso, and P. Nyholm: *Polymer Preprints,* 30(1):145 (1989).
17. Carter, F. L. (ed.): "Molecular Electronic Devices, II," Dekker, New York, 1987.
18. Botts, S. C., and G. L. Moses: *Insulation,* 7:19 (December 1961).
19. Kloeffler, R. G.: "Industrial Electronics and Control," pp. 318.326, Wiley, New York, 1949.
20. Wilson, J. L.: "Dielectric Heating," in *Enc. Polym. Sci. Eng. 2d ed.,* 5 (1) (1987).
21. Li, C., K. A. Chaffin, and A. Thakore: *Adhesives Age,* 37(7):18 (1994).
22. Saito, N., *et al.,* in F. Seitz and D. Turnbull (eds.): "Solid State Physics," vol. 14, p. 420, Academic Press, New York, 1963.
23. Reddish, W.: *Trans. Faraday Soc.,* 46:459 (1950); redrawn by A. J. Curtis in J. B. Birks and J. H. Schulman (eds.): "Progress in Dielectrics," vol. 2, p. 37, Wiley, 1960.
24. Olyphant, M., Jr.: *Insulation,* 9:23 (March 1963); 9:33 (November 1963).
25. Mills, N. J.: in *Enc. Polym. Sci. & Eng.,* 10:496 (1987).
26. Teyssier, C., and C. Tribastone: *Lasers & Optronics,* 12:50 (1990).

General References

Andrade, J. D. (ed.): "Polymer Surface Dynamics," Plenum, New York, 1988.
Aseeva, R. M., and G. E. Zaikov: "Combustion of Polymer Materials," Macmillan, New York, 1986.
Babbit, R. O. (ed.): "The Vanderbilt Rubber Handbook," 12th ed., R. T. Vanderbilt Co., Norwalk, CT, 1978.
Baer, E. (ed.): "Engineering Design for Plastics," Reinhold, New York, 1964.
Baijal, M. D. (ed.): "Plastics Polymer Science and Technology," Wiley, New York, 1982.
Bakshi, A. K.: "Electronic Structure of Biopolymers and Highly Conducting Polymers," Wiley, New York, 1993.
Barlow, F. W.: "Rubber Compounding," Dekker, New York, 1988.
Bhattacharya, S. K. (ed.): "Metal-Filled Polymers," Dekker, New York, 1986.
Bhowmick, A. K., and H. L. Stephens (eds.): "Handbook of Elastomers," Dekker, New York, 1988.
Birley, A. W., and M. J. Scott: "Plastics Materials Properties and Applications," 2d ed., Chapman & Hall, New York, 1988.

Bloor, D., and R. R. Chance (eds.): "Polydiacetylenes: Synthesis, Structure, and Electronic Properties," M. Nijhoff (Netherlands), 1985.

Blythe, A. R.: "Electrical Properties of Polymers," Cambridge, London, 1979.

Chan, C.-M.: "Polymer Surface Modification and Characterization," Hanser-Gardner, Cincinnati, Ohio, 1993.

Comyn, J. (ed.): "Polymer Permeability," Elsevier Applied Science, New York, 1985.

Cotts, D. B., and Z. Reyes: "Electrically Conductive Organic Polymers for Advanced Applications," Noyes Data, Park Ridge, NJ, 1986.

Crosby, E. G., and S. N. Kochis: "Practical Guide to Plastics Applications," Cahners, Boston, 1971.

Davies, D. K. (ed.): "Static Electrification, 1971," Institute of Physics, London, 1971.

Deanin, R. D.: "Polymer Structure-Properties-Applications," Cahners, Boston, 1971.

Farges, J.-P. (ed.): "Organic Conductors, Fundamentals and Applications," Dekker, New York, 1994.

"Fire Safety Aspects of Polymeric Materials," 10 vols., Technomic, Lancaster, Pa., 1977.1981.

Ford, W. T. (ed.): "Polymeric Reagents and Catalysts," ACS, Washington, D.C., 1986.

Frados, J. (ed.): "Plastics Engineering Handbook," 4th ed., Van Nostrand-Reinhold, New York, 1976.

Frisch, K. C., and A. V. Patsis (eds.): "Electrical Properties of Polymers," Technomic, Westport, Conn., 1972.

Goddard, E. D., and B. Vincent (eds.): "Polymer Adsorption and Dispersion Stability," ACS, Washington, D.C., 1984.

Goosey, M. T.: "Plastics for Electronics," Elsevier Applied Science, New York, 1985.

Hall, C.: "Polymer Materials," Halsted-Wiley, New York, 1981.

Hartwig, G.: "Polymer Properties at Room and Cryogenic Temperatures," Plenum, New York, 1995.

Hilado, C. J. (ed.): "Flammability Handbook for Plastics," 4th ed., Technomic, Lancaster, Pa., 1990.

Hopfenberg, H. B. (ed.): "Permeability of Plastic Films and Coatings," Plenum, New York, 1974.

Hornak, L. A. (ed.): "Polymers for Lightwave and Integrated Optics," Dekker, New York, 1995.

Hunter, R. S.: "The Measurement of Appearance," Wiley-Interscience, New York, 1975.

Ikada, Y., and Y. Uyama: "Lubricating Polymer Surfaces," Technomic, Lancaster, Pa., 1993.

Ishida, H., and J. L. Koenig (eds.): "Composite Interfaces," Elsevier Applied Science, New York, 1986.

Ishida, H., and G. Kumar (eds.): "Molecular Characterization of Composite Interfaces," Plenum, New York, 1985.

Karasz, F. E. (ed.): "Dielectric Properties of Polymers," Plenum, New York, 1972.

Kitihara, A., and A. Watanabe (eds.): "Electrical Phenomena at Interfaces," Dekker, New York, 1984.

Kovarskii, A. L. (ed.): "High-Pressure Chemistry and Physics of Polymers," CRC Press, Boca Raton, Fla., 1994.

Kroschwitz, J. I. (ed.): "Electrical and Electronic Properties of Polymers: A State-of-the-Art Compendium," Wiley, New York, 1988.

Ku, C. C., and R. Liepins: "Electrical Properties of Polymers," Macmillan, New York, 1987.

Kumins, C. A. (ed.): "Transport Phenomena Through Polymer Films," ACS, Washington, D.C., 1973.

Kuryla, W. C., and A. J. Papa: "Flame Retardancy of Polymeric Materials," 5 vols., Dekker, New York, 1973.1979.

Kuzmany, H., M. Mehring, and S. Roth (eds.): "Electronic Properties of Polymers and Related Compounds," Springer-Verlag, New York, 1985.

Labana, S. S., and R. A. Dickie (eds.): "Characterization of Highly Cross-linked Polymers," ACS, Washington, D.C., 1984.

Ladik, J. J.: "Quantum Theory of Polymers as Solids: Part 1, Quantum Theory of Polymeric Electronic Structure," Plenum, New York, 1987.

Landrock, A. H.: "Handbook of Plastics Flammability and Combustion Technology," Noyes Data, Park Ridge, N.J., 1983.

Lever, A. E., and J. A. Rhys: "The Properties and Testing of Plastics Materials," 3d ed., CRC Press, Cleveland, 1968.

Levy, S., and J. H. DuBois: "Plastics Product Design Engineering Handbook," Van Nostrand-Reinhold, New York, 1977.

Lewin, M., S. M. Atlas, and E. M. Pearce (eds.): "Flame-Retardant Polymeric Materials," Plenum, New York, 1975.

Linford, R. G. (ed.): "Electrochemical Science and Technology of Polymers," Elsevier Applied Science, New York, 1987.

Lipatov, Y. S.: "Physical Chemistry of Filled Polymers," Rubber and Plastics Research Assoc., Shrewsbury, England, 1979.

Lutz, J. T. (ed.): "Thermoplastic Polymer Additives," Dekker, New York, 1988.

Lyons, M.: "Electroactive Polymer Electrochemistry. Part 1: Fundamentals," Plenum, New York, 1994.

Margolis, J. M.: "Conductive Polymers and Plastics," Chapman & Hall, New York, 1989.

Mark, J. E., A. Eisenberg, W. W. Graessley, L. Mandelkern, E. T. Samulski, J. L. Koenig, and G. D. Wignall: "Physical Properties of Polymers," 2d ed., ACS, Washington, D.C., 1993.

McCauley, J. W., and V. Weiss (eds.): "Materials Characterization for Systems Performance and Reliability," Plenum, New York, 1986.

Meeten, G. H. (ed.): "Optimal Properties of Polymers," Elsevier Applied Science, New York, 1986.

Moiseev, Y. V., and G. E. Zaikov: "Chemical Resistance of Polymers in Aggressive Media," Plenum, New York, 1987.

Mort, J., and G. Pfister (eds.): "Electronic Properties of Polymers," Wiley, New York, 1982.

Myers, R. R., and J. S. Long (eds.): "Characterization of Coatings: Physical Techniques," Dekker, New York, Part 1, 1969; Part 2, 1973.

Nelson, G. L. (ed.): "Fire and Polymers: Hazards Identification and Prevention," ACS, Washington, D.C., 1990.

Nielsen, L. E., and R. F. Landel: "Mechanical Properties of Polymers and Composites," 2d ed., Dekker, New York, 1993.

Ogorkiewicz, R. M. (ed.): "Thermoplastics: Properties and Design," Wiley-Interscience, New York, 1974.

Park, W. R. R.: "Plastics Film Technology," Van Nostrand-Reinhold, New York, 1969.

Prasad, P. N., and D. R. Ulrich (eds.): "Nonlinear Optical and Electroactive Polymers," Plenum, New York, 1988.

Rogers, C. E. (ed.): "Permselective Membranes," Dekker, New York, 1971.

Russo, P. S. (ed.): "Reversible Polymeric Gels and Related Systems," ACS, Washington, D.C., 1987.

Samuels, R. J.: "Structured Polymer Properties," Wiley-Interscience, New York, 1974.

Scott, J. R.: "Physical Testing of Rubbers," Palmerton, New York, 1965.

Scrosati, B. (ed.): "Applications of Electroactive Polymers," Chapman & Hall, New York, 1993.

Seanor, D. A. (ed.): "Electrical Properties of Polymers," Academic, New York, 1982.

Serafini, T. T., and J. L. Koenig (eds.): "Cryogenic Properties of Polymers," Dekker, New York, 1968.

Sichel, E. K. (ed.): "Carbon-Black Polymer Composites, the Physics of Electrically Conducting Composites," Dekker, New York, 1982.

Simpson, W. G. G. (ed.): "Plastics Surface and Finish," CRC Press, Boca Raton, Fla., 1994.

Skotheim, T. A. (ed.): "Handbook of Conducting Polymers," 2 vols., Dekker, New York, 1986.

Skotheim, T. A. (ed.): "Electroresponsive Molecular and Polymeric Systems, Vol. II," Dekker, New York, 1991.

Sward, G. G. (ed.): "Paint Testing Manual," 13th ed., ASTM Spec. Tech. Publ. 500, American Society for Testing and Materials, Philadelphia, 1972.

Thomas, E. L. (ed.): "Materials Science and Technology. Structure and Properties of Polymers," VCH, New York, 1993.

Troitzsch, J.: "International Plastics Flammability Handbook," Macmillan, New York, 1983.

Urban, M. W., and C. D. Craver (eds.): "Structure-Property Relations in Polymers: Spectroscopy and Performance," ACS, Washington, D.C., 1993.

van Krevelen, D. W.: "Properties of Polymers," Elsevier, Amsterdam, 1990.

Vieth, W. R.: "Diffusion In and Through Polymers," Hanser-Gardner, Cincinnati, Ohio, 1991.

ELEVEN

DEGRADATION AND STABILIZATION
OF POLYMER SYSTEMS

Degradation is any undesirable change in properties that occurs after a material has been put into service. We can categorize degradation as affecting the polymer chemically or the polymer system physically or chemically. Examples of the latter include the wearing of tires, the loss of plasticizer from a polymer system by evaporation or migration, or the separation of polymer from rigid fillers leaving voids at the interface. To counteract the degradation of systems by various agencies, certain *stabilizers* can be added that interfere with specific reactions. In this chapter we shall consider the agencies that produce degradation, the physical and chemical consequences of degradation, and some of the preventative measures that can be taken to diminish degradation.

11.1 DEGRADING AGENCIES

The agencies that bring about changes in polymers seldom act individually except in the laboratory. We can speak about the separate effects of heat, radiation, chemicals, and mechanical energy, but in practice, all four may be present to some degree. *Weathering* is a term used for outdoor exposure. During weathering, sunlight causes damage by ultraviolet radiation absorption as well as by heat built up by infrared absorption. Atmospheric oxygen and moisture provide the most common chemical attacks. And, of course, a sample may be mounted in a strained condition to approximate a real application.

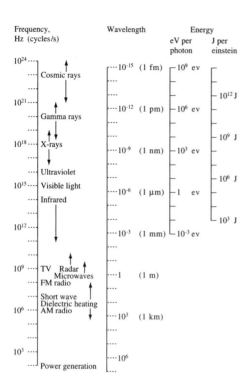

Figure 11.1 The electromagnetic spectrum. The positions of various applications are approximate, since definitions of the various bands are not exact

In the laboratory the degrading agencies are usually separated. However, in some accelerated weathering tests, a high-intensity mercury or xenon arc, water spray, and temperature and humidity control may be used. Outdoor exposure itself varies considerably from one location to another. Samples may be sent to Florida for sunlight and salt spray exposure and to Los Angeles for the effects of smog and ozone.

The spectrum of electromagnetic waves ranges from radio waves of long wavelength λ and low frequency ω_c to gamma rays (and cosmic rays) of short wavelength (Fig. 11.1). Certain radio bands are familiar because of specific applications. Almost all uses are regulated by government agencies. Industrial dielectric heating uses 1 to 100 MHz, microwave ovens use 915 to 2450 MHz, and TV and FM broadcasting bands generally fall between 100 and 915 MHz. Radar can use a wide range of frequencies, with most applications between 200 and 35,000 MHz [1].

The energy associated with a quantum or photon (the smallest unit of radiant energy) is simply related to the wavelength or frequency through Planck's constant h and the velocity of light c. Some interrelations are

$$\lambda\omega_c = c = 3.00 \times 10^8 \text{ m/s} \tag{11.1}$$

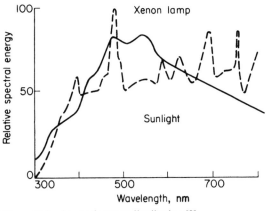

Figure 11.2   Spectral energy distribution [2]

$$\frac{\text{Energy}}{\omega_c} = h = 6.62 \times 10^{-34} \text{ J} \cdot \text{s} \tag{11.2}$$

For example, light with a wavelength of 100 nm has a frequency of 3.0×10^{15} Hz. A commonly used mercury germicidal lamp has a strong ultraviolet emission at 254 nm, corresponding to a frequency of 1.18×10^{15} Hz. The energy associated with 6.02×10^{23} (Avogadro's number) photons is called an einstein. Thus the energy in each einstein of the 254-nm radiation is 470 kJ or 112 kcal. Since the average bond energy of the carbon-carbon bond is 83 kcal/mol, it becomes apparent that absorption of ultraviolet light is more likely to cause chemical reactions in polymers than visible light (400 to 700 nm).

It can be seen from the distribution curve for the spectral energy of sunlight (Fig. 11.2) that most of the energy is in the visible region 380 to 800 nm. At the high-energy end, about $\lambda = 350$ nm, the energy per einstein is approaching or exceeding the dissociation energy of a number of covalent bonds found in polymers (see Sec. 2.1). Of course, the incident light may be transmitted, refracted, or scattered. However, if it is absorbed locally in a polymer, dissociation may result, most commonly with the loss of a hydrogen atom, leaving a free radical on the chain. When ultraviolet light or gamma radiation is used, such dissociations become much more probable. The penetrating depth also increases with increasing frequency. At the other end of the sun's spectrum, the infrared rays may be absorbed with no great localized energy but with a general increase in temperature. In the laboratory, sunlight may be simulated by a xenon lamp having the spectrum shown in Fig. 11.2.

Almost all materials are subject to attack by oxygen and by water when they are put into use. The process of oxidation is so important that it is worth noting a

Table 11.1 Relative rates of oxidation versus structure [3]

Structure	Relative rate of oxidation
CH₃ $\mid$ $\leftarrow$CH₂—C=CH—CH₂$\rightarrow$ $\uparrow$ $\uparrow$	10
CH₃ $\mid$ $\leftarrow$CH₂—CH—O$\rightarrow$ $\uparrow$	9
CH₃) $\mid$ $\leftarrow$CH₂—CH$\rightarrow$ $\uparrow$	6.5
CH₃ $\mid$ O $\mid$ $\leftarrow$CH₂—CH$\rightarrow$ $\uparrow$	2.8
CH₃CH₂—O $\mid$ C=O $\mid$ ———CH₂—CH——— $\uparrow$	1.4

few generalizations that have been made concerning it. Oxidation typically is a chain process [3]:

> Initiation:
> Production of R· or RO₂·
> Propagation:
> R· + O₂ → RO₂·
> RO₂· + RH → ROOH + R·
> Termination:
> 2R· → R—R
> RO₂· + R· → ROOR
> 2RO₂· → nonradical products

The end products are seen to consist of covalent carbon-carbon bonds, peroxide bonds, and perhaps hydroperoxides. Although any carbon-hydrogen bond might be attacked, positions that are especially vulnerable are those adjacent to a double bond, adjacent to an ether linkage, or on a tertiary carbon (indicated by arrows in Table 11.1).

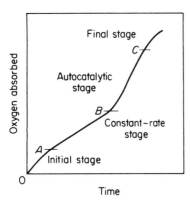

Figure 11.3 axis labels: Oxygen absorbed (vertical), Time (horizontal), with stages marked: Final stage, C, Autocatalytic stage, B, Constant-rate stage, A, Initial stage, 0

Figure 11.3 Four stages observed in oxygen-absorption measurements on elastomers [4]

In the laboratory, the effects of oxidation are measured by several kinds of test. Most commonly, samples of a polymer system are aged at an elevated temperature in nitrogen, air, or oxygen. For air, a circulating air oven is used. Exposure to atmospheres of pure nitrogen or oxygen is usually carried out in sealed tubes (bombs). Changes in structure may be inferred from changes in physical properties such as tensile modulus or elongation at break on aging. A more sophisticated measure of change is found in infrared spectral analysis. An actual measurement of the amount of oxygen absorbed is useful in studying the mechanism of oxidation. The four stages of oxidation found by Shelton and co-workers are shown in Fig. 11.3 [4]. The first two stages may not always be discernible in uninhibited polymers. In addition, one can measure the concentration of specific groups as a function of time or oxygen content (Fig. 11.4).

Moisture can be important by acting as a plasticizer or by acting as a solvent for some catalytic species. Ultraviolet light and oxygen usually will have different effects on a polymer depending on the humidity. This is especially true when the polymer has a substantial equilibrium moisture content. Nylon, for example, may vary in moisture content from 0 to over 5% depending on the humidity. Polyethylene has almost no moisture content. Water penetration at the interface between two phases can cause delamination. The hydroxyl-rich surface of glass has a great affinity for water, so that a layer of organic polymer on the glass may be displaced by water at high humidities. Another reaction with water is hydrolysis. In some of the first commercial urethane foams (1950s), it was found that the urea, urethane, and aliphatic ester groups were rather easily hydrolyzed, the rate increasing with pH above 7. Present-day formulations are much less sensitive, but water remains a factor to be considered whenever hydrolyzable groups are part of the polymeric structure.

The overall view of chemical attack is impossible to condense because of the wide variety of chemical species and polymeric substrates. Whenever a polymer is suggested as a material of construction in a chemical plant, a thorough consider-

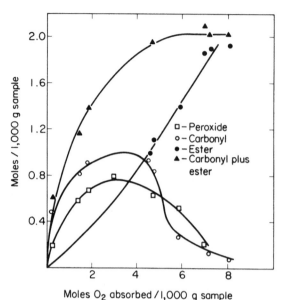

Figure 11.4 Concentration of peroxide, carbonyl, and ester groups as a function of oxygen absorbed in oxidation of 5-phenyl-2-pentene at 80°C [4]

ation of resistance to specific chemicals is desirable. Polymeric pipes and tank liners are used for acid and alkali, oxidizing and reducing agents, and all kinds of organic reactants. In each case the most economical solution will be specific, because factors of initial cost, versatility, and maintenance may enter in.

Biological attack includes alteration of polymer systems by fungi, bacteria, and larger species such as insects and animals [5–7]. Most naturally occurring polymers are biodegradable in their native state. Cellulose and natural rubber can be digested by a variety of microorganisms. Modified natural materials such as acetylated cellulose or sulfur-cross-linked rubber are quite resistant. Totally synthetic polymers, by and large, are resistant to attack when the molecular weight exceeds about 5000. Some heterochain polymers, including polyesters, polyamides, and polyurethanes, may degrade in the presence of some microorganisms [5, 8]. In this degradation the microorganisms are using the material as a source of nutriment, with enzymes generated by the organism acting as catalysts. For many years the polymer industries emphasized the development of bioresistant materials. The treatment of cellulose to prevent deterioration of cotton and linen in tropical climates has long been a major effort of military suppliers. Since the early 1970s, public concern for the problem of litter as part of the overall problem of waste disposal has led to development of polymers that are biodegradable. However, the cost and the health hazard of marketing polymer containers that are truly biodegradable has outweighed the esthetic advantages they might have. Waste disposal is also discussed in Chap. 13.

Purely mechanical agencies can include the effect of solvents on the macroscopic dimensions of a material. Swelling of rubbers and melts was mentioned in Sec. 2.5 in connection with cohesive energy density. Solvents may extract portions of polymer systems, e.g., the plasticizer from a poly(vinyl chloride)-based system. Swelling at a surface may set up a stress that is relieved by *crazing*. Craze lines consist of oriented polymer. They become evident by their opacity in glassy polymers stressed by a purely mechanical couple or by surface absorption of a liquid (see Sec. 9.7).

High-speed stirring, turbulent flow, and ultrasonic irradiation of polymer solutions can cause changes in polymer structure. The predominant response is a decrease in molecular weight, because recombination of radicals is not favored in solution. When a melt is masticated at high shear rates, a great deal of mechanical energy is dissipated in the system. Recombination of radicals can take place, as well as some other reactions depending on circumstances.

11.2 PHASE SEPARATION AND INTERFACIAL FAILURE

The degradation of a polymer system can take place without changing the structure of the whole polymer. The two common cases are where one component of a homogeneous mixture is extracted and where components of a heterogeneous assembly delaminate.

The extraction of plasticizer from a poly(vinyl chloride) composition does not always require organic solvents. Soapy water in contact long enough with a foam or film can extract low-molecular-weight plasticizers. A plasticized vinyl shower curtain has the added hazard of relatively high temperatures and humidities to contend with. When a plasticized system is put in physical contact with an unplasticized polymer, the plasticizer may *migrate* from one to the other. In designing a vinyl coaster, for example, the compounder must take into account the possibility that the item may be placed on a lacquered surface. Migration of dioctyl phthalate from the poly(vinyl chloride) coaster to a cellulose acetate surface will result in a softened, marred appearance.

Delamination is most easily illustrated by interior plywood soaked in water. The layers may come apart as the wood swells and the phenolic adhesive is displaced. The same sort of failure can occur in a glass-fiber-reinforced thermoset. Coupling agents that tie glass and polymer together can decrease this tendency (see Sec. 9.8). Systems that delaminate are not always so obviously heterogeneous to start. It is common for large injection-molded polyethylene objects to develop a layer that peels off in the vicinity of the injection point on aging. Apparently the crystal structure of the polymer that first enters the cold cavity and coats the wall is different from the structure developed by subsequent polymer cooled more slowly. The interface between the two represents a plane of weakness that is aggravated by warm water and detergent. A continued change in crystalline structure at the interface may be taking place.

Mechanical stresses often induce delamination. A tensile stress in loaded rubber may tear polymer from filler, leaving vacuoles (Sec. 9.8). Constant flexing of a conveyor belt or a tire may pull rubber from the fibrous reinforcement.

11.3 POLYMER DEGRADATION

As far as the polymers themselves are concerned, many types of "degradation" can be predicted from the reactions of low-molecular-weight analogs. Oxidation of hydrocarbons and hydrolysis of esters are often more easily studied in such analogs. However, as we consider the response of polymers that are predominantly linear, we can categorize degradation as affecting the backbone of the polymer or the side chains on the polymer. Some degradative effects are:

1. Scission
2. Depolymerization
3. Cross-linking   } Backbone effects
4. Bond changes
5. Side-group changes

In *chain scission* we have bonds broken at random within the polymer molecules. Each breaking bond creates another molecule and lowers the average molecular weight. Hydrolysis of a polyester is a good example of a *random scission* process, since the susceptibility of a bond does not depend greatly on molecular size. If we take a polyester of number-average degree of polymerization x_n in a dilute solution of concentration c (g/dl), we can calculate the concentration of ester bonds in the system $[\phi]$:

$$[\phi] = (N, \text{ concentration of monomer units in solution})$$
$$- (m, \text{ concentration of polymer molecules})$$

$$[\phi] = N - m \tag{11.3}$$

$$N = \frac{c}{M_0} \tag{11.4}$$

$$m = \frac{N}{x_n} \tag{11.5}$$

where M_0 is the molecular weight of a monomer unit.

If we postulate that the rate of bond disappearance (hydrolysis) with time

$(-d[\phi]/dt)$ will be proportional to the bond concentration $[\phi]$, we have a first-order expression,

$$-\frac{d[\phi]}{dt} = k[\phi] = k(N - m) = \frac{kc(1 - 1/x_n)}{M_0} \tag{11.6}$$

Also, differentiating the second and fourth terms of Eq. (11.6) yields

$$-k\frac{d[\phi]}{dt} = \frac{kc}{M_0}\frac{d(1/x_n)}{dt} \tag{11.7}$$

But in Eq. (11.6), $1/x_n \ll 1$; therefore, the reaction is "pseudo-zero-order" and

$$\frac{c}{M_0}\frac{d(1/x_n)}{dt} = \frac{kc(1 - 1/x_n)}{M_0} = \frac{kc}{M_0} \tag{11.8}$$

If we start at time $t = 0$ and $x_n = (x_n)_0$, we get the change in number-average degree of polymerization as a function of time.

$$\frac{1}{x_n} - \left(\frac{1}{x_n}\right)_0 = kt \tag{11.9}$$

It should be emphasized that this relationship holds only when $x_n \gg 1$. When $(x_n)_0 = 1000$, for example, breaking one bond in each molecule cuts x_n to 500. However, the available number of hydrolyzable bonds has been decreased by only 0.1%.

Polymers containing a completely substituted carbon in the chain often undergo chain scission on oxidation or exposure to ultraviolet or gamma radiation. This is because rearrangements are formed when an unpaired electron is left on a structure as in polyisobutylene:

$$R_1 - CH_2 - \underset{\underset{\overset{|}{CH_2}}{\overset{|}{\underset{\cdot}{}}}}{\overset{\overset{CH_3}{|}}{C}} - CH_2 - R_2 \rightarrow R_1 - CH_2 - \underset{\overset{|}{CH_2}}{\overset{CH_2}{C}} = CH_2 + \cdot CH_2 - R_2$$

Some other structures that favor chain scission are polypropylene, polyacrylates, and polymethacrylates. Because the industrially important process of cross-linking rubber often is carried out under oxidative conditions, and because chain scission generally is undesirable, other agencies must be employed for cross-linking when the chain contains such unstable groups.

Depolymerization also decreases molecular weight. Some common examples are merely the polymerization reactions previously considered taken in reverse. With some systems, almost quantitative regeneration of monomer is possible. In fact, depolymerization is used to recover monomers from scrap polymer when economics permit. For example, methyl methacrylate can be recovered in good yield and in high purity from the polymer by using low pressures, high temperatures, and a source of free radicals:

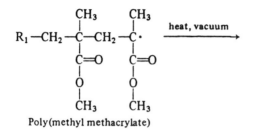

Poly(methyl methacrylate)

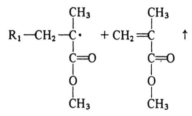

Polystyrene also depolymerizes under these conditions, but the original monomer is not expensive enough to warrant commercial regeneration from scrap polymer. Ionic reactions also may lead to depolymerization. Heating almost any polysiloxane with potassium hydroxide in a vacuum eventually will convert the entire system to volatile cyclic material, predominantly the eight-membered ring of the tetramer:

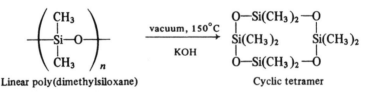

Linear poly(dimethylsiloxane)   Cyclic tetramer

Although cross-linking is a useful reaction in making a rubbery material that is dimensionally stable at high temperatures, there may be undesirable consequences when it occurs after a material is in service. The modulus increases with cross-linking, but the energy-absorbing capacity goes through a maximum and decreases thereafter. A rubbery polymer system generally becomes brittle as a result. Another typical pattern accompanying cross-linking is a decrease in compatibility, so that plasticizers exude, systems shrink, and delamination occurs. Polyethylene affords a primitive example of desirable and undesirable cross-linking. Molten polymer (150°C) may be mixed with a peroxide that does not decompose until an even higher "curing" temperature (205°C) is reached. The finished product is a cross-linked polymer very useful as high-temperature electrical insulation. Cross-linking can also be induced by gamma radiation or high-energy electron beams:

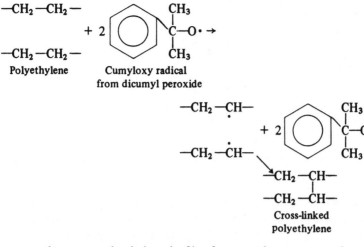

Polyethylene

Cumyloxy radical from dicumyl peroxide

Cross-linked polyethylene

However, the same polyethylene in film form outdoors may undergo slow attack from atmospheric oxygen in the presence of sunlight. Although cross-linking does take place, it is accompanied now by incorporation of oxygen to give polar groups and by chain scission. The end result is a brittle, easily ripped film with poor optical and electrical properties.

Accelerated aging tests have been mentioned in an earlier discussion of oxidation. It is desirable in analyzing such tests to be able to measure cross-linking and chain scission. For rubbery polymers this can be done by making two kinds of measurements [9]. Both depend on the proportionality of cross-link density with modulus (see Sec. 8.2). It should be obvious that if a tensile modulus E_t is measured intermittently on an aging sample, any cross-link that is formed will be reflected in an increased modulus. Chain scission, on the other hand, will decrease this modulus. In other words, the rate of change of E_t is the difference in the rates of cross-linking (dN_c/dt) and chain breaking (dN_s/dt):

$$\frac{dE_t}{dt} = K\left(\frac{dN_c}{dt} - \frac{dN_s}{dt}\right) \qquad (11.10)$$

where K is a proportionality constant.

Now in a separate test we can measure the relaxation of stress in a sample held at constant elongation. Any chain scission that occurs also decreases this relaxation modulus E_r. However, cross-linking of the strained sample does not increase the modulus, because only the cross-links that were put in the relaxed sample and were distorted from their equilibrium, random positions will contribute. For this case, we measure only the rate of chain breaking:

$$\frac{dE_r}{dt} = -K\frac{dN_s}{dt} \qquad (11.11)$$

where K is the same constant as in Eq. (11.10).

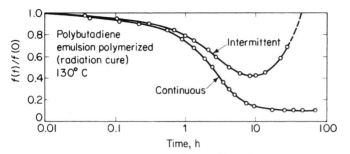

Figure 11.5 Continuous stress-relaxation and intermittent stress measurements at 130°C [9]

The simultaneous measurement of E_t and E_r for polybutadiene aged in air at 130°C can be interpreted as showing that much scission and limited cross-linking occurs up to 1 h (Fig. 11.5). After 10 h the main reaction is cross-linking, and chain scission almost ceases.

Changes can be wrought in the polymer backbone without scission or cross-linking. When poly(vinyl chloride) is heated above 200°C in the absence of stabilizing materials, copious quantities of HCl are evolved after a few minutes. Although scission and cross-linking ensue, the primary effect of heating this polymer is dehydrohalogenation, which changes the backbone structure.

$$\text{\textnormal{+CH}}_2\text{—CHCl—CH}_2\text{—CHCl\textnormal{+}} \xrightarrow{200°C} \text{\textnormal{+CH}}_2\text{—CHCl—CH}\text{=CH\textnormal{+}} + \text{HCl} \uparrow$$

Poly(vinyl chloride)

The allyl chloride structure that is left after a single abstraction of HCl is more susceptible to free-radical attack than the fully saturated material. A "zippering" occurs, leaving behind successive conjugated double bonds. This remaining structure is chromophoric and highly reactive with metal salts and with oxygen. Since the chain cannot rotate around double bonds, it is stiffened even in the absence of cross-linking. Other reactions can take place that alter the backbone stiffness. When heated in zinc metal powder, poly(vinyl chloride) can be cyclicized to a large extent:

$$\text{\textnormal{+CH}}_2\text{—CHCl—CH}_2\text{—CHCl\textnormal{+}} \rightarrow \overset{\displaystyle CH_2}{\overset{\diagup\ \diagdown}{\text{\textnormal{+CH}}_2\text{—CH}\text{——}\text{CH\textnormal{+}}}}$$
$$+ \text{Zn} \qquad\qquad\qquad + \text{ZnCl}_2$$

As with most degradation reactions, these changes in structure are sometimes channeled into useful applications. For example, the induction period for dehydrohalogenation in the presence of zinc oxide is decreased by exposure to ultraviolet light. A photograph can be reproduced using degraded polymer as the chromophores by "heat developing" the latent image from light exposure of such a system.

The exposed areas darken first on heating. A rather good halftone image can be reproduced, although the film speed is quite slow [10].

The oxidation of polyacrylonitrile to give a heat-stable, ladder polymer was mentioned in Sec. 3.5. Natural rubber is cyclicized in order to give abrasion-resistant ink and paint resins. In the presence of chlorostannic acid $(H_2SnCl_6 \cdot 6H_2O)$, *cis*-polyisoprene is cyclicized by refluxing in benzene. The new material can be reduced to a powder at room temperature, indicating that its T_g has been raised considerably over the original $-70°C$. Cross-linking and chain scissions have not taken place to any large extent, since the product redissolves completely to give solutions of high viscosity. The probable structure is [11]:

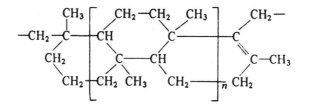

Of course, reactions of the "foliage," the side chains on polymers, can take place without altering the molecular weight or the chain stiffness. Changes in solubility, compatibility, color, and mechanical and electrical properties may result. The main chains of poly(ethyl acrylate) or poly(vinyl acetate), for example, are not affected by hydrolysis. But in either case, hydrolysis changes the water-insoluble precursors into water-soluble resins:

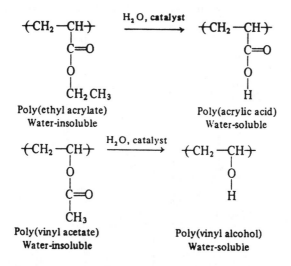

In the case of poly(vinyl alcohol), the reaction is a commercially useful one. But in either case, a drastic change in solubility has been made. It is clearly an undesirable change if the precursors are being used as water-insoluble protective coatings.

11.4 ANTIOXIDANTS AND RELATED COMPOUNDS

The most frequently employed stabilizing ingredient in plastics, fibers, rubbers, and adhesives is the antioxidant. Since oxidation is a free-radical, chain process, it is to be expected that useful molecules will be those that combine with free radicals to give stable species incapable of further reaction.

Quinones may act as free-radical stoppers by adding the radical to the ring or to the carbonyl oxygen. The most widely used classes of antioxidants are the hindered phenols and diaryl amines. Presumably these react with RO_2 radicals by giving up a hydrogen and forming a rather stable radical, which also might stop a second radical by combining with RO_2. Electron transfer and ring-addition reactions may take place also. It will be recalled that such action may be thought of as *degradative chain transfer* and that it is the principle by which inhibitors of free-radical polymerization function (see Sec. 4.4). Two important considerations in selecting antioxidants can be toxicity and color formation. The use of any chemical additive in food wrapping must be approved by the appropriate federal agency (Food and Drug Administration). Usually, only specific formulations and not general classes of materials are approved. In stabilizing rubber, antioxidants are classified as staining or nonstaining depending on whether or not they develop color in use. In a carbon black-loaded tire tread, a staining material is no drawback. But in the white sidewalls, it is important to use additives that are colorless to start with and stay colorless as they protect against oxidation.

A widely used antioxidant in food products is butylated hydroxytoluol (BHT):

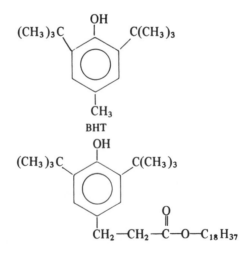

Fatty ester of hindered phenol

It is added to a variety of consumer items ranging from chewing gum and potato chips to bread wrapping and vitamin A. When modified as in the fatty derivative, an antioxidant of lower volatility and better compatibility with polyethylene is formed.

Antioxidants used in rubber compounds often are classified as "staining" or "nonstaining." Phenylenediamine derivatives are very effective antioxidants but discolor (stain) severely. This is no drawback in a rubber compound reinforced with large amounts of carbon black. Where light colors must be preserved, a sulfur-cross-linked, hindered phenol that does not discolor is preferable:

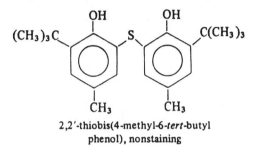

2,2'-thiobis(4-methyl-6-*tert*-butyl
phenol), nonstaining

Most unsaturated raw rubber has a small amount of antioxidant added before shipping to protect it during storage. Butyl rubber may have only four double bonds for each 1000 single bonds in the main chain, but it is usually supplied with BHT (about 0.1%). Ethylene-propylene terpolymer has a higher level of unsaturation, but it is confined to pendant groups of somewhat lower reactivity. About 0.25% of BHT is added by one manufacturer.

In contrast to the small amounts of antioxidant in raw rubber, compounded rubber often contains several percent of protective material. It seems contradictory at first that antioxidants can continue to be very effective even after the polymer system has undergone a high-temperature, oxidative cross-linking. A case in point is poly(vinyl ethyl ether) cross-linked by dicumyl peroxide:

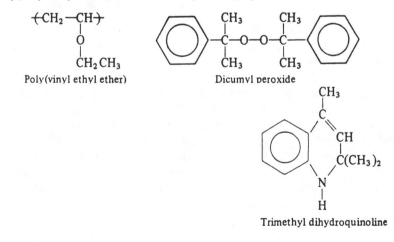

Poly(vinyl ethyl ether)   Dicumyl peroxide

Trimethyl dihydroquinoline

Table 11.2 Poly(vinyl ethyl ether) rubber [12]

Ingredient	Formulation, parts by wt
Polymer (Bakelite EDBM, Union Carbide)	100
Carbon black (Wyex, channel black, Huber)	50
Stearic acid	2
Dicumyl peroxide	4

In a primitive formula the only ingredients besides the polymer are a reinforcing filler, a lubricant, and the cross-linking agent (Table 11.2). The cross-linking reaction takes 30 min at 150°C. Adding as much as two parts of antioxidant, Agerite Resin D, a polymerized trimethyl dihydroquinoline [13], does not degrade the original tensile strength of elongation. And yet the antioxidant is quite effective in preventing degradation for a period of a week in air at 120°C (Figs. 11.6 and 11.7).

Ozone (O_3) also attacks polymers. Unlike molecular oxygen (O_2), ozone appears to add directly to a double bond [14, 15]. Subsequently, chain scission often occurs.

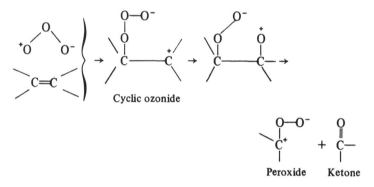

The circumstances generally calling for stabilization against ozone are the combination of a highly unsaturated material, such as diene rubber (polyisoprene, styrene-butadiene rubber), some stress, and, of course, ozone. The automobile tire, especially in some urban locations, combines these features. Although the usual concentration of ozone in the atmosphere is only 0 to 20 parts/100 million, in the Los Angeles area it may run to about 25 parts/100 million (and can reach concentrations of 80 parts in heavy smog). An even higher concentration is encountered on the surface of high-voltage insulators. *Corona discharge* is the flow of electric energy from a high-voltage conductor through ionized air on the surface or in voids or air spaces between conductor and insulation. Although most corona testing is done at voltages of over 10,000 V, the discharge may commence at less than 1000 V. The discharge is accompanied by a faint glow (hence the name "co-

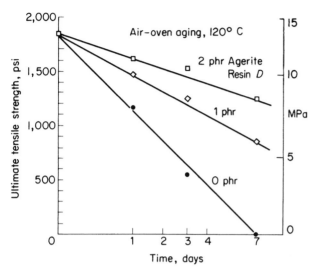

Figure 11.6 Effect of antioxidant (polymerized trimethyldihydroquinoline) on tensile strength of poly-(vinyl ethyl ether) [12]. Abscissa is square root of time

rona") and a crackling sound indicating the pulsating character of the conduction. Oxygen is converted to ozone (ozone generators are built on the principle of corona discharge) so that an insulator may be subjected to very high concentrations.

In insulators the physical changes that are noted are pitting, erosion, and discoloration; and the chemical changes are chain scission and oxidation. In rubber for tires, most tests are run on static specimens at various strains. A sample elongated 20% cracks much more rapidly than an unstrained sample. The sensitivity does not continue to increase with elongation but may go through a maximum. The particular manifestation usually studied is crack growth. Under a given set of conditions, a crack started by a razor blade will grow in linear fashion perpendicular to the direction of mechanical strain.

Some antiozonants are effective just by forming a surface barrier to diffusion of O_3. Waxes can be added that *bloom,* i.e., that exude to the surface, and form a sacrificial layer of hydrocarbon. Dialkyl-*p*-phenylenediamines (the largest chemical class) may react directly with ozone or with the ozone-olefin reaction products in such a way as to interfere with chain scission. Even the reaction products of this latter class may form an additional barrier to O_3 diffusion, since they react at the surface initially. It can be seen that addition of such an amine *increases* the rate of ozone absorption initially but affords protection after a short time (Fig. 11.8).

The proclivity of poly(vinyl chloride) toward dehydrohalogenation is shared to some extent by other chlorinated aliphatic materials including poly(vinylidene chloride), chlorinated waxes, and chlorosulfonated polyethylene. Since worldwide production of poly(vinyl chloride) exceeds 10×10^9 kg/year, the production of

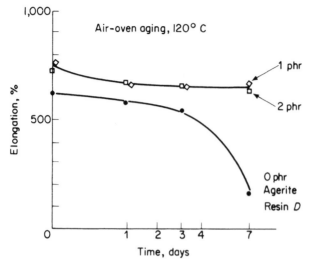

Figure 11.7   Effect of antioxidant on elongation at break for polymer of Fig. 11.6 and Table 11.2 [12]

stabilizers to retard dehydrohalogenation is itself a multimillion-dollar business. The degradation reaction is not completely understood, although it has been studied extensively for many years [16]. On a practical level, however, the species effective in stabilizing poly(vinyl chloride) typically are found to:

1. Absorb or neutralize HCl
2. Displace "active" chlorine atoms such as those on tertiary carbons
3. React with double bonds
4. React with free radicals
5. Neutralize other species that might accelerate degradation

Metal soaps of barium, cadmium, lead, zinc, and calcium obviously can react with HCl. The addition of a zinc soap (or zinc oxide) to poly(vinyl chloride) gives a material that can be heated at 175°C for 10 or 15 min with scarcely any discoloration, as opposed to a control with no zinc that turns yellow or brown as a result of the conjugated unsaturations and oxidized structures. However, once the zinc is converted in large measure to zinc chloride, there is a very rapid and copious evolution of HCl and the remaining material is blue-black and brittle. The zinc soap, which was a stabilizer, is converted to zinc chloride, a rapid and efficient degradation catalyst.

Besides the metal soaps and general antioxidants, some species used to stabilize poly(vinyl chloride) are organo-tin compounds, epoxy compounds, phosphites, and phenols.

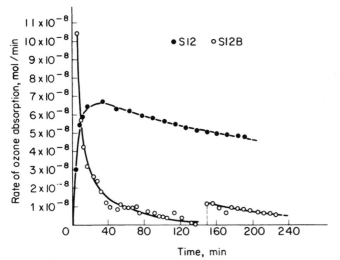

Figure 11.8 Rate of ozone absorption vs. time for a specimen under 20% elongation. Ozone flow rate = 1.05×10^{-7} mol/min. S12 is unprotected elastomer. S12B contains 2 phr of N,N'-di-*sec*-butyl-*p*-phenylenediamine [15]. The discontinuity in the curve for S12B resulted from an interruption in the test that allowed more ozone-sensitive material to diffuse to the surface

11.5 STABILIZERS FOR IRRADIATED SYSTEMS

Most *UV absorbers* protect the substrate by preferentially absorbing the harmful portion of the sun's spectrum between 300 and 400 nm. The earth's atmosphere filters out most of the shorter wavelengths. To be effective, the UV absorber should dissipate the absorbed energy by transferring it to its surroundings as heat or by reemitting it at longer wavelengths through phosphorescence, fluorescence, or infrared radiation. Some other desirable properties for a UV absorber are compatibility with the substrate, lack of visible color, low vapor pressure, low reactivity with other materials in the system, and low toxicity.

Carbon black can be tolerated in a limited number of formulations where color is not a criterion. The pigment not only absorbs light, it is reactive with those free-radical species that might be formed. Wire and cable insulation and pipe for outdoor applications can be protected this way. Films for mulching and water containment in ponds may be made from polyethylene containing carbon black. Five UV absorbers and their corresponding transmission curves are shown in Table 11.3 and Fig. 11.9. The common feature of these UV absorbers is conjugated unsaturation enhanced by stable aryl groups. A variety of groups may be substituted on these basic structures in order to decrease solubility in water and volatility or to increase compatibility.

Phenyl salicylate has the common name "salol." In addition to its use as a stabilizer in polymers, salol is used at concentrations of about 1 to 10% in suntan

Table 11.3 Structures of five different UV absorbers [17]

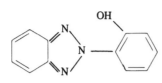

2-(Hydroxyphenyl)benzotriazole

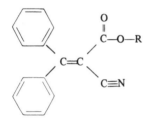

2-Hydroxybenzophenone

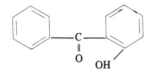

Alkyl-2-cyano-3-phenyl cinnamate
(substituted acrylonitrile)

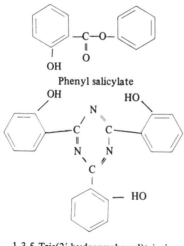

Phenyl salicylate

1,3,5-Tris(2'-hydroxyphenyl)triazine

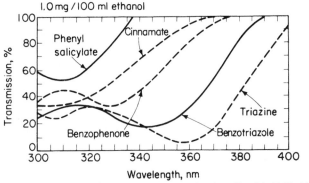

1.0 mg / 100 ml ethanol

Figure 11.9 Transmission of UV light by absorbers listed in Table 11.3 [17]

oils and ointments, as are other absorbers. The function of the stabilizer in such a situation is the same as in the polymer, to decrease the concentration of skin-damaging radiation. Toxicity and reactivity take on a personal and obvious importance for this application. Suntan oils illustrate one way in which stabilizers can be used. If the UV absorber is in a thin layer on the surface, the article is protected to a greater extent than if it is dispersed throughout. Occasionally this can be done, as when multiple coats of lacquer are applied to furniture. All the UV absorber can be put in the top coat. When articles are to be molded, however, it is not economical to coat separately, so the UV absorber is mixed in with the other ingredients.

Despite the addition of stabilizers, many polymer systems reflect less visible light at short wavelengths (400 nm) than at long ones (700 nm). To compensate for this nonuniformity, a blue dye can be added that absorbs at long wavelengths but not at short ones. For years *laundry bluing* has been added to fabrics to give an overall whiteness. In doing this the total reflectance is decreased. *Brighteners,* also called *optical bleaches,* operate by absorbing invisible light below 400 nm and then fluorescing in the visible wavelengths [18]. The effect is seen in Fig. 11.10. A violet dye decreases total reflectance in a poly(vinyl chloride) plaque. The fluorescent whitening agent absorbs light at less than 400 nm and reemits it between 400 and 500 nm. Some typical structures (Table 11.4) incorporate the stilbene group:

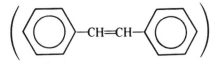

In Fig. 11.11, the absorption and emission spectra for structure 1 of Table 11.4 have been superimposed. Brighteners have been used in soaps and detergents extensively since 1948 and undoubtedly gave rise to the various extravagant advertising claims for whiteness with which the soap-consuming public has been bom-

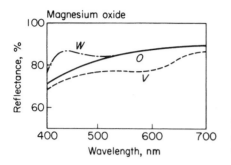

Figure 11.10 Reflectance curves of poly(vinyl chloride) alone (*O*), with violet dye (*V*), and with brightener (*W*) [19]

barded ever since. Textile and paper usage of brighteners followed. Typical concentrations in plastics are of the order of 0.001 and 0.05%. Because the brighteners "feed" on ultraviolet light, their effect is most noticeable by sunlight and hardly perceptible by incandescent light. Fluorescent lights give an intermediate effect. Also, more brightener has to be added if a UV absorber is incorporated in the same system to maintain a given intensity of whiteness.

As work with ionizing radiation (x-rays, gamma rays, electron beams) has increased, some means of protecting systems has been sought to interfere with degradation by radiation. In rubber, aromatic amines and phenolic compounds similar to the antioxidants and antiozonants already described are effective in decreasing the production of cross-links during gamma radiation [20, 21]. In aqueous poly-(acrylic acid) solutions, Sakurada [22] reports that as little as 1 molecule of thiourea per 3 molecules of polymer will increase the dose of gamma radiation needed for gelation to occur by a factor of 5.

11.6 ABLATION

When a space vehicle reenters the earth's atmosphere (or any other planet's atmosphere), it adiabatically compresses the air in front of it. Aerodynamic heating also

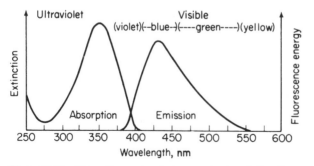

Figure 11.11 Absorption and emission spectra in solution for a compound of structure 1 in Table 11.4 [18]

Table 11.4 Derivatives of (di)aminostilbene(di)sulfonic acid [18]

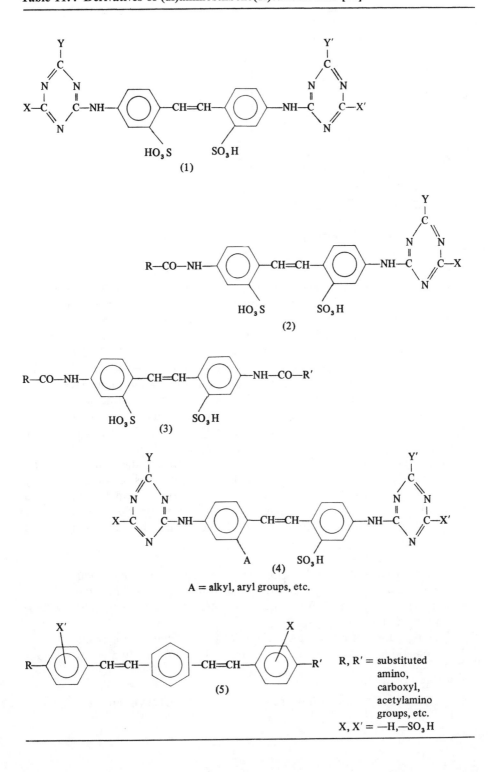

(1)

(2)

(3)

(4)

A = alkyl, aryl groups, etc.

(5)

R, R' = substituted amino, carboxyl, acetylamino groups, etc.

X, X' = —H, —SO₃H

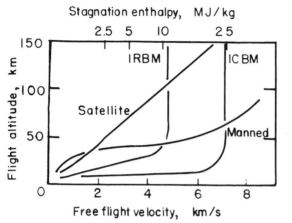

Figure 11.12   Representative reentry trajectories for various flight vehicles [23]

results from the viscous dissipation of energy in the layer of air around the vehicle. Using a reentering satellite as a frame of reference, the situation is the same as if air at a velocity $u = 6$ km/s encountering the nose of the satellite suddenly had its kinetic energy entirely converted to enthalpy (J/g of air). With an appropriate conversion of units, the result is a change in enthalpy of about 18 kJ/g air (8×10^3 Btu/lb). To interpret this in terms of a temperature is rather difficult. It can be appreciated that measures must be taken to protect the contents of a space vehicle from the heat. The actual heat load will vary with vehicle geometry, entry and flight path angles, velocity, and altitude. Some typical trajectories with their corresponding *stagnation enthalpies* are shown in Fig. 11.12. The heat load may be handled by heating up a large mass acting as a *heat sink,* by energy transfer by *radiation,* and by *transpiration,* where a coolant is pumped to the surface and evaporated much as the body is cooled by perspiration. In terms of cooling capacity and reliability a fourth method has been favored. *Ablation* can be defined as a sacrificial loss of material accompanied by transfer of energy. Polymers have been used in *ablative* systems for a combination of reasons. These have been summarized by Schmidt [23] as:

1. High heat absorption and dissipation per unit mass expended, which ranges from several hundred to several thousand Btu/lb of ablative material
2. Excellent thermal insulation, which eliminates or reduces the need for an additional internal cooling system
3. Useful performance in a wide variety of hyperthermal environments
4. Automatic temperature control of surface by self-regulating ablative degradation
5. Light weight
6. Increase in efficiency with the severity of the thermal environment

7. Tailored performance by varying the individual material components and construction
8. Design simplicity and ease of fabrication
9. Low cost and nonstrategic

Limitations listed by the same author for polymers are:

1. The loss of surface material and attendant dimensional changes during ablation must be predicted and incorporated into the design.
2. Service life is greatly time-dependent. Present uses are for transitory periods of several minutes or less at very high temperatures and heating rates.
3. Thermal efficiency and insulative value are generally lowered by combined conditions of low incident flux and long-time heating.

The heat of ablation (the ability of the material to absorb and dissipate energy per unit mass) can be measured in various ways. In general it will be made up of the energy to heat the material up to a reaction temperature, the energy to decompose or depolymerize, the energy to vaporize, and the energy to heat the gases. Any exothermic reactions such as oxidation of hydrocarbons work in the opposite direction and lower the heat of ablation.

In addition to a high heat of ablation, it is desirable that the material possess good strength even after charring. Any sloughing off of chunks of material represents poor usage, because the pieces have absorbed only enough heat to come to temperature and have not decomposed or vaporized. The various stages of heating, charring, and melting are profiled in Fig. 11.13 for two composites. It can be noted that the glass-reinforced system produces a molten glass surface, which can be an advantage mechanically. On the other hand, the nylon-reinforced system may have the higher thermal efficiency.

11.7 CONTROLLED DRUG DELIVERY SYSTEMS

One application that takes advantage of controlled polymer degradation is a novel drug delivery system. In general, drugs that are delivered into the body in discrete doses by injection or ingestion are present at concentrations that vary widely with time. The more time that elapses between doses, the greater is the danger that the concentration in the body will go from being too concentrated (or nearly toxic) initially to being too dilute (ineffective) just before the next dose. There are various ways in which a relatively constant flow of medicine can be introduced into the body. Skin patches have been used in which a low-molecular-weight material diffuses through a membrane at a controlled rate. Scopolamine patches for control of motion sickness can be effective for up to 3 days. Another familiar example is a skin patch for overcoming nicotine addiction.

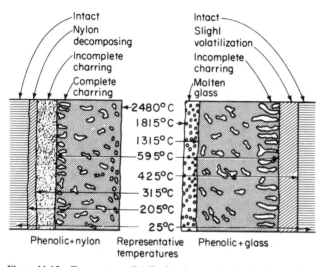

Figure 11.13   Temperature distribution in two plastics during steady-state ablation [23]

With a high-molecular-weight drug, diffusion may be too slow through any practical membrane. On the other hand, if such a drug can be mixed with a surface-degradable polymer, the drug can be made available at the same rate as the polymer erodes. Langer [24–26] has described the design of various systems. One such system is used for the delivery of a powerful chemotherapy agent for brain cancer [26]. The agent, BCNU [1,3-bis-(2-chlorethyl)-1-nitrosourea], has a systemic half-life of only 15 min when it is introduced into the body. Moreover, the agent has various deleterious side effects. However, when the drug is incorporated into a proper polymer matrix, release can be controlled so that a wafer implanted into the site of an excised tumor will release an effective flow of BCNU over a period of days or even weeks.

Proper design calls for a polymer which will degrade by surface erosion much as a bar of soap wears away. Unfortunately, most polymers decompose by penetration of solvent into the interior of a wafer, resulting in "bulk erosion." Also, the rate should be linear and predictable. The polyanhydrides form a class which degrade by hydrolysis and which can be tailored accurately to rigid specifications. For the BCNU delivery system, Langer and his collaborators chose a polyanhydride copolymer (20:80) of CPP:SA.

$$HO-CO\!\!\bighexagon\!\!O-(CH_2)_3-O\!\!\bighexagon\!\!CO-OH \qquad\qquad HO-CO-(CH_2)_8-CO-OH$$

1,3-bis(*p*-Carboxyphenoxy)propane (CPP)   Sebacic acid (SA)

$$-CO-R_1-CO-O-CO-R_2-CO-O-$$

Polyanhydride structure

In order to produce the wafers, a cosolution of BCNU with the copolymer was spray-dried to make uniform spheres which could be compressed into wafers about 1.4 cm in diameter and 1.0 mm thick. They were sterilized by gamma radiation before implantation. The properties of this system include the desired rate of surface erosion, proper release of the BCNU, and no ill effects from the products of hydrolysis (the mechanism for degradation). Of course, the tests for safety and efficacy went through many stages before human patients received treatment. Other diseases which can benefit from implanted degradable systems include Alzheimer's disease, osteomyelitis, and diabetes.

KEYWORDS FOR CHAPTER 11

Section

1. Degradation
   Weathering
   Electromagnetic spectrum
   Oxidation
2. Plasticizer migration
   Delamination
3. Chain scission
   Hydrolysis
   Depolymerization
   Dehydrohalogenation
4. Antioxidant
   Corona discharge
   Antiozonant
   Bloom
5. UV absorber
   Brightener
6. Ablation
   Stagnation enthalpy
7. Skin patch
   Chemotherapy
   Surface erosion
   Bulk erosion
   Polyanhydride

PROBLEMS

11.1 Why does a suspension of poly(vinyl chloride) powder in dioctyl phthalate on heating first become clear and then start to turn yellow?

11.2 In mechanical degradation of a polymer in solution, the rate of chain scission may be faster for higher-molecular-weight chains. If in Eq. (11.6) we have $k = k'x_n^a$, where a is a power greater than zero, derive an expression for x_n as a function of time.

11.3 Cite instances in which each of these "degradation" reactions is desirable: (*a*) hydrolysis, (*b*) cross-linking, and (*c*) depolymerization.

11.4 If a heat shield 5 ft in diameter is to be made from glass-reinforced polystyrene, how much heat can be removed by the depolymerization of a layer 1 in thick of polymer (70 vol% polymer, 30 vol% glass)? $\Delta H_{vap} = 153$ Btu/lb for monomer.

REFERENCES

1. Fink, D. G., and D. Christiansen (eds.): "Electronics Engineers' Handbook," 3d ed, sec. 1, McGraw-Hill, New York, 1989.
2. Brighton, C. A., in S. H. Pinner (ed.): "Weathering and Degradation of Plastics," p. 51, Columbine, London, 1966.
3. Norling, P. M., T. C. P. Lee, and A. V. Tobolsky: *Rubber Chem. Technol.*, 38:1198 (1965).
4. Shelton, J. R.: *Rubber Chem. Technol.*, 30:1251 (1957).
5. Rodriguez, F.: *Chem. Tech.*, 1:409 (1971).
6. Sharpley, J. M., and A. M. Kaplan (eds.): "Proceedings of the Third International Biodegradation Symposium," Applied Science Publs., London, 1976.
7. "Degradability of Polymers and Plastics" (Conference Reprints), The Plastics Institute, London, 1973.
8. Fields, R. D., F. Rodriguez, and R. K. Finn: *J. Appl. Polym. Sci.*, 18:3571 (1974).
9. Tobolsky, A. V.: "Properties and Structure of Polymers," chap. V., Wiley, New York, 1960.
10. Kosar, J.: "Light-Sensitive Systems," p. 361, Wiley, New York, 1965.
11. Stern, H. J.: "Rubber: Natural and Synthetic," 2d ed., p. 49, Palmerton, New York, 1967.
12. Rodriguez, F., and S. R. Lynch: *Ind. Eng. Chem. Prod. Res. and Develop.*, 1:206 (1962).
13. Winspear, G. G. (ed.): "Vanderbilt Rubber Handbook," p. 278, R. T. Vanderbilt, New York, 1968.
14. Bateman, L. (ed.): "The Chemistry and Physics of Rubber-like Substances," chap. 12, Wiley, New York, 1963.
15. Ambelang, J. C., R. H. Kline, O. M. Lorenz, C. R. Parks, and J. R. Shelton: *Rubber Chem. Technol.*, 36:1497 (1963); also E. R. Erickson et al., *Rubber Chem. Technol.*, 32:1062 (1959).
16. Koleske, J. V., and L. H. Wartman: "Poly(Vinyl Chloride)," Gordon and Breach, New York, 1969.
17. Cipriani, L. P., and J. F. Hosler: *Mod. Plastics*, 45(1A, Encycl. Issue):406 (September 1967).
18. Zweidler, R., and H. Hausermann: in "Encyclopedia of Chemical Technology," 2d ed., vol. 3, p. 738, Wiley-Interscience, New York, 1964.
19. Villanne, F. G.: *Mod. Plastics*, 41(1A, Encycl. Issue):428 (September 1963).
20. Bauman, R. G., and J. W. Born: *J. Appl. Polymer. Sci.*, 1:351 (1959); R. G. Bauman, *J. Appl. Polymer Sci.*, 2:328 (1959).
21. Shelberg, W. E., and L. H. Gevantman: *Rubber Age*, 87:263 (May 1960).
22. Sakurada, I., and Y. Ikada: *Bull. Inst. Chem. Res. Kyoto Univ.*, 41:123 (1963).
23. Schmidt, D. L.: *Mod Plastics*, 37:131 (November 1960), 147 (December 1960).
24. Langer, R.: *Science*, 249:1527 (1990).
25. Tamada, J., and R. Langer: *J. Biomater. Sci. Polymer Ed.*, 3:315 (1992).
26. Tamada, J., and R. Langer: *Proc. Natl. Acad. Sci. USA*, 90:552 (1993).

General References

Albertsson, A.-C., and S. J. Huang (eds.): "Degradable Polymers, Recycling, and Plastics Waste Management," Dekker, New York, 1995.

Allara, D. L., and W. L. Hawkins (eds.): "Stabilization and Degradation of Polymers," ACS, Washington, D.C., 1978.

Allen, N. S. (ed.): "Developments in Polymer Photochemistry—1," Elsevier Applied Science, New York, 1980.

Allen, N. S. (ed.): "Developments in Polymer Photochemistry—2," Elsevier Applied Science, New York, 1981.

Allen, N. S. (ed.): "Developments in Polymer Photochemistry—3," Elsevier Applied Science, New York, 1982.

Allen, N. S. (ed.): "Degradation and Stabilisation of Polyolefins," Elsevier Applied Science, New York, 1983.

Allen, N. S., and M. Edge: "Fundamentals of Polymer Degradation and Stabilization," Chapman & Hall, New York, 1993.

Allen, N. S., and J. F. Rabek (eds.): "New Trends in the Photochemistry of Polymers," Elsevier Applied Science, New York, 1985.

Allen, N. S., and W. Schnabel (eds.): "Photochemistry and Photophysics in Polymers," Elsevier Applied Science, New York, 1984.

Aseeva, R. M., and G. E. Zaikov: "Combustion of Polymer Materials," Macmillan, New York, 1986.

Bamford, C. H., and C. F. H. Tipper (eds.): "Degradation of Polymers," Elsevier, New York, 1975.

Billingham, N. C., and D. M. Wiles (eds.): "Polymer Stabilization Mechanisms and Applications," Elsevier, New York, 1991.

Brostow, W., and R. Corneliussen (eds.): "Failure of Plastics," Macmillan, New York, 1986.

Buist, J. M., S. J. Grayson, and W. D. Woolley (eds.): "Fire and Cellular Polymers," Elsevier Applied Science, New York, 1987.

Casale, A., and R. S. Porter: "Polymer Stress Reactions," Academic Press, New York, vol. 1, 1978; vol. 2, 1979.

Conley, R. T.: "Thermal Stability of Polymers," Dekker, New York, vol. 1, 1970; vol. 2, 1974.

Davis, A., and D. Sims: "Weathering of Polymers," Elsevier Applied Science, New York, 1983.

Dickie, R. A., and F. L. Floyd (eds.): "Polymeric Materials for Corrosion Control," ACS, Washington, D.C., 1986.

Eby, R. K. (ed.): "Durability of Macromolecular Materials," ACS, Washington, D.C., 1979.

El-Nokaly, M. A., D. M. Piatt, and B. A. Charpentier (eds.): "Polymeric Delivery Systems: Properties and Applications," ACS, Washington, D.C., 1993.

Emanuel, N. M., and A. L. Buchachenko: "Chemical Physics of Polymer Degradation and Stabilization," VNU Science, Utrecht (Netherlands), 1987.

"Fire Safety Aspects of Polymeric Materials," 10 vols., Technomic, Lancaster, Pa., 1977–1981.

Fouassier, J. P., and J. F. Rabek (eds.): "Radiation Curing in Polymer Science and Technology," 4 vols., Chapman & Hall, New York, 1994.

Gächter, R., and H. Müller (eds.): "Plastics Additives Handbook," 4th ed., Hanser-Gardner, Cincinnati, Ohio, 1993.

Geuskens, G. (ed.): "Degradation and Stabilization of Polymers," Halsted (Wiley), New York, 1975.

Grassie, N. (ed.): "Developments in Polymer Degradation—1," Elsevier Applied Science, New York, 1977.

Grassie, N. (ed.): "Developments in Polymer Degradation—2," Elsevier Applied Science, New York, 1979.

Grassie, N. (ed.): "Developments in Polymer Degradation—3," Elsevier Applied Science, New York, 1981.

Grassie, N. (ed.): "Developments in Polymer Degradation—4," Elsevier Applied Science, New York, 1982.

Grassie, N. (ed.): "Developments in Polymer Degradation—5," Elsevier Applied Science, New York, 1984.

Grassie, N. (ed.): "Developments in Polymer Degradation—6," Elsevier Applied Science, New York, 1985.

Griffin, G.: "Chemistry and Technology of Biodegradable Polymers," Chapman & Hall, New York, 1993.

Guillet, J. (ed.): "Polymers and Ecological Problems," Plenum, New York, 1973.

Guillet, J.: "Polymer Photophysics and Photochemistry," Cambridge Univ. Press, Cambridge, 1986.

Halim Hamid, S., M. B. Amin, and A. G. Maadhah (eds.): "Handbook of Polymer Degradation," Dekker, New York, 1992.

Hilado, C. J.: "Flammability Handbook for Plastics," 3d ed., Technomic, Lancaster, Pa., 1982.

Hoyle, C. E., and J. F. Kinstle (eds.): "Radiation Curing of Polymeric Materials," ACS, Washington, D.C., 1990.

Hoyle, C. E., and J. M. Torkelson (eds.): "Photophysics of Polymers," ACS, Washington, D.C., 1987.

Ivanov, V. S.: "Radiation Chemistry of Polymers," VSP, Zeist, The Netherlands, 1992.

Jellinek, H. H. G.: "Aspects of Degradation and Stabilization of Polymers," Elsevier, New York, 1978.

Jellinek, H. H. G., and H. Kaachi (eds.): "Degradation and Stabilization of Polymers: A Series of Comprehensive Reviews," vol. 2, Elsevier, New York, 1989.

Kaplan, D. L., E. L. Thomas, and C. Ching (eds.): "Biodegradable Polymers and Packaging," Technomic, Lancaster, Pa., 1993.

Kelly, J. M., C. B. McArdle, and M. J. F. Maunder (eds.): "Photochemistry and Polymeric Systems," CRC Press, Boca Raton, Fla., 1993.

Klemchuk, P. P. (ed.): "Polymer Stabilization and Degradation," ACS, Washington, D.C., 1985.

Krongauz, V. V., and A. D. Trifunac (eds.): "Processes in Photoreactive Polymers," Chapman & Hall, New York, 1994.

Kumar, G. S.: "Biodegradable Polymers," Dekker, New York, 1986.

Kuryla, W. C., and A. J. Papa: "Flame Retardancy of Polymeric Materials," Dekker, New York, vol. 1, 1973; vol. 2, 1973; vol. 3, 1975; vol. 4, 1978; vol. 5, 1979.

Labana, S. S. (ed.): "Ultraviolet Light Induced Reactions in Polymers," ACS, Washington, D.C., 1976.

Landrock, A. H.: "Handbook of Plastics Flammability and Combustion Technology," Noyes Data, Park Ridge, N.J., 1983.

McKellar, J. F., and N. S. Allen: "Photochemistry of Man-Made Polymers," International Ideas, Philadelphia, 1979.

Minsker, K. S., S. K. Kolesov, and G. E. Zaikov (eds.): "Degradation and Stabilization of Vinyl Chloride-Based Polymers," Pergamon, New York, 1988.

Mobley, D. P. (ed.): "Plastics from Microbes," Hanser-Gardner, Cincinnati, Ohio, 1994.

Murakami, K., and K. Ono: "Chemorheology of Polymers," Elsevier, New York, 1979.

Owen, E. D. (ed.): "Degradation and Stabilisation of PVC," Elsevier Applied Science, New York, 1984.

Park, K., W. S. W. Shalaby, and H. Park: "Biodegradable Hydrogels for Drug Delivery," Technomic, Lancaster, Pa., 1993.

Phillips, D. (ed.): "Polymer Photophysics," Chapman & Hall (Methuen), New York, 1985.

Pinner, S. H. (ed.): "Weathering and Degradation of Plastics," Gordon & Breach, New York, 1968.

Platzer, N. A. J. (ed.): "Irradiation of Polymers," ACS, Washington, D.C., 1967.

Platzer, N. A. J. (ed.): "Stabilization of Polymers and Stabilizer Processes," ACS, Washington, D.C., 1968.

Popov, A., N. Rapoport, and G. Zaikov: "Oxidation of Stressed Polymers," Gordon & Breach, New York, 1991.

Rabek, J. F.: "Polymer Photodegradation: Mechanisms and Experimental Methods," Chapman & Hall, New York, 1994.

Rånby, B., and J. F. Rabek: "Photodegradation, Photo-oxidation, and Photostabilization of Polymers," Wiley, New York, 1975.

Rånby, B., and J. F. Rabek: "Long-Term Properties of Polymers and Polymeric Materials," Wiley, New York, 1979.

Randell, D. R.: "Radiation Curing of Polymers, II," CRC Press, Boca Raton, Fla., 1992.

Reich, L., and S. Stivala: "Elements of Polymer Degradation," McGraw-Hill, New York, 1971.

Reichmanis, E., and J. H. O'Donnell (eds.): "The Effects of Radiation on High-Technology Polymers," ACS, Washington, D.C., 1989.

Reichmanis, E., C. W. Frank, and J. H. O'Donnell (eds.): "Irradiation of Polymeric Materials: Processes, Mechanisms, and Applications," ACS, Washington, D.C., 1993.

Rodriguez, F.: "The Prospects for Biodegradable Plastics," *Chem. Tech.,* 1:409 (1971).

Rosato, D. V., and R. T. Schwartz (eds.): "Environmental Effects on Polymeric Materials" (vol. 1, Environments; vol. 2, Materials), Wiley-Interscience, New York, 1968.

Schnabel, W.: "Polymer Degradation," Macmillan, New York, 1982.

Schweitzer, P.: "Corrosion Resistance of Elastomers," Dekker, New York, 1990.

Scott, G. (ed.): "Developments in Polymer Stabilization—1," Elsevier Applied Science, New York, 1979.

Scott, G. (ed.): "Developments in Polymer Stabilization—2," Elsevier Applied Science, New York, 1980.

Scott, G. (ed.): "Developments in Polymer Stabilization—3," Elsevier Applied Science, New York, 1980.

Scott, G. (ed.): "Developments in Polymer Stabilization—4," Elsevier Applied Science, New York, 1981.

Scott, G. (ed.): "Developments in Polymer Stabilization—5," Elsevier Applied Science, New York, 1982.

Scott, G. (ed.): "Developments in Polymer Stabilization—6," Elsevier Applied Science, New York, 1983.

Scott, G. (ed.): "Developments in Polymer Stabilization—7," Elsevier Applied Science, New York, 1984.

Scott, G. (ed.): "Mechanisms of Polymer Degradation and Stabilization," Elsevier, New York, 1990.

Shalaby, S. W. (ed.): "Biomedical Polymers: Designed-to-Degrade-Systems," Hanser-Gardner, Cincinnati, Ohio, 1994.

Shlyapintokh, V. Y.: "Photochemical Conversion and Stabilization of Polymers," Macmillan, New York, 1984.

Singh, A., and J. Silverman (eds.): "Radiation Processing of Polymers," Hanser-Gardner, Cincinnati, Ohio, 1992.

Struik, L. C. E.: "Physical Aging in Amorphous Polymers and Other Materials," Elsevier, New York, 1978.

Tötsch, W., and H. Gaensslen: "Polyvinylchloride: Environmental Aspects of a Common Plastic," Elsevier, New York, 1992.

Troitzsch, J. (ed.): "International Plastics Flammability Handbook," 2nd ed., Hanser-Gardner, Cincinnati, Ohio, 1990.

Vert, M., J. Feijen, A. Albertsson, G. Scott, and E. Chiellini: "Biodegradable Polymers and Plastics," CRC Press, Boca Raton, Fla., 1992.

Wypych, J.: "Polyvinyl Chloride Degradation," Elsevier Applied Science, New York, 1985.

Yehaskel, A.: "Fire and Flame Retardant Polymers," Noyes Data, Park Ridge, N.J., 1979.

Zlatkevich, L.: "Radioluminescence and Transitions in Polymers," Springer-Verlag, New York, 1987.

TWELVE

FABRICATION PROCESSES

12.1 THE FABRICATOR

It is only in general terms that we can differentiate the fabricator from other participants who carry out the steps that change a fraction of natural gas into a sweater, a petroleum fraction into a garbage can, or linseed oil into a layer of paint. In every case, the first step is the production of a suitable monomer by a combination of physical separation processes such as distillation and extraction together with chemical reactions such as dehydrogenation and cyanoethylation. Polymerization of acrylonitrile and ethylene for two of the products mentioned is not considered to be *fabrication,* whereas the polymerization of linseed oil is. However, converting the polyacrylonitrile into a fiber, which may then be compounded, crimped, dyed, and twisted, does comprise fabrication. Likewise, taking the polyethylene pellets from the end of the polymerization process and injection-molding the can is a process of fabrication. For our purposes we shall consider the further processes of weaving or knitting cloth and the assembly of apparel as being beyond the scope of the fabricator. It can be seen from these examples that no unambiguous definition comes to hand. The fabricator sometimes starts with monomers or prepolymers, but more often with polymers. He changes the ingredients physically and perhaps chemically. His customer may be the ultimate user, as with the garbage can or paint, or his customer may be someone who changes the physical form again, as with the fiber for the apparel manufacturer. Several fabricators may add to the value of materials they process in sequence. One company may produce a rubbery polymer, a second may compound it, a third may convert it into a sheet, and a fourth finally may form the sheet into a gasket. Liaison between the polymer producer and the ultimate user is a difficult and demanding facet of polymer technology. Because the ultimate user often has limited technical resources and is preoccupied with immediate problems, most polymer and monomer producers have built up substantial *technical service* organizations to generate techniques and materials (together with immense quantities of data) that guide fabricators and provide a basis for selecting the best polymers and fabrication conditions for a given application.

12.2 RAW MATERIAL FORMS

Liquid raw materials may be received in 55-gal drums or 8000-gal (30-m³) tank cars. Smaller units such as carboys and cans may be used. Individual tank trucks and tank cars vary in capacity from one to another. Monomers and prepolymers often come as liquids. The large-scale producer of urethane foams receives polyols and diisocyanates in tank cars and uses them directly from the siding or perhaps transfers them to storage tanks. On a different scale, the familiar two-part epoxy adhesive can be bought as a pair of tubes in most hardware stores. Once again, the two liquids represent a liquid prepolymer and another reactive ingredient, in this case an amine hardener. Both epoxy systems and solutions of unsaturated polyesters in styrene monomer are used as liquids by fabricators of glass-reinforced structures, among which are auto bodies, boat hulls, and filament-wound rocket casings or storage tanks.

Most latexes are the result of emulsion polymerization processes. For the coatings industry no chemical change is necessary, and the compounder adds value to the product by mixing the latex with a stable dispersion of pigments, stabilizers, and other ingredients to make a protective or decorative coating. On the other hand, the latex may be coagulated and cross-linked in the process of making a foam or dipped glove. It is important that the polymerization process leave no objectionable residues in the latex, because only very volatile ingredients can be removed economically. Not all latexes come from emulsion polymerization. Natural rubber latex is recovered from the tree. Polymers that are best produced by bulk or solution techniques can be converted into latexes by *postemulsification*. In a typical process a polymer solution is emulsified in water with a surfactant, then the solvent is evaporated. Latexes of butyl rubber and polyethylene are available, even though neither is usually polymerized in emulsion.

It is rare that a polymer is supplied in the same solvent in which it was polymerized. For example, alkyd resins may be polymerized in the presence of toluene or xylene, but they are supplied to the paint compounder diluted with mineral spirits. A liquid dispersion of poly(vinyl chloride) in a plasticizer (Sec. 3.6) may be the raw material for a slush molder (Sec. 12.6).

Massive solid forms are common in the rubber industry. Bales of natural rubber may weigh 100 kg. Synthetic rubber is often sold in bales (blocks) that are wrapped in polyethylene or dusted with talc and weigh 20 to 50 kg. Polychloroprene is available in batons or smaller pieces that weigh less than a pound and are packaged in bags. For ease in dissolving rubber in solvents for coatings and adhesives applications, both natural and synthetic rubbers have been put in crumb form. The porous particles have a much greater surface area for solvent attack. Most thermoplastics for molding and extrusion applications are sold as pellets (often called "powder"), which are cylinders about 3 mm in diameter and length. In this form the polymer flows freely in automatic dispensing devices and in bins and hoppers. Thermoplastics are used as fine powders in rotational and fluid-bed coating systems (Secs. 12.4 and 12.6). Because the fabricator may wish to compact a specific unit weight of a resin for compression-molding himself, most thermoset resins are supplied as powders that can be compressed at room temperature into cakes or "preforms." Poly(vinyl chloride) from suspension or emulsion polymerization is supplied as a powder. Suspension polymer particles range from 50 to a few hundred micrometers in diameter and are quite porous. They are dry-blended with plasticizers, yielding a dry, flowing, plasticized power. The emulsion-polymerized resin is smaller in particle diameter and denser so as to be suitable for dispersion in plasticizers that are mobile liquids.

For many fabricators the raw form of the polymers they use is the result of a previous fabrication step. This is true of the sheet formers, since the molder of blister packages or refrigerator door liners seldom forms his own sheet but buys it from a separate converter. Builders of ductwork, storage tanks, and luggage commonly buy their raw materials as sheets, tubes, or other extruded shapes.

12.3 MIXING

Mixing low-viscosity liquids or blending dry powders can be accomplished with conventional equipment. Turbine and propeller agitators are used to mix low-viscosity liquids, and twin-cone blenders or tumblers are used for dry powders. These devices have low power requirements and rather low residence times. At the other extreme, the dispersion of a finely divided solid such as carbon black in a rubber with a viscosity of about a million poise presents a situation requiring more power.

The degree of mixing, the uniformity of dispersion of one phase in another, can be measured in several ways. One could take a series of samples of uniform size from a large batch and determine in each the concentrations of various ingredients. A statistical analysis coupled with a mathematical model can yield a number that can be fitted into a scale of uniformity or degree of mixedness. It is more common to judge the uniformity by some performance test. Mechanical failure tests such as tensile, tear, and impact strength are used. Optical tests are conventional for judging pigment dispersion. When an entire cable is tested for electrical

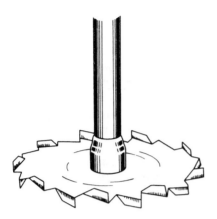

Figure 12.1   The impeller of the Cowles dissolver [1]

strength through the insulation, a very sensitive measure of uniformity is obtained, since it takes only one pinpoint of poorly dispersed filler to bring about failure at a low impressed voltage.

For mixing liquids with viscosities less than about 10 poise (1 Pa·s), paddles, turbines, and propellers are suitable. Power requirements vary from 0.5 to 15 hp/ 1000 gal (0.1 to 3 kW/m³). If a solid polymer is to be dissolved in a solvent, a mixing impeller with a saw-toothed edge is available that can cut the swollen solid ribbons in addition to circulating the contents of the tank (Fig. 12.1). On the other hand, the high speeds characteristic of some agitators can decrease the molecular weight of the polymer by a mechanical shearing action.

Mixing viscoelastic materials involves techniques different from those used with low-viscosity fluids. In the latter case, the normal procedure is to induce turbulence to a degree such that molecular diffusion can proceed rapidly to establish a uniform concentration of ingredients. With polymers it is necessary to use high local shear rates and rather gross mechanical "folding" to get the same result. A simple analogy to the two processes is the comparison between mixing cream in coffee and dispersing a spoonful of caraway seeds in a batch of bread dough. The power and time requirements are much more stringent for the dough, of course.

Pastes and liquids with viscosities higher than 100 poise (10 Pa·s) are handled in machinery resembling that familiar in the baking industry. Kneaders and dough mixers are used to mix pigments in printing inks, to disperse poly(vinyl chloride) in plastisols, and to blend together the more viscous prepolymers and other ingredients for urethane foams. The agitator speeds are lower and the power requirements higher (10 to 1000 hp/1000 gal, or 2.5 to 250 kW/m³) than those for turbines and propellers.

Amorphous polymers that are 50 to 200°C above T_g and have molecular weights in the range of 10^4 to 10^6 are most often mixed with solids or other polymers in intensive mixers or on mixing rolls. The intensive mixer is distinguishable from the dough mixers and kneaders by two features. One is the sturdier construction, which allows a higher power input per unit volume. The second is pressurized mixing, usually with temperature control. In the Banbury mixer two interrupted spirals (Fig. 12.2*a*) rotate in opposite directions at 30 to 40 rpm. Polymers, fillers, and other ingredients are charged through the feed hopper (Fig. 12.2*b*) and then held in the mixing chamber under the pressure of the hydraulic ram. Both the rotors and the walls of the mixing chamber can be cooled or heated by circulating fluid. In the 270-liter model shown in Fig. 12.3, a 1200-kW motor (not shown) might be used for mixing 250 kg of a tire compound. This amounts to almost 4.8 W/g or about 1.1 cal/g·s. Even in a mixing cycle of a few minutes with cooling water, the temperature increase can be 30 to 60°C.

The mainstays of the rubber industry for over 50 years have been the two-roll mill and the Banbury mixer. Roll mills were first used for rubber mixing over 100 years ago! The plastics and adhesives industries adopted these tools later on.

The two-roll mill shears the material in the nip between the rolls (Fig. 12.4), one of which is rotating faster than the other. Mixing is a function of:

1. Roll speeds
2. Amount of turnover (folding) between passes
3. Number of passes
4. Gap between rolls

The value of (1) may be fixed in any machine; (2) and (3) are functions of the amount of material in a batch and time of milling. They are also functions of the viscoelastic properties of the material being mixed, since multiple passes often depend on the material adhering to the roll and being carried around back to the nip. The value of (4) is usually changed fairly easily on most machines and is often decreased during the course of a single mixing operation.

In the Banbury mixer the material is subjected to less shearing than in roll mills, since the gap between scroll and wall is not adjustable. Also, heat transfer is more difficult, so that a temperature rise invariably is produced. However, incorporation of solids is more rapid because the mixing is done under the pressure of a hydraulic (or manual) ram. Also, the mixing depends much less on the viscoelastic properties of the material than does mixing on a roll mill. White has reviewed the history of internal mixers in fascinating detail [3].

Powders can be blended together by a simple tumbling action. In the ribbon blender, double-cone mixer, or twin-shell blender (Figs. 12.5 and 12.6) a simple rotational motion causes particles to be raised together and then to fall, splitting into small domains that are rejoined at random in the next cycle. Rolling and folding of the charge occur also. The same apparatus is used to blend porous poly(vinyl chloride) with plasticizers. Liquid plasticizer is poured or sprayed onto the powdered resin. During the tumbling, capillary action draws the liquid into the pores

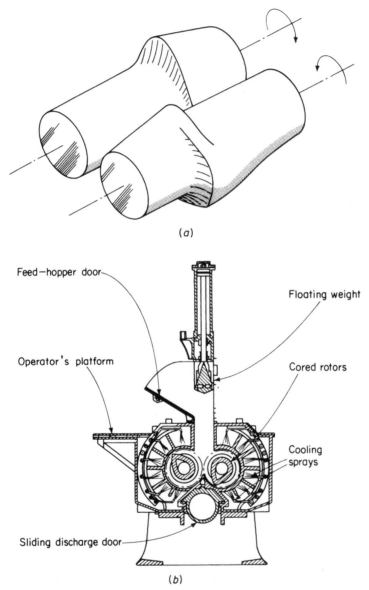

(a)

Feed—hopper door

Floating weight

Operator's platform

Cored rotors

Cooling
sprays

Sliding discharge door

(b)

Figure 12.2 (*a*) Roll mixing blades in the Banbury mixer. (*b*) Cross section of Banbury mixer (registered trademark of Farrel Co.) [2]

Figure 12.3 This Banbury internal mixer has a gross capacity of 270 liters for a typical tire batch of 225 kg and is driven by a motor (not shown) with up to 2000 hp at 60 rpm. (Photograph of model F270 courtesy of Farrel Corp., Ansonia, Conn.)

Figure 12.4 Both rolls of this mill are 26 in in diameter, 84 in long, and are independently driven. The spacing between the rolls also is variable. Two 125-hp motors (not shown) would be attached to the speed reducers at the far end. (Photograph courtesy of Farrel Corp., Ansonia, Conn.)

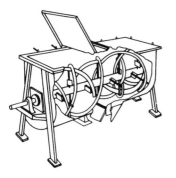

Figure 12.5   Ribbon blender [2]

of the resin, so that in a few minutes the dry, powdery character is restored and the resin-plasticizer combination can be charged to some other equipment (extruder or calendar) in which it will be fused into a continuous product.

Roll mills have been mentioned as functioning to mix polymers with fillers. The same equipment can be used not only to disperse a solid but to attrite it as well. Roll mills with up to five rolls, each rotating faster than the preceding one, are used to disperse pigments in paint master batches. A slowly rotating cylinder half-filled with ceramic balls is a familiar sight in the paint industry. The falling action of the balls (Fig. 12.7) decreases the size of solids and disperses them in a liquid at the same time. As opposed to the roll mills, which disperse the pigment in a few passes with an elapsed machine time of a minute or so, the ball mills usually operate on a cycle of hours or days.

It is obvious that less energy (and expense) is required to disperse solids in low-viscosity media than in rubbery polymer melts. In applications where a standard filler and plasticizer are being used in large amounts, the producer of a rubber may find it expedient to add the carbon black and oil to the latex before coagulating and baling it. About half the styrene-butadiene rubber supplied in recent years

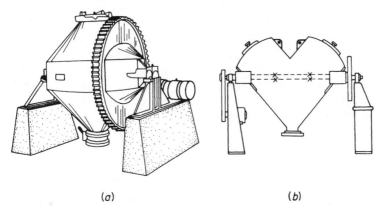

(a)   (b)

Figure 12.6   Tumbler mixers: (a) double-cone mixer; (b) twin-shell blender [2]

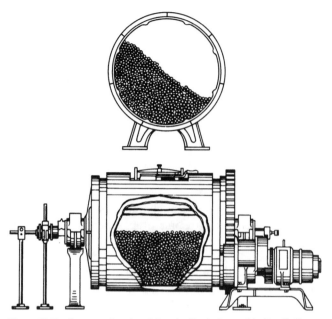

Figure 12.7 Cross-sectional and longitudinal views of ball mill (Patterson-Ludlow Division, Banner, Ind.) [2]

has had oil added to it before coagulation (about a gram of oil for each 3 g of rubber in a typical formulation). Half of the oil-extended rubber has had carbon black added also (see Sec. 14.3). Master batching undoubtedly will continue to grow in popularity whenever a market justifies it to the polymer producer.

12.4 ONE-DIMENSIONAL PROCESSES

In the application of coatings and adhesives the only important dimension usually controlled is the thickness. There are many other parameters to be considered, of course. Adhesion is of overwhelming concern. In coatings, adhesion to one surface must be controlled. Adhesives often represent a more complicated situation because two unlike surfaces may be involved. On the other hand, the coating must face the challenge of its environment with its hazards of chemical, bacteriological, and mechanically abrasive attack over a high surface area. The adhesive seldom has to contend with these to the same extent.

Coatings

A coating is a thin layer of material intended to protect or decorate a substrate. Most often it is expected to remain bonded to the surface permanently, although there are strippable coatings that afford a temporary protection. We can subdivide

Table 12.1 Peelable hot-melt coating [4]

Material	Parts by weight
Ethyl cellulose (50 cps grade)	25
Castor oil	10
Mineral oil	61
Paraffin wax	3
Stabilizer (antioxidant)	1

coatings by considering whether or not they contain a volatile solvent or diluent and whether or not a chemical reaction is involved in film formation.

Nondiluent, nonreactive. The oldest recorded use of polymers was for a coating of this type. The Lord told Noah, "Make yourself an ark of gopher wood; make rooms in the ark, and cover it inside and out with pitch" (Genesis 6:14). The pitch may have been a naturally occurring bitumen or wood rosin applied as a hot melt. A more recent application in which moisture impermeability is sought is the wax- or polyethylene-coated milk container now so familiar. The hot melt can be applied from a roll on a continuous basis. More commonly, the polyethylene might be extruded through a slot (Sec. 12.5, Extrusion) as a film layer and laminated with the cardboard. The formulation of a strippable hot-melt coating with ethyl cellulose as the binder is given in Table 12.1. The ingredients are melted together at about 190°C. Metal pieces such as drill bits or other tools and gears can be dipped in the molten mass and cooled in air with a thick (2-mm) coating. Castor oil is used as a plasticizer. Mineral oil and wax are less compatible than the castor oil, but they serve as lubricants as well as plasticizers. Exuding to the metal surface, they can function to prevent corrosion as well as mechanical abrasion during shipping and handling. The coating is easily stripped when the tool or other article is to be put into use.

Another application of nonreactive melt coating is *fluidized-bed* coating. In a typical setup (Fig. 12.8), a bed of powdered polyethylene in the range of 200 mesh (75 μm in diameter) is fluidized by the passage of air through a porous plate. A metal object to be coated is heated to well above the fusion point of the resin. When it is dipped into the bed and moved laterally as well, a layer of powder adheres to its surface and melts to form a continuous coating. The process is self-limiting because the metal cools as the polymer melts, and the layer of polymer is a poor conductor of heat. After being removed from the bed, the object may be further heated to ensure the integrity of the film. There is nothing in the technique that limits it to thermoplastic coatings. The bed is not heated; only the surface of the object to be coated is hot. A mixture of powders that react when heated can be fluidized and a thermoset coating produced in the same equipment.

The powder can also be applied to a hot surface as a charged spray. Powder is

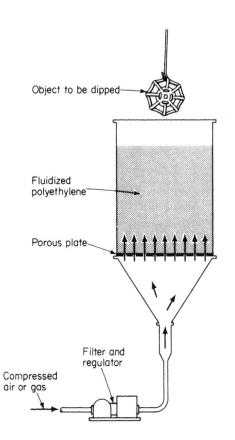

Object to be dipped

Fluidized polyethylene

Porous plate

Filter and regulator

Compressed air or gas

Figure 12.8 Fluidized-bed coating with powdered polyethylene [5]

carried by an air stream from a fluidized reservoir and through an orifice, where it is charged by a high-voltage power supply (Fig. 12.9). The parts to be coated are grounded, so the charged powder is applied very efficiently to the desired locations. The part may be hot when it is coated, or it can be heated subsequent to coating. Some applications of the electrostatic spray process are automobiles, appliances, metal furniture, luggage hardware, and even glass bottles [7].

Nondiluent, reactive. The fluidized-bed method can be used when the polymer is a reactive powder. A mixture of an epoxy resin and an acid "hardener" can be fluidized and deposited simultaneously. Reaction occurs only on the hot surface.

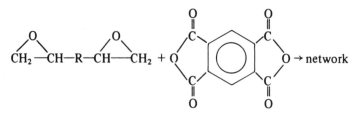

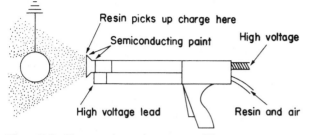

Figure 12.9   Electrostatic powder spray gun [6]

A general technique has been developed whereby a reactive monomer is sprayed on a surface and forms a polymer immediately. In one application the monomer is di-*p*-xylylene formed by the pyrolysis of *p*-xylene at 950°C in the presence of steam [8]:

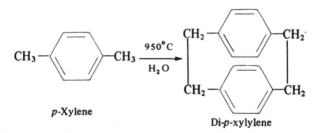

When the purified monomer is heated to about 550°C at a reduced pressure, a diradical results in the vapor phase, which when deposited on a surface below 50°C polymerizes instantaneously to a molecular weight of about 500,000:

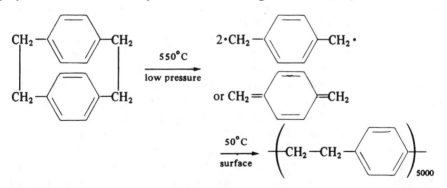

A thin, adherent coating, Parylene N (Union Carbide), can be applied to metals or to other substrates. A major application has been production of miniature capacitors having the polymer as a dielectric. The monochloro and dichloro derivatives (Parylene C and Parylene D) have slightly higher dielectric constants and dissipation factors but much lower gas permeabilities. The melting temperatures are 405, 280, and 340°C for the N, C, and D modifications, respectively.

The process of producing polymers on a surface through the use of ionized, gaseous plasmas has been used with hydrocarbons, fluorocarbons, and other systems [9]. In glow-discharge polymerization, the surface to be coated can be made the anode for an electrical discharge over a potential of several thousand volts. Monomers adsorbed on the surface can form highly adherent, and often cross-linked, films. Another technique that makes use of plasma technology uses polymer powders rather than monomers. The plasma in this case consists of an inert gas such as argon passed through an electric arc so that its temperature is raised to 5000 to 8000°C. Poly(tetrafluoroethylene) powder is blown at high velocity through the plasma jet and onto the surface to be coated. As the polymer impinges on the surface, it is sintered to a continuous coating.

In a special class of nondiluent, reactive systems are those that react because of high-energy radiation. Coatings that are useful are usually based on unsaturated molecules. They may contain reactive, nonvolatile monomers as well as a photo-initiator. It has been pointed out that Egyptian mummy wrappings were made from linen saturated with oil of lavender and bitumen of Judea and then exposed to sunlight [10]. The nontacky composite that results is quite durable, as visitors to the British Museum can attest. As a modern-day example, a photocurable printing ink might include an acrylated urethane polymer (containing pendant unsaturation), several monomers such as neopentyl glycol diacrylate and *N*-vinyl pyrroli-done, a sensitizer such as a benzoin ether or benzoquinone, and various pigments. The ink is transferred to a surface by conventional printing press processes. When such an ink is applied to a metal surface under an inert atmosphere, conversion to a dry, smudge-proof film can take place in seconds under a strong UV lamp rather than by evaporation of a solvent or by oxidation of a drying oil, which would take much longer. The monomers act as diluents in the system, but they react to become part of the film and are not evaporated.

Photoresists have been used for many years in lithography for the production of printing plates. Strictly speaking, these generally are applied from a solution. However, since the reaction takes place well after the film has been dried, they are included here in the nondiluent category. Chromated gelatin is one classical example that still finds some use in making masks for silkscreens. Somewhat newer are films made from poly(vinyl alcohol) containing bichromate. For a printing process the film might be deposited on a metal surface from an aqueous solution. The dry film on exposure to UV light through a mask is insolubilized by oxidative cross-linking. Patient washing with warm water removes the unexposed portions of the film. Next the film is baked to increase adherence of the cross-linked portions. Then an etchant is applied to attack the metal and to leave a raised pattern where the polymer was irradiated. The polymer "resists" the etching of the metal. In some applications etching is not needed, as when ink receptivity is different for bare metal compared to a polymer surface. Another application of the poly(vinyl

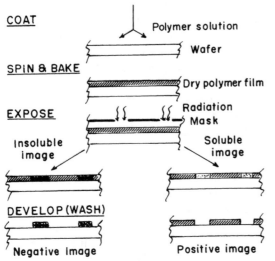

COAT

Polymer solution

Wafer

SPIN & BAKE

Dry polymer film

EXPOSE

Radiation

Mask

Insoluble
image

Soluble
image

DEVELOP (WASH)

Negative image

Positive image

Figure 12.10 The silicon single crystal wafer (with oxidized surface) is coated with a polymer film by spinning and baking and then selectively exposed to UV light or x-rays, or to an electron beam. The latent image is developed into a positive or negative pattern on the wafer by dissolution of the appropriate areas of the polymer "resist" film

alcohol)-bichromate system has been in the production of color television tubes. The screen of the tube requires a pattern of three types of phosphor dots. Each of the three phosphors can be applied in separate stages by exposure through a carefully registered set of masks. In each of three stages a single phosphor is suspended in the polymer film and the dots are insolubilized. The un-cross-linked polymer with its phosphor content is washed away. Etching is not required. The screen is baked between stages to oxidize the polymer film and to leave the phosphor firmly attached to the glass. The glass screen is fused onto the front of the TV tube with the phosphors on the inside surface. Because of environmental concerns, chromate-based systems are no longer recommended.

The production of semiconductor devices, especially integrated circuits, represents the most demanding set of conditions for polymeric resists. This operation still is a form of lithography (*litho* = stone, *graph* = image) although it is more often termed *microlithography*. The "image" being created is the circuit, and the silicon (or its surface layer of silicon oxide) is the "stone." Many silicon chips used in calculators and computers are produced by some variation of the following sequence of operations (Fig. 12.10):

1. A silicon wafer about 152 mm in diameter has one surface oxidized to a controlled depth.
2. Polymer is applied by placing a few drops of solution at the center of the oxide side of the wafer and then spinning the wafer to obtain a thin, uniform coating

that when dry is about 1 μm thick. The polymers used are those that will change in solubility upon irradiation. Polymers such as poly(vinyl cinnamate) (see Sec. 14.5) and cyclized rubber (see Sec. 11.3) cross-link on exposure to UV light and are termed negative resists (Fig. 12.10). Others, such as the novolak-diazoketone mixtures (see below), change from oil-soluble to base-soluble as a result of irradiation and are termed positive resists.

3. In either case, exposure to UV light via a mask (in contact or with a projected image) changes the solubility of exposed polymer film.

4. In the case of negative resists, the un-cross-linked polymer is washed off, leaving the cross-linked image behind. For positive resists, it is the exposed film that is washed off.

5. The bare substrate parts are etched through the oxide layer down to the silicon layer by a fluoride solution in water or by a plasma containing reactive ions.

6. The chemical compositions of the etched regions are altered. Ion implantation can be used to introduce "dopants," that is, impurities that make semiconductors of the diffused-base transistor type. Other operations can include depositing a layer of aluminum to act as a conductor or other materials to act as insulators.

7. All polymer is removed by solvents, plasmas, or baking.

8. The wafer is recoated and a new pattern imposed and processed. This entire sequence may be repeated a number of times to give integrated circuits with many layers and amazing complexity. The minimum line width on chips produced by photolithography has decreased steadily from year to year. As a rule of thumb, the minimum feature size on a DRAM (dynamic random access memory) silicon chip has changed by a factor of 2 about every six years. From 1985 to 1995, many commercial DRAM chips went from minimum line widths of 1 μm down to 0.35 μm. The mercury lamps used as light sources have also come down in wavelength from G line, 436 nm, to H line, 405 nm, to I line, 365 nm. Further reductions in wavelength to deep UV (248 and 193 nm) probably will require excimer laser sources. For example, a KrF excimer laser operating at 248 nm has been used to achieve resolution better than 0.25 nm and should be capable of 0.18 nm [11]. A single wafer is processed as a unit but then is cut up to make the individual chips. Wafers in use in 1995 typically were 100 to 200 mm in diameter, but 300-mm wafers were being tested. The chips themselves range from several centimeters on a side (for a complex chip such as Intel's Pentium) to a few millimeters on a side (for a wristwatch chip). Attaching electrical leads to the tiny conducting spots on the chip is a separate, demanding task.

In the past, the most popular photoresist type has been a physical combination of a phenol-formaldehyde novolak resin with a diazoquinone. The novolak (molecular weight of about 500 to 2000) does not dissolve in dilute alkali when 20 to 50% of a diazoketone is present in the dry film. However, UV light converts the alkali-

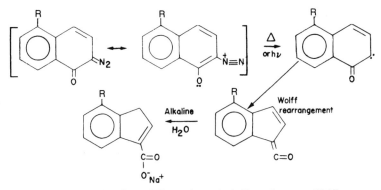

Figure 12.11 Photochemistry of diazoquinone including subsequent Wolff rearrangement to carboxyl

insoluble diazoquinone into the hydrophilic carboxylate form via the Wolff rearrangement (Fig. 12.11). Without the diazoquinone "dissolution inhibitor" in the exposed areas, the novolak now dissolves in dilute alkali.

Further demands for increased speed and finer resolution have given rise to "chemically amplified" resists that are usable in the deep UV range and also are very sensitive. A very small amount of acid produced by a light-sensitive compound becomes the catalyst for a subsequent chemical reaction that amplifies the change in resist solubility [12]. For example, a diaryl iodonium salt can act as the photoacid generator in a matrix of an acid-sensitive polymer such as the "protected" form of poly(p-hydroxystyrene). Following exposure, heating the mixture causes loss of the *tert*-butoxycarboxyl protecting group to occur in the exposed regions (Fig. 12.12). Now the film is soluble in alkali where it has been exposed, but insoluble where it is still protected. In this way, the small amount of energy used in exposure is amplified by the subsequent chemical reaction.

Electron beam lithography is used to produce masks for UV photolithography and, in some places, to produce the chips themselves. A scanning electron beam similar to a scanning electron microscope can be programmed to produce patterns on a very thin polymer layer (less than 1 μm thick). The electron beam may cause cross-linking or may solubilize the polymer. Both poly(methyl methacrylate) and poly(butene-1-sulfone) respond to electrons by chain scission. The lower-molecular-weight polymer produced is more soluble than the original. Another type of radiation that has been used for lithography is high-energy x-rays.

Diluent, nonreactive. The term *lacquer* usually connotes a solution of a polymer. Once used for solutions of naturally occurring polymers such as shellac in alcohol, the term now is applied to all solutions in which film formation results from mere evaporation of the solvent and does not require that a chemical reaction take place. Another name sometimes used as a synonym for lacquer is *dope*. Airplane dope, for example, is a lacquer applied to fabric that has been stretched over airframes

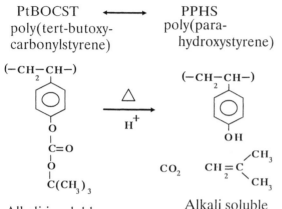

PtBOCST ⟶ PPHS
poly(tert-butoxy-carbonylstyrene)   poly(para-hydroxystyrene)

Alkali insoluble

Alkali soluble

Figure 12.12 In a chemically amplified resist, the acid generated by exposure to light is used to catalyze the thermal deprotection of the alkali-insoluble polymer, rendering it soluble in alkali

and other surfaces to increase tautness and decrease permeability. Film formation from a lacquer requires only the evaporation of solvents. Mixtures of solvents are usually needed in order to achieve a smooth, coherent film. Compatibility, often guided by the solubility parameter concept (Sec. 2.4), obviously is important. Volatility is adjustable by using solvents of different molecular weights or by using mixtures of solvents and *diluents* (extenders that may not be good solvents when used alone). As the coating dries, dimensions change, and the change is more rapid at the surface. A proper balance of volatility and compatibility assures that the film will not "skin over" on its surface and create a barrier to further diffusion of solvent from the film. If a good solvent is too volatile, a poor one left behind in the film may reach a concentration where the polymer precipitates, giving a dull or wrinkled film. When the humidity is high, the surface may cool from evaporation of solvent and condense water. The water in turn may cause precipitation of the polymer, giving the film an undesirable opacity (*blushing*). A typical lacquer formulation (Table 12.2) contains polymer, pigment, plasticizer (nonvolatile solvent), and volatile solvents. The vinyl resins used here are copolymers of vinyl chloride with about 10 to 20% vinyl acetate. The VMCH also contains a few percent of maleic anhydride as a termonomer. This lacquer is designed for steel surfaces, and the acid functionality results in good adhesion to the iron surface.

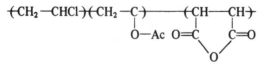

(random distribution)

Vinyl chloride, vinyl acetate, maleic anhydride terpolymer (VMCH)

Table 12.2 White vinyl topcoat [13]

Material	Pounds
Titanium dioxide	100.0
Vinyl resin (VAGH)	66.7
Vinyl resin (VMCH)	66.7
Tricresyl phosphate	25.0
Methyl isobutyl ketone	288.0
Toluene	288.0
Total	834.4
Weight per gallon 8.34 lb	

It should be noted that we have classified this as a lacquer because reaction is not essential to film formation even though a reaction between the anhydride and the metal surface is possible. Titanium dioxide is widely used as a white pigment because of its excellent *hiding power* (ability to mask the substrate per unit weight of pigment). The tricresyl phosphate is used as the plasticizer in sufficient volume to lower the glass transition temperature to about 40°C. Thermoplastic (un-cross-linked) binders for coatings usually are formulated to have T_g somewhat above room temperature. Too high a T_g gives a brittle film that may flake off when the temperature is cycled because the coefficient of expansion for metals is so much smaller than that for polymers (Sec. 10.5). Too low a T_g gives a soft film that will not resist abrasion. Vinyl chloride polymers, cellulose derivatives (acetate and nitrate), and acrylic copolymers can be formulated to remain tough over a wider range of temperatures than most homopolymers, so they have been favored for lacquers.

Some homely examples of lacquers include the spray cans of "touch-up paint" sold to the auto owner. Most of these are pigmented acrylic resins in solvents together with a very volatile solvent that acts as a propellant (usually a low-molecular-weight hydrocarbon). Model airplane dope may be a solution of cellulose acetate-butyrate in a mixture of ketones and aromatic solvents.

The latex paints are another class of coatings that form films by loss of a liquid and deposition of a polymer layer. During World War II the latexes resulting from emulsion polymerization of styrene and butadiene for rubber were found to be well suited to film formation. The fast drying, low odor, and, above all, the ease of cleaning up spots and applicators when water is the diluent contributed to a do-it-yourself movement of tremendous proportions. A parallel rise in the wage level of painters and other craftsmen accelerated the process. The formulation of a modern latex paint is not a matter of merely dispersing a pigment in a latex, however. Some ingredients listed in Table 12.3 for an exterior, light-toned paint based on an acrylic-vinyl copolymer (Ucar Latex 180) include pigments (TiO_2, mica, and talc), surfactants for dispersing the pigment in addition to the surfactant remaining in the latex from polymerization, a mercuric salt as a mildew preventative, a defoamer

Table 12.3 White latex paint for exterior application [14]

Material		Parts by weight
Vinyl-acrylic polymer latex (Ucar Latex 180) (55% solids)		400
Pigments: Rutile TiO$_2$	175 ⎫	
Anatase TiO$_2$	50 ⎪	375
Mica, water-ground	25 ⎬	
Talc	125 ⎭	
Wetting agents, dispersants, surfactants		14
Phenyl mercuric acetate (30% solids)		5
Defoamer		5
Carbitol (diethylene glycol monoethyl ether)		15
2-Ethylhexyl acetate		5
Ethylene glycol		20
Water (added to latex during formulation)		293
Total		1132
PVC = 36%; lb/gal = 11.3		

(needed during mixing and during application to decrease trapped air bubbles), and plasticizers. The ethylene glycol also functions as an antifreeze, since freezing and thawing can break a suspension and bring about coagulation.

In 1990, paint manufacturers in the United States voluntarily agreed to eliminate mercurial biocides from interior latex paints. Organotin polymers and cuprous oxide compositions were available as substitutes. The term PVC in Table 12.3 refers to the *pigment volume concentration,* which is the volume percent of pigment in the nonvolatile portion of the formulation (film ingredients). The pounds per gallon is important to the formulator, since ingredients are bought by the pound, but paint is sold by the gallon.

As in the case of lacquers, the T_g of a latex paint polymer usually is adjusted by copolymerization or plasticization to the range of 30 to 50°C. Film formation proceeds differently, however. One can picture the latex particles as so many marbles being deposited on a surface. As the water evaporates or, on wood and plaster surfaces, diffuses into the substrate, the spherical particles approach each other more and more closely. The minimum temperature at which the particles will coalesce to form a continuous layer depends mainly on the T_g. The water may have some plasticizing action, so the actual T_g may be slightly lower than one that might be measured in dry test. The driving force tending to push the particles together is related to the capillary pressure of water in the system. Smaller particles and higher surface tensions should result in lower film-forming temperatures. In practice it is found that the ratio of monomers or addition of plasticizers is a better control for film formation than particle size or surface tension [15].

While most latex house paints fall in the nonreactive class, many applications

Table 12.4 Brown house paint [13]

Material		Pounds
Iron oxide brown		64.0
Acicular ZnO		254.0
Calcium carbonate		600.0
Soya lecithin		4.6
Raw linseed oil		280.0
Kettle-bodied linseed oil (X viscosity)		93.0
24% lead naphthenate		9.3
6% manganese naphthenate		1.1
Mineral spirits		108.9
Total		1414.9
Pigment volume	35.8%	
Consistency	83 KU	
Weight per gallon	14.1 lb	
Add puffing agent to increase viscosity		

use a subsequent cross-linking reaction to enhance durability. In a simple example, pendant hydroxyl and carboxyl groups on a polymer chain may not react with each other while the polymer is suspended as a latex. But when the film is dried and heated, esterification can take place. It may be possible to wash off a fresh film with soap and water, but not one that has aged and become cross-linked. The advantages in an exterior house paint are obvious because cleaning up is easy but permanence (water resistance) increases with time.

Another broad application of latex films involving very low pigment volumes is the whole field of water-based polishes for industrial and home use. A typical material for vinyl tile floors consists of a polystyrene or acrylic copolymer latex of fine particle size together with plasticizers, waxes, and hard resins totaling almost the same weight as the polymer. The total solids content in the final product might be only 10 to 20%.

Latexes have also been used as paper coatings, as wet-strength promoters in paper, asphalt dispersions, and in concrete. In each case it is the binding action of the continuous film formed when the water is gone that is sought.

Diluent, reactive. Although the term "paint" is used for latex-based as well as many other systems, some favor its use to describe one of the oldest coating systems known, that of a pigment combined with a drying oil and a solvent. The brown house paint of Table 12.4 does not differ in principle from coatings used hundreds of years ago. Linseed oil is a *drying oil* by virtue of its multiple unsaturation. In effect, the oil is a polyfunctional monomer that can polymerize ("dry") by a combination of oxidation and free-radical propagation. Modern methods of recovery by pressing and extraction remove almost all of the oil content (about 40%) of the flax seeds. The oil is a triglyceride of a mixture of fatty acids. About half the fatty acid content is linolenic acid, another 20% is linoleic, and another 20% is oleic:

$$CH_2-O-CO(CH_2)_7CH=CH-CH_2-CH=CH-CH_2-CH=CH-CH_2CH_3 \quad \text{linolenic acid}$$
$$CH-O-CO(CH_2)_7CH=CH-CH_2-CH=CH(CH_2)_4CH_3 \quad \text{linoleic acid}$$
$$CH_2-O-CO(CH_2)_7CH=CH(CH_2)_7CH_3 \quad \text{oleic acid}$$

Typical triglyceride from linseed oil

One can distinguish three stages in the film formation of a drying oil by oxidation. In the first, *autoxidation,* the oil reacts with oxygen to form peroxy compounds. In the second, *network formation,* the peroxy compounds react to create covalent bonds between the original drying oil molecules. In the third, much slower, stage, *aging,* the polymer film continues to react with oxygen, forming additional cross-links and also some volatile products.

In oils with isolated double bonds, autoxidation proceeds by the formation of hydroperoxides adjacent to the unsaturation:

$$R-CH=CH-CH_2-R' + O_2 \rightarrow R-CH=CH-CH-R'$$
$$|$$
$$O$$
$$|$$
$$OH$$

Hydroperoxide

Oils with conjugated unsaturation, such as tung oil with about 80% eleostearic acid, react more rapidly with oxygen, forming cyclic peroxides together with other oxygenated species.

$$CH_3(CH_2)_3CH=CH-CH=CH-CH=CH(CH_2)_7\overset{\displaystyle O}{\overset{\displaystyle \|}{C}}-OH$$

Eleostearic acid

Oil-soluble metallic soaps are used to catalyze the oxidation process. The time necessary to change a film of linseed oil to an insoluble network is decreased from days to hours by the addition of 0.05% cobalt metal insoluble form. Cobalt, manganese, and lead are commonly used as salts (soaps or *driers*) of naphthenic acid. The acid, a by-product of naphthenic lubricating oil refining, is predominantly composed of saturated, cyclic, monocarboxylic species. With a typical molecular weight of 150 to 250, the acid yields soaps with excellent solubility in aliphatic solvents and good drying activity. The metallic soaps accelerate the rate of oxygen uptake and network formation and may also alter the details of the mechanism. In Fig. 12.13, linseed oil by itself increases in acid content much more slowly than it does with added metallic driers. The gain in weight shown is the net gain. It has been demonstrated that linseed oil may react with oxygen to the extent of 48% of

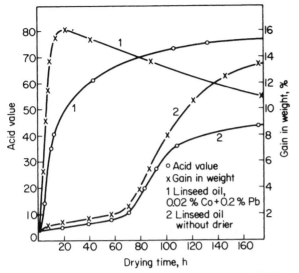

Figure 12.13 Acid value and gain in weight of linseed oil film on drying [16]

its original weight over a period of several months. The net gain may be only 15 to 20% because volatile products, mainly CO_2, are lost.

The pigments in Table 12.4 include calcium carbonate and zinc oxide, which would give a white appearance of moderate hiding power. The iron oxide gives a durable brown color, which is stable over a long period of time. The soya lecithin is a substituted phosphoric acid ester of a diglyceride that acts as a dispersing and stabilizing agent for the pigments. Kettle-bodied linseed oil has been heated with enough oxygen to cause some cross-linking, which increases the molecular weight and viscosity.

In the slow process of aging or weathering, the gradual erosion of the surface that accompanies further oxidation is not always undesirable. A *chalking* paint, one that erodes at the surface, is self-cleaning. In a house paint this means that atmospheric dirt will be removed from the surface along with some of the film by ordinary wind and rain or by deliberate washing.

Oleoresinous *varnishes,* as the name implies, are combinations of resins and drying oils. Prior to the introduction of alkyd resins in the 1920s (discussed next below), the term "varnish" usually denoted the reaction product of a heat-treated natural resin with drying oil. The product often was dissolved in a solvent, called a *thinner* because it decreased the viscosity of the varnish, together with appropriate driers. Addition of a pigment to a varnish yields an *enamel,* although the terminology is far from standardized. The natural resins are not widely used any longer. The synthetic resins that are used often are somewhat complex. A simple example is a phenolic *resole,* a reactive, soluble material. Reaction of formaldehyde with a para-substituted phenol under alkaline conditions yields the resin, which can be further reacted with an oil to form an air-drying coating.

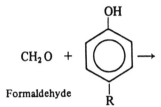

CH₂O + (p-Substituted phenol)

Formaldehyde

p-Substituted phenol

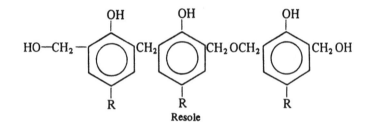

Resole

    The reaction at 200 to 250°C with a triglyceride such as tung oil is accompanied by the evolution of water, so presumably the resole reacts with free acid groups from oxidation of tung oil as well as by ester interchange with the terminal hydroxyls of the resole. Other reactions involving the phenolic hydroxyl also are probable. The combination of hard, wear-resistant resin with softer, energy-absorbing drying oil films can be designed to give products with a wide range of gloss and durability. The terms long oil and short oil are used to denote varnishes that are higher and lower in the ratio of drying oil to resin, respectively.

    Another route to balancing drying properties, durability, gloss, and hardness is the alkyd resins. These esters are formed from *al*cohols and ac*ids,* hence the coined name "alkyd." Originally the procedure resembled the cooking of an oleoresinous varnish in that a drying oil would be transesterified with a mixture of polyfunctional acids and alcohols. A common mode of operation today is to start with the free fatty acids from the drying oil rather than with the triglycerides. Another source of unsaturated fatty acids is the sulfate pulping process for making cellulose from pine chips. When the chips are digested at 170°C with sodium hydroxide and sulfide, a dark, odoriferous, pasty mass is extracted, which when evaporated and neutralized is crude tall oil, a mixture of rosin and fatty acids. Distillation separates the acids into fractions, one of which is high in unsaturation. A typical analysis might show equal amounts of oleic and linoleic acids. The formulations of three typical alkyd resins based on tall oil fatty acids are indicated in Table 12.5.

    The layout of a typical alkyd resin plant is shown in Fig. 12.14. The heart

Table 12.5 Typical alkyd resin formulations*

Ingredient	Formula	Molecular weight	Long oil, parts by wt	Medium oil, parts by wt	Short oil, parts by wt
Tall oil fatty acids	—	ca. 280	123.2	49.1	39.6
Phthalic anhydride	(phthalic anhydride ring structure)	148	50.0	28.2	58.7
Pentaerythritol	$C(CH_2OH)_4$	136	40.5	18.6	
Trimethylol propane	$CH_3CH_2C-CH_2OH$ with CH_2OH and CH_2OH branches	134	—	—	51.7
Ethylene glycol	$HO-CH_2CH_2-OH$	62	—	4.1	
Final concentration in mineral spirits, % solids			70	50	60
Total cooking time, h			6.5	7	5.5
Cooking schedule			Charge all ingredients. Raise temp to 250°C. Hold with inert gas and removal of water for acid number less than 6.	Charge all plus 6.5 parts xylene. Heat to reflux temp 230°C, and hold for acid number less than 8.	Charge as with long oil. Hold at 230°C for acid number less than 10.
Application			Trim and trellis paints. Air-dry in ½ h with driers.	Industrial enamels. Air-dry in 40 min with driers and TiO_2 pigment.	Appliance enamel. Bake 20 min at 300°F (no driers).

*Glidden-Durkee Div. of SCM Corp.

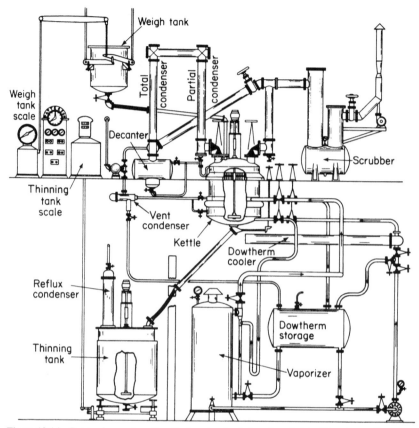

Figure 12.14 Processing system for alkyd resins heated and cooled by Dowtherm (Patterson-Ludlow Division, Banner, Ind.) [17]

of the process is a jacketed 200- to 4000-gal kettle, usually of stainless steel. For temperatures of 200 to 300°C the circulating fluid in the jacket may be Dowtherm, a proprietary mixture of diphenyl and diphenyl oxide. To protect the drying oils from oxygen and to remove water, an inert gas such as carbon dioxide or nitrogen is sparged through the kettle. In the medium-oil example of Table 12.5, xylene is added, and since it is the only volatile ingredient, it refluxes and controls the temperature. When the vaporized xylene is condensed in a side arm, water that condenses with it can be removed in the decanter and only the xylene is returned to the kettle. This "azeotropic" process is somewhat faster and gives more uniform product than straight fusion. However, one has to contend with slightly more complicated equipment together with the fire hazard of the volatile solvent.

The *acid number* is defined as the weight of KOH in milligrams needed to neutralize 1 g of sample to a phenolphthalein end point, usually in a water-methanol mixed solvent. Because an excess of hydroxyl over carboxyl is used, the acid number is a good indication of the approach to complete reaction.

The finished resin is very viscous and flows with difficulty from the kettle even at the elevated temperature. As indicated in Fig. 12.14, the batch is dropped to a *thinning tank,* where it is diluted with a volatile solvent, often mineral spirits, a petroleum cut. Even in solution, the final product has a viscosity of 10 to 100 poise (1 to 10 Pa·s). Although the large paint producers make most of their own resins, there is a substantial market in resins that are sold to paint formulators who, in turn, serve industrial and household consumers.

The drying mechanism for alkyd resins should resemble that described for linseed oil. It should be noted that some applications (short oil, Table 12.5) do not use driers but react in a short time at an elevated temperature.

A major influence on the coatings industry is the legislation enacted over the past several decades limiting the discharge of solvents into the atmosphere [18]. It is seldom economical to recover or incinerate solvents even in concentrated industrial applications. The smog problem in Los Angeles County, California, led to the adoption of "Rule 66" in 1966. This rule, which limits the discharge of "photochemically reactive solvents," was widely copied in other locales, so the term "Rule 66" has acquired international acceptance despite local variations. Besides defining certain solvents and combinations of solvents as photoreactive, the rule also limits total emissions of nonreactive solvents. The basic premise is that some aromatic compounds, branched-chain ketones, and unsaturated compounds react with nitrogen oxides in UV light, giving rise to smog. Smog, in turn, is manifested by poor visibility, eye irritation, and plant damage. Although the particular conditions of Los Angeles are not found in many other places, Rule 66 has been adopted in some form in many locations, especially urban areas where automobile exhaust raises the nitrogen oxide levels. The obvious answer to Rule 66 was, at first, reformulation with the allowed solvents. However, with the limitation on solvent vapor discharges of all kinds, there has been a trend toward water-borne coatings, powder coatings, and UV-curable (100% reactive) coatings.

Adhesives

In the broad sense of the word, adhesion is important in every heterophase system including the coatings just described. However, we can look at the process more narrowly as involving a layer of material (the *adhesive*) between two surfaces (*adherends* or *substrates*). *Mechanical* adhesion is easy to visualize. When two porous surfaces such as paper are glued together with a liquid adhesive, the interlocking action of the liquid that penetrates the pores acts as a mechanical fastener. On the other hand, gluing together two smooth, impervious surfaces must involve other forces. In *specific* adhesion the secondary forces between adhesive and surface bond the two. Actual covalent bonds may be formed, but this is hard to demonstrate in most cases. In any kind of adhesion, wetting of the surface by the

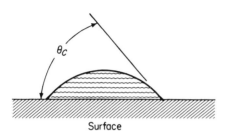

Figure 12.15 Definition of contact angle, θ_c, for liquid droplet on a surface

Surface

adhesive is important. One way of characterizing a surface is to measure the *contact angle* θ_c made by a drop of liquid on the surface at the point where the two phases meet (Fig. 12.15). Perfect wetting occurs when $\cos \theta_c = 1$, that is, when $\theta_c = 0$. When $\cos \theta_c$ is plotted for a series of liquids on a single surface, the intercept at $\cos \theta_c = 1$ is γ_c, the critical surface tension of that surface (Fig. 12.16). The values obtained for some common polymers are summarized in Table 12.6. In general, liquids will spread when the surface tension of the liquid is lower than the critical surface tension of the surface. Thus glass and metals are easily wetted by organic materials when they are clean. On the other hand, some liquids, upon contacting a surface, orient in such a way that the surface is converted to one more or less easily wetted than the original. Poly(dimethyl-siloxane) liquids have a surface tension of 19 to 20 dyne/cm (mN/m). However, a closely packed, adsorbed monolayer of the same polymer exhibits a surface of methyl groups with a critical surface tension of about 24 dyne/cm. Thus, silicones spread on almost all surfaces, since each surface, once wetted, is converted to one of higher surface tension than that

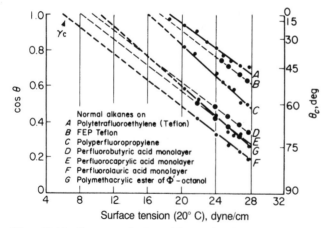

Figure 12.16 Contact angles formed by a series of liquid *n*-alkanes on various fluorinated low-energy solid surfaces [19]

Table 12.6 Critical surface tensions of polymeric solids [19] at 20°C

Polymeric solid	γ_c, dyne/cm (mN/m)
Polymethacrylic ester of Φ'-octanol*	10.6
Poly(hexafluoropropylene)	16.2
Poly(tetrafluoroethylene)	18.5
Poly(trifluoroethylene)	22
Poly(vinylidene fluoride)	25
Poly(vinyl fluoride)	28
Polyethylene	31
Poly(trifluorochloroethylene)	31
Polystyrene	33
Poly(vinyl alcohol)	37
Poly(methyl methacrylate)	39
Poly(vinyl chloride)	40
Poly(vinylidene chloride)	40
Poly(ethylene terephthalate)	43
Poly(hexamethylene adipamide)	46

*Φ'-Octanol is $CF_3 \cdot (C_2 F_2)_3 \cdot CH_2 OH \cdot CH_2 OH$.

of the liquid itself. Normal octyl alcohol also orients at a high-energy surface to convert it to a methyl surface. But now the surface tension of the alcohol (25 dyne/cm) is higher than that of the layer on the surface. Such a system is *autophobic;* i.e., spreading does not occur because the first molecules antagonize the surface.

If surfaces were perfectly smooth and clean, no adhesive would be needed. Almost any material cleaved in a high vacuum can be fused back together without an adhesive or high temperature. However, surfaces are rough on an atomic scale, and one of the functions of an adhesive is to fill in the asperities. Therefore all adhesives are applied as liquids. We can classify adhesives as being permanently liquid, solidifying by physical processes, or solidifying by chemical processes.

Water is a good example of a liquid adhesive for almost any two surfaces. Its volatility, low viscosity, and high surface tension make it a poor adhesive, although when it freezes, it can be a very strong adhesive—as someone who has put his tongue on a frozen sled runner can attest. A plasticized rubbery polymer has low volatility, high viscosity, and low surface tension and makes a good adhesive. The common cellophane tape is a typical application of such a *pressure-sensitive* adhesive. The material remains permanently liquid but forms a strong bond between the surface and the cellophane backing when slight pressure is applied to cause flow. When the surface is paper, presumably both specific and mechanical adhesion may be involved.

Tar and sealing wax are examples of adhesives that solidify by cooling after being applied as hot liquids. Some animal glues and thermoplastic resins are applied similarly. Most adhesives are counterparts of the coating systems we have already discussed. Model airplane cement is a polymer solution, often cellulose nitrate in a mixture of ketones and aromatic solvents. As a coating we would call it a lacquer. Aqueous solutions of natural and synthetic gums are used in library paste. Some of the popular white glues for paper and wood are simply poly(vinyl acetate) emulsions with a small amount of plasticizer. All of these materials solidify after contacting the surface as a liquid by loss of solvent or diluent. Evaporation or diffusion into a porous substrate may be involved.

The stronger adhesives generally are those that are thermosetting. They are applied as liquids but form network polymers by chemical reaction. It may be necessary to heat the liquid to cause the reaction to occur. A common system involves mixing two ingredients that will react after an induction period. The *pot life* of such a system is the time between mixing and conversion to a high viscosity unsuitable for spreading. Plaster of paris is a slurry of calcined calcium sulfate in water. The hardening reaction involves both hydration and crystallization. The epoxy resins can be formulated so that the reaction between the epoxy prepolymer and the amine "hardener" (both polyfunctional) takes place 15 to 30 min after the mixing. Not every reactive adhesive is made by mixing two ingredients. The room-

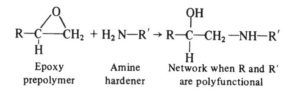

temperature-vulcanizing (RTV) silicones are stable pastes for months as long as they are protected from the atmosphere. Exposure to a humid atmosphere brings about hydrolysis of acetate ester groups. In the presence of certain catalysts, the SiOH groups condense to form Si—O—Si bonds.

$$R-Si(CH_3)_2-O-Si(CH_3)_2-O-\overset{\overset{\displaystyle O}{\|}}{C}-CH_3 + H_2O \xrightarrow{\text{Sn catalyst}}$$

Acetate end-blocked silicone

$$R-Si(CH_3)_2-O-Si(CH_3)_2-OH + HOOCCH_3$$

The cyanoacrylate adhesives were introduced in the late 1950s. Unlike the epoxy or silicone adhesives, these materials contain a monomer that undergoes rapid polymerization at room temperature. The monomer ethyl cyanoacrylate is stable with trace amounts of an acid inhibitor such as polyphosphoric acid. The commercial formulations may contain methyl and higher cyanoacrylate esters, thickeners such as organic polymers, and plasticizers for increased final film toughness. A slightly basic surface is needed for ionic polymerization to occur. Even the surface moisture of an ostensibly "dry" metal or ceramic surface is sufficiently

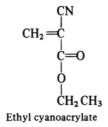

Ethyl cyanoacrylate

basic. A really acid surface does require an alkaline pretreatment. The reaction time for polymerization is less than 10 s when a thin layer of monomer is applied to a clean, dry, smooth surface. A variety of cyanoacrylate adhesives are on the market with numerous brand names that convey the rapidity of bonding, the "superstrength" of the bond, or merely the imagination of a sales organization. Repairs of china, glass, plastics, and metals all can be quite dramatic, as demonstrated in TV ads. However, the cyanoacrylates also have been used as sutures to glue together skin or other body tissues, as dental adhesives, and in industrial applications such as attaching ornaments in automobiles and in assembling electronic devices.

12.5 TWO-DIMENSIONAL PROCESSES

Extrusion in General

In *extrusion* a material is forced through a *die* that shapes the profile. Continuous flow of material results in a long shape of constant cross section. Unlike the extrudate from a toothpaste tube or a meat grinder, plastic extrudates generally approach truly continuous formation. They may attain dimensional stability in various ways. Like the usual meat grinder, the extruder (Fig. 12.17) is essentially a screw conveyer carrying cold plastic pellets forward and compacting them in the compression section with heat from external heaters and from the friction of viscous flow. The pressure is highest right before the plastic enters the die that shapes the extrudate. The screen pack and breaker plate between the screw and the die filter out dirt and unfused polymer lumps. When thermoplastics are extruded, it is necessary to cool the extrudate below T_m or T_g in order to gain dimensional stability. In some cases this can be done by simply running the product through a tank of water or, even more simply, by air cooling. When rubber is extruded, dimensional stability comes about by cross-linking (vulcanization). Special attachments for continuous vulcanization are described later. Actually, rubber extrusion for wire coating was the first application of the screw extruder. The first extruder in the United States was built in 1880 by John and Vernon Royle, although some machines may have been in use in England as early as 1866 [21].

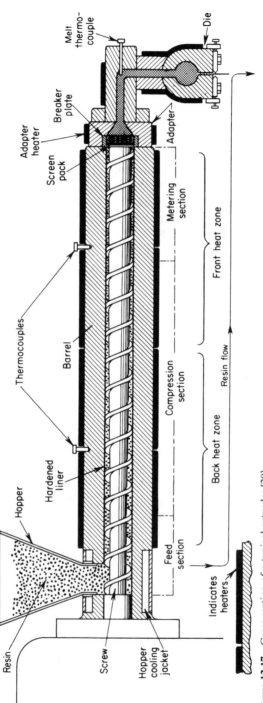

Figure 12.17 Cross section of a typical extruder [20]

452

Standard sizes of single-screw extruders (specified by the inside diameter of the barrel) are $1\frac{1}{2}$, 2, $2\frac{1}{2}$, 3, $3\frac{1}{2}$, $4\frac{1}{2}$, 6, and 8 in. Smaller sizes are available for laboratory evaluations. As a rough guide, extruder capacity Q_e in pounds per hour, varies with the barrel diameter D_b, in inches, to the 2.2 power [22]:

$$Q_e = 16D_b^{2.2} \tag{12.1}$$

Another guide to extruder capacity stems from the fact that most of the energy that melts the thermoplastic results from mechanical work. The barrel heaters serve mainly to insulate the material. Allowing for an efficiency from drive to screw of about 75%, the capacity then depends on the power supplied H_p (horsepower), the heat capacity of the material c_p (Btu/lb·°F), and the temperature change from feed to extrudate ΔT (°F):

$$H_p = 5.3 \times 10^{-4} Q_e c_p \, \Delta T \tag{12.2}$$

For example, a 10-hp motor on a 2-in extruder might be used for poly(methyl methacrylate), for which c_p is about 0.6 Btu/lb·°F. Equation (12.1) gives $Q_e = 74$ lb/h. Equation (12.2) indicates that ΔT_{max} would be 430°F. In actual practice, a ΔT of 350°F would usually be adequate for this polymer. The same calculation is, of course, much simpler in SI units. For example, if a 10-kW motor delivers 75% of its energy to a throughput of 10 g/s of polymer with a heat capacity of 2.5 J/g·°C, the temperature change is

$$\Delta T = \frac{7500 \text{ J/s}}{(10 \text{ g/s})(2.5 \text{ J/g·°C})} = 300°C$$

Obviously, these are very rough guides, since both polymers and machines are variable. Also, we have ignored the heat of melting and other thermal effects.

Besides barrel diameter, the length to diameter ratio L/D of an extruder is most often specified. Ratios of 20:1 and 24:1 are common for thermoplastics. Lower values are used for rubber. A long barrel gives a more homogeneous extrudate, which is especially important when pigmented materials are handled. A popular variant of the standard single-screw extruder is the vented model. Even when polymer pellets are predried, the compacting process may trap air bubbles in the melt. By placing an obstruction in the barrel (the reverse flights in Fig. 12.18), the pressure on the melt can be reduced suddenly to that of the atmosphere. Since the polymer does not have to be remelted, the feed section does not have to be replicated in the remainder of the screw. Valves can be used to adjust the pressure in various sections of the barrel. Although many screw designs are available, most have three zones, which differ in channel depth but not in pitch (which is synonymous with lead, since we are dealing with a single thread) (Fig. 12.19). The ratio of channel depth in the feed section to that in the metering section, the *compression ratio*, probably would best be adjusted to each polymer and each temperature used in a given machine. However, it is seldom feasible to stock more than two or three

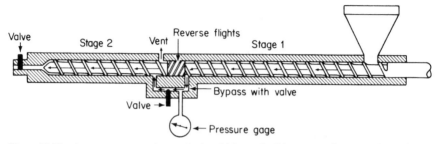

Figure 12.18 A two-stage vented extruder in which a valved bypass section steps down the pressure developed in the first stage to atmospheric pressure for venting; valve is for adjustment [22]

screws for a single extruder, so a compromise is made. Some considerations would be (1) the form in which polymer is to be fed to the machine, i.e., pellets, powder, continuous ribbon; (2) the melt viscosity; and (3) thermal degradation. One screw might be suitable for polyethylene and ABS polymers, another for polyamides with their characteristic low melt viscosities, and another for poly(vinyl chloride) or poly(vinylidene chloride), each of which degrades easily at elevated temperatures. In laboratory-sized equipment one might extrude a filled rubber by using still another screw. However, since rubber usually is fed as a warm ribbon, it has to be cooled to prevent cross-linking in the barrel, and then must be heated to cross-link subsequent to extrusion; it becomes apparent that production machinery that attempts to do all this and still be versatile enough to handle thermoplastics probably will be inefficient for both.

Twin-screw extruders may have the two parallel screws rotating in the same direction. Another form has counterrotating screws with some meshing to give a conveying action similar to a positive displacement pump. Some advantages generally claimed for twin-screw over single-screw extruders are better control, lower possible melt temperatures, and better mixing. These advantages must be weighed against a higher initial cost. Some other features of extruders are best described in terms of specific shapes.

Flat Film, Sheet, and Tubing

By definition, the term *film* is used for material less than 0.010 in thick and *sheet* for that which is thicker. Since the material will issue from the extruder as a thin, unsupported web of molten polymer, it is obvious that a high melt viscosity and some elasticity are desirable. The most direct control for these parameters is the molecular weight distribution. Often a polymer blend will perform better than a single component. *Flat-film* extrusion uses a T-shaped die (Fig. 12.20). The die opening for polyethylene may be 0.015 to 0.030 in, even for films that are less than 0.003 in thick. The speed of taking up the film (various driven rolls) is high enough

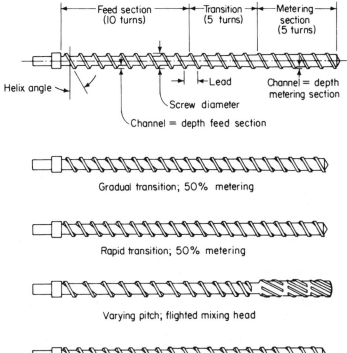

Figure 12.19 The various zones or sections of a standard constant-pitch, rapid-transition, metering extruder screw are labeled in the above drawing [22]. Some other designs are shown also

to draw down the film with a concurrent thinning. Dies may be as wide as 10 ft. On leaving the extruder the film may be chilled below T_m or T_g by passing through a water bath or contacting a water-cooled roll. A schematic drawing of a *chill-roll* (also called *cast-film*) operation (Fig. 12.21) shows how the hot web is made dimensionally stable by contacting several chrome-plated chill rolls before being pulled by the driven carrier rolls and wound up. The edges of the sheet may be trimmed off, since some curling occurs there. The draw-down process (followed by cooling before relaxation can occur) results in an oriented film whether the polymer is glassy or crystalline. Thicker sheets generally are not drawn down as much and exhibit much less residual orientation. *Biaxially oriented* film can be produced from a flat web by using a *tenter* (Fig. 12.22). Polypropylene, for example, might be extruded as a web about 0.005 in thick from a slot die. Inside a temperature-controlled box, the web is grasped on either side by tenterhooks which can exert a drawing tension as well as a widening tension. Where hooks disengage, the final sheet might be two or three times as wide as it was originally and, of course, consid-

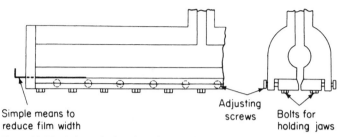

Simple means to
reduce film width

Adjusting
screws

Bolts for
holding jaws

Figure 12.20 Schematic drawing of a manifold-type flat-film die with adjustable jaws [20]

erably thinner. Biaxial orientation can be equal in two directions or unbalanced. In either case, molecules are now randomly oriented in two dimensions, just as fibers would be in a random mat. With polyethylene, biaxial orientation often can be achieved in *blown-film extrusion*. The molten polymer (Fig. 12.23) is extruded through a ring-shaped die around a mandrel. The tube or sleeve so formed is expanded around an air bubble, is cooled to below T_m, and then is rolled into a flattened tube and wound up. Since the bubble is sealed at one end by the mandrel and at the other by the nip where the tube is flattened, air cannot escape and the bubble acts like a permanent shaping mandrel once it has been injected. Cooling of the film is controlled by an air ring below the bubble. When polyethylene cools, the crystalline material is cloudy compared to the clear amorphous melt, so the transition line, which coincides with dimensional stabilization, is called the *frost line*. The *blowup* ratio (bubble diameter/die diameter) may range as high as 4 or 5, although 2.5 is a more typical figure. Orientation occurs in the hoop direction during blowup. Orientation in the machine direction (in the direction of extrudate flow from the die) can be induced by tension from the nip roll. The method pro-

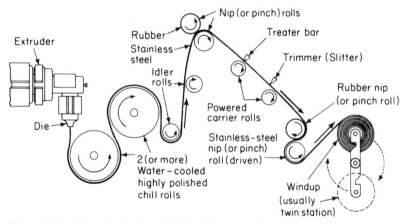

Figure 12.21 Schematic drawing of chill-roll film extrusion equipment [20]

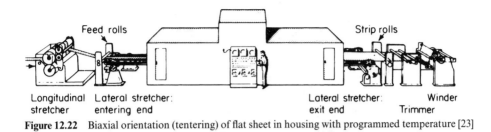

Feed rolls   Strip rolls

Longitudinal Lateral stretcher: Lateral stretcher: Winder
stretcher   entering end   exit end   Trimmer

Figure 12.22 Biaxial orientation (tentering) of flat sheet in housing with programmed temperature [23]

duces wide films used in construction trades as wind or moisture barriers. One example is reported as involving a 10-in extruder, a 5-ft-diameter die, and a blowup ratio of 2.5 to give 1100 lb/h of polyethylene film, which when slit is 40 ft wide [24]. Such films in thicknesses of 0.004 to 0.008 in are readily available from hardware firms and mail-order companies for farm and construction applications. Nylon and poly(vinylidene chloride) are also produced by blow extrusion.

For some applications, more than one polymer layer is desirable. For example, a food tray might need a strong, abrasion-resistant surface layer such as polypropylene combined with an inner layer such as poly(vinylidene chloride) to contribute low oxygen permeability. House siding of poly(vinyl chloride) can benefit from a

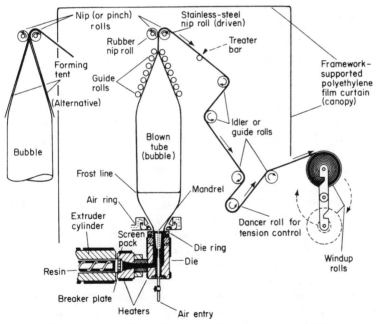

Figure 12.23 Schematic drawing of the blown-film extrusion process [20]

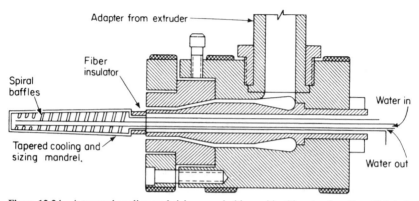

Figure 12.24 A tapered cooling and sizing mandrel is used in this extrusion die, which is designed for use in producing either pipe or tubing [22]. (Courtesy of Union Carbide Corp.)

surface layer of an ABS-type polymer with superior weathering properties. In order to produce laminated, *multilayer film* by *coextrusion,* several extruders can be coupled. Each will contribute a separate stream of polymer melt to be combined without mixing in a *feed-block* from which a layered melt issues. Much better adhesion between film layers often can be achieved in this way than could be attained by coating one melt onto a previously extruded and cooled second film.

Pipe or tubing is extruded through a ring-shaped die around a mandrel. Because heavier walls are involved than in blown-film extrusion, it is advantageous to cool by circulating water through the mandrel (Fig. 12.24) as well as by running the extrudate through a water bath. The extrusion of rubber tubing differs from thermoplastic tubing, since dimensional stability results from a cross-linking reaction at a temperature above that in the extruder rather than by cooling below T_g or T_m. The high melt viscosity needed to ensure a constant shape during the cross-linking may be an inherent property of the rubber being extruded. More often, however, the addition of a filler will induce the same behavior. For example, 20 parts of finely divided silica added to 100 parts of a poly(dimethyl siloxane) will increase the apparent viscosity only about twofold at a rate of shear equal to 100 s^{-1}. The viscosity at a stress corresponding to the unsupported polymer is measurable for the unfilled material but essentially infinite for the filled polymer. It is possible to extrude a filled rubber through a ring die to give a tube with a 12-mm outside diameter and a wall thickness of 3 mm and to cross-link it continuously by carrying it through a radiant-heated tunnel on a steel conveyer belt. With a residence time of 10 s in a 6-m-long tunnel, the tubing emerges with very little sagging of the original circular cross section.

Almost any cross section can be extruded. A complication is the swelling of the polymer when the elastic energy stored in capillary flow is relaxed on leaving the die. The flat sheet or circular cross sections discussed so far are not sensitive

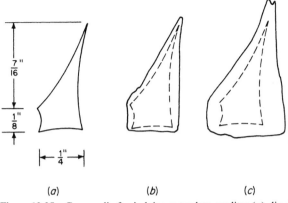

$\frac{7}{16}$"

$\frac{1}{8}$"

$\frac{1}{4}$"

(*a*)   (*b*)   (*c*)

Figure 12.25 Garvey die for judging extrudate quality: (*a*) die dimensions (and dimensions of ideal extrudate); (*b*) marginal extrudate showing swelling and some distortion; (*c*) badly distorted extrudate

to this *die swell* (Sec. 7.8), since the shape remains symmetrical even though the extrudate dimensions differ from those of the die. Also, the tension exerted on the extrudate can be used to counteract the swelling. Unsymmetrical cross sections may be distorted as the result of die swell. A standard test in the rubber industry gives a qualitative answer to the question of suitability for a given compound and machine setting. Extrusion through the peculiarly shaped Garvey die (Fig. 12.25) is sensitive both to die swell (shown by rounded corners and convex surfaces) and to other irregularities that may occur because of poor cohesion and poor surface lubricity [25]. The maximum rate at which even a symmetrical cross section can be extruded generally will correspond to the flow rate region of "melt fracture" or some other kind of periodic flow distortion. Die entrance geometry and temperature control may affect the extrudate as well as the usual variables of composition.

A particularly clever mechanical arrangement permits the extrusion of a continuous sleeve of loosely knitted fibers suitable for being made into bags of the sort used for onions and potatoes. Two elements that rotate concentrically (Fig. 12.26) bear hemicylindrical slots through which the polymer melt is pumped. When the slots coincide, the filaments being extruded converge and are welded together. With a die like that in Fig. 12.26, 16 threads are extruded, 8 of them convex outward and 8 in toward the axis about which the elements rotate.

Another intriguing process also uses counterrotating dies. Two polymers are fed between the dies in the manner suggested by Fig. 12.27. As the dies rotate, the two polymers are stretched out laterally into a multiplicity of layers. In one example [27], poly(methyl methacrylate) and polystyrene are made into a tube with over 100 layers. Combined with blown-film extrusion, a film only 25 μm thick can be made with individual layers having thicknesses near the wavelength of visible light. Very attractive diffraction patterns are produced, which suggest the use of

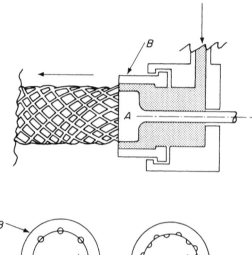

Figure 12.26 Rotating dies for extrusion of net material [26]

the film in decorative applications. An iridescent film based on polypropylene and polystyrene has been available commercially since 1976 [28]. Some uses for it include overwraps for cosmetics and toiletries, gift wrapping, and wallcoverings.

The application of insulation to a continuous length of conductor represented one of the first uses of extruders for rubber about 100 years ago and also one of the first uses with thermoplastics in the 1930s. The process resembles that used for pipe with extrusion around a mandrel except for the drawing of the conductor through the mandrel on a continuous basis. For thermoplastics such as polyethylene, nylon, and plasticized poly(vinyl chloride), a slight downward pitch from the extruder into a water trough is enough to cool the insulation below T_m or T_g. At the far end of the trough the wire or cable with its hardened insulation can be withdrawn under tension around a pulley or drum, dried, and wound up. The conductor is not always a single metal strand. It can be a multiple strand or even a bundle of previously individually insulated wires. Cloth or paper may have been wrapped around the bundle in a previous stage also. With rubbers that have to be cross-linked by heating subsequent to extrusion, several routes are available. If the insulation has high dimensional stability, the un-cross-linked extrudate may be coiled up and the entire assembly heated in a pressurized, steam-heated tank. The hot-air vulcanization mentioned previously for silicone tubing is feasible. Higher heat transfer rates and less porosity than the hot-air tunnel are obtained by vulcan-

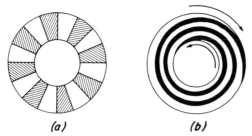

(a)   *(b)*

Figure 12.27   Production of multilayered sheet by extrusion. (*a*) Head-on view of two polymers entering an annular die like spokes on a wheel. (*b*) Spiral formation of multiple layers by a single spoke between counterrotating dies

izing in high-pressure steam. The continuous vulcanizing (CV) process puts the coated conductor through as much as several hundred feet of pipe containing steam at up to 250 psig (1.7 MPa). By extruding vertically downward, the sagging of the insulated wire is avoided (Fig. 12.28). The seal at the upper end is obtained by butting the pipe against the extruder head. The bottom 3 m of the pipe contains cooling water, which hardens the now cross-linked insulation somewhat and allows the cable to be withdrawn through a seal without damaging the coating.

Pultrusion is a continuous molding process resembling the application of insulation to wire and cable (Fig. 12.29). Typically, a glass fiber or fabric reinforcement is fed continuously from a spool through a resin bath. The resin-impregnated reinforcement is then pulled through a die that forms the desired cross section and heats the material to start a "curing" reaction. If the resin is a mixture of styrene monomer and unsaturated polyester with a peroxide initiator, heating causes a free-radical copolymerization to occur. After emerging from the die, the pultruded shape may be further heated by radio-frequency energy (see Sec. 10.6). Some standard shapes produced by this process are angles, channels, I-beams, and tubes, as well as table legs and platforms [30].

Extruders are used as auxiliary equipment in many other kinds of forming operations. Pellets for thermoplastic molding are made by chopping the cylindrical "spaghetti" extruded through a multidie extruder head. The feed for a calender (see below) may come from an extruder. The thin film for a flat-film extruder may be laminated directly onto cardboard or another polymer film. Finally, a vented extruder can be used to dry a product or remove residual monomer from a polymer in an intermediate stage of production.

Two methods of producing flat films that do not involve extrusion should be mentioned. Most of the cellulose acetate (or acetate-butyrate) film that is coated with a gelatin emulsion for photographic purposes is produced by depositing a polymer solution (dope, lacquer) on a large drum and evaporating the solvent. In a typical installation the drum might be 4 ft wide and 18 ft in diameter. The solu-

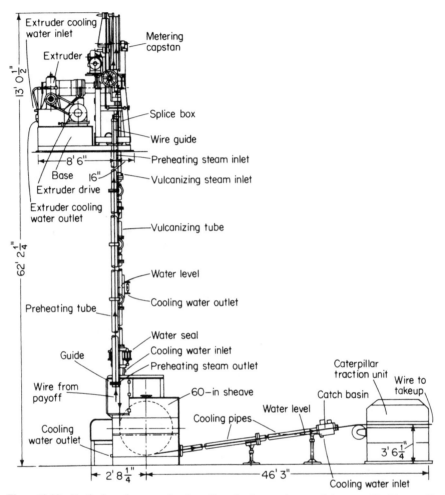

Figure 12.28 Vertical continuous vulcanizer. Extruder is near the top of the assembly (Fawcett Davis-Standard, Davy Plastics Machinery, Ltd.) [29]

tion is introduced at the top from a flow box, and by the time the wheel has turned about 300° (say 15 or 20 s), enough solvent has evaporated that the web is self-supporting and can be carried on to further drying and conditioning [31]. The alignment and polishing of the drum are very critical, because the thickness of the photographic emulsion to be added later on will reflect any variations in the film base material. Continuous metal belts also have been used, although control of film thickness is more difficult. Associated with this kind of operation is an elaborate solvent-recovery system, without which it is uneconomical.

Calenders resemble the two-roll mill (Fig. 12.4), except that three to five rolls

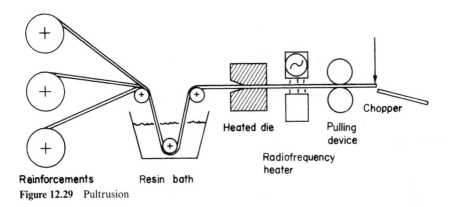

Figure 12.29 Pultrusion

may be used and they all rotate at the same speed. The action is essentially that of a rolling pin or washer wringer. The material fed to the calender is usually preheated and may come directly from an internal mixer or an extruder. The calender squeezes the mass into a film or sheet, usually at an elevated temperature, and imparts a surface gloss or texture. Embossed or printed patterns are added on subsequent rolls. Some variations are shown in Fig. 12.30. Thick sections of rubber can be made by applying one layer of polymer upon a previous layer (*double plying*) or onto cloth fabric. Most plasticized poly(vinyl chloride) film and sheet from the 0.1-mm film for baby pants to the 2.5-mm "vinyl" tile for floor coverings is calendered. With thermoplastics the cooling below T_g or T_m can be accomplished on the rolls with good control over dimensions. With rubber, cross-linking can be carried out with the support of the rolls, once again with good control. It is only fair to point out that despite the simple appearance of the calender compared to the extruder, the close tolerances involve and other mechanical problems bring the cost of a calendering line to about a million dollars.

Fibers [33–35]

The term *spinning* as used with the natural fibers (and with the spinning wheel of colonial days) refers to the twisting of short fibers into continuous lengths. In the modern synthetic fiber industry the term is used to denote the process of producing continuous lengths by any means. First, we should define several other terms. A *fiber* is the fundamental unit of textiles and fabrics and can be defined as a unit of matter having a length at least 100 times its width or diameter. A *filament* is an individual strand of continuous length. *Yarn* may be formed by several processes. Continuous filament yarn, as the name implies, is made by twisting together filaments into a strand. If the filaments are assembled in a loose bundle, we have *tow* or *roving*, which can be chopped into lengths of *staple* (an inch to several inches long). *Spun yarn* is made by twisting (plying) the lengths of staple into a single

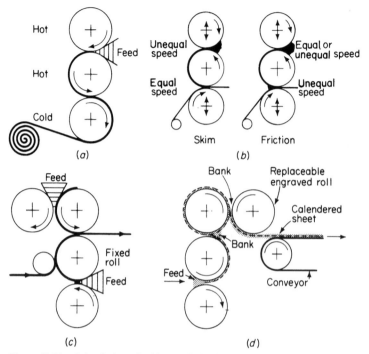

Figure 12.30 Calendering of rubber and plastics [32]: (*a*) single-ply sheeting; (*b*) applying rubber to fabrics; (*c*) double-ply sheeting; (*d*) profiling—four-roll engraving calender

continuous strand. Sewing *thread* is a variety of yarn that is plied and then finished to give a smooth, compact strand. *Cord* is formed by twisting together two or more yarns. In the production of synthetic fibers the primary fabrication process of interest is the formation, "spinning," of filaments. In every case a polymer melt or solution is put in filament form by extrusion through a die. Usually the *spinneret,* the element through which the polymer is extruded, has a multiplicity of holes. Some rayon spinnerets have as many as 10,000 holes in a 15-cm-diameter platinum disc. Textile yarns might be spun from spinnerets with 10 to 120 holes and industrial yarns such as tire cord from spinnerets with up to 720 holes. Very soft fabrics for apparel are produced from various polymers by using small-diameter fibers. The term *microfiber* is used to denote fibers with a denier less than unity [34]. Three major categories of spinning processes are used: melt, dry, and wet spinning [35].

Melt spinning of fibers is fundamentally the same as melt extrusion of gaskets and tubing. However, the actual machinery for carrying out the processing steps may differ greatly. Polypropylene may require an extruder together with a gear pump, whereas nylon, which has a lower melt viscosity, may require only the gear pump (Fig. 12.31). For producing fully oriented polypropylene yarn, a compact

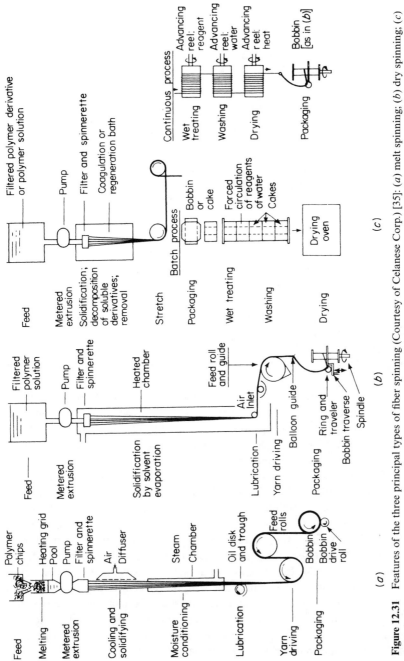

Figure 12.31 Features of the three principal types of fiber spinning (Courtesy of Celanese Corp.) [35]: (a) melt spinning; (b) dry spinning; (c) wet spinning

465

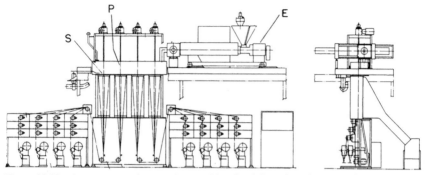

Figure 12.32 A compact melt extrusion machine for fully oriented yarn (polypropylene, nylon, or polyester). Key components are the extruder (E), the metering pumps (P) and spinnerets (S), and the take-up apparatus for controlled drawing and finishing of product. (Courtesy of Hills Research and Development, Inc., West Melbourne, Fla.)

melt extrusion apparatus combines a screw extruder with gear (metering) pumps (Fig. 12.32). The polymer chips fed to the extruder are melted, mixed, and pressurized. A color concentrate can be added to the extruder feed. After filtration, the melt goes to a "spin beam" containing four metering pumps. Each pump controls the flow of melt through spinnerets (76 holes per spinneret) to as many as four threadlines (individually wound spools of product) per pump. Thus one extruder and four pumps can combine to feed 16 take-up stations. When producing 400-denier yarn, such an apparatus can handle 84 kg/h with final yarn take-up speeds of 2000 m/min [36].

Dry spinning usually uses a pump to feed a polymer solution to a spinneret. Dimensional stability now involves evaporation of solvent as in the film-casting process. Even when dried under some tension, the diffusion of solvent out of the partly dried fiber results in an uneven fiber formation. The skin that forms first on the fiber gradually collapses and wrinkles as more solvent diffuses out and decreases the diameter. The cross section of a dry-spun fiber characteristically has an irregularly lobed appearance (Fig. 12.33). Solvent recovery is important to the economics of the process. Cellulose diacetate dissolved in acetone and polyacrylonitrile in dimethyl formamide are two current examples.

Wet spinning also involves pumping a solution to the spinneret. Now, however, the polymer is precipitated in an immiscible liquid. Polyacrylonitrile in dimethyl formamide, for example, can be precipitated by passing a jet of the solution through a bath of water, which is miscible with the solvent but causes the polymer to coagulate. Cellulose triacetate can be wet-spun from a methylene chloride-alcohol mixture into a toluene bath, where it precipitates. In other fibers the precipitation can involve a chemical reaction. Viscose rayon is made by regenerating cellulose from a solution of cellulose xanthate.

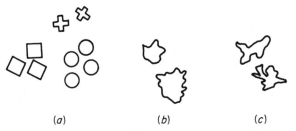

(a)	(b)	(c)

Figure 12.33 Typical cross sections of (a) melt-spun nylon from various shaped orifices, (b) dry-spun cellulose acetate from round orifice, and (c) wet-spun viscose rayon from round orifice

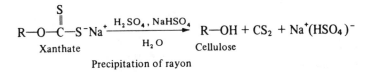

Precipitation of rayon

This process could be carried out with a slot die rather than a spinneret, in which case *cellophane* would be produced. A ring die would give a continuous tube used as sausage casing for the ubiquitous American hot dog. The casing is removed before final packaging of the "skinless" frankfurter.

When "liquid crystal" polymers are spun, the molecules are highly oriented. Some, such as copolyarylesters of 4-hydroxybenzoic acid, are melt-spun, while others, such as the polyarylamide based on *p*-phenylene terephthalamide, are wet-spun. In either case, the tensile strength can exceed that of steel (Table 12.7). A flexible polymer such as ultrahigh-molecular-weight, linear polyethylene (UHM-WPE), can also be drawn into an extended-chain configuration when fibers are formed by "gel spinning." In one example [37], a hot, dilute (5 to 15%) solution of

Table 12.7 Comparative properties of selected fibers [37]
(These values are only typical; actual values vary)

Material (fiber)*	Tensile modulus, GPa	Tensile strength, GPa	Compressive strength, GPa	Density, g/cm³
Steel	200	2.8	—	7.8
Carbon	585	3.8	1.67	1.94
Polyarylamide	185	3.4	0.39	1.47
Polyarylester	65	2.9	—	1.4
UHMPE	172	3	0.17	1
Nylon	6	1	0.1	1.14
Polyester	12	1.2	0.09	1.39

*Polyarylamide = Kevlar; Polyarylester = Vectra; UHMPE = Spectra.

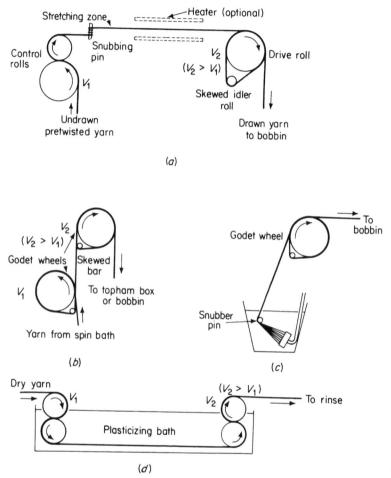

Figure 12.34 Four types of drawing processes used for synthetic fibers, V_1 and V_2 indicate relative velocities of take-up rolls (courtesy of Celanese Corp.) [35]. The processes involve (a) cold (or hot) draw, (b) godet-controlled wet stretch of nascent yarn, (c) snubber-controlled wet stretch of nascent yarn, and (d) roll-controlled wet stretch of replasticized yarn

the polymer (molecular weight exceeding 1 million) in a solvent such as decalin is extruded through a spinneret and immediately cooled to form a gel. The extrudate is quenched and extracted in a hydrocarbon liquid bath and dried under vacuum. Then the fiber, retaining the conformation it had in the gel, is drawn 30 to 100 times its original length at a controlled temperature of 100 to 150°C. The highly oriented structure that results gives the fibers a modulus and strength in tension much higher than that of the ordinary nylon and poly(ethylene terephthalate) used in textiles (Table 12.7). Both the polyarylamides and the UHMWPE fibers have

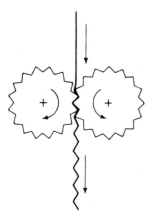

Figure 12.35 Mechanical crimping

found many uses. Probably the most exotic application is in the manufacture of light-weight, bullet-proof clothing for police protection.

Almost all synthetic yarns are so drawn that the crystalline structure is oriented in the direction of the fiber axis. Usually yarns are drawn at a temperature between T_g and T_m. The T_g may be depressed by the presence of a plasticizer, particularly water (Secs. 3.3 and 3.6). Some mechanical arrangements for continuously drawing yarn are shown in Fig. 12.34. In each case the drawing is accomplished by winding the yarn around a drum or wheel driven at a faster surface velocity than a preceding one. Polyethylene could be drawn, as indicated in Fig. 12.34a, at room temperature, whereas nylon 66 would have to be heated or humidified to be drawn. Fluted glass feed rolls called *godets* are used in Figs. 12.34b and 12.34c to control the speed of yarn. Many fibers are drawn as an integral part of the spinning process. In some cases it is preferable to store the fibers or carry out some other operation first, such as dying or twisting, and then to draw, in which case a plasticizing bath may be used to facilitate the process. In any case, the drawing does orient crystallites in the direction of the applied strain so that the modulus in that direction is increased and elongation at break is decreased.

The smooth cylindrical shape may not be the optimum one for apparel use. The seeming futility of chopping tow or roving into staple, only to respin the staple into a yarn, makes more sense when one sees the difference the operation makes in the apparent bulkiness of fibers in a sweater or blouse. Another means of increasing the bulkiness of yarn is to *crimp* the fibers by mechanical or other means. A simple means of inducing crimp (waviness) is to pass the fiber between two heated, fluted rolls or between two meshing gears (Fig. 12.35). More elaborate means of texturing yarns include twisting and untwisting, knitting and deknitting, and chemical treatments. Staple fibers often are textured before being chopped into staple.

In some applications it is desirable to have individual fibers that are very stiff.

A column with a cruciform cross section is stiffer than one with a circular cross section (Fig. 12.33), and it is quite feasible to melt-extrude such shapes in polypropylene for rugs, for example. Hollow fibers can be produced in various ways. Discontinuous air cells increase the bulk of the fiber and can give some interesting handling properties, causing some synthetics to resemble fine wool. By extruding around a mandrel, a continuous void can be formed inside a fiber. So-called hollow fibers are really small-size tubes. Bundles of such fibers are assembled into separators that resemble shell-and-tube heat exchangers. The tubes act as semipermeable membranes. Such devices have been used for water desalination and as artificial kidneys. For separating hydrogen from gas streams, over 10,000 fibers (0.8-mm OD, 0.4-mm ID) may be packed into a shell that is 20 cm in diameter and 3 m long [38].

For many of the mechanical operations with fibers such as knitting and spinning, lubricants or other agents may be needed to prevent fouling of machinery. Printing and dying of textile fabrics is another large field in itself.

In recent years the application of thermosetting resins to fibers as wash-and-wear treatments has become a major industry. The general idea is to enhance the "memory" of the fabric for creases put in before setting the resin and to minimize subsequent wrinkling. Melamine and urea resins applied to cotton comprise one large class of such materials. Due to their comparatively hydrophobic nature, nylon and poly(ethylene terephthalate) fibers do not require such treatment. While low moisture absorption is good for dimensional stability, it is often accompanied by low electrical conductivity. This brings on a second problem, that of static electricity. The charges on hydrophobic fibers do not leak off, and they cause apparel to cling uncomfortably. Also, most people have had the experience of walking across a deep-pile rug on a low-humidity, winter day and finding on reaching for the doorknob that a substantial difference in electrical potential has been induced between themselves and the knob. Finishes are available to counteract that static buildup too. In apparel the solution most often is to blend the nonconducting synthetics with the more conductive cotton or wool.

Laminates

In industrial terminology, laminates include many types of heterogeneous materials whether in layers or not. Here we shall restrict the term to standard shapes that are made in flat sheets of the same or different materials. A familiar example that illustrates the method and the advantages of lamination is plywood. Thin sheets of *veneer* are cut by knives whose blades stretch the length of a log and cut at an angle almost tangential to the curvature of the log surface. Such sheets are very strong in the direction of the tree axis and comparatively weak at 90° to that axis in the plane of the veneer. By plying the layers of veneer at right angles (usually with an odd number of layers), the strength can be made nearly equal in both directions. This makes the plywood superior to a board cut from the same wood, provided the plys can be held together strongly. Phenol-formaldehyde resins are used to bond, and to a certain extent impregnate, the veneers. The conversion of the prepolymers to the final thermosetting form takes place when the assembly is in a press under high pressure. There are several other advantages of plywood over single boards besides equalizing the directional strength. The impregnation decreases the moisture pickup of the wood, so that there is less swelling in humid atmospheres. For decorative purposes, plywood paneling can be made from inexpensive wood such as fir except for the top layer on one side, which might be a specially decorated material or an expensive, rather rare wood. It is also possible to build moisture barriers, flameproofing materials, or insulating layers into such panels by modifying the inner layers.

Paper layers are laminated with various resins to make materials useful in electrical applications as well as the decorative laminates that are best known by their trade names such as Formica and Micarta. Kraft paper, about the weight used in shopping bags, is taken directly from rolls and run through a solution of melamine-formaldehyde prepolymer in a typical application. Driving off the solvent or drying out water leaves an impregnated sheet that can be handled easily, since the brittle prepolymer does not leave the surface sticky. As many as a dozen or more layers of impregnated paper are laid up. For decorative purposes, a printed rag or cloth paper is put on top and covered with a translucent paper layer. The entire assembly is heated between smooth plates in a high-pressure press to carry out the thermosetting reaction that binds the sheets together into a strong, solvent-resistant, heat-resistant surfacing material. For use in tabletops, countertops, and so on, the laminate, which is only about 1.5 mm thick, can be glued to a plywood base.

Numerous standardized classes of industrial laminates are available. They vary in the reinforcement and in the binder. Besides paper, the reinforcement might be woven or knitted cloth, most often cotton or glass but possibly wool, asbestos, nylon, or rayon. The binder can be phenolic, melamine, polyester, epoxy, or any one of a variety of thermosetting resins. One of the more important industrial applications of such laminates is as the base material for printed circuits.

The impregnated sheet materials can be wound on themselves or on a mandrel and heated in a press with curves surfaces to make rods or tubes. In any case, ultimate use of the laminates often involves machining operations. For this reason the ease with which a particular laminate can be sawed, drilled, or punched often has a bearing on its suitability for a given application.

Sandwich construction usually connotes assembly with one material for both surfaces and a second material in between. A core of foamed polymer covered by plywood or metal layers is very stiff without being very heavy and is a good insulator. Plywood surface with a foamed polystyrene core is available in 4- by 8- and 4- by 10-ft panels for quick assembly into small housing units. Surface and core can be varied to impart abrasion resistance, weathering resistance, flame resistance, or radiation resistance for particular applications.

When a thin layer of a tough thermoplastic such as plasticized poly(vinyl chloride) is laminated to a metal sheet, the bond may be strong enough that the laminate can be punched, cut, or shaped without parting the two layers. A pattern can be printed or embossed on the plastic before subsequent fabrication. Appliance cabinets and typewriter cases have been produced by shaping operations carried out on such materials.

Laminates can be produced during extrusion. A multilayered laminate made using rotating dies has already been mentioned. Coextruded film and sheet may combine two or three different polymers in a single product. Adhesion is enhanced by cooling the laminate directly from the melt rather than in a separate operation after the film components have been formed and cooled separately. In one form, flows from individual extruders are combined in a *flowblock* and then conveyed to a single manifold die. Because laminar flow must be maintained, all the polymer streams have to have about the same viscosity. Systems with up to seven layers are reported to be in operation [39]. There are various reasons for wanting multilayer films and sheets. A chemically resistant sheet can be combined with a good barrier to oxygen or water diffusion. A decorative, glossy sheet can be placed over a tough, strong material. Not all combinations adhere equally well, so there are limits to the design of such structures. One commercial example is a four-layered sheet (ABS, polyethylene, polystyrene, and rubber-modified polystyrene) for butter and margarine packages. Another is an ABS, high-impact-polystyrene sheet, which can be thermoformed (see Sec. 12.6, Sheet Forming) to make the inside door and food compartment of a refrigerator [39].

12.6 THREE–DIMENSIONAL PROCESSES

Molding

In almost all molding operations, some pressure and usually heat is applied to make the polymer flow into a preferred form. Dimensional stability then is conferred by cooling or further heating. Thermoplastic materials have to be cooled below T_m or T_g before being removed from the mold. This means a sequence of heating, pressurized flow, and pressurized cooling. Thermoset materials and rubbers have to be heated long enough for some chemical reaction to occur that brings about a stable, network structure. The sequence here is heating, pressurized flow, and pressurized reaction. Although some thermosets are cooled before ejection from the mold, most can be ejected hot.

Compression molding of combinations of shellac, straw, and various waxes and pitches goes back at least to the 1860s. The simplest compression mold is the flash mold. The only pressure on the material remaining in the mold when it is closed (Fig. 12.36) results from the high viscosity of the melt, which does not allow it to escape. Since most rubbers have high melt viscosities, especially when filled, the flash mold is widely used for mechanical goods such as gaskets and grommets and also for shoe heels, tub and flask stoppers, doormats, and other familiar items.

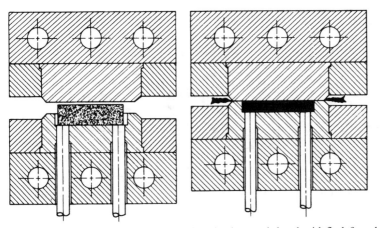

Figure 12.36 Flash mold: open, with preform in place, and closed, with flash forced out between mold halves [40]

Most thermoset prepolymers (phenolic and melamine, for example) are brittle and powdery at room temperature but have a low viscosity when heated, although the viscosity rapidly increases as cross-linking occurs. A primitive *positive-pressure* mold can be made from a ring and two plates (Fig. 12.37). The final thickness of the molded disc varies with the amount of polymer charged initially. An *internally landed* semipositive mold (Fig. 12.38) exerts more pressure on the melt than the flash mold, but the final dimensions are controlled by the meeting of the plunger and a land, or shelf, on the cavity.

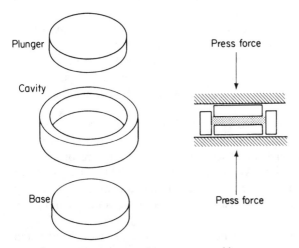

Figure 12.37 Elementary positive-pressure mold

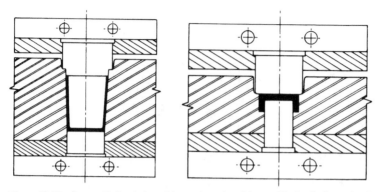

Figure 12.38 Internally landed positive and semipositive molds (both shown in closed position) [40]

In either of the compression molds, the prepolymer (molding powder) is melted and forced into the cavity simultaneously. If the polymer is being molded around a metal insert such as an electrical lug, the prepolymer may resist flow initially and exert a high force on the insert. In such a case it is desirable to melt the polymer in a separate operation and then inject it into the cavity. Because only seconds are available, both operations are carried out in the same equipment by *transfer molding.* The prepolymer is put into a pot in the mold (Fig. 12.39), from which it flows under the force of the plunger only after it has become hot enough to be rather low in viscosity. The material then is pushed into the cavity where it is held for the thermosetting reaction. When the mold is disassembled, note that the part is pushed out of the mold by *ejector pins* that operate automatically. The *sprue* is withdrawn, because of a slight undercut section, along with the *cull,* the uninjected remainder. The *gate* is the point where the sprue meets the *runners* that carry resin from the sprue bushing to the individual cavities.

Because compression molding of a thermoplastic resin involves heating and then cooling the mold, most such materials are more efficiently molded in the injection machines described next. One exception is the production of high-quality phonograph records, usually of compounded poly(vinyl chloride). The thin section, coupled with preheating of resin "biscuits," allows a relatively fast heating, forming, cooling cycle.

Preformed biscuits may be heated to 150 ± 5°C on a steam table adjacent to the molding machine. The molding press is a flash mold that resembles a waffle iron. The faces of the mold are nickel negatives of an original disc recording that have been made by electrodeposition. At 150 to 170°C and 1800 to 2000 psi (12 to 14 MPa), the record is formed and cooled with a total cycle time of only 40 s. Less expensive records such as children's 7-in discs may be injection-molded from polystyrene.

Injection molding accomplishes the cyclic operation by carrying out each step in a separate zone of the same apparatus. A plunger (also called a ram or piston)

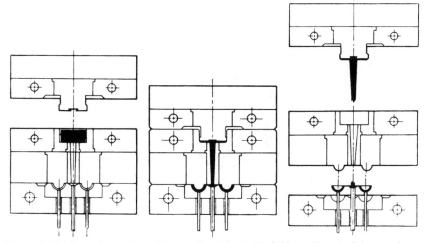

Figure 12.39 Typical transfer-mold operation. Material is fed into the pot of the transfer mold and then forced under pressure when hot through an orifice and into a closed mold. After the polymer has formed, the part is lifted by ejector pins. The sprue remains with the cull in the pot [40]

pushes pellets, which fall from the hopper into the barrel when the plunger is withdrawn (Fig. 12.40) into a heated zone. By spreading the polymer around a *torpedo* into a thin film, a rapid heating is possible. The already molten polymer displaced by this new material is pushed forward through the *nozzle,* the sprue bushing, past the gate, down the runner, and into a cooled cavity. The thermoplastic must be cooled under pressure below T_m or T_g before the mold is opened and the part is ejected. The plunger is withdrawn, new pellets drop down, the mold is clamped shut, and the entire cycle is repeated. Mold pressures of 8000 to 40,000 psi (55 to 275 MPa) and cycles as low as 15 s are achieved on some machines. Injection-molding machines are rated by their capacity to mold polystyrene in a single shot. A 2-oz machine can heat and push 2 oz of general-purpose polystyrene into a mold. Plunger diameter, plunger travel, and heating capacity all are involved. Although the plunger-type machines are inherently simple, the limited heating rate has caused a decline in popularity, so that most machines being sold today in the United States are of the reciprocating-screw type. By virtue of a check valve on the forward end of the screw (Fig. 12.41), it can push forward without rotation and act just as the plunger in an ordinary injection-molding machine. The screw may remain forward while the melt cools or, in the case of rubber, the polymer in a hot mold cross-links. The screw then turns and threads its way back to the rear of the barrel. The check valve is open during rotation, allowing molten polymer to flow around the screw. It is in this part of the cycle that the screw is valuable, because it increases heat transfer at the walls and also does considerable heating by the conversion of mechanical energy into heat in an adiabatic system. When the

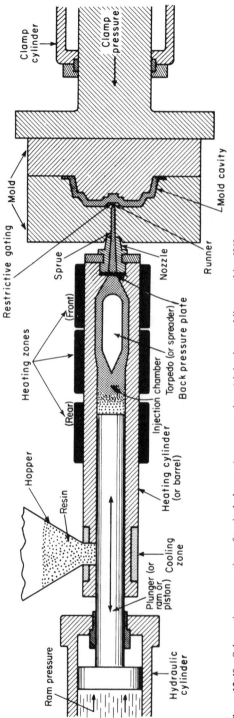

Figure 12.40 Schematic cross section of typical plunger (or ram or piston) injection molding machine [20]

476

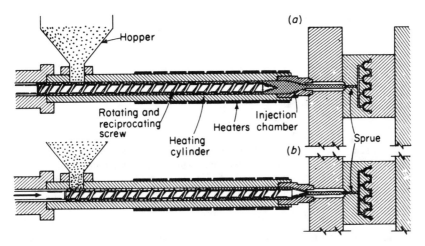

Figure 12.41 Cross section of a typical screw injection molding machine, showing the screw in the retracted (*a*) and forward (*b*) position [20]

finished part is ejected from the mold, the entire cycle can be repeated. Another advantage of the screw machine is its mixing and homogenizing action. Another way of taking advantage of the screw feature is by *preplasticizing* the melt with a screw arrangement which then feeds to a barrel from which the molten polymer can be injected by a ram (Fig. 12.42).

As with extrusion, injection molding of thermoset resins is difficult but possible. In effect, a kind of automated transfer molding is used. Injection molding of rubber (with a heated mold and ejection of hot parts) gives faster cycles and pieces with less porosity than flash compression molding. A side benefit is that rubber molded and cross-linked at a high pressure has built in residual stresses that in-

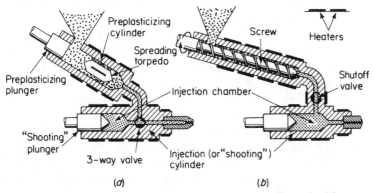

Figure 12.42 Drawings of a plunger-type (*a*) and screw-type (*b*) preplasticizer atop plunger-type injection molding machines [20]

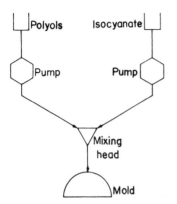

Figure 12.43 Reaction injection molding. Polyols and isocyanates are pumped at high pressure from storage reservoirs into the mixing head. The mixed streams are then injected into the mold at a much lower pressure

crease resilience. A cross-linked polybutadiene rubber ball will bounce back up to about 85% of its original height when molded at pressures of less than 1000 psi (7 MPa). However, molding at about 10 times that pressure can increase the bounce to as much as 92%. Since this particular property is not of great importance except to toy manufacturers, the more often cited benefits of injection molding or rubber are the faster cycle time and lower scrap level (less flash, sprues, etc.) compared to compression molding.

The rapidity with which isocyanate-polyol reactions take place has been used to advantage in reaction injection molding (RIM). In the typical process (Fig. 12.43), two components are mixed by high-pressure injection into a special chamber and injected almost immediately into a closed, low-pressure mold. The finished parts can be cellular or solid depending on the amount of material injected and on the formulation used. In the automobile industry, the need for light-weight, cushioning parts has been met by RIM. Front and rear bumpers and the "fascia" (face part, which can include grill and headlight housings) are made as single units.

The mixing head is the critical point in the system. Volume in the mixing chamber is kept small (1 to 5 cm³), and impingement pressures for mixing are 100 to 200 atm. There is a self-cleaning feature so that when one mold is filled, the head will remain clear until another mold is ready. Mold filling may take 2 to 3 s at mold pressures of only 3 to 4 atm. Typical conditions are summarized in Table 12.8. A single metering pump may feed up to 12 mixing heads. In another modification called liquid injection molding (LIM), the entire shot for a mold is mixed in a chamber before injection so that it resembles transfer molding.

Rigid structural foams with densities about 0.5 g/cm³ have been used for furniture and for housings for electronic equipment. Low-modulus elastomers such as the material described in Table 12.8 have been used for bumpers. With two mixing machines and eight molds, a production rate can be maintained of 1400 6-kg bumpers per day. Each cycle involves a 3-s injection time and a 2-min demolding time [42]. The addition of milled glass or mineral fillers to RIM formulations has led to high-modulus materials that may find use as automobile fenders.

Table 12.8 Typical conditions for reaction injection molding [41]

Formulation	Parts by weight
Resin stream:	
Polyols	150
Dibutyltindilaurate (catalyst)	0.045
Amine cocatalyst	0.6
Isocyanate stream:	
Polyfunctional isocyanate	96
Blowing agent (Freon type)	3
Methylene chloride	1

Processing parameter	Condition
Resin temperature	54°C
Isocyanate temperature	21°C
Mold temperature	57 ± 3°C
Resin pressure	120 atm
Isocyanate pressure	100 atm
Throughput	1.1 kg/s
Demolding time	1 to 2 min
Post-cure	1 h at 120°C

Physical property	Value
Density	0.99 g/cm^3
Hardness, Shore D	51
Tensile strength	18.6 MPa (2.7 × 10^3 psi)
Elongation at break	175%
Flexural modulus at	
-29°C	830 MPa (120 × 10^3 psi)
$+21$°C	150 MPa (22 × 10^3 psi)
$+65$°C	70 MPa (10 × 10^3 psi)

Molding with little or no pressure is practical if the melt viscosity is very low so that the polymer can be *cast* (poured) into a mold and allowed to set by reaction or cooling. Most polymers with molecular weights low enough to be pourable have poor physical strength. Paraffin wax is a good example of a kind of polyethylene with low enough melt viscosity to be poured into a mold. When a cotton string insert is included, such a casting operation produces a candle. Since mold casting of candles goes back at least to colonial times, it ranks as one of the more antique fabricating methods still in use. Casting a monomer in a mold to make a sheet of acrylic plastic was described in Sec. 5.2. Some other polymers that are cast are the epoxies and the silicones. A highly filled epoxy resin with good cured strength and abrasion resistance can be cast inexpensively to make a metal stamping die. One takes a pattern (same dimensions as final piece), places it in a box, and pours in the resin-hardener mixture (usually with a filler), which sets to a network polymer

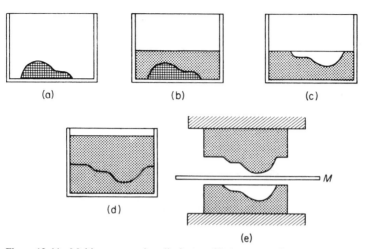

Figure 12.44 Making a stamping die from a filled epoxy resin: (*a*) pattern placed in box; (*b*) resin added and hardened to make lower half of die; (*c*) pattern removed, lower half of die inverted and coated with release agent; (*d*) resin added and hardened to make upper half of die; (*e*) two halves of die mounted in press to form metal sheet, *M*

on heating (Fig. 12.44). After a mold-release compound is sprayed on the surface, another layer is poured and set. The two pourings are attached to opposite platens of a stamping press and form a matched mold for the forming of metal sheets. Compression and injection molds can be made in the same manner.

It is possible to produce three-dimensional prototype parts using photopolymerization. In one system [43], a UV laser light source is moved in two dimensions (*x* and *y*) over a pool of photosensitive prepolymer in a programmed fashion, producing a thin film of cross-linked polymer on the surface wherever the light strikes. A fresh layer of prepolymer is added and the next "slice" of the part is cross-linked selectively. By repeating this operation a number of times with increasing amounts of prepolymer, and with careful control of the cross-linking in a thin surface layer each time (by limiting the amount of transmitted light through the top layer), a final part can be made with a thickness (*z* axis) of 10 to 20 cm. Pouring off the un-cross-linked polymer leaves a part which can be quite complex. The prototyping process has been used in automotive, electronics, aerospace, medical, and consumer-product companies.

Plastisols represent a special heat-convertible liquid category that is made possible by the unique properties of poly(vinyl chloride). The polymer is quite insoluble in a large class of high-boiling liquids including di-2-ethylhexylphthalate. It is possible to disperse finely divided polymer particles in this liquid and have a stable, free-flowing suspension (a *plastisol*) at a weight ratio of 3:5 or higher, liquid to polymer. The polymer particles preferably have been recovered from an emulsion polymerization to aid in the plastisol formation. Such polymers are sometimes called plastisol, dispersion, paste, or stir-in resins. When the plastisol is heated to

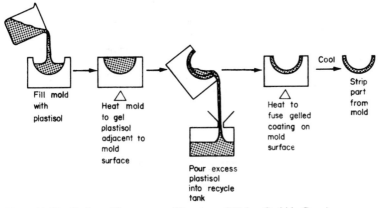

Figure 12.45 Slush molding process. (Courtesy of Union Carbide Corp.)

about 175°C (the temperature varies with resin and plasticizer), the resin rapidly dissolves in the plasticizer. Only a small amount of plasticizer is volatilized at this temperature. On cooling, the solution of resin in plasticizer remains stable indefinitely. It will be recalled that the major effect of a plasticizer is to change the T_g of a system. Examples were given in Sec. 3.6. The utility of such a liquid, castable, heat-convertible system is obvious. Although other polymers have been proposed as plastisol resins, only polymers and copolymers of vinyl chloride seem to have the combination of low-temperature insolubility, high-temperature solubility, and high tensile strength when plasticized that is needed [44, 45]. While hundreds of liquids are sold as plasticizers, esters of phthalic and adipic acid, along with epoxidized drying oils, chlorinated waxes, and esters of phosphoric acid, dominate the market. Primary consideration may be given to the final properties of the system such as T_g, modulus, strength, stability, and odor. On the other hand, the viscosity of the plastisol initially and on aging often becomes a controlling factor in plasticizer selection. The problems of physical and chemical stability (plasticizer migration and thermal breakdown) are discussed in Chap. 11.

In the laboratory the resin can be dispersed in the liquid by stirring in a beaker with a spatula. Industrially, a dough mixer adapted from the baking industry may be used. The unmodified plastisol may have a viscosity of 1 to 100 poise (0.1 to 10 Pa·s). To get the best dispersion, the mixture may be refined on a three-roll mill or some other high-shear device. Because of the rather high viscosity and surfactant present, air bubbles are trapped. An arrangement for cyclic deaeration of plastisols is often used.

Some articles, such as fishing lures, washers, and stoppers, are made by casting in open molds. A sizable application involves *slush molding*. A typical produce made by this technique is an overshoe. There is a great resemblance to fluidized-bed casting (Sec. 12.4). A preheated, hollow metal mold is filled with plastisol and then inverted (Fig. 12.45). The hot wall solvates some resin so that on inversion of

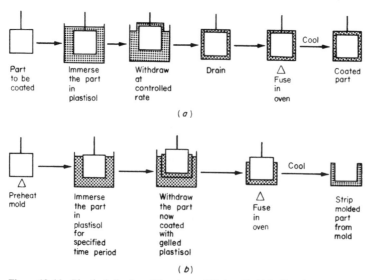

(a)

(b)

Figure 12.46 Plastisol dipping. (Courtesy of Union Carbide Corp.)

the mold the core flows out but the skin remains. Further heating completes the fusion process. Some cooling is necessary before the boot can be stripped from the mold. The mold might be split to facilitate removal of the molded object, which now bears on its outer side the pattern from the inner side of the metal mold. Squeezable dolls or parts of dolls are most often made by slush molding.

Even closer to the fluidized-bed technique is the process of *dipping* (Fig. 12.46). Wire dish drainers, coat hangers, and other industrial and household metal items can be coated with a thick layer of flexible vinyl plastic by simple dipping and fusion. A more elaborate artifact is a plastic glove, which could be produced either by dipping a hand-shaped mandrel or by slush molding in a hollow mold for the same shape. Several variations of the plastisol have been developed to alter the properties of the liquid as well as the final solid.

The viscosity of the plastisol increases as the ratio of resin to plasticizer increases. Where a thin coating of rigid product is required and the plasticizer content is so low that the plastisol will not flow easily, a nonsolvating diluent can be added. The process of solidification is more complex now, since the diluent (usually an aliphatic or naphthenic liquid in the mineral spirits range) must evaporate before the fusion is attempted. *Organosols,* as these diluted plastisols are termed, are used to advantage in dipping processes as well as in impregnating cloth or rope with vinyl polymer. There are some applications where it is desirable to have an infinite viscosity at low shear stresses. *Plastigels* are made by adding to the resin-plasticizer suspension a metallic soap or finely divided filler as a gelling agent. Aluminum stearate is an effective gelling agent, especially if a petroleum oil is being used as a secondary (extending) plasticizer. The suspension of resin in gelled plasticizer can be cold-molded, placed on a pan, and heated to fusion without discernible flow. The whole operation resembles baking cookies, with molding and baking being done in two separate steps.

There are two routes to making fluid plastisols that become rigid when fused. One is to extend the easily dispersed emulsion process resin with a larger-particle-size, dense, suspension-process resin. The latter may be 10 to 50 μm in diameter compared with the ultimate particles of the former, which are in the 0.1- to 1-μm-diameter range. The larger particles remain suspended because of the thickening action of the small ones. A slightly longer fusion time may be required to get complete solvation. Very rigid, glassy solutions can be made when the plasticizer can be polymerized during or right after fusion. For example, a *rigidsol* might be made from 100 parts of poly(vinyl chloride) and 100 of triethylene glycol dimethacrylate, together with one part of di-*tert*-butyl peroxide. The viscosity of the mixture is only 3 poise compared to 24 poise for a phthalate-plasticized material. However, on fusion for 10 min at 175°C, the resin solvates and the plasticizer forms a network almost simultaneously. Back at room temperature, the baked product is a hard, rigid glass with a flexural modulus of over 2.5×10^5 psi (1.7 GPa).

$$(CH_2 = \underset{\underset{O}{\overset{\|}{\underset{}{C}}}}{\overset{\overset{CH_3}{|}}{C}} - C - O - CH_2 - CH_2 - O - CH_2 \,)_2$$

Triethylene glycol dimethacrylate

Slush molding has been mentioned as one method of producing a hollow object. Several others have been developed. *Rotational* molding can be used with plastisols, but it can also be used with finely divided powders of polyethylene and other polymers (Fig. 12.47). In this case the hollow, split mold is charged only with enough material to make the final object. The mold is closed and heated while being rotated in an eccentric manner so that all surfaces on the inside of the mold become coated with a fused material. Cooling, opening the mold, and removing the hollow object go rather rapidly, since heat transfer is to a thin skin through a metal wall. Large containers, suitcases, and display novelties are inexpensively produced from a lightweight mold that undergoes no great stress during the process.

The need for a rapid production of bottles with uniform thickness for soft drinks, milk, and myriad household products is best met by *blow molding.* A tube of molten polymer, the *parison,* is sealed at both ends and held in place while two halves of a hollow mold surround it. Air is injected into the parison, which blows up as if it were a rubber balloon. When the polymer surface meets the cold metal wall of the mold, it is cooled rapidly below T_g or T_m. Once the product is dimensionally stable, the mold is opened, the bottle is ejected, a new parison is introduced, the mold is closed, and so on. The parison can be produced continuously

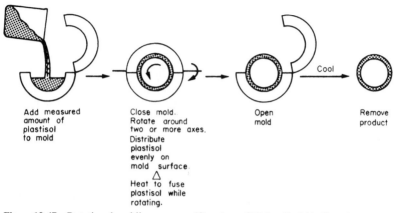

Add measured amount of plastisol to mold

Close mold. Rotate around two or more axes. Distribute plastisol evenly on mold surface. △ Heat to fuse plastisol while rotating.

Open mold

Cool

Remove product

Figure 12.47 Rotational molding process. (Courtesy of Union Carbide Corp.)

from a screw extruder and converted to a bottle on a rotating platform (Fig. 12.48). In the ram-extrusion method, the parison is formed in a cyclic manner by a plunger forcing a charge out from an accumulated molten mass as in the preplasticizer injection-molding machine (Fig. 12.49). In the production of large bottles, efficiency demands a rather uniform wall thickness and no dangerous thin spots. Inherently, the weight of the bottom of a parison that is suspended vertically from a machine will cause the wall to be thinner toward the top. Unless corrected, a long parison may give a bottle with a wall thickness near the bottom twice that near the top. One way to correct the situation is to program the position of a mandrel around which the parison is being extruded so that the die opening increases during

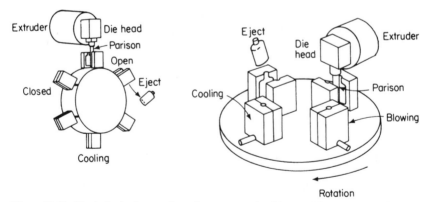

Figure 12.48 Vertical wheel type of continuous extrusion blow molding machine; molds are carried to the die head by rotation of the wheel. In a horizontal table type, all steps of the process are carried out simultaneously also [46]

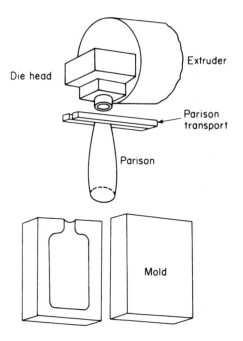

Die head

Extruder

Parison
transport

Parison

Mold

Figure 12.49 In this continuous-extrusion blow molding system, a parison transfer device is used to lower the extruded parison from the die head into the waiting mold. The transport arm cuts and holds the parison during the transfer operation [46]

a shot (Fig. 12.50). The parison now has a thicker wall toward the top, which will be evened out as it hangs waiting a second or two to be molded. Cycles are fast. Small plastic vials can be molded 10 at a time from one machine with multiple cavities using a cycle time of 10 s. For many food-packaging applications, multilayered parisons are *coextruded* in order to produce multilayered bottles. As many as seven layers may be sandwiched together to get the proper combination of appearance, strength, and barrier properties for foods such as ketchup or salad dressings. Even fuel tanks for automobiles are blow-molded from high-density polyethylene. The inner layer of the tank may be treated with SO_3 in order to enhance barrier properties. One trade source estimated that 35% of automobiles being produced in 1995 were equipped with plastic fuel tanks [49].

In *stretch-blow* molding, biaxial orientation in the bottle is increased by longitudinal stretching during radial blowing. This is accomplished by an internal rod (Fig. 12.51). Better clarity, increased impact strength, and improved barrier properties (water and gas diffusion) are claimed. Beverage bottles made from poly(ethylene terephthalate) have used the technique.

Large moldings of glass fiber-reinforced polyester often are made using low pressures. A number of variations exist. A typical resin formulation might contain 30 to 40% styrene monomer, with the balance being an unsaturated polyester (based on maleic anhydride, for example). The glass reinforcement can be in chopped strands, mats of strands, woven cloth, or continuous filaments.

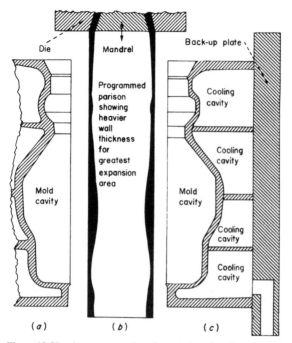

Figure 12.50 A programmed parison designed to fit a particular mold configuration [47]

For the production of a boat hull or an automobile body, layers of glass cloth or mat can be placed on a form and then impregnated by pouring "activated" resin from a pail. The activation usually consists of a peroxide initiator with an accelerator such as cobalt naphthenate. The resin is spread out with rollers and polymerization takes place after an induction period ("pot life") of an hour or so at room temperature. If production volume justifies the added cost, a matched-metal compression mold can be used to give a denser, stronger finished product. The hand layup may still be used. A more uniform product is obtained if the glass fibers are sprayed onto the form or if resin and glass are sprayed on simultaneously.

Another means of enhancing uniformity is to combine glass and resin beforehand. Sheet molding compound (SMC) is one approach [50]. When the liquid polyester (actually polyester plus styrene) is thickened by less than 1% of calcium or magnesium oxide, the gel has a yield stress and can be handled without dripping. SMC is made by combining thickened polyester, fillers, peroxides, and chopped glass fiber between layers of polyethylene. The mat can be produced continuously and uniformly. Molding requires stripping off the polyethylene film. The amount of SMC can be tailored to the requirements of the piece being molded. It is usually molded under some pressure and with heating.

Bulk molding compound (BMC) is a puttylike mixture of polyester and

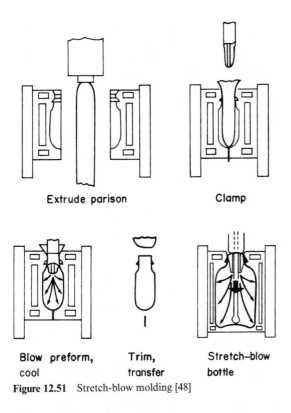

Extrude parison   Clamp

Blow preform,   Trim,   Stretch–blow
cool   transfer   bottle

Figure 12.51   Stretch-blow molding [48]

chopped fiber glass strands [51]. Fillers such as calcium carbonate are the factor that changes the liquid polyester to a putty that can be applied with a trowel.

Foams

Foamed polymeric materials have several inherent features that combine to make them economically important. Any foamed material is a good heat insulator by virtue of the low conductivity of the gas, usually air, contained in the system. In a rigid foam, the gas contained in the cells may not be exchanged with air for a long time. One advantage of fluorocarbons as expanding agents is their low thermal conductivity compared to air. For example, the thermal conductivities of some common gases in W/m°C at 27°C are: air, 0.0262; n-pentane, 0.0144; and CCl_2F_2, 0.0125. The reluctance of some manufacturers to replace the fluorocarbons in rigid insulation is understandable.

A glassy polymer, when foamed, will have a much higher ratio of flexural modulus to density than when unfoamed. A beam of foam deflects less than an unfoamed beam of the same length, width, and weight under the same load. Finally, any foam will have energy-storing or -dissipating capacity operating through a greater distance than the unfoamed material. The resilient foam made when the polymer is rubbery can store most of the energy reversibly and makes a good cushion for upholstery and packaging applications. A compressible foam that breaks down on impact can absorb a great deal of energy and is widely used for packaging. The mere bulk of highly foamed polymers, along with good machinability, makes them economical for some decorative purposes where massiveness without great weight is demanded. One example is foamed polystyrene used as an artistic medium. A block of foam can be sculptured much as the artist might use clay for three-dimensional figures.

Foams can be classified according to several schemes. The nature of the cells is one. Each cell may be completely enclosed by a membrane, or it may be interconnected with its neighbors. *Closed-cell* foams typically are produced in processes where some pressure is maintained during the cell-formation process. *Open-cell* foams often are made during a free expansion. Most processes produce both kinds. In some applications there is an advantage to converting all cells to the open variety. A closed cell tends to store energy reversibly, as if it were a small rubber balloon. In cushioning applications it is desirable to have compression cause the flow of air from cell to cell as if they were a series of orifices. This dissipates energy that is not entirely regained on unloading. An open-cell foam acts as a sponge, soaking up liquid by capillary action. A closed-cell foam makes a better buoy or life jacket because the cells do not fill with liquid.

Foams differ in the process conditions under which they are generated. A material that requires a high temperature or an external pressure for foam formation cannot be easily *foamed in place,* as could a system that can be mixed and poured at room temperature. And, of course, the polymer from which the foam is derived can be a thermoplastic or a thermoset, rubbery or glassy.

Foamed polyurethanes, rubber, and poly(vinyl chloride) dominate the upholstery market, so it is convenient to consider the rather different processes that are used for each. The earlier consideration of polyurethane foams (Sec. 4.7) showed only diisocyanate, water, and polyhydroxy compound as the ingredients. An actual foam formulation is more complex (Table 12.9). A combination of a diol and a triol is used to give a partially cross-linked structure in the final form. The diisocyanate is a mixture if two isomers that differ slightly in reaction rate. The mixture is less expensive than either pure compound. The isocyanate group reacts slowly with the secondary hydroxyl groups that these polymers have on one end. The tin compound catalyzes the reaction, especially in the presence of a small amount of tertiary amine. Note that the amine has no reactive hydrogen. The silicone surfactant is an essential ingredient that regulates the cell size, uniformity, and nature to a large extent, controlling the viscosity and surface tension of the cell membranes as they are stretched during foaming. The last ingredient is a low-boiling liquid (bp 23.8°C) that volatilizes when the urethane reaction starts. The "blowing agent," as this fluorinated hydrocarbon is termed, increases the total volume as indicated

Table 12.9 Urethane one-shot foam system [52]

Ingredient		Parts by weight
Poly(propylene oxide) $M_n = 2000$, 2 OH/molecule	$\begin{array}{c}CH_3\\|\\HO\text{-}(CH_2CHO)\text{-}H\end{array}$	35.5
Poly(propylene oxide) started on tri- functional alcohol, $M_n = 3000$, 3 OH/molecule	$\begin{array}{c}HO\text{—}R\text{—}OH\\|\\OH\end{array}$	35.5
Toluene diisocyanate (80% 2,4- and 20% 2,6-substituted)	$(C_6H_3)(CH_3)(NCO)_2$	26.0
Dibutyl tin dilaurate	$(C_4H_9)_2Sn(O_2C_{12}H_{23})_2$	0.3
Triethyl amine	$(CH_3\text{—}CH_2)_3N$	0.05
Water		1.85
Surfactant (silicone)		0.60
Trichloromonofluoromethane, CCl_3F		12
Final density = 1.4 lb/ft³ (2.0 lb/ft³ if CCl_3F omitted).		

without itself participating in the chemical reactions that are occurring. Concern for environmental air pollution by fluorocarbon vapors is an incentive to replace this particular ingredient with other volatile liquids.

The foaming reaction takes place so rapidly that large-scale production demands automatic mixing equipment. In one arrangement, mixing and reaction start in a traveling head that cycles back and forth to make a continuous slab (Fig. 12.52).

The urethane foam is a typical polymer system in that the number of variables that can significantly affect important properties is very large. The optimization of the relative quantities of only the specific ingredients in Table 12.9 is in itself a time-consuming, expensive exercise, even with the aid of a computer and multiple-regression analysis. We must add to that problem the possibilities of varying each

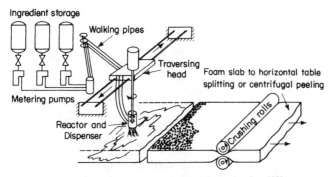

Figure 12.52 Continuous urethane foam slab production [53]

ingredient chemically. The polyol can vary in functionality, with as many as five or six hydroxyls per molecule. Although polyols based on propylene oxide dominate the field, other polyethers are used based on tetrahydrofuran, ethylene oxide, or styrene oxide. Polyesters are widely used in rigid foams and, to a certain extent, in flexible foams. Such polyesters may be made with an excess of diol or by growth

$$HO(CH_2)_4O[CO(CH_2)_4CO-O(CH_2)_4O]_nH$$
Polyester diol based on butanediol and adipic acid

$$(HO(CH_2)_5CO-O(CH_2)_5CO-O-CH_2)_2$$
Polyester diol based on caprolactone "grown" on ethylene glycol

of a hydroxy acid on a diol. The diisocyanate most often used in foams is the toluene derivative, but the ratio of isomers can be changed. The remaining ingredients present similar opportunities for exploration. Even a single company may offer a line of a half-dozen to a dozen surfactants, catalysts, blowing agents, or auxiliary materials. Water is not an essential ingredient. Some urethane foams are made with only the blowing agent. Another parameter that affects foam properties is the manner in which mixing takes place and the time scale of the process. All these things not only can change the physical appearance of the foam, cell size, size distribution, openness, and strength but also can encourage or diminish subsidiary chemical reactions depending on the concentration of reactants, the temperature, and viscosity of the system. Some groups that form in a typical foam, in addition to urethane and urea which were mentioned before, are the allophanate and biuret. It is seen that these represent secondary reactions of the previously formed groups that still contain an available hydrogen. However, the hydrogen is not as easily reacted as the hydrogen of an alcohol or primary amine. The groups occur less frequently than the urethane and urea and can probably be dissociated more readily under conditions of higher temperature or hydrolytic attack. Nonetheless, they are covalent cross-links and make a definite contribution to modulus and permanent set.

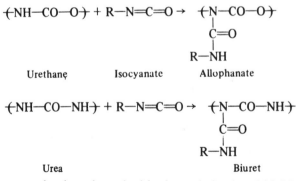

Rigid foams result when the polyol is short-chained and highly functional so that the distance between cross-links is small or when the polyol backbone contains bulky groups that raise the glass-transition temperature. Fillers such as fibers or finely divided silica can be used to stiffen a foam, but they are seldom used because they increase the density of the system. Polyisocyanates of high functionality can be employed to increase the cross-link density. Some of these are condensation products of the commonly used toluene diisocyanate.

One of the largest markets for flexible urethane foams is in upholstery. Danish Modern furniture, with its straight lines and flat cushions, is a large-scale consumer product eminently suited for using polyurethane foam. The foam is made in continuous loaves several feet in width and height and then sliced into slabs of the required thickness. Some other products are crash pads for automobiles and packaging. Weather stripping is made from a thin tape of foam about 6 mm $\times$ 3 mm that is attached to a door frame by means of paper tape that has adhesive on both surfaces. Rigid urethane foam can be foamed in place. One of the early commercial applications (1940s) was the filling of certain aircraft components with rigid foam. The advantages of the urethane foam are that it can be foamed in place, say in the rudder or aileron, it adheres well to metal surfaces (as a rule), it adds little weight, and it greatly reduces the flexing of the element on stressing. Hollow walls in chambers such as boxcars can be filled with foam to provide strength and good thermal insulation. Completely open-cell foams can be produced directly in the foaming process or by chemical or mechanical treatment of a previously formed material. These "reticulated" foams are widely used as air filters.

Although foamed rubber has many of the same properties as foamed urethanes, the foaming and the cross-linking of the polymer are accomplished in a radically different manner. To natural rubber latex is added a solution of a soap that yields a froth on beating. In addition, antioxidants, cross-linking agents, and a foam stabilizer are added as aqueous dispersions. Foaming can be done with an eggbeater type of wire whip, but most processes today use automatic mixing and foaming machines that combine agitation and aeration. The foam would collapse rapidly if it were not for the stabilizer, which usually is sodium silicofluoride, Na_2SiF_6. The hydrolysis of this salt yields a silica gel that increases the viscosity of (gels) the aqueous phase and prevents the foam from collapsing. The cross-linking agent has to act at 100°C or lower in order to react before the water is removed from the foam. One such combination is sulfur with the zinc salt of mercaptobenzothiazole:

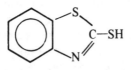

Mercaptobenzothiazole

Cross-linking (curing or vulcanization) takes place in 30 min at 100°C. In a large article such as a mattress, a metal mold is filled with the foam during the few minutes between the time foaming is complete and the silica gel has formed. The entire mold is then heated by steam at atmospheric pressure. After being removed from the mold, the foam may be dewatered by compression between rolls or centrifuging followed by drying with hot air in a tunnel dryer. Part of the natural latex can be replaced by a synthetic rubber such as the latex from emulsion copolymerization of styrene and butadiene. The overall composition of such a combination in Table 12.10 illustrates again the complexity of simple "model" systems in the real world.

Although poly(vinyl chloride) foams cannot be made as inexpensively as those from rubber or urethanes, their mechanical properties can sometimes be of value. Most vinyl foams are stronger and are torn less easily than rubber or urethane foams of comparable density and are less flammable. Two processes for making foamed poly(vinyl chloride) use plastisols (see Sec. 12.6, Molding). The first is a very generally applicable method that can be used whenever a system can be mixed at a lower temperature than the decomposition temperature of a *blowing agent*. Three important characteristics of a blowing agent are the decomposition temperature, the volume of gas generated per unit weight, and the nature of the decomposed residue. Several examples (Table 12.11) are especially recommended for vinyl plastisols [55]; in each case the gas generated is nitrogen. Carbon dioxide can be generated by heating some inorganic carbonates. To produce uniform cells, the blowing agent must be uniformly dispersed or dissolved, uniformly nucleated, and rapidly, smoothly decomposed over a narrow temperature range, which is matched to the attainment of a high viscosity or gelation of the polymer system. In plastisols, gelation involves the solvation of resin in plasticizer at 150 to 200°C, depending on the particular ingredients employed. If gelation occurs before gas release, large fissures or holes may form. Cells formed too soon before gelation may collapse and give a coarse, weak, spongy material. Closed cells result when the decomposition and gelation are carried out in a closed mold almost filled with plastisol. After the heating cycle, the article is cooled in the mold still under pressure until it is dimensionally stable (about 50°C). It expands slightly on removal from the mold because of the compressed gas it now contains. Final expansion is accomplished by heating the free article somewhat below the previous molding temperature. Buoys, life jackets, floats, and protective padding are some items made by this process.

The same blowing agents can be used in making foamed rubber. The step corresponding to fusion of plastisols is the cross-linking reaction to give a stable network (vulcanization). Some thermoplastics such as polyethylene and cellulose derivatives can also be foamed by blowing agents even though they do not undergo an increase in dimensional stability at an elevated temperature. As long as the polymer melt is viscous enough to slow down collapse of the cells and the system is cooled below its T_m or T_g, a reasonably uniform cell structure can be built in.

When the plastisol containing a blowing agent is spread on a substrate and fused without a second confining surface, the cell structure is usually open. Fabrics with a bonded layer of insulation are made this way. Extrusion can also be used

Table 12.10 Foamed rubber formulation [54]

Ingredient	Parts by weight
Styrene-butadiene latex (65% solids)	123
Potassium oleate	0.75
Natural rubber latex (60% solids)	33
Sulfur	2.25
Accelerators:	
Zinc diethyl dithiocarbamate	0.75
Zinc salt of mercaptobenzothiazole	1.0
Reaction product of ethyl chloride, formaldehyde	
and ammonia ("Trimene Base")	0.8
Antioxidant (phenolic)	0.75
Zinc oxide	3.0
Na_2SiF_6	2.5

with a variety of thermoplastic and rubber compounds. Foamed gaskets and weather stripping are made by decomposing the agent in the barrel of the extruder and allowing the extrudate to expand freely. The product is predominantly open-celled.

Plastisols can be used in a mechanical foaming process. A substantial amount of carbon dioxide will dissolve in the plasticizer phase when mixed at about 7 atm and below room temperature. Release of pressure through a nozzle yields a stream of foamed, unfused plastisol. In order to convert this to the stable, fused plasticized vinyl product, two innovations have been introduced. A gelling agent, an aluminum soap, is used to make the liquid plastisol into a highly pseudoplastic fluid, which flows easily at high stresses but is extremely viscous at low stresses. Even with the gelling agent, the cells would collapse on fusing the resin-plasticizer mixture in an oven that depended on slow heating by conduction through the foam, which is an excellent thermal insulator. Dielectric heating can be used to heat uniform cross sections and results in very rapid fusion throughout the foamed mass. The whole process of cooling and dissolving carbon dioxide in the plastisol using a scraped-surface heat exchanger, pressure release through a nozzle, gelation, and fusion can be carried out continuously, since the fusion can be accomplished in seconds by dielectric heating.

The foams based on urethanes, rubber, and vinyls discussed so far are flexible and useful for upholstery and packaging. Increasing the degree of cross-linking will make any one of them more rigid. Of the three, only the urethanes are important in the commercial applications of rigid foams. Another way to achieve rigidity is to use a polymer that is below T_g at normal use temperatures. The various forms of polystyrene foams illustrate techniques that can be applied to other thermoplastic polymers.

A solution of polystyrene in methyl chloride can be heated well above the

Table 12.11 Commercial foaming (blowing) agents [55]

Chemical description	Structure	Decomposition temperature in air, °C	Decomposition range in plastics, °C	Gas yield ml(STP)/g	Literature reference
A. Azo compounds					
Azobisformamide (Azodicarbonamide)	$H_2N-CO-N=N-CO-NH_2$	195–200	160–200	220	BIOS Final Report 1150, 23 German Patent 871,835
Azobisisobutyronitrile	$NC-C(CH_3)_2-N=N-C(CH_3)_2-CN$	115	90–115	130	German Plastics Practice, DeBell-Richardson p. 456 (1946) German Patent 899,414
Diazoaminobenzene	C₆H₅–NH–N=N–C₆H₅	103	95–100	115	U.S. Patent 2,299,593
B. N-Nitrosocompounds					
N,N'-Dimethyl-N,N'-dinitrosoterephthalamide	$H_3C-N(NO)-OC-C_6H_4-CO-N(NO)-CH_3$	105	90–105	126*	U.S. Patent 2,754,326
N,N'-Dinitrosopentamethylenetetramine	(structure)	195	130–190	265†	U.S. Patent 2,491,709

C. Sulfonyl hydrazides

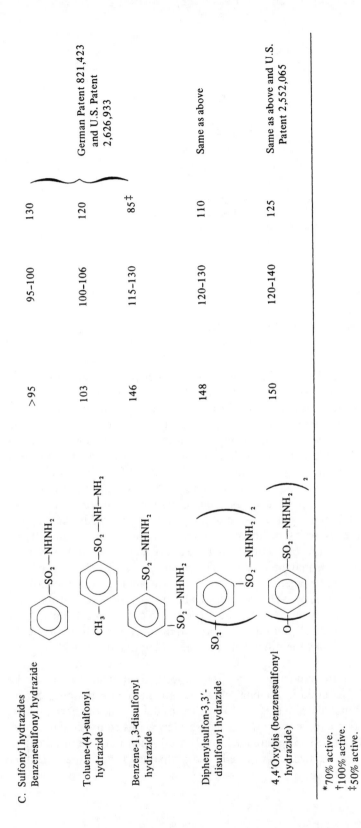

Benzenesulfonyl hydrazide	>95	95–100	130	German Patent 821,423 and U.S. Patent 2,626,933
Toluene-(4)-sulfonyl hydrazide	103	100–106	120	
Benzene-1,3-disulfonyl hydrazide	146	115–130	85‡	
Diphenylsulfon-3,3'-disulfonyl hydrazide	148	120–130	110	Same as above
4,4'Oxybis(benzenesulfonyl hydrazide)	150	120–140	125	Same as above and U.S. Patent 2,552,065

*70% active.
†100% active.
‡50% active.

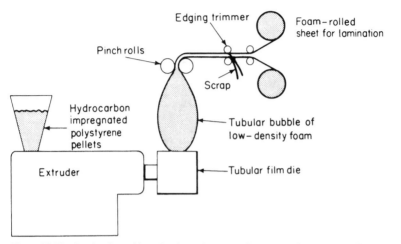

Figure 12.53 Production of low-density polystyrene foam sheet from tubular film die [57]

normal boiling point as long as the system is kept under pressure as in the barrel of an extruder [56]. Upon issuing from the extruder, the solvent evaporates, simultaneously foaming the plastic and cooling it below the T_g. Foaming and dimensional stabilization thus are carried out in one step. Rather extensive facilities are required to mix and convey the concentrated solution and to extrude the foamed "logs." Because it is uneconomical to ship low-density foams a great distance, a modification of the solution process has been introduced [57]. Pellets chopped from an ordinary melt extruder or beads from a suspension polymerization are impregnated with a hydrocarbon such as pentane. As long as the beads or pellets are stored below 20°C, the vapor pressure of the pentane dissolved in the polymer is low enough that pressures of less than 1 atm build up in a closed container. Even so, manufacturers do not recommend storing the impregnated material more than a few months. The pellets may be used in an extrusion process to give logs or thin sheets. Usually a nucleating agent is added, such as citric acid or sodium bicarbonate [58], to control the cell size. This process has been especially successful in producing thin sheet that can subsequently be formed into containers. Tubular blow extrusion is used (Fig. 12.53). Egg cartons and cold drink cups can be made by the thermoforming techniques mentioned in the next section starting with foamed sheet.

The beads often are molded in a two-step process. Prefoaming almost to the density that will be required in the final object is similar to corn-popping. Steam heat is used in an agitated drum (Fig. 12.54) with a residence time of a few minutes. The expanded beads can be stored for as much as a few weeks before final molding. In a typical mold (Fig. 12.55) prefoamed beads are loaded into the mold cavity. After the mold is closed, steam is injected. The prefoamed beads expand further

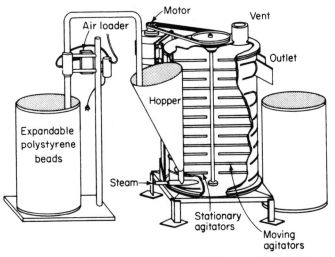

Figure 12.54 Prefoaming of polystyrene beads in a steam-heated device [59]

and fuse together as the temperature exceeds T_g. The article must be cooled in the mold to a temperature below T_g before removal. Packages shaped to fit their contents, picnic coolers, drinking cups, toys and small sailboats, are common consumer items fabricated from beads. The thin-walled cups demand a small bead for easy flow into the molding cavity.

Some typical properties of rigid foams are listed in Table 12.12. Comparing the two figures for the polystyrene, it can be seen that convection within the larger cells actually causes the overall thermal conductivity to be higher at the lower density. On the other hand, for polystyrene foam densities in the 30- to 100-kg/m³ range, thermal conductivity does not change much. The lower thermal conductivity for the polyurethane foam vs. the polystyrene is due mainly to the lower thermal conductivity of the gas in the cells. For most foams, an increase in temperature of

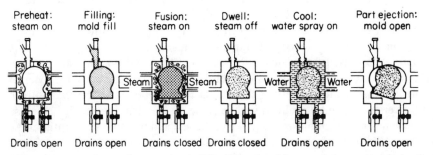

Figure 12.55 Molding polystyrene beads [60]. (Reproduced by permission, copyright by the Dow Chemical Co., 1966.)

Table 12.12 Some typical properties of expanded plastics [61]

	Polystyrene		Polyurethane	
Blowing agents	Beads expanded with pentane		Expanded using fluorocarbon	
Density, kg/m³	15	30	20	45
Tensile strength, kPa	175	350	100	650
Compressive stress at 10% comp., kPa	85	250	100	400
Thermal conductivity, W/m°C	0.040	0.035	0.016	0.025

about 30°C is enough to increase thermal conductivity by about 10%. This is due almost entirely to the normal temperature coefficient of thermal conductivity for gases.

Foams of phenol-formaldehyde resins can be made from a dispersion of a volatile diluent (isopropyl ether dispersed with the aid of a surfactant) in an aqueous solution of an incomplete phenol-formaldehyde reaction product [62]. Addition of an acid catalyst such as hydrochloric or sulfuric acid causes further condensation of phenol and formaldehyde to give a dimensionally stable, network structure. At the same time the heat of reaction volatilizes the diluent, yielding a foam. The foaming can be done in place. Phenolic foams are used as heat-stable, flame-retardant, thermal insulation.

Another way to make a foam is to embed hollow spheres of glass or phenolic resin in another polymer. Usually the spheres are mixed with low-viscosity liquids that can be converted to solids. The spheres are not easily dispersed without breaking when extremely viscous rubbers and melts are used.

The structural foam process gets its name from the application for its product rather than the mechanics of the process itself [63]. In a manner directly opposite to the vented extruder (Fig. 12.18), a blowing agent, often nitrogen, is injected into the melt in the extruder (Fig. 12.56). Polymer melt that has already been injected with gas goes to the accumulator. When a sufficient charge has accumulated it is transferred into the mold. The melt foams and fills the mold at a relatively low pressure (200 to 400 psi, that is, 1.3 to 2.6 MPa) compared to the much higher pressure in the accumulator. The lower pressures in the molds make the molds cheaper than those used for conventional injection molding. However, the cycle times are longer because the foam, being a good insulator, takes longer to cool. A major market for structural foams is in furniture. ABS and high-impact polystyrene are used to replace wood in seat panels, legs, chairs, and even some exterior panels for pianos. A wide variety of thermoplastics can be used. The trade-offs in deflection versus strength are illustrated in Table 12.13. Structural foams can also be made using a chemical blowing agent like those listed in Table 12.11 rather than an inert gas. A change in pressure or temperature on entering the mold triggers gas formation.

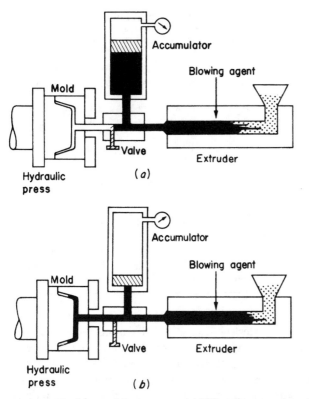

Figure 12.56 Structural foam process. (*a*) Filling the accumulator. The blowing agent—usually nitrogen—is injected into the melt in the extruder. The melt is passed into the accumulator, where it is maintained at a pressure and temperature high enough to prevent foaming until the predetermined charge is collected. (*b*) Filling the mold. The accumulator ram then injects the charge into the mold, where the reduced pressure allows the gas to foam the resin and drive it throughout the mold

Sheet Forming

Often it is more economical to fabricate an article from a previously formed sheet than to start from raw pellets or slabs. In the case of thermoplastic materials, the sheet is heated to a temperature at which it will deform under stress but not flow easily. This corresponds to the rubbery plateau region above T_g (and above T_m for a predominantly crystalline material). In this state the sheet readily deforms elastically. However, if the sheet is cooled below T_g (or T_m) in the deformed condition, it will remain there permanently. The process has many variations, as seen in Fig. 12.57. In each case the sheet is first heated to a rubbery condition while firmly held in a frame. With thin sheets a few seconds exposure to a radiant heating panel may suffice. *Vacuum forming* can be used for making something like a contour map.

Table 12.13 Typical properties of three plastics for structural foam

Property	General-purpose polypropylene		High-density polyethylene		High-impact polystyrene	
	Solid	Foam	Solid	Foam	Solid	Foam
Sample thickness, cm	0.38	0.635	0.38	0.635	0.44	0.635
Sample density, g/cm³	0.902	0.6	0.962	0.6	1.04	0.7
Flexural strength, MPa	45	22	34	19	50	31
Flexural modulus, GPa	1.30	0.83	1.10	0.83	1.86	1.45
Deflection of simple beam with 10-cm span loaded with 2.3 kg, cm	0.36	0.23	0.50	0.32	0.20	0.11
Tensile strength, MPa	33	14	28	9.0	24	12
Coefficient of linear expansion, cm/cm·°C $\times$ 10⁶	9.4	9.4	12.1	12.1	9.0	9.0
Heat distortion temperature at 1.8 MPa load, °C	56	55	45	34	76	80
Mold shrinkage, %	0.018	0.015	0.020	0.018	0.010	0.008
Skin thickness, cm	0.38	0.080	0.38	0.080	0.44	0.080

The heated sheet is held above a plaster mold and sealed around the edges by a gasket. When a vacuum is applied through small holes in the mold, the sheet conforms to the mold and quickly cools to a temperature below T_g, since the mold surface, even of a poor conductor, has a heat capacity much greater than that of the thin sheet of plastic. The other methods illustrated use various combinations of pressure and vacuum as well as plugs and pads.

If the sheet has been subjected to biaxial orientation during extrusion (see Sec. 12.5, Flat Film, Sheet, and Tubing), heating above the deformation temperature will cause the sheet to shrink. A tight package can be made by wrapping an article in biaxially oriented sheet followed by a brief heat treatment. Such *shrink-packaging* with polypropylene film is used for lettuce and meats.

Postforming of thermoset sheets resembles metal forming. As opposed to the various sheet-forming methods, the dies are rugged, the stresses high, and the temperature often fairly high. Because thermoset materials already possess a dimensionally stable structure, it is necessary actually to break some covalent bonds and to form some new ones during postforming. The curvature is limited, since a sharp bend may crack the sheet. Paper laminates and vulcanized fiber are shaped by this method. Some common examples are trays, tote boxes, and countertops.

Some thermoplastic sheets can be *cold-formed*. It is possible to form sheets of some ABS polymers into dishes, louvers, and housings by application of sufficient pressure.

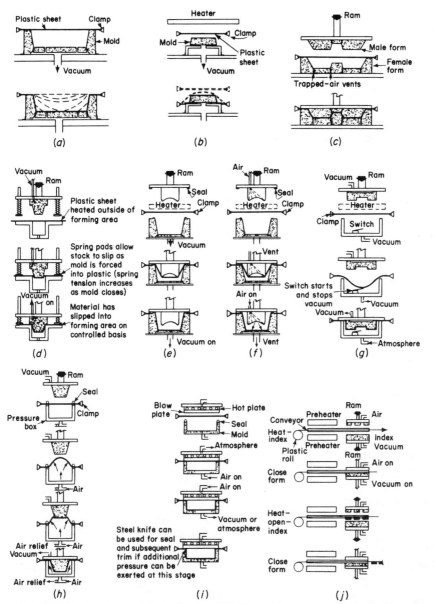

Figure 12.57 Ten methods of thermoforming [64]: (*a*) straight vacuum; (*b*) drape; (*c*) match-mold; (*d*) slip-ring; (*e*) plug-assist, vacuum; (*f*) plug-assist, pressure; (*g*) snap-back, vacuum; (*h*) pillow; (*i*) trapped-sheet, contact-heat, pressure; (*j*) preheat, plus-assist, pressure forming

KEYWORDS FOR CHAPTER 12

Section

1. Fabrication
3. Roll mill
   Internal mixer
   Blender
4. Fluidized-bed coating
   Electrostatic spray coating
   Photocurable coating
   Photoresists
   Microlithography
   Lacquer
   Dope
   Blushing
   Hiding power
   Paint
   Drying oil
   Autoxidation
   Drier
   Varnish (oleoresinous)
   Enamel
   Thinner, diluent
   Alkyd
   Adhesive
   Adherend
   Mechanical adhesion
   Specific adhesion
   Contact angle
   Critical surface tension
5. Extrusion
   Die (die swell)
   Film
   Sheet
   Biaxial orientation
   Blown film extrusion
   Frost line
   Coextrusion
   Continuous vulcanization
   Pultrusion
   Calender
   Spinning
   Spinneret

Fiber
Melt spinning
Dry spinning
Wet spinning
Crimp
Laminate
Sandwich construction
6. Compression mold
Flash mold
Positive-pressure mold
Internally landed mold
Transfer molding
Injection (screw) molding
Reaction injection molding
Casting
Plastisol
Slush molding
Dip molding
Rotational molding
Blow molding
Parison
Stretch-blow molding
Foam
Expanded plastic
Closed-cell foam
Open-cell foam
Blowing agent
Expanded-bead foam
Structural foam
Sheet forming
Vacuum forming

PROBLEMS

12.1 How might you expect the critical surface tension of a polymer to vary with the cohesive energy density? What factor could change the expected correlation? Compare the values of γ_c (Table 12.6) with values of δ_2 and δ_h (Table 2.5).

12.2 The statement is often made that one cannot glue anything to polyethylene. Show that this is incorrect by applying a strip of cellophane tape to a film of polyethylene (a bread wrapper or freezer bag will do). From the force necessary to pull the tape off in shear, estimate the shear strength of the bond. Is the strength dependent on the time scale of the test? How?

12.3   A method for making rubber "thread" is to extrude and continuously vulcanize a flat sheet of rubber. The flat sheet is cut into individual strands with a rectangular cross section. For a sheet 0.030 in thick, strands are cut 0.060 in wide. Calculate the denier and the strength in grams per denier if the rubber compound has a density of 0.85 g/cm³ and a tensile strength of 3500 psi.

12.4   What are the similarities between transfer and injection molding? What are the differences? Injection molding has been used for rubber and thermosets recently. What major problems must be overcome? What relationship between flow temperature and cross-linking or polymerization temperature is desirable?

12.5   A candlemaker has decided to automate production of small wax candles by using a continuous extrusion-coating process such as that used to insulate wire. Cotton string will be used instead of wire, and the extrudate will be chopped into candles 6 in long and ½ in in diameter. What problems might she expect? How might they be overcome? Should she hire a butcher and a baker as consultants? What experience have they had with extrusion?

12.6   Obtain a plastic bottle (an old bleach, shampoo, or beverage bottle will do). Using a razor blade, cut two cross sections through the length of the bottle at right angles to each other (one down the front, one down the side). Observe and explain any variation in thickness.

12.7   Obtain a foamed polystyrene article made from expanded beads. Industrial packaging or a picnic hot-drink cup will do. From the weight and dimensions of a large piece, estimate the density. Break off some of the individual beads and estimate what the *original* bead diameter must have been before foaming.

12.8   How would you go about making rubber bands from vulcanized rubber?

12.9   A 60:40 emulsion copolymer of styrene and ethyl acrylate is proposed as a coating material (a latex paint). What difficulty is likely to be encountered? How might it be overcome?

12.10   Outline a suitable method of manufacture for (*a*) 100,000 ft of garden hose from plasticized poly(vinyl chloride), (*b*) 50,000 pocket combs from polystyrene, and (*c*) two boat hulls, each 15 ft long, from glass cloth and a solution of an unsaturated polyester in styrene monomer.

12.11   Describe the role of each of these ingredients within each system and classify each system as a paint, enamel, varnish, lacquer, or putty.

System *A:*
  Linseed oil
  Mineral spirits
  Cobalt naphthenate
  Calcium carbonate

System *B:*
  Cellulose nitrate
  Methyl ethyl ketone
  Titanium dioxide

System *C:*
  Tall oil fatty acids
  Ethylene glycol
  Glycerol
  Phthalic anhydride
  Mineral spirits
  Lead naphthenate

} cooked together but
  describe each separately

System *D:*
  Shellac
  Ethyl alcohol

REFERENCES

1. Nylen, P., and E. Sunderland: "Modern Surface Coatings," pp. 534 and 543, Wiley Interscience, New York, 1965.
2. McCabe, W., and J. C. Smith: "Unit Operations of Chemical Engineering," 2d ed., pp. 851–859, McGraw-Hill, New York, 1967.
3. White, J. L.: *Rubber Chem. Technol.,* 65:527 (1992).
4. Parker, D. H.: "Principles of Surface Coating Technology," p. 762, Wiley, New York, 1965.
5. "Coating Metals and Other Rigid Materials with Microthene Polyethylene Powder," 2d ed., U.S.I. Chemicals, New York, 1965.
6. "Microthene F Microfine Polyolefin Powders," U.S.I. Chemicals, New York, 1976.
7. Glover, E.: *Modern Plastics Encyl.,* 56:250 (October 1979).
8. Gorham, W. F.: *Mod. Plastics,* 45(1A, Encycl. Issue): 173 (September 1967).
9. Shen, M., and A. T. Bell (eds.): "Plasma Polymerization," ACS, Washington, 1979.
10. Thompson, L. F., and R. E. Kerwin: *Ann. Rev. Matls. Sci.,* 6:267 (1976).
11. *Laser Focus World,* 31(3):9 (1995).
12. Reichmanis, E., F. M. Houlihan, O. Nalamasu, and T. X. Neenan, in L. F. Thompson, C. G. Willson, and S. Tagawa (eds.): "Polymers for Microelectronics," chap. 1, ACS, Washington, D.C., 1994.
13. Parker, D. H.: "Principles of Surface Coating Technology," pp. 563 and 752, Wiley, New York, 1965.
14. *Ucar Latex 180,* booklet F-41315, Union Carbide Corp., 1966.
15. Brodnyan, J. G., and T. Konen: *J. Appl. Polymer Sci.,* 8:687 (1964).
16. Nylen, P., and E. Sunderland: "Modern Surface Coatings," p. 123, Wiley-Interscience, New York, 1965.
17. Nylen, P., and E. Sunderland: "Modern Surface Coatings," p. 167, Wiley-Interscience, New York, 1965.
18. Tess, R. W., in J. K. Craver and R. W. Tess (eds.): "Applied Polymer Science," ACS, Washington, D.C., 1975.
19. Zisman, W. A.: *Ind. Eng. Chem.,* 55:18 (October 1963).
20. "Petrothene Polyolefins: A Processing Guide," 3d ed., U.S. Industrial Chemicals Co., New York, 1965.
21. Hovey, V. M.: *Wire,* 192 (February 1961).
22. Van Ness, R. T., G. R. De Hoff, and R. M. Bonner: *Mod. Plastics,* 45(14A, Encycl. Issue): 672 (October 1968).
23. "Plastic Film Orienting," Bull. 1-6, Marshall and Williams, Providence, R.I., 1967.
24. Schenkel, G.: "Plastics Extrusion Technology and Theory," pp. 323, American Elsevier, New York, 1966.
25. Garvey, B. S., Jr., M. H. Whitlock, and J. A. Freese, Jr.: *Ind. Eng. Chem.,* 34:1309 (1942).
26. Schenkel, G.: "Plastics Extrusion Technology and Theory," pp. 383–385, American Elsevier, New York, 1966.

27. Schrenk, W. J., and T. Alfrey, Jr.: *Polym. Eng. Sci.*, 9:393 (1969).
28. "Mearl Iridescent Film," The Mearl Corporation, Peekskill, N.Y., 1976.
29. Stern, H. J.: "Rubber: Natural and Synthetic," 2d ed., p. 388, Palmerton, New York, 1967.
30. Ewald, G. W.: *Mod. Plastics Encycl.*, 56:379 (October 1979).
31. Winding, C. C., and G. D. Hiatt: "Polymeric Materials," p. 110, McGraw-Hill, New York, 1961.
32. Winspear, G. G. (ed.): "Vanderbilt Rubber Handbook," p. 392, R. T. Vanderbilt Co., 1958.
33. Mark, H. F., S. M. Atlas, and E. Cernia (eds.): "Man-made Fibers: Science and Technology," 3 vols., Wiley-Interscience, New York, 1967, 1968.
34. Warner, S. B.: "Fiber Science," Prentice-Hall, Englewood Cliffs, N.J., 1995.
35. Riley, J. L., in C. E. Schildknecht (ed.): "Polymer Processes," chap. XVIII, Wiley-Interscience, New York, 1956.
36. Fredericks, S. D.: Paper presented at the Polypropylene Technology Conference, Clemson University, September 1987.
37. Jiang, H., W. W. Adams, and R. K. Eby, in R. W. Cahn, P. Haasen, and E. J. Kramer (eds.): "Materials Science and Technology: Structure and Properties of Polymers," chap. 13, VCH, New York, 1993.
38. MacLean, D. L., and T. E. Graham: *Chem. Eng.*, 87:54 (February 25, 1980).
39. Johnson, J. E.: *Mod. Plastics Encycl.*, 56:256 (October 1979).
40. Vaill, E. W.: *Mod. Plastics*, 40(1A, Encycl. Issue): 767 (September 1962).
41. "Urethanes for Liquid Reaction Molding," Union Carbide Corp., 1975.
42. Bonk, H. W.: *Mod. Plastics Encycl.*, 56:368 (October 1975).
43. Belforte, D. A.: *Laser Focus World*, 29(6):126 (1993).
44. Todd, W. D., in C. E. Schildknecht (ed.): "Polymer Processes," Wiley-Interscience, New York, 1956.
45. Randolph, A. F. (ed.): "Plastics Engineering Handbook," 3d ed., chap. 16, Reinhold, New York, 1960.
46. Morgan, B. T., D. L. Peters, and N. R. Wilson: *Mod. Plastics*, 45(1A, Encycl. Issue): 797 (September 1967).
47. "Polyethylene Blow Molding," U.S.I. Chemicals, New York, 1970.
48. *Mod. Plastics*, 53:36 (February 1976).
49. *Mod. Plastics*, 72(2):13 (1995).
50. Wood, A. S.: *Mod. Plastics*, 57:56 (January 1980).
51. Lichtenberg, D. W.: *Mod. Plastics Encycl.*, 56:147 (October 1979).
52. "One-Step Urethane Foams," Bull. F-40487, Union Carbide Corp., 1959.
53. Eldib, I. A.: *Hydrocarbon Process. Petrol. Refiner*, 42(12):121 (December 19, 1963).
54. Stern, H. J.: "Rubber: Natural and Synthetic," p. 450, Palmerton, New York, 1967.
55. Lasman, H. R.: *Mod. Plastics*, 45(1A, Encycl. Issue):368 (September 1967).
56. McIntire, O. R. (Dow Chemical): U.S. Patent 2,515,250, 1950.
57. Collins, F. H.: *SPE J.*, 16:705 (July 1960).
58. Miles, D. C., and J. H. Briston: "Polymer Technology," p. 187, Chemical Publishing, New York, 1965.
59. "PELASPAN Expandable Polystyrene," *Form* 171-90, Dow Chemical Co., 1958.
60. "PELASPAN Expandable Polystyrene," *Form* 171-414, Dow Chemical Co., 1966.
61. *Mod. Plastics*, 65(11):656 (1989).
62. Randolph, A. F. (ed.): "Plastics Engineering Handbook," 3d ed., chap. 12, Reinhold, New York, 1960.
63. Cargile, H. M.: *Mod. Plastics Encycl.*, 56:296 (October 1979).
64. Zelnick, D.: *Plastics Design Proces.*, 8(5):12 (May 1968).

General References

(The entries under each heading often are not exclusively on one subject, but they are grouped for convenience according to major emphasis.)

Compounding, molding, and extrusion

Agassant, J. F., P. Avenas, J. Sergent, and P. J. Carreau: "Polymer Processing: Principles and Modeling," Hanser-Gardner, Cincinnati, Ohio, 1991.

Astarita, G., and L. Nicolais (ed.): "Polymer Processing and Properties," Plenum, New York, 1984.

Baird, D. G., and D. M. Collias: "Polymer Processing: Principles and Design," Butterworth-Heinemann, Newton, Mass., 1995.

Beck, R. D.: "Plastic Product Design," 2d ed., Van Nostrand-Reinhold, New York, 1980.

Becker, W. E. (ed.): "Reaction Injection Molding," Van Nostrand-Reinhold, New York, 1979.

Belofsky, H.: "Plastics: Product Design and Process Engineering," Hanser-Gardner, Cincinnati, Ohio, 1995.

Birley, A. W., B. Haworth, and J. Batchelor: "Physics of Plastics: Processing, Properties and Materials Engineering," Hanser-Gardner, Cincinnati, Ohio, 1992.

Brewis, D. M. (ed.): "Surface Analysis and Pretreatment of Plastics and Metals," Elsevier Applied Science, New York, 1982.

Bruins, P. F.: "Basic Principles of Rotational Molding," Gordon & Breach, New York, 1971.

Bruins, P. F. (ed.): "Basic Principles of Thermoforming," Gordon & Breach, New York, 1973.

Bruins, P. F. (ed.): "Unsaturated Polyester Technology," Gordon & Breach, New York, 1976.

Chanda, M., and S. K. Roy: "Plastics Technology Handbook," 2d ed., Dekker, New York, 1992.

Charrier, J. M.: "Polymeric Materials and Processing: Plastics, Elastomers, and Composites," Hanser-Gardner, Cincinnati, Ohio, 1990.

"Chemical Additives for the Plastics Industry, Properties, Applications, Toxicologies," Noyes Data, Park Ridge, N.J., 1987.

Cheremisinoff, N. P.: "Polymer Mixing and Extrusion Technology," Dekker, New York, 1987.

Cheremisinoff, N. P. (ed.): "Product Design and Testing of Polymeric Materials," Dekker, New York, 1990.

Collyer, A. A. (ed.): "Rubber Toughened Engineering Plastics," Chapman & Hall, New York, 1994.

Corish, P. J. (ed.): "Concise Encyclopedia of Polymer Processing & Applications," Pergamon, New York, 1991.

Cracknell, P. S., and R. W. Dyson: "Handbook of Thermoplastics Injection Mould Design," Chapman & Hall, New York, 1993.

Crawford, R. J.: "Plastics Engineering," 2d ed., Pergamon, New York, 1987.

DiStasio, J. I. (ed.): "Epoxy Resin Technology (Developments since 1979)," Noyes Data, Park Ridge, N.J., 1982.

Domininghaus, H.: "Plastics for Engineers: Materials, Properties, Applications," Hanser-Gardner, Cincinnati, Ohio, 1993.

DuBois, J. H. (ed.): "Plastics Mold Engineering Handbook," 3d ed., Van Nostrand Reinhold, New York, 1979.

DuBois, J. H., and F. W. John: "Plastics," 5th ed., Van Nostrand Reinhold, New York, 1974.

Dym, J. B.: "Injection Molds and Molding: A Practical Manual," Van Nostrand Reinhold, New York, 1979.

Enikoloyan, N. S. (ed.): "Filled Polymers I. Science and Technology," Springer, Secaucus, N.J., 1990.

Farnham, S. E.: "A Guide to Thermoformed Plastic Packaging," Cahners, Boston, 1972.

Flick, E. W.: "Industrial Synthetic Resins Handbook," Noyes Data, Park Ridge, N.J., 1985.

Flick, E. W.: "Plastics Additives, An Industrial Guide," Noyes Data, Park Ridge, N.J., 1986.

Flick, E. W.: "Epoxy Resins Curing Agents, Compounds, and Modifiers: An Industrial Guide," Noyes Data, Park Ridge, N.J., 1987.

Florian, J.: "Practical Thermoforming," Dekker, New York, 1987.

Folkes, M. J. (ed.): "Processing, Structure and Properties of Block Copolymers," Elsevier Applied Science, New York, 1985.

Fouassier, J. P., and J. F. Rabek (eds.): "Radiation Curing in Polymer Science and Technology," 4 vols., Chapman & Hall, New York, 1994.

Frados, J. (ed.): "Plastics Engineering Handbook," 4th ed., Van Nostrand Reinhold, New York, 1976.

Gächter, R., and H. Müller (eds.): "Plastics Additives Handbook," 4th ed., Hanser-Gardner, Cincinnati, Ohio, 1993.

Gillies, M. T. (ed.): "Stabilizers for Synthetic Resins," Noyes Data, Park Ridge, N.J., 1981.

Gomez, I. L. (ed.): "Engineering with Rigid PVC," Dekker, New York, 1984.

Gomez, I. L.: "High Nitrile Polymers for Beverage Container Applications," Technomic, Lancaster, Pa., 1990.

Goodman, I. (ed.): "Developments in Block Copolymers—1," Elsevier Applied Science, New York, 1982.

Goodman, I. (ed.): "Developments in Block Copolymers—2," Elsevier Applied Science, New York, 1985.

Goodman, S. H. (ed.): "Handbook of Thermoset Plastics," Noyes Data, Park Ridge, N.J., 1986.

Griff, A. L.: "Plastics Extrusion Technology," 2d ed., Krieger, Melbourne, Fla., 1976.

Griskey, R. G.: "Polymer Process Engineering," Chapman & Hall, New York, 1994.

Gruenwald, G.: "Thermoforming: A Plastics Processing Guide," Technomic, Lancaster, Pa., 1987.

Harper, C. A. (ed.): "Handbook of Plastics, Elastomers, and Composites," 2d ed., McGraw-Hill, New York, 1992.

Hensen, F. (ed.): "Plastics Extrusion Technology," Hanser (Oxford Univ. Press), New York, 1988.

Hermansen, R. D.: "Formulating Plastics and Elastomers by Computer," Noyes, Park Ridge, N.J., 1991.

Holmes-Walker, W. A.: "Polymer Conversion," Halsted-Wiley, New York, 1974.

Hoyle, C. E., and J. F. Kinstle (eds.): "Radiation Curing of Polymeric Materials," ACS, Washington, D.C., 1990.

Isayev, A. I. (ed.): "Injection and Compression Molding Fundamentals," Dekker, New York, 1987.

Isayev, A. I. (ed.): "Modeling of Polymer Processing," Hanser-Gardner, Cincinnati, Ohio, 1991.

Ivanov, V. S.: "Radiation Chemistry of Polymers," VSP, Zeist, The Netherlands, 1992.

Janssen, L. P. B. M.: "Twin Screw Extrusion," Elsevier, New York, 1978.

Kia, H. G. (ed.): "Sheet Molding Compounds," Hanser-Gardner, Cincinnati, Ohio, 1993.

Kircher, K.: "Chemical Reactions in Plastics Processing," Macmillan, New York, 1987.

Kresta, J. E. (ed.): "Polymer Additives," Plenum, New York, 1984.

Kresta, J. (ed.): "Reaction Injection Molding," ACS, Washington, D.C., 1985.

Lee, N. C.: "Plastic Blow Molding Handbook," Chapman & Hall, New York, 1990.

Levy, S., and J. H. DuBois (eds.): "Plastics Product Design Engineering Handbook," 2d ed., Chapman & Hall (Methuen), New York, 1985.

Lin, S.-C.: "High-Performance Thermosets," Hanser-Gardner, Cincinnati, Ohio, 1993.

Lutz, J. T. (ed.): "Thermoplastic Polymer Additives," Dekker, New York, 1988.

MacDermott, C. P.: "Selecting Thermoplastics for Engineering Applications," Dekker, New York, 1984.

Macosko, C.: "Fundamentals of Reaction Injection Molding," Hanser (Oxford Univ. Press), New York, 1988.

Manas-Zloczower, I., and Z. Tadmor (eds.): "Mixing and Compounding of Polymers," Hanser-Gardner, Cincinnati, Ohio, 1994.

Manzione, L. T. (ed.): "Application of Computer-Aided Engineering in Injection Molding," Hanser, New York, 1987.

Mascia, L.: "Thermoplastics: Materials Engineering," Elsevier Applied Science, New York, 1982.

Matthews, G.: "Polymer Mixing Technology," Elsevier Applied Science, New York, 1982.

Meyer, R. W.: "Handbook of Pultrusion Technology," Chapman & Hall (Methuen), New York, 1985.

Meyer, R. W.: "Handbook of Polyester Molding Compounds and Molding Technology," Chapman & Hall (Methuen), New York, 1987.

Michaeli, W.: "Plastics Processing: An Introduction," Hanser-Gardner, Cincinnati, Ohio, 1995.

Middleman, S.: "Fundamentals of Polymer Processing," McGraw-Hill, New York, 1977.

Miller, E. (ed.): "Plastics Products Design Handbook, Part B, Processes and Design for Processes," Dekker, New York, 1983.

Miller, R. L. (ed.): "Flow-induced Crystallization in Polymer Systems," Gordon & Breach, New York, 1979.

Morton-Jones, D. H.: "Polymer Processing," Chapman & Hall, New York, 1989.

Morton-Jones, D. H., and J. W. Ellis: "Polymer Products," Chapman & Hall (Methuen), New York, 1986.

O'Brien, K. T. (ed.): "Applications of Computer Modeling for Extrusion and Other Continuous Polymer Processes," Hanser-Gardner, Cincinnati, Ohio, 1992.

Pearson, J. R. A.: "Mechanics of Polymer Processing," Elsevier Applied Science, New York, 1985.

Pearson, J. R. A., and S. M. Richardson (eds.): "Computational Analysis of Polymer Processing," Elsevier Applied Science, New York, 1983.

Pinner, S. H., and W. G. Simpson: "Plastics: Surface and Finish," Butterworth, New York, 1971.

"Plastics Processing, Technology and Health Effects," Noyes Data, Park Ridge, N.J., 1986.

Plueddemann, E. P.: "Silane Coupling Agents," 2d ed., Plenum, New York, 1990.

Potter, W. G.: "Uses of Epoxy Resins," Chemical Publishing, New York, 1975.

Powell, P. C.: "Engineering with Polymers," Chapman & Hall (Methuen), New York, 1983.

Progelhof, R. C., and J. L. Throne: "Polymer Engineering Principles: Properties, Processes and Tests for Design," Hanser-Gardner, Cincinnati, Ohio, 1993.

Provder, T. (eds.): "Computer Applications in the Polymer Laboratory," ACS, Washington, D.C., 1986.

Randell, D. R. (ed.): "Radiation Curing of Polymers," CRC Press, Boca Raton, Fla., 1987.

Randell, D. R.: "Radiation Curing of Polymers, II," CRC Press, Boca Raton, Fla., 1992.

Rauwendaal, C. J. (ed.): "Mixing in Polymer Processing," Dekker, New York, 1991.

Rauwendaal, C.: "Polymer Extrusion," 3rd ed., Hanser-Gardner, Cincinnati, Ohio, 1995.

Rees, H.: "Understanding Injection Molding Technology," Hanser-Gardner, Cincinnati, Ohio, 1994.

Reichmanis, E., and J. H. O'Donnell (eds.): "The Effects of Radiation on High-Technology Polymers," ACS, Washington, D.C., 1989.

Reichmanis, E., C. W. Frank, and J. H. O'Donnell (eds.): "Irradiation of Polymeric Materials: Processes, Mechanisms, and Applications," ACS, Washington, D.C., 1993.

Richardson, P. N.: "Introduction to Extrusion," Society of Plastics Engineers, Greenwich, Conn., 1974.

Ritchie, P. D., S. W. Critchley, and A. H. Hill: "Plasticizers, Stabilizers, and Fillers," Butterworth, New York, 1972.

Rosato, D. V.: "Rosato's Plastics Encyclopedia and Dictionary," Hanser-Gardner, Cincinnati, Ohio, 1993.

Rosato, D. V., and D. V. Rosato (eds.): "Blow Molding Handbook," Hanser-Gardner, Cincinnati, Ohio, 1989.

Rosato, D. V., and D. V. Rosato: "Injection Molding Handbook," 2d ed., Chapman & Hall, New York, 1994.

Rubin, I. I.: "Injection Molding," Wiley-Interscience, New York, 1973.

Ryan, M. E. (ed.): "Fundamentals of Polymerization and Polymer Processing Technology," Gordon & Breach, New York, 1983.

Saechtling, H.: "International Plastics Handbook," Macmillan, New York, 1983.

Sarvetnick, H. A.: "Plastisols and Organosols," Van Nostrand-Reinhold, New York, 1971.

Satas, D. (ed.): "Plastics Finishing and Decoration," Van Nostrand-Reinhold, New York, 1986.

Schwartz, S., and S. Goodman: "Plastics Materials and Processes," Van Nostrand-Reinhold, New York, 1982.

Seferis, J. C., and P. S. Theocaris (eds.): "Interrelations between Processing, Structure and Properties of Polymeric Materials," Elsevier Applied Science, New York, 1984.

Starr, T. F. (ed.): "Databook of Thermoset Resins for Composites," Elsevier, New York, 1993.

Stevens, M. L.: "Extruder Principles and Operation," Elsevier Applied Science, New York, 1985.

Suh, N. P., and N.-H. Sung (eds.): "Science and Technology of Polymer Processing," MIT Press, Cambridge, Mass., 1979.

Sweeney, F. M.: "Introduction to Reaction Injection Molding," Technomic, Westport, Conn., 1979.

Sweeney, F. M.: "Reaction Injection Molding Machinery and Processes," Dekker, New York, 1987.

Sweeting, O. J. (ed.): "The Science and Technology of Polymer Films." Wiley-Interscience, New York, vol. 1, 1968; vol. 2, 1971.

Tadmor, Z., and C. G. Gogos: "Principles of Polymer Processing," Wiley, New York, 1979.

Tadmor, Z., and I. Klein: "Engineering Principles of Plasticating Extrusion," Van Nostrand-Reinhold, New York, 1970.

Tess, R. W., and G. W. Poehlein (eds.): "Applied Polymer Science," 2d ed., ACS, Washington, D.C., 1985.

Throne, J. L.: "Plastics Process Engineering," Dekker, New York, 1979.

Throne, J. L.: "Thermoforming," Macmillan, New York, 1987.

Tucker, C. L., III, and E. C. Bernhardt (eds.): "Fundamentals of Computer Modeling for Polymer Processing," Hanser-Gardner, Cincinnati, Ohio, 1989.

Valcarcel, M., and M. D. Luque de Castro: "Flow Injection Analysis." Wiley, New York, 1987.

Weir, C. L.: "Introduction to Injection Molding," Society of Plastics Engineers, Greenwich, Conn., 1975.

Whelan, A., and J. A. Brydson (eds.): "Developments with Thermosetting Plastics," Applied Science, London, 1975.

Whelan, A., and J. L. Craft (eds.): "Developments in PVC Production and Processing—1," Applied Science, London, 1977.

Whelan, A., and J. L. Craft (eds.): "Developments in Injection Molding—1," International Ideas, Philadelphia, 1978.

Whelan, A., and J. L. Craft (eds.): "Developments in Injection Molding—2," Elsevier Applied Science, New York, 1981.

Whelan, A., and J. L. Craft (eds.): "Developments in Plastics Technology—2," Elsevier Applied Science, New York, 1985.

Whelan, A., and J. L. Craft (eds.): "Developments in Plastics Technology—3," Elsevier Applied Science, New York, 1986.

Whelan, A., and D. Dunning (eds.): "Developments in Plastics Technology—1," Elsevier Applied Science, New York, 1982.

Whelan, A., and J. P. Goff (eds.): "Developments in Plastics Technology—4," Elsevier Applied Science, New York, 1989.

Whelan, A., and J. P. Goff (eds.): "Developments in Injection Molding—3," Elsevier Applied Science, New York, 1985.

Whelan, T.: "Polymer Technology Dictionary," Chapman & Hall, New York, 1994.

White, J. L.: "Twin-Screw Extrusion," Hanser-Gardner, Cincinnati, Ohio, 1990.

White, J. L.: "Rubber Processing," Hanser-Gardner, Cincinnati, Ohio, 1995.

Wickson, E. J. (ed.): "Handbook of PVC Formulating," Wiley, New York, 1993.

Wittfoht, A. M.: "Plastics Technical Dictionary," 3 vols., Macmillan, New York, 1983.

Wright, R. E.: "Injection Transfer Molding of Thermosets," Hanser-Gardner, Cincinnati, Ohio, 1995.

Wright, R. E.: "Molded Thermosets: A Handbook for Plastics Engineers, Molders, and Designers," Hanser-Gardner, Cincinnati, Ohio, 1991.

Xanthos, M. (ed.): "Reactive Extrusion Principles and Practice," Hanser-Gardner, Cincinnati, Ohio, 1992.

Coatings

Allen, N. S. (ed.): "Photopolymerization and Photoimaging Science and Technology," Elsevier, New York, 1989.

Ash, M., and I. Ash (eds.): "A Formulary of Paints and Other Coatings," Chemical Publishing, New York, 1978.

Athey, R. D., Jr.: "Emulsion Polymer Technology," Dekker, New York, 1991.

Benkreira, H. (ed.): "Thin Film Coating," CRC Press, Boca Raton, Fla., 1993.

Biederman, H., and Y. Osada: "Plasma Polymerization Processes," Elsevier, New York, 1992.

Calbo, L. J. (ed.): "Handbook of Coatings Additives," Dekker, New York, 1987.

Calbo, L. J. (ed.): "Handbook of Coatings Additives," vol. 2, Dekker, New York, 1992.

Calvert, K. O. (ed.): "Polymer Latices and Their Applications," Elsevier Applied Science, New York, 1981.

Colbert, J. C. (ed.): "Modern Coating Technology," Noyes Data, Park Ridge, N.J., 1982.

Daniels, E. S., E. D. Sudol, and M. S. El-Aasser (eds.): "Polymer Latexes: Preparation, Characterization, and Applications," ACS, Washington, D.C., 1992.

Dören, K., W. Freitag, and D. Stoye: "Water-Borne Coatings," Hanser-Gardner, Cincinnati, Ohio, 1994.

Dubin, P. (ed.): "Microdomains in Polymer Solutions," Plenum, New York, 1985.

Flick, E. W.: "Water-Based Paint Formulations," Noyes Data, Park Ridge, N.J., 1975.

Flick, E. W.: "Solvent-Based Paint Formulations," Noyes Data, Park Ridge, N.J., 1977.

Flick, E. W.: "Contemporary Industrial Coatings," Noyes Data, Park Ridge, N.J., 1985.

Flick, E. W.: "Printing Ink Formulations," Noyes Data, Park Ridge, N.J., 1985.

Gaynes, N. I.: "Testing Organic Coatings," Noyes Data, Park Ridge, N.J., 1977.

Gillies, M. T. (ed.): "Solventless and High Solids Industrial Finishes," Noyes Data, Park Ridge, N.J., 1980.

Gillies, M. T. (ed.): "Water-Based Industrial Finishes," Noyes Data, Park Ridge, N.J., 1980.

Gooch, J. W.: "Lead-based Paint Handbook," Plenum, New York, 1993.

Holmberg, K.: "High Solids Alkyd Resins," Dekker, New York, 1987.

Kelly, J. M., C. B. McArdle, and M. J. F. Maunder (eds.): "Photochemistry and Polymeric Systems," CRC Press, Boca Raton, Fla., 1993.

Koleske, J. (ed.): "Paint and Coating Testing Manual (Gardner-Sward Handbook, 14th ed.)," ASTM, Philadelphia, 1995.

Konstandt, F.: "Organic Coatings: Properties and Evaluation," Chemical Publishing, New York, 1985.

Krongauz, V. V., and A. D. Trifunac (eds.): "Processes in Photoreactive Polymers," Chapman & Hall, New York, 1994.

Lambourne, R. (ed.): "Paints and Surface Coatings," Wiley, New York, 1987.

Licari, J. J., and L. A. Hughes: "Handbook of Polymer Coatings for Electronics," 2d ed., Noyes Data, Park Ridge, N.J., 1990.

Marrion, A. R. (ed.): "The Chemistry and Physics of Coatings," CRC Press, Boca Raton, Fla., 1994.

Morgans, W. M.: "Outlines of Paint Technology," 2 vols., Wiley, New York, 1984.

Myers, R. R., and J. S. Long (eds.): "Film-Forming Compositions," Dekker, New York, part 1, 1967; part 2, 1968; part 3, 1972.

Myers, R. R., and J. S. Long (eds.): "Pigments," Dekker, New York, 1975.

Parfitt, G. D., and A. V. Patsis (eds.): "Organic Coatings," Dekker, New York, 1984.

Patton, T. C. (ed.): "Pigment Handbook," 3 vols., Wiley-Interscience, New York, 1973.

Paul, S.: "Surface Coatings Science and Technology," Wiley, New York, 1985.

Ranney, M. W.: "Powder Coatings Technology," Noyes Data, Park Ridge, N.J., 1975.

Ranney, M. W.: "Specialized Curing Methods for Coatings and Plastics," Noyes Data, Park Ridge, N.J., 1977.

Robinson, J. S. (ed.): "Paint Additives," Noyes Data, Park Ridge, N.J., 1981.

Roffey, C. G.: "Photopolymerization of Surface Coatings," Wiley, New York, 1982.

Seymour, R. B., and H. F. Mark (eds.): "Organic Coatings: Their Origin and Development," Elsevier, Amsterdam, 1989.

Solomon, D. H., and D. G. Hawthorne: "Chemistry of Pigments and Fillers," Wiley, New York, 1983.

Turner, G. P. A.: "Introduction to Paint Chemistry," 3d ed., Chapman & Hall, New York, 1988.

Adhesives

Adams, R. D., and W. C. Wake: "Structural Adhesive Joints in Engineering," Elsevier Applied Science, New York, 1984.

Allen, K. W. (ed.): "Adhesion," vols. 1–5, Elsevier Applied Science, New York, 1977–1978.

Allen, K. W. (ed.): "Adhesion," vols. 6–11, Elsevier Applied Science, New York, 1983–1987.

Anderson, G. P., S. J. Bennett, and K. L. Devries: "Analysis of Adhesive Bonds," Academic Press, New York, 1977.

Andrade, J. D. (ed.): "Polymer Surface Dynamics," Plenum, New York, 1987.

Bateman, D. L.: "Hot Melt Adhesives," Noyes Data, Park Ridge, N.J., 1978.

Cagle, C. V. (ed.): "Handbook of Adhesive Bonding," McGraw-Hill, New York, 1973.

Dunning, H. R.: "Pressure Sensitive Adhesives," 2d ed., Noyes Data, Park Ridge, N.J., 1977.

Evans, R. M.: "Polyurethane Sealants—Technology and Applications," Technomic, Lancaster, Pa., 1993.

Feast, W. J., and H. S. Munro (eds.): "Polymer Surfaces and Interfaces," Wiley, New York, 1987.

Flick, E. W.: "Construction and Structural Adhesives and Sealants: An Industrial Guide," Noyes Data, Park Ridge, N.J., 1988.

Gutcho, M. (ed.): "Adhesives Technology (Developments since 1979)," Noyes Data, Park Ridge, N.J., 1983.

Hartshorn, S. R. (ed.): "Structural Adhesives," Plenum, New York, 1986.

Herman, B. S.: "Adhesives—Recent Developments," Noyes Data, Park Ridge, N.J., 1976.

Kinloch, A. J.: "Adhesion and Adhesives," Chapman & Hall, New York, 1988.

Kinloch, A. J. (ed.): "Developments in Adhesives—2," Elsevier Applied Science, New York, 1981.

Kinloch, A. J. (ed.): "Durability of Structural Adhesives," Elsevier Applied Science, New York, 1983.

Kinloch, A. J. (ed.): "Structural Adhesives: Developments in Resin and Primers," Elsevier Applied Science, New York, 1986.

Landrock, A. H.: "Adhesives Technology Handbook," Noyes Data, Park Ridge, N.J., 1985.

Lee, L.-H. (ed.): "Adhesive Chemistry," Plenum, New York, 1985.

Lee, L.-H. (ed.): "Adhesives, Sealants, and Coatings for Space and Harsh Environments," Plenum, New York, 1988.

Minford, J. D. (ed.): "Treatise on Adhesion and Adhesives," vol. 7, Dekker, New York, 1991.

Padday, J. F. (ed.): "Wetting, Spreading, and Adhesion," Academic Press, New York, 1978.

Patrick, R. L. (ed.): "Treatise on Adhesion and Adhesives," Dekker, New York; vol. 1, "Theory," 1967; vol. 2, "Materials," 1969; vols. 3–5, "Special Topics," 1973–1981.

Patrick, R. L., K. L. DeVries, and G. P. Anderson (eds.): "Treatise on Adhesion and Adhesives, Vol. 6," Dekker, New York, 1988.

Patsis, A. V. (ed.): "Advances in Organic Coatings Science and Technology," vol. 12, Technomic, Lancaster, Pa., 1990.

Pizzi, A. (ed.): "Wood Adhesives, Chemistry and Technology," Dekker, New York, 1983.

Pizzi, A.: "Advanced Wood Adhesives Technology," Dekker, New York, 1994.

Pizzi, A., and K. L. Mittal (eds.): "Handbook of Adhesive Technology," Dekker, New York, 1994.

Sanchez, I. C. (ed.): "Physics of Polymer Surfaces and Interfaces," Butterworth, Oxford, U.K., 1992.

Sellers, T., Jr.: "Plywood and Adhesive Technology," Dekker, New York, 1985.

Shields, J.: "Adhesives Handbook," 2d ed., Butterworth, New York, 1976.

Skeist, I. (ed.): "Handbook of Adhesives," 2d ed., Van Nostrand Reinhold, New York, 1979.

Torrey, S. (ed.): "Adhesive Technology," Noyes Data, Park Ridge, N.J., 1980.

Vacula, V. L., and L. M. Pritykin: "Physical Chemistry of Polymer Adhesion," Prentice-Hall, Engle-
wood Cliffs, N.J., 1991.
Wake, W. C.: "Adhesion and the Formulation of Adhesives," Applied Science, London, 1976.
Wake, W. C. (ed.): "Developments in Adhesives—1," Applied Science, London, 1977.
Wake, W. C. (ed.): "Synthetic Adhesives and Sealants," Wiley, New York, 1987.
Wegman, R. F., and T. R. Tullos: "Handbook of Adhesive Bonded Structural Repair," Noyes Data,
Park Ridge, N.J., 1992.
Wool, R. P.: "Polymer Interfaces Structure and Strength," Hanser-Gardner, Cincinnati, Ohio, 1995.
Wu, S.: "Polymer Interface and Adhesion," Dekker, New York, 1982.

Fibers

Economy, J. (ed.): "New and Specialty Fibers," Wiley, New York, 1976.
Hongu, T., and G. O. Phillips: "New Fibers," Prentice-Hall, Englewood Cliffs, N.J., 1990.
Lewin, M., and E. M. Pearce (eds.): "Handbook of Fiber Science and Technology, Fiber Chemistry,"
Dekker, New York, 1985.
Lewin, M., and J. Preston (eds.): "Handbook of Fiber Science and Technology, High Technology Fi-
bers," Part A, Dekker, New York, 1984.
Lewin, M., and J. Preston (eds.): "Handbook of Fiber Science and Technology, vol. III, High Technol-
ogy Fibers, Part C," Dekker, New York, 1993.
Lewin, M., and S. B. Sello (eds.): "Handbook of Fiber Science and Technology, Functional Finishes,"
Part A, Dekker, New York, 1983.
Lewin, M., and S. B. Sello (eds.): "Handbook of Fiber Science and Technology, Chemical Processing:
Fundamentals and Preparation," Part A, Dekker, New York, 1983.
Lewin, M., and S. B. Sello (eds.): "Handbook of Fiber Science and Technology, Functional Finishes,"
Part B, Dekker, New York, 1984.
Lewin, M., and S. B. Sello (eds.): "Handbook of Fiber Science and Technology, Chemical Processing:
Fundamentals and Preparation," Part B, Dekker, New York, 1984.
Masson, J. C. (ed.): "Acrylic Fiber Technology and Applications," Dekker, New York, 1995.
Militky, J., V. Vanicek, J. Krystufek, and V. Hartych: "Modified Polyester Fibres," Elsevier, Amster-
dam, 1991.
Robinson, J. S. (ed.): "Fiber-Forming Polymers," Noyes Data, Park Ridge, N.J., 1980.
Schwartz, P., T. Rhodes, and M. Mohamed: "Fabric Forming Systems," Noyes Data, Park Ridge,
N.J., 1982.
Turbak, A. F. (ed.): "Solvent Spun Rayon, Modified Cellulose Fibers and Derivatives," ACS, Washing-
ton, D.C., 1977.
Walczak, Z. K.: "Formation of Synthetic Fibers," Gordon & Breach, New York, 1976.
Warner, S. B.: "Fiber Science," Prentice-Hall, Englewood Cliffs, N.J., 1995.
Ziabicki, A.: "Fundamentals of Fibre Formation," Wiley-Interscience, New York, 1976.

Composites

Ashbee, K. H. G.: "Fundamental Principles of Fiber Reinforced Composites," 2d ed., Technomic, Lan-
caster, Pa., 1993.
Béland, S.: "High Performance Thermoplastic Resins and Their Composites," Noyes Data, Park Ridge,
N.J., 1990.
"Carbon Fibers, Technology, Uses, and Prospects," Noyes Data, Park Ridge, N.J., 1986.
Cheremisinoff, N. P., and P. N. Cheremisinoff: "Fiberglass-Reinforced Plastics Deskbook," Ann Arbor
Science Publishers, Ann Arbor, Mich., 1978.
DiBenedetto, l. Nicolais, and R. Watanabe (eds.): "Composite Materials," Elsevier, New York, 1992.
Donnet, J.-B., and R. C. Bansal: "Carbon Fibers," Dekker, New York, 1989.

Finlayson, K. M. (ed.): "Carbon Reinforced Epoxy Systems, Parts VI and VII," Technomic, Lancaster, Pa., 1989.

Garbo, S. P. (ed.): "Composite Materials: Testing and Design," ASTM, Philadelphia, 1990.

Grayson, M. (ed.): "Encyclopedia of Composite Materials and Components," Chemical Publishing, New York, 1983.

Halpin, J. C.: "Primer on Composite Materials Analysis," 2d ed., Technomic, Lancaster, Pa., 1992.

Harris, B. (ed.): "Developments in GRP Technology," Elsevier Applied Science, New York, 1983.

Jones, R. F. (ed.): "Handbook of Short Fiber Reinforced Plastics," Chapman & Hall (Methuen), New York, 1987.

Jones, F. (ed.): "Interfacial Phenomena in Composite Materials," Butterworth, Oxford, U.K., 1989.

Kaelble, D. H.: "Computer-Aided Design of Polymers and Composites," Dekker, New York, 1985.

Katz, H. S., and J. V. Milewski (eds.): "Handbook of Fillers for Plastics," Van Nostrand Reinhold, New York, 1987.

Kausch, H. -H. (ed.): "Advanced Thermoplastic Composites," Hanser-Gardner, Cincinnati, Ohio, 1992.

Kelly, A. (ed.): "Concise Encyclopedia of Composite Materials," Pergamon, New York, 1994.

Lee, J. A., and D. L. Mykkanen: "Metal and Polymer Matrix Composites," Noyes Data, Park Ridge, N.J., 1987.

Lee, S. M. (ed.): "Reference Book for Composites Technology," vols. 1 & 2, Technomic, Lancaster, Pa., 1989.

Lubin, G.: "Handbook of Composites," Van Nostrand Reinhold, New York, 1982.

Mallick, P. K.: "Fiber-Reinforced Composites; Materials, Manufacturing, and Design," Dekker, New York, 1987.

Mallick, P., and S. Newman (eds.): "Composite Materials Technology," Hanser-Gardner, Cincinnati, Ohio, 1990.

Margolis, J. M. (ed.): "Advanced Thermoset Composites," Van Nostrand Reinhold, New York, 1986.

Matthews, F. L., and R. D. Rawlings: "Composite Materials: Engineering and Science," Chapman & Hall, New York, 1994.

Mayer, R.: "Design with Reinforced Plastics: A Guide for Engineers and Designers," Chapman & Hall, New York, 1993.

Mayer, R., and N. Hancox: "Design Data for Reinforced Plastics: A Guide for Engineers and Designers," Chapman & Hall, New York, 1993.

Meyer, R. W.: "Handbook of Polyester Molding Compounds and Molding Technology," Chapman & Hall, New York, 1987.

Milewski, J. V., and H. S. Katz (eds.): "Handbook of Reinforcements for Plastics," Van Nostrand Reinhold, New York, 1987.

Powell, P. C.: "Engineering with Fibre-Polymer Laminates," Chapman & Hall, New York, 1993.

Pritchard, G. (ed.): "Developments in Reinforced Plastics—1," Elsevier Applied Science, New York, 1980.

Pritchard, G. (ed.): "Developments in Reinforced Plastics—2," Elsevier Applied Science, New York, 1982.

Pritchard, G. (ed.): "Developments in Reinforced Plastics—3," Elsevier Applied Science, New York, 1984.

Pritchard, G. (ed.): "Developments in Reinforced Plastics—4," Elsevier Applied Science, New York, 1984.

Pritchard, G. (ed.): "Developments in Reinforced Plastics—5," Elsevier Applied Science, New York, 1986.

Ranney, M. W.: "Reinforced Plastics and Elastomers," Noyes Data, Park Ridge, N.J., 1977.

"Reinforced Plastics Handbook," Elsevier, Cambridge, 1994.

Serafini, T. T. (ed.): "High Temperature Polymer Matrix Composites," Noyes Data, Park Ridge, N.J., 1987.

Sheldon, R. P.: "Composite Polymeric Materials," Elsevier Applied Science, New York, 1982.

Sichel, E. K. (ed.): "Carbon-Black Polymer Composites," Dekker, New York, 1982.

Strong, A. B.: "High Performance and Engineering Thermoplastic Composites," Technomic, Lancaster, Pa., 1993.

Titow, W. V., and B. J. Lanham: "Reinforced Thermoplastics," Applied Science, London, 1975.

"Tough Composite Materials," Noyes Data, Park Ridge, N.J., 1985.

Vigo, T. L., and B. J. Kinzig (eds.): "Composite Applications. The Role of Matrix, Fiber, and Interface," VCH, New York, 1992.

Wake, W. C. (ed.): "Textile Reinforcement of Elastomers," Elsevier Applied Science, New York, 1982.

Yosomiya, R., K. Morimoto, A. Nakajima, Y. Ikada, and T. Suzuki: "Adhesion and Bonding in Composites," Dekker, New York, 1989.

Zinoviev, P., and Y. N. Ermakov: "Energy Dissipation in Composite Materials," Technomic, Lancaster, Pa., 1994.

Elastomers

Babbit, R. O. (ed.): "The Vanderbilt Rubber Handbook," 12th ed., R. T. Vanderbilt Co., Norwalk, CT, 1978.

Barlow, F.: "Rubber Compounding: Principles, Materials, and Techniques," 2d ed., Dekker, New York, 1993.

Bhowmick, A. K., M. M. Hall, and H. A. Benarey (eds.): "Rubber Products Manufacturing Technology," Dekker, New York, 1994.

Bhowmick, A. K., and H. L. Stephens (eds.): "Handbook of Elastomers," Dekker, New York, 1988.

Cheremisinoff, N. P. (ed.): "Elastomer Technology Handbook," CRC Press, Boca Raton, Fla., 1993.

Donnet, J.-B. and R. C. Bansal, and M.-J. Wang (eds.): "Carbon Black," 2d ed., Dekker, New York, 1993.

Eirich, F. R. (ed.): "Science and Technology of Rubber," Academic Press, New York, 1978.

Evans, C. W.: "Practical Rubber Compounding and Processing," Elsevier Applied Science, New York, 1981.

Evans, C. W. (ed.): "Developments in Rubber and Rubber Composites—1," Elsevier Applied Science, New York, 1980.

Evans, C. W. (ed.): "Developments in Rubber and Rubber Composites—2," Elsevier Applied Science, New York, 1983.

Franta, I. (ed.): "Elastomers and Rubber Compounding Materials: Manufacture, Properties and Applications," Elsevier, Amsterdam, 1989.

Hofmann, W.: "Rubber Technology Handbook," Hanser-Gardner, Cincinnati, Ohio, 1989.

Kohudic, M. A. (ed.): "Advances in Elastomers, Vol. 1: Advances in the Chemistry and Processing of Various Elastomers," Technomic, Lancaster, Pa., 1994.

Lynch, W.: "Handbook of Silicone Rubber Fabrication," Van Nostrand Reinhold, New York, 1978.

Morton, M. (ed.): "Rubber Technology," 3d ed., Van Nostrand Reinhold, New York, 1987.

Nutt, A. R.: "Toxic Hazards of Rubber Chemicals," Elsevier Applied Science, New York, 1984.

Penn, W. S.: "Injection Molding of Elastomers," Gordon & Breach, New York, 1969.

Roberts, A. D. (ed.): "Natural Rubber Science and Technology," Oxford Univ. Press, New York, 1988.

Schweitzer, P.: "Corrosion Resistance of Elastomers," Dekker, New York, 1990.

Whelan, A., and K. S. Lee (eds.): "Developments in Rubber Technology—1," Elsevier Applied Science, New York, 1979.

Whelan, A., and K. S. Lee (eds.): "Developments in Rubber Technology—2," Elsevier Applied Science, New York, 1981.

Whelan, A., and K. S. Lee (eds.): "Developments in Rubber Technology—3," Elsevier Applied Science, New York, 1982.

Foams

Berlin, A. A., F. A. Shutov, and A. K. Shitinkin: "Foam Based on Reactive Oligomers," Technomic, Lancaster, Pa., 1982.

Klempner, D., and K. C. Frisch (eds.): "Handbook of Polymeric Foams and Foam Technology," Hanser-Gardner, Cincinnati, Ohio, 1992.

Lichtenberg, F. W. (ed.): "CFCs & the Polyurethane Industry," vol. 2, Technomic, Lancaster, Pa., 1989.

Lichtenberg, F. W. (ed.): "CFCs & the Polyurethane Industry," vol. 3, Technomic, Lancaster, Pa., 1990.

Lichtenberg, F. W. (ed.): "CFCs & the Polyurethane Industry," vol. 4, Technomic, Lancaster, Pa., 1991.

Lichtenberg, F. W. (ed.): "CFCs & the Polyurethane Industry," vol. 5, Technomic, Lancaster, Pa., 1993.

McBrayer, R. L., and D. C. Wysocki (eds.): "Advances in Polyurethane Foams Formulations," Technomic, Lancaster, Pa., 1994.

Meltzer, Y. L.: "Foamed Plastics," Noyes Data, Park Ridge, N.J., 1976.

Meltzer, Y. L.: "Expanded Plastics and Related Products (Developments since 1978)," Noyes Data, Park Ridge, N.J., 1983.

Wendle, B. C.: "Structural Foam," Dekker, New York, 1985.

Woods, G.: "Flexible Polyurethane Foams," Elsevier Applied Science, New York, 1982.

Electronics

Bowden, M., and S. R. Turner (eds.): "Polymers for High Technology, Electronics and Photonics," ACS Symp. Series 346, ACS, Washington, D.C., 1987.

Bowden, M. J., and S. R. Turner (eds.): "Electronic and Photonic Applications of Polymers," ACS, Washington, D.C., 1988.

Chilton, J. A., and M. Goosey (eds.): "Special Polymers for Electronics and Optoelectronics," Chapman & Hall, New York, 1995.

Davidson, T. (ed.): "Polymers in Electronics," ACS, Washington, D.C., 1984.

Feit, E. D., and C. Wilkins, Jr. (eds.): "Polymer Materials for Electronic Applications," ACS, Washington, D.C., 1982.

Flick, E. W.: "Adhesives, Sealants, and Coatings for the Electronics Industry," 2d ed., Noyes Data, Park Ridge, N.J., 1992.

Goosey, M., and J. A. Chilton: "Special Polymers for Electronics and Optoelectronics," Chapman & Hall, New York, 1994.

Ito, H., S. Tagawa, and K. Horie (eds.): "Polymeric Materials for Microelectronic Applications: Science and Technology," ACS, Washington, D.C., 1995.

McArdle, C. B.: "Side Chain Liquid Crystal Polymers," Chapman & Hall, New York, 1989.

McArdle, C. B.: "Applied Photochromic Polymer Systems," Chapman & Hall, New York, 1991.

Moreau, W. M.: "Semiconductor Lithography," Plenum, New York, 1988.

Prasad, P. N., and D. R. Ulrich (eds.): "Nonlinear Optical and Electroactive Polymers," Plenum, New York, 1988.

Soane, D. S., and Z. Martynenko: "Polymers in Electronics: Fundamentals and Applications," Elsevier, New York, 1989.

Thompson, L. F., C. G. Willson, and M. J. Bowden (eds.): "Introduction to Microlithography," 2d ed., ACS, Washington, D.C., 1994.

Thompson, L. F., C. G. Willson, and J. M. J. Frechet (eds.): "Materials for Microlithography," ACS, Washington, D.C., 1984.

Thompson, L. F., C. G. Willson, and S. Tagawa (eds.): "Polymers for Microelectronics: Resists and Dielectrics," ACS, Washington, D.C., 1994.

Wong, C. P. (ed.): "Polymers for Electronic and Photonic Applications," Academic Press, San Diego, Calif., 1992.

Biomedical applications

DeRossi, D., K. Kajiwara, Y. Osada, and A. Yamauchi (eds.): "Polymer Gels: Fundamentals and Biomedical Applications," Plenum, New York, 1991.

Donaruma, L. G., and O. Vogl: "Polymeric Drugs," Academic Press, New York, 1978.

Dunn, R. L., and R. M. Ottenbrite (eds.): "Polymeric Drugs and Drug Delivery Systems," ACS, Washington, D.C., 1991.

El-Nokaly, M. A., D. M. Piatt, and B. A. Charpentier (eds.): "Polymeric Delivery Systems: Properties and Applications," ACS, Washington, D.C., 1993.

Gebelein, C. G.: "Biotechnology Polymers—Medical, Pharmaceutical and Industrial Applications," Technomic, Lancaster, Pa., 1993.

Gebelein, C. G. (ed.): "Polymeric Materials and Artificial Organs," ACS, Washington, D.C., 1984.

Gebelein, C. G., and C. E. Carraher, Jr. (eds.): "Biotechnology and Bioactive Polymers," Plenum, New York, 1994.

Gebelein, C. G., and R. L. Dunn (eds.): "Progress in Biomedical Polymers," Plenum, New York, 1990.

Kramer, O. (ed.): "Biological and Synthetic Polymer Networks," Elsevier Applied Science, New York, 1988.

Migliaresi, C., L. Nicolais, P. Giusti, and E. Chiellini (eds.): "Polymers in Medicine, III," Elsevier, New York, 1988.

Ottenbrite, R. M. (ed.): "Polymeric Drugs and Drug Administration," ACS, Washington, D.C., 1994.

Ottenbrite, R. M., and E. Chiellini (eds.): "Polymers in Medicine: Biomedical and Pharmaceutical Applications," Technomic, Lancaster, Pa., 1992.

Park, K., S. W. Shalaby, and H. Park: "Biodegradable Hydrogels for Drug Delivery," Technomic, Lancaster, Pa., 1993.

Paul, D. R., and F. W. Harris (eds.): "Controlled Release Polymeric Formulations," ACS, Washington, D.C., 1976.

Peppas, N. K. (ed.): "Hydrogels in Medicine and Pharmacy," CRC Press, Boca Raton, Fla., vol. 1, "Fundamentals," 1986; vol. 2, "Polymers," 1987; vol. 3, "Properties and Applications," 1987.

Shalaby, S. W., Y. Ikada, R. Langer, and J. Williams (eds.): "Polymers of Biological and Biomedical Significance," ACS, Washington, D.C., 1994.

Stoy, V. A., and C. K. Kliment: "Hydrogels: Specialty Plastics for Biomedical, Pharmaceutical and Industrial Applications," Technomic, Lancaster, Pa., 1990.

Tarcha, P. J. (ed.): "Polymers for Controlled Drug Delivery," CRC Press, Boca Raton, Fla., 1990.

Miscellaneous applications

Benning, C. J.: "Plastic Films for Packaging," Technomic, Lancaster, Pa., 1983.

Brannon-Peppas, L., and R. S. Harland (eds.): "Absorbent Polymer Technology," Elsevier, Amsterdam, 1990.

Bruins, P. F. (ed.): "Packaging with Plastics," Gordon & Breach, New York, 1974.

Buchholz, F. L., and N. Peppas (eds.): "Superabsorbent Polymers: Science and Technology," ACS, Washington, D.C., 1994.

Buscall, R., T. Corner, and J. F. Stageman (eds.): "Polymer Colloids," Elsevier Applied Science, New York, 1985.

Dickie, R. A., and F. L. Floyd (eds.): "Polymeric Materials for Corrosion Control," ACS, Washington, D.C., 1986.

Dickie, R. A., S. S. Labana, and R. S. Bauer (eds.): "Cross-Linked Polymers: Chemistry, Properties, and Applications," ACS, Washington, D.C., 1988.

Finlayson, K. M. (ed.): "Plastic Film Technology, Vol. 1: High Barrier Plastic Films for Packaging," Technomic, Lancaster, Pa., 1989.

Finlayson, K. M. (ed.): "Plastic Film Technology, Vol. 2: Extrusion of Plastic Film & Sheeting," Technomic, Lancaster, Pa., 1993.

Goddard, E. D., and K. P. Ananthapadmanabhan (eds.): "Interactions of Surfactants with Polymers and Proteins," CRC Press, Boca Raton, Fla., 1993.

Guenet, J.-M.: "Thermoreversible Gelation of Polymers and Biopolymers," Academic Press, San Diego, Calif., 1992.

Hara, M. (ed.): "Polyelectrolytes," Dekker, New York, 1992.

Jenkins, W. A., and J. P. Harrington (eds.): "Packaging Foods with Plastics," Technomic, Lancaster, Pa., 1991.

Jenkins, W. A., and K. R. Osborn: "Plastic Films: Technology and Packaging Applications," Technomic, Lancaster, Pa., 1992.

Kesting, R. E.: "Synthetic Polymeric Membranes," 2d ed., Wiley, New York, 1985.

Kesting, R. E., and A. K. Fritzche: "Polymeric Gas Separation Membranes," Wiley, New York, 1993.

Klempner, D., and K. C. Frisch (eds.): "Advances in Interpenetrating Polymer Networks." vol. 1, Technomic, Lancaster, Pa., 1989.

Klempner, D., and K. C. Frisch (eds.): "Advances in Interpenetrating Polymer Networks." vol. 2, Technomic, Lancaster, Pa., 1990.

Klempner, D., and K. C. Frisch (eds.): "Advances in Interpenetrating Polymer Networks." vol. 3, Technomic, Lancaster, Pa., 1991.

Klempner, D., L. H. Sperling, and L. A. Utracki (eds.): "Interpenetrating Polymer Networks," ACS, Washington, D.C., 1994.

Kohudic, M. A., and K. M. Finlayson (eds.): "Advances in Polymer Blends and Alloys Technology," vol. 2, Technomic, Lancaster, Pa., 1989.

Koros, W. J. (ed.): "Barrier Polymers and Structures," ACS, Washington, D.C., 1990.

Lee, L.-H. (ed.): "Advances in Polymer Friction and Wear," 2 vols., Plenum, New York, 1974.

Lloyd, D. R. (ed.): "Materials Science of Synthetic Membranes," ACS, Washington, D.C., 1985.

MacCallum, J. R., and C. A. Vincent (eds.): "Polymer Electrolyte Reviews—2," Elsevier, New York, 1989.

Matsuura, T.: "Synthetic Membranes and Membrane Separation Processes," CRC Press, Boca Raton, Fla., 1993.

McGreavy, C. (ed.): "Polymer Reaction Engineering," Chapman & Hall, New York, 1993.

Montella, R.: "Plastics in Architecture, a Guide to Acrylic and Polycarbonate," Dekker, New York, 1985.

Nass, L. I., and C. A. Heiberger (eds.): "Encyclopedia of PVC, Resin Manufacture and Properties," 2d ed., Dekker, New York, 1986.

Noble, R. D., and S. A. Stern (eds.): "Membrane Separations Technology: Principles and Applications," Elsevier, Amsterdam, 1995.

Noble, R. D., and J. D. Way (eds.): "Liquid Membranes," ACS, Washington, D.C., 1987.

Osborn, K. R., and W. A. Jenkins: "Plastic Films: Technology and Packaging Applications," Technomic, Lancaster, Pa., 1992.

Oswin, C. R.: "Plastic Films and Packaging," Applied Science, London, 1975.

Paul, D. R., and Y. Yampol'skii (eds.): "Polymeric Gas Separation Membranes," CRC Press, Boca Raton, Fla., 1994.

Piirma, I.: "Polymer Surfactants," Dekker, New York, 1992.

Riew, C. K., and A. J. Kinloch (eds.): "Toughened Plastics I: Science and Engineering," ACS, Washington, D.C., 1993.

Scrosati, B. (ed.): "Applications of Electroactive Polymers," Chapman & Hall, New York, 1993.

Selke, S. E. M.: "Packaging and the Environment: Alternatives, Trends, and Solutions," Technomic, Lancaster, Pa., 1994.

Seymour, R. B.: "Plastics versus Corrosives," Wiley, New York, 1982.

Shalaby, S. W., C. L. McCormick, and G. B. Butler (eds.): "Water Soluble Polymers: Synthesis, Solution Properties, and Applications," ACS, Washington, D.C., 1991.

Simpson, W. G. G. (ed.): "Plastics Surface and Finish," CRC Press, Boca Raton, Fla., 1994.

Stahl, G. A., and D. N. Schulz (eds.): "Water-Soluble Polymers for Petroleum Recovery," Plenum, New York, 1988.

Stroeve, P., and E. I. Franses (eds.): "Molecular Engineering of Ultrathin Polymeric Films," Elsevier Applied Science, New York, 1987.

Szycher, M.: "High Performance Biomaterials," Technomic, Lancaster, Pa., 1991.

Tipp, G., and V. J. Watson: "Polymeric Surfaces for Sports and Recreation," Elsevier Applied Science, New York, 1982.

Titow, W. V.: "PVC Technology," 4th ed., Elsevier Applied Science, New York, 1984.

Vergnaud, J. M.: "Liquid Transport Processes in Polymeric Materials: Modeling and Industrial Applications," Prentice Hall, Englewood Cliffs, N.J., 1991.

RECYCLING AND RESOURCE RECOVERY

13.1 WASTE GENERATION AND DISPOSAL

Waste disposal would be a problem even if the amount of refuse per person stayed constant over the years. However, an increased standard of living and other changes in living style have aggravated the situation. In the United States, the population increased from 180 million in 1960 to 250 million in 1990, an increase of 39%. In the same period (Table 13.1), municipal solid waste (MSW) increased at a faster rate so that the total waste *per person* increased by 62%. On the other hand, the per-capita generation of waste remained about the same from 1990 to 1993 and was projected to decrease slightly in the 1994–2000 period [2, 3]. MSW includes material from commercial, industrial, and institutional sources as well as residences. Construction and demolition wastes are not included, nor is sludge from sewage treatment.

Disposal

In 1960 almost all solid waste was dumped or buried in landfills or incinerated. Open dumps are breeding grounds for rats and other carriers of pestilence. They have been largely replaced by "sanitary landfills," in which layers of refuse are covered at regular intervals by layers of soil. Many communities have run out of space for landfills and are seeking alternatives. One of the largest landfills is the Fresh Kills landfill on Staten Island, New York. Originally a marshland, the volume of material deposited there since it opened in the 1950s is 25 times the volume

Table 13.1 Selected data on waste generation and disposal (millions of tons) [1, 2]

Municipal Solid Waste Generation					
Item	1960	1970	1980	1990	1993
Waste, total	88.0	122.0	152.0	196.0	206.9
Paper products	30.0	44.0	55.0	73.0	77.8
Ferrous metals	10.0	13.0	12.0	12.0	12.9
Aluminum	0.4	0.8	1.8	2.7	3.0
Other nonferrous metals	0.2	0.7	1.1	1.2	1.2
Glass	6.7	12.7	15.0	13.0	13.7
Plastics	0.4	3.1	7.9	16.2	19.3
Rubber and leather	2.0	3.2	4.2	4.7	6.2
Textiles	1.7	2.0	2.6	5.7	6.1
Wood	3.0	4.0	6.7	12.3	13.7
Food wastes	12.2	12.8	13.2	13.1	13.8
Yard wastes	20.0	23.2	27.6	35.0	32.8
Waste, total	88.0	122.0	152.0	196.0	206.9
Other wastes	1.4	2.7	5.1	6.1	6.4

Municipal Solid Waste Disposal					
Method	1960	1970	1980	1990	1993
Landfill	55.0	88.0	123.0	130.0	129.0
Combustion/energy recovery	0.0	0.4	2.7	29.7	32.9
Other combustion	27.0	24.7	11.0	2.2	nil
Recovered materials, total	5.9	8.6	14.5	33.4	38.5
Paper products	5.4	7.4	11.9	20.9	26.5
Ferrous metals	0.1	0.1	0.4	1.9	3.4
Aluminum	0.0	0.0	0.3	1.0	1.1
Other nonferrous metals	0.0	0.3	0.5	0.8	0.8
Glass	0.1	0.2	0.8	2.6	3.0
Plastics	0.0	0.0	0.0	0.4	0.7
Other	0.3	0.6	0.6	5.8	3.1

of the Great Pyramid of Khufu at Giza. The prediction was made in 1992 that it would eventually be the highest geographical feature on the U.S. Atlantic seaboard [4].

Much of the focus on waste has concentrated on the disposal of packaging. Unlike many other applications, packaging represents a short-term usage. Long-term usage also feeds products into the waste stream, but in a less obvious manner. Construction materials, clothing, automobiles, tires, batteries, and appliances all appear in the waste stream eventually, but in a whole spectrum of shapes, sizes, and conditions. For many concerned groups, waste has become almost synonymous with packaging made of paper, glass, aluminum, and plastics.

A concern expressed in some quarters is that polymers in the waste stream represent a massive dissipation of irreplaceable fossil fuel resources. In order to put this in perspective, it is well to note that production of all synthetic polymers (plastics, rubber, and textiles) in the United States amounted to about 41×10^9 kg in 1994. Less than a fifth of the total was used for packaging. In the same year, fuels for heating and transportation based on petroleum and natural gas consumed over 1000×10^9 kg. A typical shopping experience provides an illustration. The average American grocery bag is made from less than 6 g of high-density polyethylene. As a liquid fuel, the same amount of hydrocarbon would run the engine of a compact automobile for less than 10 s.

Along the same lines, there is an ongoing controversy over the merits of paper packaging versus plastics. A rather elegant comparison of paper cups vs. foamed polystyrene cups [5] showed that the choice between the two is complex and depends on various subjective criteria. A similar analysis compared grocery bags of paper and polyethylene [6]. Neither case demonstrated a clear-cut superiority for one material over the other for a container of the same capacity. The analyses included the costs of material and energy for production and ultimate disposal of each alternative.

Several terms should be distinguished. *Postconsumer waste* includes materials which have been gathered from a multiplicity of sources, often by a community recycling program. A somewhat more restrictive term often used to describe recovered materials that may potentially be recycled is *postconsumer recyclate*. *Postindustrial waste* is material which has not gone into consumer products. The trimmings from edges of sheet materials and returned bulk containers are often easier to handle, since they are not geographically scattered. The explanatory legends on most packaging that makes use of recycled materials (cereal boxes, for example) usually distinguish between the two major categories of recycled waste.

Some governments have passed laws specifying minimum standards for recycling or have issued outright bans on certain products. Most of these have been at the local level, but national governments have also set specific targets. In Germany, for example, a law was passed to the effect that 64% of all plastics packaging was to be recycled by 1995 [7]. Similar laws have set goals for recycling of entire automobiles in some countries [8]. A U.S. Highway Bill passed in 1991 mandated the use of "rubberized asphalt" for paving 5% of all new roads starting in 1994, increasing to 20% by 1997 [9]. Rubberized asphalt contains 1% by weight of crumb rubber, mainly from ground-up, discarded tires. It was estimated that the program would consume between 35 and 50% of the 260 million tires discarded annually in the United States. However, the program has encountered resistance in the road construction industry.

An obvious way to minimize waste is to reduce the sources of waste. The downsizing of the polyethylene grocery bag provides a successful example. The 6-g grocery bag in current use is capable of containing a volume of about 15 liters. The wall thickness is only 0.013 mm (0.0005 in). Not so very long ago, a plastic bag of the same capacity required a wall thickness at least five times greater. The increased yield strength of the newer bag material has been brought about by changes in polymer structure and by improvements in the fabrication process.

Recycling of Paper, Glass, and Aluminum

The recycling of all materials has increased dramatically in recent years (Table 13.1). Much of this is due to community-based recycling programs which mandate collection of certain classes of material. Paper products, which are a large fraction of the waste stream, are successfully recycled, although the supply of recycled paper appears to exceed the demand. Consumer products containing recycled paper often sell at a premium over the same products with non-recycled material. There are lessons that polymer recyclers can learn from older programs. Recycling of aluminum beverage cans is a true success. Over 50% of all cans were recycled in 1993, mostly back to the original application [3]. Several factors leading to this situation are worth considering for polymers.

1. The cans are all aluminum. Older cans which had some parts of ferrous alloys were not easily recycled.
2. All vendors use the same material. If one used a modified alloy that was not compatible with that used by the others, the cans would have to be segregated before recycling.
3. The value of scrap aluminum is high (ca. $0.70/lb).

On the other hand, only about 28% of all aluminum goes into cans. Wire, foil, and appliances are not as easily recycled. Also, most nonbeverage metal containers (foods, personal care products, etc.) are not aluminum. Almost 29% of the metal cans used in 1987 were steel (both tin-plated and tin-free) [10]. Not since World War II days has there been a booming market for "tin can" scrap.

The replacement of glass containers by aluminum and plastics is apparent in the statistics from 1970 to 1990. Glass is easily distinguished from other containers, which makes recycling easier. About 25% of glass containers used in 1993 were recycled [3]. Since less than one-fourth of all silica-based glass goes into other products (flat glass, fiberglass, etc.), the 25% figure is rather impressive. Most recyclers (and most municipalities that recycle) require that glass be separated into clear and colored portions.

Paper, particularly newspapers and office paper, was recycled at a rate of 34% in 1993 [2]. It has been observed that in one year a single subscription to *The New York Times* uses about 520 lb of paper [11]. Newsprint itself does not make large use of reclaimed paper because fibers are degraded in structure and length during processing, giving rise to problems in high-speed printing [10]. The use of recycled newsprint in cardboard and building applications goes back many years. The degradation of properties and use in downgraded products are factors in polymer use also. The separation problems with paper are simpler than those with polymers,

Table 13.2 Plastics indentification code [12]

Number and letters (polymer type)	U.S. Sales in billions of kg (1993)			Typical uses	Density, g/cm³
	Total	Containers	Film		
♻1 PETE (polyethylene terephthalate)	1.2	0.8	0.4	Beverages	1.35
♻2 HDPE (high-density polyethylene)	4.8	1.8	0.5	Milk, bags	0.95
♻3 V (poly(vinyl chloride))	4.7	0.1	0.1	Oil, shampoos	1.32–1.38
♻4 LDPE (low-density polyethylene)	5.9	0.2	1.9	Bags	0.93
♻5 PP (polypropylene)	4.1	0.2	0.3	Foods	0.90
♻6 PS (polystyrene)	3.1	0.5	0.1	Foam	1.08
7 Other	7.5	0.3	0.1		
Totals	31.3	3.8	3.4		

since ink and adhesives are not very much like paper pulp. On the other hand, polyethylene and polystyrene have much more in common. Separation of fillers (kaolin and other inorganic materials) from glossy papers is unattractive economically and generally avoided.

13.2 IDENTIFICATION AND SEPARATION

Gathering individual components from the waste stream into well-labeled and economically viable masses is the most difficult part of the recycling process. In many communities, some form of recycling is mandated by law. Residents may be required to separate their trash into separate containers, or they may be allowed to commingle (mix) all the "recyclables." In some places, bottle deposit laws encourage consumers to return beverage containers to the place of purchase. Aluminum, glass, and polyester soft-drink and beer containers are collected in stores and shipped to collection centers where they can be compacted. High-density polyethylene milk containers also are easily identified and collected.

Identification

In order to facilitate identification of containers, the Society of Plastics Industries introduced a code which has been used by most manufacturers since the 1990s. At or near the bottom of almost all plastic containers sold in the United States appears a triangular symbol with a number and an abbreviated name (Table 13.2)

[12]. While this is a help when consumers participate in a recycling program, the symbols are not very visible and do not enhance automatic sorting. Most containers and almost all film materials are potentially included in the six categories. There are several cautionary points to be observed. Except for poly(ethylene terephthalate) (PET), containers are not the predominant use for any one polymer. For poly(vinyl chloride) (PVC), consumer-identifiable containers are less than 10% of the total. The identification on products is not always very prominently displayed. It can take some effort to ascertain that a given grocery bag is either No. 2 (HDPE), high-density polyethylene, or No. 4 (LDPE), low-density polyethylene. And there is a problem with composite products. Even the symbol on the base of a 2-liter soda bottle is somewhat ambiguous. The symbol is No. 1 (PETE). While it is true that the clear bottle is, indeed, PET, often a base made of No. 2 (HDPE) is placed over the round bottom of the bottle. Also, the bottle usually will contain small amounts of paper (label) and poly(vinyl acetate) (adhesive) in addition to a metal or polypropylene cap with a copolymer liner.

There are other limitations on the system. PET from bottles is of a higher molecular weight than that used for fibers, so PET mixtures will not always be the same on recycling. Whether or not the identification code will be applicable for consumer sorting of more long-lived products such as garden hose (PVC) and pantyhose (nylon) remains to be seen. Multicomponent products that do not come apart easily (laminated food trays, some food-product bottles) end up in the "other" category, No. 7.

On the whole, consumer sorting is a real step forward and greatly simplifies fuller treatment of any type. The environmentally aware consumer is able to take an active part in a process of great symbolic and some real significance. Because they are such bulky items, removal of aluminum, glass, and plastic containers from the municipal waste stream has a very perceptible impact on any disposal system. The recycling of PET bottles reached over 41% in 1993 [14]. Bottle deposit laws are especially effective in gathering the bottles into central depots.

The ease with which materials can be identified and separated varies with the source of supply. Most automobile batteries are made with polypropylene (PP) casings. Since the metal in the batteries is reclaimed, the casings represent a centralized and relatively homogeneous source of one polymer. In 1990 it was reported that almost 150 million pounds of PP were recovered annually in the United States, representing as much as 95% of all discarded batteries. About 40% of the recovered PP went into the next generation of batteries, with the balance going into other automotive products and miscellaneous consumer products [15]. Automobile tires, on the other hand, contain several different polymers in addition to the metal bead and fabric reinforcement. The reuse of tires by separating all the components is economically unsound.

It has been suggested that automobiles, along with many other consumer products, should be designed for easy disassembly so that components can be separated and identified more easily. A drawback of the plan is the possibility of encouraging theft and vandalism. Nonetheless, manufacturers in Europe and the United States appear to be investigating the idea. Automatic methods of separating commingled plastics into homogeneous streams have been suggested. Manual sorting of recovered containers is still the rule for most processors.

Size Reduction

After any presorting, the scrap must be reduced to a manageable processing size. Shredders and granulators are employed to do this. The standard chip size is ¼ in. Chips larger than this size have been determined to cause bridging and plugging within the process. The price for the smaller particle size is a decreasing granulator capacity. If shredded bottles are purchased from a supplier, as sometimes happens, only the granulation step is needed. These steps are industry standard.

The granulation of bottles and other scrap can be responsible for loss of physical strength. The application of mechanical shear in the shredders and granulators breaks the polymer chains into smaller fragments. The decrease in chain size, and thus the molecular weight, can cause a corresponding degradation of the mechanical properties of the polymer. Mechanical properties are responsible for a substantial part of value of most plastics used in films and containers. As long as mechanical reduction of particle size is employed, this is a consideration.

Direct Flotation

Once a mixture of plastics is reduced to chips of the order of a centimeter in diameter, liquids with intermediate densities can be used to differentiate polymers, assuming that the polymers do not contain fillers that alter their density. A mixture of the six plastics listed in Table 13.2 can be separated into two groups using water. The polyolefins (numbers 2, 4, and 5) will float and the others (numbers 1, 3, and 6) will sink.

Further differentiation of the polyolefins requires liquids that are lighter than water and that do not swell the polymers. Aqueous solutions of methyl or ethyl alcohol are applicable. A solution of sodium chloride in water can be used to separate polystyrene from the two denser polymers. However, separation of polyester (PET) from poly(vinyl chloride) (PVC) is complicated by the fact that many variations in compounding PVC result in densities that can be greater or less than that of PET.

There are other complications in using float-sink separations. The adhesion of a small bit of label made of a different material alters the apparent density of a chip. Fillers and dirt may increase the apparent density. The method is obviously of greatest use when the overall composition of the mixed plastic stream remains relatively uniform.

Froth Flotation

A variation of direct flotation is froth flotation. In froth flotation, a stirred tank is generally employed, with air sparged in from the bottom. The feed also enters from the bottom. Originally a tool of the mining industry, it operates on the principle of the affinity of different materials for air bubbles. Materials with lower wettabilities will cling to air bubbles and rise to the surface. Frothing agents provide a stable froth in which floated materials remain suspended [16]. In the case of PET recycling, though, the wettabilities of the HDPE and PET would be too close for effective separation. Indeed, surfactants are usually added to the direct flotation column to ensure that froth flotation does not occur. Disposal of these chemicals is a problem. At present, direct flotation is the preferred approach to HDPE removal.

Water Removal

The presence of water will not only impair the electrostatic separation of plastics and aluminum but also severely interfere with any extrusion operations downstream. The maximum water level for extrusion is about 1 ppm [17]. Water levels of less than 0.5% must exist for effective electrostatic separation. As washing, rinsing, and flotation are key steps in recovery processes, some mechanism for water removal must be present.

One process uses "spin-drying" followed by a dessicant bed to achieve drying [18]. Spin-drying presumably refers to centrifugal drying, as is employed by a household washer. The spin-drying could also be combined sequentially with heat drying methods. A countercurrent direct-heat rotary dryer is a possibility. In this device, feed enters the top of a rotating drum which is tilted at an angle. Hot gases pass up the drum, drying the feed. This is the most thermally efficient of the rotary dryers [19]. Centrifugal drying by itself is more efficient in terms of energy but cannot remove as much water as hot-air dryers.

Paper Removal

Paper labels are often attached to plastic products. The paper is generally removed first. Two approaches are currently practiced. One uses a fluidized bed, and the other uses a cyclone. Both of these approaches are gravimetric. Thus, it should be kept in mind that product loss is possible if polymer particles are of the same mass as the paper. Grinding to too fine a size must be avoided if a clean separation is to be achieved.

In the fluidized bed, air is fed into the bottom and the feed enters near the middle of the bed. When the lifting force of the air overcomes the force of gravity, the particles rise. The lightest particles are pushed out the top and the heaviest sink to the bottom, where they are removed. The particles exist in a gradient throughout the column. Several cuts can be taken from the bed.

A cyclone is generally conical in shape, with the feed entering tangentially at the top, the widest part. The outlets are at the bottom and through a pipe whose mouth is lower than the feed entry point. Having the feed enter off-center causes the linear velocity of the stream to be translated into angular velocity. Acceleration rates of several hundred gravities may be easily achieved this way. A vortex is formed, with the lightest material in the center and the heaviest toward the outside. Due to the nature of the vortex formed, material inside the vortex moves upward and out the top of the device. Material outside the vortex moves downward. In this way separation is simply and effectively accomplished. The cyclone offers simplicity to the process and thus lower capital costs and easier operation.

Fluidized beds offer more flexibility than cyclones in that more than one cut of material may be taken. This is valuable if partial recycling of the product stream is necessary. On the other hand, a second cyclone may be used if the first is not effective enough. Indeed, the use of a sequence of cyclones is a standard approach to many separation operations. Being more complex, fluidized beds are more expensive. There is a trade-off between price and efficiency.

Metal Removal

Two cases must be considered in metal removal: One may assume that aluminum is the only metal present or one may assume that some ferrous contaminants are present as well. If ferrous components are assumed to be present, magnetic separation must be employed [20]. This is generally performed in a drum separator. This would be the case only if very impure feedstocks were being used.

To separate aluminum from plastics, gravimetric, densiometric, or electrostatic means can be employed. Gravimetric methods have been less favored, mainly because separation depends on particle size as well as density. It is quite likely that some metal particles will settle more slowly than the plastic, if only because of their small size. On the other hand, metal particles will always be denser. One firm which has used this approach uses an "eddy current" separator combined with a fluidized bed to effect removal [21].

Electrostatic separation takes advantage of the difference in the conductivities of materials to sort them. Conducting materials will lose a charge more quickly than insulators. If charged pieces of both materials pass over a grounded drum and then fall past a rotating charged object, the weaker conductor will be more strongly attracted because it will retain more of its original charge. Thus, separation is achieved by the deflection of the insulating fraction away from the conducting fraction. The typical electrostatic separator is based on ion bombardment. An electric discharge forms a corona in which air is ionized. These ions contact the surface of the particles, transferring their charge [22].

Dissolution

A second general method of separation is by selective dissolution and precipitation. In principle, this method is capable of purifying bonded, blended, and filled plastics. Dissolution of the polymer releases the impurities, which can be removed by filtration, adsorption, or flotation/sedimentation [23, 24]. This is capable of yielding a polymer of high purity. The major drawback of a solvent system is the increased expense due to the complexity of equipment and the higher energy requirements.

Research on solvent-based polymer separation processes is by no means in its infancy. One of the first studies on mixed plastics was conducted by Sperber and Rosen [25, 26] in the mid-1970s. These investigators used a blend of xylene and cyclohexanone to separate a mixture of polystyrene (PS), poly(vinyl chloride) (PVC), high-density polyethylene (HDPE), low-density polyethylene (LDPE), and polypropylene (PP) into three separate phases. In addition, many U.S. and foreign patents dating from the 1970s were granted for the solvent recovery of thermoplastic polymers. The interest in solvent processes waned in the late 1970s as the oil crisis eased, but the growing need to develop solutions to the solid waste problem has renewed the research effort.

Early investigators of solvent processes for mixed plastics [25, 26] could not foresee the penetration of poly(ethylene terephthalate) (PET) into the packaging industry, and so their processes did not address the separation of PET from mixtures. The RPI process [27], a much later research project on mixed plastics, did include the separation of PET. This process utilizes a single solvent (either tetrahydrofuran or xylene) to separate the polymers in batch mode based on the temperature-dependent dissolution rates of each resin. The polymer solution is then exposed to elevated temperatures and pressures before flash-devolatilizing the solution. In addition to the high pressures required in the feed to the flash stage, elevated pressures are required in the dissolution stage of the RPI process because the normal boiling point of the solvent is below the temperature required to achieve PET dissolution. The use of elevated temperatures increases the risk of thermally degrading the polymers, while the elevated pressures translate into higher energy and equipment costs. An additional limitation of the RPI process lies in the use of a single solvent capable of dissolving all of the polymers. This reduces the purity of the final polymer streams due to the partial dissolution of polymers which were intended to be dissolved at a higher temperature and the carryover of undissolved polymers to a higher-temperature-dissolving fraction. These limitations can be overcome by using multiple solvents, with each solvent being compatible with only a limited number of polymers at reasonable temperatures, and by incorporating a less energy-intensive recovery stage. In addition, lower-cost processes could be employed to make initial separations.

Combined Technology

Each separation process has its merits and limitations. It would seem logical that a process combining various separation technologies could take advantage of the inherent strengths of an individual process while attenuating the limitations. For example, a combined-technology process could utilize a flotation system to produce segregated polymer streams which could then be further purified in solvent-processing trains. The strength of the flotation system is cost-effectiveness, but it is limited by the relatively low purity of the final products, as is true of all compositionally blind separators. Selective dissolution processes, on the other hand, can yield polymers of high purity but at an increased cost.

13.3 DOWN-CYCLING

The recycling of polymers is generally considered to be the most environment-friendly alternative for dealing with the disposal problem. Even if consumers can be persuaded to segregate polymer products before putting them into the municipal waste stream, there will always be the matter of separating (1) inadvertant mixtures and (2) multicomponent products. One process which seems economically attractive in the current market is to separate "impurities" from a stream which is predominantly one polymer by using differences in density. The poly(ethylene terephthalate) beverage bottle with a polyethylene base is a case in point. Of course, there is the alternative of not separating the components of a waste stream, but to make a "plastic lumber" out of a mechanical mixture of polymers.

In general, a polymer recovered from a waste stream will have inferior physical properties compared to the original material. The exception to this rule is polymers produced by repolymerization of monomers recovered from waste. Virtually all polymers deteriorate after undergoing production processes involving heating and shearing, and weathering processes including oxidation, hydrolysis, or exposure to sunlight. Moreover, almost all commodity polymers are mutually immiscible. This means that any mixture of polymers will be likely to fail mechanically under less stress than pure materials. For this reason, recycling often takes the form of *down-cycling*. That is, the polymer is put into a less demanding application than the one from which it was derived. For example, 60 to 70% of the 400 million pounds of poly(ethylene terephthalate) recovered from beverage bottles in 1992 was used to make fibers for carpet yarn and for the fibrous packing (fiberfill) that goes into pillows and bedding [28]. Another characteristic of down-cycling shown by this example is that it converts a material from a short-term usage (typically packaging) into a long-term usage such as construction, furniture, or decorative items.

Nonetheless, a great deal of work has been done to devise consumer products that can be made from mixtures. "Plastic lumber" is a term often used to describe forms which can be converted into park benches, shipping pallets, and flooring. Even when a single source of plastic is used, a mixture may be economically attractive. One plastics producer combines recovered HDPE from grocery sacks and stretch-wrapping with an equal weight of sawdust (from discarded wooden shipping lumber) to make a product said to compete with top-grade treated marine lumber and clear cedar [29]. Factory floors are a suggested use. Some ways in which mixtures can be modified is by the addition of compatibilizing resins (often block copolymers) and by fillers.

A general approach to using unsorted plastic mixtures involves granulation, mastication, heating, and molding. Dark colors usually are chosen in order to minimize the nonuniform appearance. Some advantages cited for plastic lumber over wood include resistance to chemicals, salt water, termites, and bacteria [30]. In a lumber process developed at the Center for Plastics Recycling Research at Rutgers University, the lower-melting plastics in a mixture act as a flowable matrix in which as much as 40% of nonflowing materials such as paper, metal, glass, dirt, and crosslinked plastics can be accommodated [31].

Film Reclamation Process

In any reclamation process, certain steps are essential. For example, recovery of polyethylene from soiled agricultural film involves size reduction, washing, drying, and regranulating [32]. Agricultural film from mulch and packaging may be 25 to 150 μm thick and may contain as much as an equal weight of soil. Similar materials may be recovered from construction operations. A schematic flow sheet for a process owned by Herbold, a German machinery supplier (Fig. 13.1), starts with bales of the soiled film being fed to a shredder which operates at a slow speed and has specially toughened, blunt blades which will not be worn rapidly by the soil particles. The coarse pieces of film are conveyed to an air separator followed by a grinding and washing tank. Most of the water and dirt is removed in the dewatering screw and is filtered or centrifuged to yield a sludge ("schlamm") and a stream which goes to a water treatment plant. The washed product goes to a mechanical drier followed by a cyclone separator to recover fines and, finally, a hot-air drying section leading to storage. Not shown are an extruder and pelletizer for converting the polymer into a form suitable for return to a film production line. It is obvious that even a relatively homogeneous entering stream involves a sizable capital investment and running costs for water and power. For a plant with a throughput of 500 kg/h (4 million kg/yr), the energy needed for size reduction, washing, and drying is about 0.3 kW·hr/kg. Additional energy would be required for the pelletizer.

13.4 MONOMER RECOVERY

In principle, recovery of monomers and repolymerization should result in a product that is indistinguishable from the original polymer. Polystyrene and poly(methyl methacrylate), when heated with a free-radical source under vacuum, give

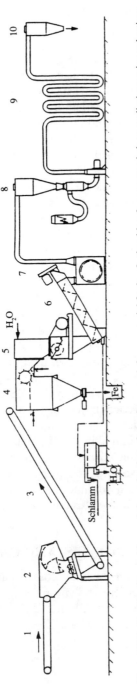

Figure 13.1 Facility for washing and size reduction of plastic scrap: 1, base material feeder; 2, shredder; 3, conveyor belt to preliminary size reduction; 4, separation of coarse matter; 5, wet grinder; 6, dewatering screw; 7, mechanical drier; 8, cyclone separator; 9, hot-air dryer; 10, cyclone separator (to storage) [32]

high yields of the respective monomers (see Sec. 11.3). Poly(dimethyl siloxane) heated with KOH in a vacuum also produces a cyclic monomer. All of these processes are used for in-plant scrap treatment, but economics do not yet favor applying them to postconsumer plastic waste.

When poly(ethylene terephthalate) is heated with an excess of methanol, a mixture of dimethyl terephthalate and ethylene glycol is produced [33]. In one example, a 1:4 mixture of polymer and methanol is said to produce a 99% yield of monomers after 1 h at 160 to 240°C and 20 to 70 atm. The process is somewhat sensitive to contaminants such as some of the dyes used in beverage bottles. Since polyester bottles are very difficult to distinguish from PVC, chlorine contamination at a low level is hard to avoid. The chlorine can interfere with subsequent repolymerization. A number of variants on the methanolysis route have been described.

A second route [33] to depolymerization is glycolysis. Transesterification of PET with an excess of propylene glycol in the presence of a catalyst such as an amine, alkoxide, or metal acetate is used to obtain low-molecular-weight fragments with hydroxyl end groups. Glycolysis for 8 h at reflux (200°C) with a 1.5:1 ratio of glycol to PET results in a polyol with a number-average molecular weight of 480. The polyols are recommended for use in making polyurethanes (by reaction with isocyanates; see Secs. 4.7 and 15.6).

Polyurethane scrap from industrial processes such as reaction injection molding, foam production, and recovered auto parts can also be treated with steam or with glycols [34]. Yields of 60 to 80% of polyols and diamines have been reported from polyurethane foam scrap using steam under pressure at 288°C in a continuous vertical reactor with a residence time of 10 to 28 min. Glycolysis as an alternative to steam hydrolysis produces a mixture of polyols. Digesting ground-up polyurethanes in a 90/10 mixture of di(alkylene)glycol/diethanolamine for several hours at 190 to 210°C gives a mixture which, after treatment with propylene oxide, can be used as part of the polyol charge in making rigid polyurethane foam.

Somewhat similar is the devulcanization of rubber, in which a sulfur crosslinked network is converted to a system of more-or-less thermoplastic components. Rubber reclaiming has fallen on hard times for various reason, including the difficulty of making an environmentally attractive process.

13.5 HYDROGENATION, PYROLYSIS, AND GASIFICATION

Straightforward reversion to monomers is not convenient with some of the commodity polymers, namely polyethylene, polypropylene, and poly(vinyl chloride). An omnibus solution is to use some sort of nonoxidizing thermal treatment which will produce a low-molecular-weight product. The terms *chemical recycling* and *plastics-to-chemicals* are used.

Hydrogenation (Liquid Product, Hydrogen Atmosphere)

A plant in Germany was modified to accept ground-up plastic wastes as a partial feed. The plant, originally built to process coal to synthetic fuel, had been converted to oil sludge upgrading. The process has been described as follows [7]:

> The unit will [be capable of] recycling up to 40,000 metric tons per year of mixed plastic containing up to 10% PVC. Up to 20% of its feed will be milled plastic particles less than 8 mm in diameter. [The] unit operates at 150–300 bars and about 470°C in a hydrogen atmosphere, producing a "syncrude" containing 60% paraffins, 30% naphtha, 9% aromatics and 1% olefins. Chlorine . . . is converted to HCl in the reactor, then neutralized with calcium carbonate.

Pyrolysis (Gaseous Product, Inert Atmosphere)

In a pyrolysis plant tested in Great Britain, a sand bed is fluidized using some of the cracked gases produced by thermal cracking of plastics introduced as chips or strips. A mixture of polyolefins and up to 10% PVC is cracked at 600°C in an inert atmosphere to yield a naphtha-type gas. Mass yields are about 90%, and 10% of the gas is used to heat the bed. Chlorine is adsorbed within the bed by calcium oxide. Sand beds can be quite insensitive to the size of material fed. Kaminsky graphically describes one reactor which produced a mixture of gas, liquid, and carbon black from scrap tires [35]:

> The whole tires roll through a gas tight lock into the reactor. The tire, landing on the fluid bed gradually sinks into the sand. The material heats up and softens, its surface becoming covered with hot sand grains. Through the shearing forces of the fluid bed and "exchange" of sand grains takes place at the surface of the softened material; the abrasion of small particles and their decomposition begins. A methane and ethylene rich gas as well as a condensable liquid with a high percentage of aromatics are produced. The main part of the filler materials (carbon black and zinc oxide) is blown out of the the fluid bed and can be separated in a cyclone. After 2 or 3 minutes the tire is completely pyrolyzed. What remains in the sand bed is a mass of twisted steel wires, which are removed by a tiltable grate extended into the fluid bed.

Gasification (Gaseous Product, Oxidizing Atmosphere)

Gasification proceeds under higher temperatures (above 900°C) than pyrolysis and often under oxidizing conditions. It is essentially a two- or three-step incineration. An Italian pilot plant [7] has the following operating parameters:

Input per hour: 4200 kg of municipal solid waste, compacted and degassed
Output per kilogram charged:
   660 kg syngas
   220 kg slag
   23 kg metals
   18 kg salts

The compacted waste is pyrolyzed at 600°C and then fed to a gasifier at 2000°C and pressures of 20 to 65 bar. The syngas after cleaning is used to drive a gas turbine producing 300 kW of electricity.

13.6 INCINERATION WITH ENERGY RECOVERY

A one-step incineration with energy recovery is possible under a variety of circumstances. Among the many possibilities, three specific kinds of facilities can be differentiated: power plants, incinerators, and cement kilns. In a power plant, dry polymer waste is an additive to a standard fuel (coal, oil, or gas). In an incinerator, the usual feed is a mixture of polymers, paper, and other components of the municipal solid waste stream. Because paper and other cellulosic materials have a lower energy value and because water is evaporated, the total energy recovery is lower than from plastics or tire-burning power plants. In both power plants and incinerators, energy is recovered in the form of steam, which is used in turn to generate electricity. Also, in both, the disposition of gases and of solid residues must be dealt with. On the other hand, burning waste in a cement kiln makes use of the energy to produce the cement, and the solid residues may be incorporated in the product.

Power Plants

Most power plants have automatic feed mechanisms which will not accommodate odd shapes of solid feed. A mixture of chipped or powdered waste with coal or oil is preferred. For example, tires are combined with bituminous coal at the Jennison generating station near Bainbridge, New York [36]. The two generating units were built in 1945 and 1950 with a total capacity of 73 MW. Scrap tires are delivered to the power plant, where a chipper reduces them into 5 × 5 cm chips. The tires are delivered from various sources such as businesses, service groups, and municipalities that pay a fee to get rid of the waste. Before the unit could operate, state and local regulations presented hurdles to be overcome. The stoker boilers are fed by traveling chain grates, which have proved to be quite adaptable for conveying the mixture of chipped rubber and coal in a 1:4 ratio. With a boiler flame temperature of 1650°C, no problems have been reported with soot or odors. As in any power plant, fly ash is collected and sold to a cement manufacturer as a raw material. The metal from the tire chips melts into small clumps in the bottom ash and is removed by magnetic separation before storage. The bottom ash accumulates and is eventually used by local municipalities as a road traction agent, a necessity amidst the ice and snow of upstate New York winters. The average heat value of a 20-lb passenger car tire (13,000 Btu/lb) is equivalent to 30 lb of coal. Although the tires contain sulfur, the overall composition is favorable for clean combustion. When mixed plastics are used, a question often arises about HCl generation from PVC, with a consequent corrosion problem in older power plants. New plants can be designed with proper controls to minimize corrosion problems.

A California plant was built in 1987 to generate 14 MW using only tires as fuel [37]. The grate in this plant feeds whole tires into the boiler. The emission controls system is a vital part of the operation including removal of fly ash and bottom ash as well as flue gas desulfurization and thermal treatment of NO_x. The selling price of power in 1989 was 8.3 cents per kilowatt-hour, giving total sales of about $10 million from about 4.9 million tires.

Incineration

Although incinerators have been in use for many years, the modern incinerator is a waste-to-energy facility with many features that older models do not possess [38]. In a typical plant (Fig. 13.2), trucks deliver waste to a storage pit, from which it is loaded into a hopper above the charging chute. Hydraulic rams located at the bottom of the chute are used to push the fuel onto moving grates. On the grates, three successive stages are drying, combustion of volatiles, and a completion of the burning of solids. Steam is generated in the tubes lining the boiler. The three main streams leaving the incinerator are the ash from the bottom of the grate, the fly ash collected by precipitation or filtration, and the gases discharged after being scrubbed. Since the average heating value of MSW is less than half that of coal, an incinerator is not nearly as efficient a generator of power as a dedicated power plant. Arguments still rage over the desirability of building large, centrally located plants for the "mass burn" of municipal solid waste. A rather negative view of MSW incineration has been detailed by the Environmental Defense Fund [39]. The arguments cited include:

1. *Expense.* To benefit from the economy of scale, most proposed incinerators are designed to handle refuse from a wide area. Construction costs typically are in the hundreds of millions of dollars. Local governments may have to participate in bond issues, special taxes, and tipping fees in order to build and maintain the operation.
2. *Solid residue (ash) disposal.* If the plant handles unsorted MSW, as much as 20 to 30% by weight of the charge is noncombustible material. The resulting residue must be disposed of by land filling. There are concerns about leaching of heavy metals from the landfill into potable water sources.
3. *Air emissions.* Aside from CO_2, about which there is concern over the "greenhouse" effect, there are fears that other objectionable gases may be discharged.
4. *Effect on conservation.* A major objection by environmentalists is that massburn incineration removes any incentive by consumers and governments to get on with recycling, which is regarded as the most environment-friendly alternative.

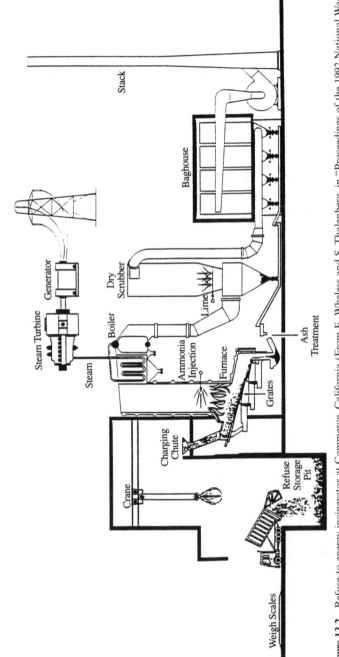

Figure 13.2 Refuse-to-energy incinerator at Commerce, California (From E. Wheless and S. Thalenberg, in "Proceedings of the 1992 National Waste Processing Conference," 1992, p. 165. Copyright © The American Society of Mechanical Engineers, 345 East 47th Street, N.Y., N.Y. 10017. Used with permission.)

Cement Kilns

Burning tires in a cement kiln is attractive, since the tires replace more expensive fuels. Usually the tires have to be reduced to pieces that are 5 to 15 cm long, in which form they are referred to in the trade as *tire-derived fuel.* A cement plant in California derives about 25% of its energy needs from shredded tires, the balance being coal [40]. The temperatures in cement kilns (1450°C), long residence times, and an excess of oxygen assure complete combustion. Moreover, even the steel from the tire beads can become part of the cement product and does not have to be removed as a separate stream.

13.7 DEGRADABLE POLYMERS

Many people, including producers, consumers, and municipal waste collectors, would be very happy if postconsumer polymers would simply disappear without any particular effort on anyone's part. This is the attraction of the degradable polymer. Unfortunately, the reality is that even paper and food products do not "disappear," as has become abundantly clear to those who have excavated landfills and found readable newspapers that are decades old [4]. There certainly are polymers which are truly degradable and there certainly are applications for which degradability is a valuable property. But there seems very little likelihood that the commodity resins used in packaging today are going to be replaced by degradable materials very soon.

As a prescient observer [41] remarked a third of a century ago: "One must remember that the paramount feature sought by polymer engineers has been stability. Biodegradability has not been an attraction. Few workers have designed their experiments by a technique of steepest descent, so to speak, to the depths of degradation." Indeed, packaging, especially for food, seems to be the last place one would want to have microbial growth.

There are many ways in which polymers may degrade (see Chap. 11). However, in developing "disposable" polymers, three main pathways can be distinguished, photooxidation, hydrolysis, and biodegradation.

Photooxidation

A combination of sunlight and oxygen is capable of changing the structure of most hydrocarbons [42]. Although polyethylene is relatively stable under such circumstances, the incorporation of carbonyl groups in the main chain can increase the absorption of photonic energy in the ultraviolet range and also can provide a site at which chain scission will be favored. Additives have been used which will promote photooxidation even when they are not part of the polymer. One such additive is Fe(III)-acetylacetonate:

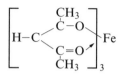

The additive may actually delay the apparent effects of oxidation while hydroperoxide groups are formed. However, eventually the chain-scissioning process predominates and molecular weight decreases. The decrease in molecular weight leads to an increased vulnerability to biological attack. Thus, one often encounters photodegradation being recommended as an adjunct to biodegradation. Photodegradable plastics at one time seemed to be the answer to the problem of litter or, at least, to the problem of certain classes of litter. Some products that have been marketed include beverage rings, agricultural mulch, and hot-drink lids.

Hydrolysis

Most polyesters, polyamides, and polyurethanes are susceptible to hydrolysis, with a consequent decrease in molecular weight. Aliphatic polymers often hydrolyze more rapidly than aromatic polymers. Once again, the lower-molecular-weight materials are subject to biological attack. The hydrolysis itself may be part of enzymatic attack on the main chains. Some polyurethanes based on polyester polyols are easily hydrolyzed, especially if they are in the form of a foamed structure with a very large surface-to-volume ratio. On the other hand, most of the condensation polymers used in films, fibers, and castings to not hydrolyze on casual contact with water. Elevated temperatures and catalysts will accelerate the process.

Microbial Degradation (Biodegradation)

The term *microbial degradation,* as used to describe the fate of plastics in the environment, embraces any change in physical properties brought about by burial in microbially active soil. Soil burial suffers as a laboratory technique for judging degradability because large samples are required and the conditions of exposure are not always reproducible. The "clear zone" technique [43] can be applied to water-immiscible polymers and gives relatively rapid (days rather than months) indications of change. The polymer is suspended in a nutrient-agar medium poured into Petri dishes. The dilute suspension is hazy. As a colony of cells grows on the surface of the gel, a clear zone may occur in the medium surrounding the colony. From this, one can infer that the polymer has been metabolized. Usually this would occur by the excretion of an extracellular enzyme (or more than one) by the organisms. In the case of a polyester, the steps of metabolism might be hydrolysis followed by digestion with growth of cells or oxidation to carbon dioxide and water.

Another, more quantitative technique can be used to measure actual digestion rates. A thin layer of polymer is deposited on the bottom of a Petri dish (Fig. 13.3)

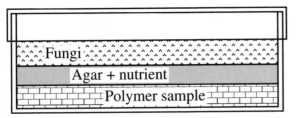

Figure 13.3 Quantitative measurement of polyester degradation [43]

[43]. The dry film is weighed and then a layer of nutrient agar medium is poured over the polymer. The plate is then inoculated with a suspension of the microorganism and incubated. The polymer is recovered after the incubation period and the loss in weight is measured. Sterile plates also are prepared for comparison. These distinguish nonmicrobial degradation such as hydrolysis from the truly microbe-mediated effects. In one test [44], several polyesters were compared after being exposed to the organism *Pullularia pullulans* for 21 days at 30°C (Fig. 13.4). The polymers were similar in structure, but the rate of erosion differed significantly. At intrinsic viscosities above about 0.4 dl/g, the weight loss was hardly measurable.

Commercial and Near-Commercial Degradable Polymers

A variety of products have been made from mixtures of photodegradable polyethylene and starch. A pro-oxidant based on transition metals and lipids might be

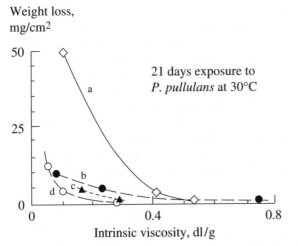

Figure 13.4 Degradation of polyesters: (a) copolymer of tetramethylene adipate and sebacate, 1:1 mole ratio; (b) homopolymer of adipate; (c) homopolymer of sebacate; (d) polycaprolactone [44]

Table 13.3 Some biodegradable polymers

Monomer name	Polymer repeat unit
Glycolic acid	$-CH_2-CO-O-$
Lactic acid	$-CH(CH_3)-CO-O-$
Caprolactone	$-CH_2-CH_2-CH_2-CH_2-CH_2-CO-O-$
β-Hydroxybutyrolactone	$-CH(CH_3)-CH_2-CO-O-$
β-Hydroxyvalerolactone	$-CH(CH_2-CH_3)-CH_2-CO-O-$
Vinyl alcohol	$-CH_2-CH(OH)-$

added. It would appear that the starch part of the composite does degrade, but the polyethylene portion survives much longer, albeit in a weak and flimsy condition.

A review article [45] cited the following kinds of products as being marketed with claims to biodegradability (see also Table 13.3):

Entirely starch-based polymers
Polylactic acid and copolymers containing lactic acid
Polycaprolactone
Water-soluble polymers; poly(vinyl alcohol) is said to biodegrade after dissolving
Unmodified, naturally occurring polymers

In the last category, a polyester produced by fermentation has received a great deal of attention. Poly-β-hydroxybutyrolactone (PHB) is made in high yield by fermentation of glucose [46–49]. A copolymer with hydroxyvalerolactone, PHBV, results when the fermentation is carried out in the presence of some propionic acid.

The properties of PHB are often compared with polypropylene (PP) and poly-(ethylene terephthalate) (PET) (Table 13.4). PHBV is more easily processed than PHB, but has a lower melting temperature. The lower crystallinity of the PHBV copolymer is reflected in the lower energy of melting. The unnotched impact strength of the copolymer is much higher than that for the homopolymer. Uses for PHBV have included blow-molded bottles and packaging film. Yields of 1 kg of polymer for 3 or 4 kg of glucose and fermentation times of 4 to 5 days have made the production of PHB expensive. Commodity applications appear to be distant.

Cellulose (cotton) and natural rubber are biodegradable polymers that have had many uses over the years, almost none of them benefiting from biodegradability. In fact, the treatment of cotton to combat mildew and the addition of biocides to latex compounds is much more in the tradition of commercial uses than working in the opposite direction.

Applications

Almost all producers of degradable polymers aim for large-scale commodity markets such as packaging. Trash bags, grocery bags, and nonfood containers have

Table 13.4 Typical properties of poly (β-hydroxybutyrolactone) (PHB) compared with those of a copolymer (14% hydroxyvalerolactone), polypropylene (PP), and poly(ethylene terephthalate) (PET) [48]

Polymer	PHB	PHB/HV*	PP	PETE
$M_w \times 10^{-3}$	292	258	—	—
M_w/M_n	2.75	3.5	—	—
T_g, °C	6	2	−18	80
T_m, °C	191	177	176	265
Heat of cryst., J/g	127	45	—	—
Density, g/cm³	1.256	1.2	0.905	1.35
Modulus, GPa	3.25	1.47	1.4	2.5
Tensile st., MPa	40	25	40	55
Impact st., notched, J/m	66	65	20–75	35–55
Impact st., unnotched, J/m	115	463	—	—

*Mechanical properties of PHB/HV are for 18 mol% copolymer.

been explored with marginal results. Certain niche applications, however, are much more successful. In the medical area, degradable sutures have been sold for many years. "Absorbable" sutures have been made from collagen (extracted from calf skins) and from synthetic polymers, notably the copolymer of glycolic and lactic acids (see Table 13.3).

A polymer which erodes can be used as the host for a drug delivery system. As the polymer is removed from the surface, usually by biodegradation, the drug dispersed in the polymer becomes available in the body [see sec. 11.7]. An agricultural application with the same idea uses the slowly degrading polymer as a host to deliver pesticides, herbicides, or fertilizers, especially when combined in a tree planting container [50].

13.8 PROSPECTS

Certain aspects of conservation present paradoxical alternatives. The effort to reduce the amount of material in packaging, for example, often requires the use of a material which is more difficult to recycle. The use of an easily recyclable polymer may require the use of expensive raw materials or less efficient processing. A case in point is the modern automobile tire, which makes use of steel, inorganic or organic fibers, several different rubbery polymers, and a variety of compounding additives (see Sec. 10.2). A passenger car tire composed of one polymer without reinforcement or compounding ingredients has not yet been designed for highway use.

Energy recovery is attractive since it does not usually involve separation of components. In the case of some consumer items, the curb-side separation of polymers may be extended to all six currently identified plastics. This could make re-

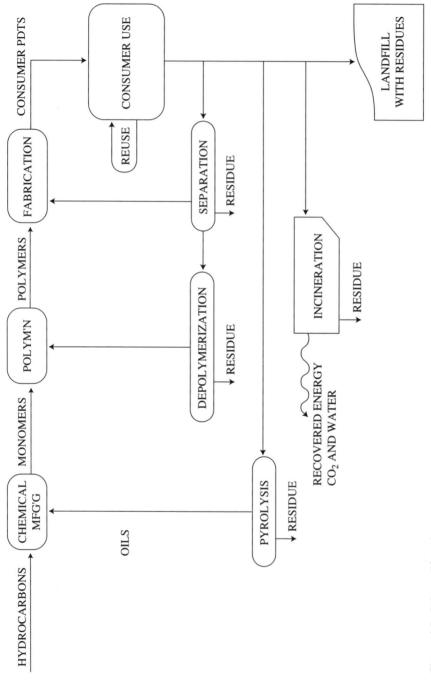

Figure 13.5 Polymer life cycle

cycling processes for LDPE, PVC, PP, and PS as attractive as current programs for HDPE and PET.

It is worth considering a hierarchy of aims that consumer groups, polymer producers, and politicians might agree on with regard to polymer waste disposal:

Reduce usage of polymers at the source by more efficient design.
Reuse items.
Recycle without separation (by melt processes).
Recycle after separation (and design for ease of separation).
Depolymerize.
Convert to liquid fuels.
Incinerate to recover energy.

When the entire life cycle of polymers is considered, it becomes obvious that no matter what steps are taken to reuse or recycle materials, the end products of the hydrocarbons (and other materials that are used to make polymers useful), are CO_2, water, energy, and some solid residues (Fig. 13.5). Moreover, Fig. 13.5 describes only some of the options that apply to polymers, primarily plastics, rubber, and fibers. Not included are many other contributors to the waste stream, such as metal, glass, paper, yard wastes, manufacturing wastes, and sewage.

Every step in the process has inefficiencies which can be diminished but not eliminated. Also, every option has associated social and economic costs. Probably the one factor that overwhelms all other considerations is population growth. While some will think of this only in terms of their own country or community, the worldwide exponential growth of population together with the improvement in the standard of living should put the possible control of disposal by technical advances in perspective.

KEYWORDS FOR CHAPTER 13

Section

1. Municipal solid waste (MSW)
   Sanitary landfill
   Postconsumer waste
   Postindustrial waste
   Rubberized asphalt
2. SPI container code
   Flotation
   Froth flotation
   Cyclone
   Electrostatic separation
   Selective dissolution

3. Down-cycling
   Plastic lumber
4. Repolymerization
   Devulcanization
5. Chemical recycling
   Plastics-to-chemicals
   Hydrogenation
   Pyrolysis
   Gasification
6. Incineration
   Mass-burn
   Tire-derived fuel
7. Biodegradability
   Photodegradability
   Clear-zone technique
   Pro-oxidant
   Absorbable suture

PROBLEMS

13.1 Suppose that a mausoleum were constructed in which the departed would be entombed in caskets made from compacted postconsumer polymer wastes with a density of 1000 kg/m^3. Each casket would have a length of 3 m and an internal volume of 1 m^3. The other two dimensions would be identical. If each of the 250 million people in the United States accumulates half of the total per-capita output of polymers from 1994 (Table 1.1) for 80 years, what would be the necessary dimensions of the casket?

13.2 A small automobile averages about 30 mi/gal (0.425 km/liter) of fuel. Assuming that the density of polyethylene is about 1.2 times that of gasoline and has a similar fuel value, the equivalent of how many shopping bags (6.0 g each) are required to move the automobile 1 mi (and also, 1 km)? How many 1.75-g polystyrene cups would be needed assuming that the density of polystyrene is 1.3 times that of gasoline, and the fuel value is slightly higher (say, 1.2 times more than gasoline)?

13.3 A mixture of wet postconsumer plastic chips is received after removal of paper and metal. All of the six categories in Table 13.2 are included. Draw a flow sheet showing how the chips could be separated into four streams using three liquids. Include mixers, separators (filters or centrifuges), and dryers. Indicate any recycle streams. Liquids: water (1.00 g/cm^3), 20% NaCl in water (1.15 g/cm^3), and tetrahydrofuran (THF) (0.89 g/cm^3). THF is a solvent for polystyrene and PVC at room temperature, but not PET or polyolefins. Output streams: mixed polyolefin chips, polystyrene chips, PET chips, and PVC as a solution in THF.

13.4   For the 14-MW California power plant mentioned in Sec. 13.6, what is the effectiveness of the plant expressed as power produced divided by the heat value of the fuel (tires)? The 4.9 million tires/year can be modeled as each weighing 20 lb with a heat value of 13,000 Btu/lb (1 Btu = 1055 J).

13.5   A plant is to be designed to produce styrene monomer from polystyrene scrap. Current prices of polystyrene and styrene monomer can be found in the *Chemical Marketing Reporter*. Suppose a plant to produce 10 million kg/yr can be built for $20 million and that plant cost scales with production to the 0.6 power. What is the minimum capacity plant (at the breakeven point) that can produce monomer economically?

   Assumptions:

1. Cost of clean polystyrene scrap is one-third that of crystal molding polystyrene.
2. Monomer recovery is 70% of scrap input.
3. Energy and disposal of residues costs $0.08/kg scrap input.
4. Capital depreciation is 10% of the plant cost per year (10-year, straight-line depreciation).

If no current prices are available, use the April 1995 prices of $0.53 and $0.47/lb for polymer and monomer, respectively.

13.6   If a product were made from the low-molecular-weight copolyester (Fig. 13.4) with an intrinsic viscosity of 0.2 dl/g, how long would it take for article which is 1 mm thick to be completely digested? Remember that it would be attacked from both surfaces. Assume a polymer density of 1.1 g/cm^3.

REFERENCES

1. "Statistical Abstract of the United States," p. 227, U.S. Dept. of Commerce, Washington, D.C., 1993.
2. "Characterization of Municipal Solid Waste in the United States: 1994 Update, Executive Summary," U.S. Environmental Protection Agency, Washington, D.C., 1994.
3. Felton, M. K.: *Resource Recycling,* 14(1): 50 (1995).
4. Rathje, W., and C. Murphy: "Rubbish! The Archaeology of Garbage," p. 4, Harper Collins, New York, 1992.
5. Hocking, M. B.: *Science,* 504:251 (1991).
6. Allen, D. T., and N. Bakshani: *Chem. Eng. Educ.,* 26:82 (1992), reprinted in *ChemTech,* 22 (Nov. 1993) and 53 (Dec. 1993).
7. Fouhy, K.: *Chem.Eng.,* 100:30 (Dec. 1993).
8. Culp, E.: *Mod. Plastics,* 69:10 (1992).
9. Cain, M. E.: *Rubber Developments,* 45:66 (1992).

10. Leaversuch, R., *Mod. Plastics,* 65:65 (June 1988).
11. Rathje, W., and C. Murphy: "Rubbish! The Archaeology of Garbage," p. 102, Harper Collins, New York, 1992.
12. "Plastic Container Code System," The Plastic Bottle Information Bureau, Washington, D.C.
13. *Mod. Plastics,* 70:73 (Jan. 1994).
14. *Recycling Today,* 32(8):38 (1994).
15. *Plast. Eng.,* 46(11):14 (1990).
16. "Froth Flotation," in "Kirk-Othmer Encyclopedia of Chemical Technology," Vol. 18, 3d ed., Wiley, New York, 1982.
17. *Chem. & Eng. News,* 63:26 (Oct. 21, 1985).
18. Smoluk, G. R., *Mod. Plastics,* 65:87 (Feb. 1988).
19. "Environmental Impact of Nitrile Barrier Containers, LOPAC: A Case Study," p. 24, Monsanto Co., St. Louis, Mo., 1973.
20. *Chem. Eng.,* 91:22 (June 25, 1984).
21. *Mod. Plastics,* 57:82 (April 1980).
22. Perry, R. H., and D. W. Green, "Perry's Chemical Engineers Handbook, 6th ed., pp. 21-42 through 21-44, McGraw-Hill, New York, 1986.
23. Rodriguez, F., L. M. Vane, J. J. Schlueter, and P. Clark: in G. F. Vandegrift, D. T. Reed, and I. R. Tasker (eds.): "Environmental Remediation," ACS Symp. Ser. 509, ACS, Washington, D.C., 1992.
24. Vane, L. M., and F. Rodriguez, in G. D. Andrews and P. M. Subramanian (eds.): "Emerging Technologies in Plastics Recycling," ACS Symp. Ser. 513, ACS, Washington, D.C., 1992.
25. R. J. Sperber and S. L. Rosen, *Polym. Eng. Sci.,* 16:246 (1976).
26. R. J. Sperber and S. L. Rosen, SPE ANTEC Tech. Papers, 21:521 (1975).
27. Lynch, J. C., and E. B. Nauman: Presented at SPE RETEC: New Developments in Plastics Recycling, Oct. 30–31, 1989; also U.S. Patent 5,278,282 (Jan. 11, 1994).
28. *Mod. Plastics* 70(10):79 (1993).
29. Leaversuch, R. D.: *Mod. Plastics,* 70(1):20 (1993).
30. Nir, M. M.: *Plastics Eng.,* 46(9):29 & 46(10):21 (1990).
31. Thayer, A. M.: *Chem. & Eng. News,* 67(5):7 (1989).
32. Herbold, K.: *Kunstoffe: German Plastics,* 1989.
33. Milgrom, J., in R. J. Ehrig (ed.): "Plastics Recycling," Hanser, New York, 1992.
34. Farissey, W. J. in R. J. Ehrig (ed.): "Plastics Recycling," Hanser, New York, 1992.
35. Kaminsky, W., and H. Sinn: *Hydrocarbon Proc.,* 59(11):187 (1980).
36. M. R. Tesla: *Power Eng.,* 98(5):43 (1994).
37. Clark, C., K. Meardon, and D. Russell: "Scrap Tire Technology and Markets," Noyes Data, Park Ridge, N.J., 1993, p. 53.
38. Wheless, E., and S. Thalenberg: The Design and Operation of an Ash Treatment System at the Commerce Refuse to Energy Facility," in "Proceedings of the 1992 National Waste Processing Conference," Am. Soc. Mech. Eng., New York, 1992.
39. Denison, R. A., and J. Ruston (eds.): "Recycling and Incineration," Island Press, Washington, D.C., 1990.
40. Clark, C., K. Meardon, and D. Russell: "Scrap Tire Technology and Markets," p. 59, Noyes Data, Park Ridge, N.J., 1993.
41. Rodriguez, F., *Chem Tech,* 1:409 (1971).
42. Schnabel, W.: "Polymer Degradation," p. 124, Hanser, New York, 1981.
43. Fields, R. D., F. Rodriguez, and R. K. Finn: *J. Appl. Polym. Sci.,* 18:3571 (1974).
44. Fields, R. D., and F. Rodriguez, in J. M. Sharpley and A. M. Kaplan (eds.): "Proceedings of the Third International Biodegradation Symposium," p. 775, Applied Science, London, 1976.
45. Nir, M. M., J. Miltz, and A. Ram: *Plast. Eng.,* 49(3):75 (1993).
46. Holmes, P. A., L. F. Wright, and S. H. Collins: "Beta-hydroxybutyrate polymers," European Patent Appl., EP 52 459 A1, May 26, 1982.

47. Dawes, E. A. (ed.): "Novel Biodegradable Microbial Polymers," Kluwer, New York, 1990.
48. Holmes, P. A.: *Phys. Tech.,* 16:32 (1985).
49. Baptist, J. N.: "Process for Preparing Poly-β-Hydroxybutyric Acid," U.S. Patent 3,044,942 (1960).
50. Lindsay, K. F.: *Mod. Plastics,* 69(2):62 (1992).

General References

Albertsson, A.-C., and S. J. Huang (eds.): "Degradable Polymers, Recycling, and Plastics Waste Management," Dekker, New York, 1995.

Andrews, G. D., and P. M. Subramanian (eds.): "Emerging Technologies in Plastics Recycling," ACS, Washington, D.C., 1992.

Bisio, A. L., and M. Xanthos (eds.): "How to Manage Plastics Waste," Hanser-Gardner, Cincinnati, Ohio, 1995.

Clark, C., K. Meardon, and D. Russell: "Scrap Tire Technology and Markets," Noyes Data, Park Ridge, N.J., 1993.

Dawes, E. A. (ed.): "Novel Biodegradable Microbial Polymers," Kluwer Academic Publishers, New York, 1990.

Denison, R. A., and J. Ruston (eds.): "Recycling and Incineration," Island Press, Washington, D.C., 1990.

Doi, Y., and K. Fukada (eds.): "Biodegradable Plastics and Polymers," Elsevier, Amsterdam, 1994.

Ehrig, R. J. (ed.): "Plastics Recycling: Products and Processes," Hanser-Gardner, Cincinnati, Ohio, 1992.

Glass, J. E., and G. Swift (eds.): "Agricultural and Synthetic Polymers: Biodegradability and Utilization," ACS, Washington, D.C., 1990.

Griffin, G.: "Chemistry and Technology of Biodegradable Polymers," Chapman & Hall, New York, 1993.

Hegberg, B. A., G. R. Brenniman, and W. H. Hallenbeck: "Mixed Plastics Recycling Technology," Noyes, Park Ridge, N.J., 1992.

Jellinek, H. H. G., and H. Kaachi (eds.): "Degradation and Stabilization of Polymers: A Series of Comprehensive Reviews," vol. 2, Elsevier, New York, 1989.

Kaplan, D. L., E. L. Thomas, and C. Ching (eds.): "Biodegradable Polymers and Packaging," Technomic, Lancaster, Pa., 1993.

La Mantia, F. P. (ed.): "Recycling of Plastic Materials," Chemtec, Toronto, 1993.

Milgrom, J., R. Stephenson, G. Mavel, and G. Vanhaeren: "Recycling Polyethylene, PET, and PVC," Technomic, Lancaster, Pa., 1992.

Mobley, D. P. (ed.): "Plastics from Microbes," Hanser-Gardner, Cincinnati, Ohio, 1994.

Mustafa, N. (ed.): "Plastics Waste Management: Disposal, Recycling, and Reuse," Dekker, New York, 1993.

Rathje, W., and C. Murphy: "Rubbish! The Archaeology of Garbage," Harper Collins, New York, 1992.

Shalaby, S. W. (ed.): "BioPolymers: Designed-to-Degrade-Systems," Hanser-Gardner, Cincinnati, Ohio, 1994.

Swartzbaugh, J. T., D. S. Duvall, L. F. Diaz, and G. M. Savage: "Recycling Equipment and Technology for Municipal Solid Waste," Noyes Data, Park Ridge, N.J., 1993.

Tötsch, W., and H. Gaensslen: "Polyvinylchloride: Environmental Aspects of a Common Plastic," Elsevier, New York, 1992.

Vandegrift, G. F., D. T. Reed, and I. R. Tasker (eds.): "Environmental Remediation," ACS, Washington, D.C., 1992.

CARBON CHAIN POLYMERS

14.1 INTRODUCTION

In considering manufacturing processes for individual polymers, it is convenient to group them in two broad classes, those with backbones containing only carbon and those with backbones containing additional atoms such as oxygen, nitrogen, sulfur, or silicon. Polymers in the first class are made by addition reactions at the double bond (or 1,4-polymerization at a conjugated double bond system). Polymers in the second class can be made by a variety of reactions, including condensation and ring scission.

Two of the major polymer-based industries are found to depend more on one class than the other. The rubber industry uses both carbon chain and heterochain polymers. A generic classification of elastomers with the abbreviations suggested by the American Society for Testing and Materials (ASTM) indicates no bias toward carbon chain polymers (Table 14.1). However, the seven heterochain rubbers listed probably constituted less than 5% of the total rubber produced in 1986. The reason for this situation appears to be economic and historical rather than fundamentally physical or chemical. Heterochain polymers of excellent strength, resilience, and durability have been designed, but never at a price that attracted widespread use. The generally conservative nature of an industry founded on one natural polymer has been another strong deterrent to innovation.

The fiber industry presents a bias toward heterochain polymers. Once again, a listing of the generic names of the man-made fibers (Table 14.2) includes both classes. Since rayon, acetate, nylon, and polyester are the fibers produced in largest

Table 14.1 Generic classification of some commercial elastomers [1]

ASTM designation	Common or trade name	Chemical designation
IR	Synthetic natural rubber	Synthetic polyisoprene
IIR	Butyl, Chloro-butyl	Isobutylene-isoprene
ACM	Acrylic	Polyacrylate
AU*	Urethane (UR)	Polyurethane (polyester)
EU*	Urethane (UR)	Polyurethane (polyether)
BR	CBR, PBd	Polybutadiene
CO*	Hydrin (CO, ECO)	Polyepichlorohydrin
CR	Neoprene	Chloroprene
CSM	Hypalon (HYP)	Chlorosulfonyl polyethylene
EPM	EP elastomer	Ethylene propylene copolymer
EPDM	EP elastomer	Ethylene propylene terpolymer
ET*	Thiokol A	Ethylene polysulfide
EOT*	Thiokol B	Ethylene ether polysulfide
FKM	Viton, Fluorel, Kel-F	Fluorinated hydrocarbon
NBR	Buna N, Nitrile	Butadiene-acrylonitrile
SBR	GR-S, Buna S	Styrene-butadiene
VMQ*	Silicone	Poly(dimethylsiloxane) with vinyl groups
PVMQ*	Silicone	Silicone rubber having methyl, phenyl, and vinyl groups
FVMQ*	Silastic LS	Silicone rubber having methyl, vinyl, and fluorine groups
CM	Plaskon CPE	Chlorinated polyethylene
NR	Natural rubber	Natural polyisoprene
YSBR	Kraton, Solprene	Block copolymers of styrene and butadiene

*Heterochain polymers.

volume, heterochains apparently dominate. In the case of fibers, one can see indications of change, because the acrylics, modacrylics, and olefins are growing in importance. The reason for the dominance by heterochain polymers does have a physical and chemical basis. The *sine qua non* of fibers is a high T_m together with a reasonable degree of crystallinity on stretching (drawing). The heterochain structure is higher in polarity than the carbon chain, and it can often be tailored to a suitable chain stiffness. The high polarity of the nitrile group in acrylics and the remarkable regularity of isotactic polypropylene permit these carbon chain polymers to compete successfully.

Inevitably, some polymers refuse to fit in one class or the other. The polyxylylenes (Sec. 12.4) are carbon chain polymers except that the phenylene group is in the backbone. Another exception is the ladder polymer from polyacrylonitrile (Sec. 3.5), which starts as a carbon chain polymer but is converted by oxidation to a polycyclic heterochain structure.

Table 14.2 Classification and properties of principal man-made fibers† [2]

Fibers	Characteristic properties
Cellulosic fibers	
Rayon—composed of regenerated cellulose from wood and cotton fiber*	Easy to dye and finish, rayon offers high moisture absorption, flexibility, soft hand, good drape, minimum tendency to develop static charges
Acetate and triacetate—composed of regenerated cellulose from wood and cotton fiber that has been treated with acetic acid*	Easy to dye, supple, fast drying, good drape, wrinkle-resistant; heat-set triacetate has high resistance to shrinking, stretching and wrinkling
Synthetic polymer fibers (noncellulosic)	
Acrylic—at least 85% by weight acrylonitrile-based polymer.	Warmth without weight, durable, soft hand, shape retention, resistant to sunlight, weather, oil and chemicals
Modacrylic—at least 35% but not more than 85% acrylonitrile-based polymer	Resemble acrylic fibers; self-extinguishing when exposed to flame
Anidex—a manufactured fiber in which the fiber-forming substance is any long-chain synthetic polymer composed of at least 50% by weight of one or more esters of a monohydric alcohol and acrylic acid	Imparts permanent stretch and recovery properties of fabrics
Aramid—a manufactured fiber in which the fiber-forming substance is a long-chain synthetic polyamide in which at least 85% of the amide linkages are attached directly to two aromatic rings*	High strength and high temperature stability
Novoloid—a manufactured fiber containing at least 85% by weight of a cross-linked novolac	Converts to carbon fiber at high temperatures; used in flame-protective garments
Nylon—a manufactured fiber in which the fiber-forming substance is a long-chain synthetic polyamide in which less than 85% of the amide linkages are attached directly to two aromatic rings*	Great strength, chemical resistance, high elasticity, durability, and range of dyeability
Nytril—a manufactured fiber containing at least 85% of a long-chain polymer of vinylidene dinitrile where the vinylidene dinitrile content is no less than every other unit in the polymer chain	Soft and resilient, used in sweaters
Olefin—at least 85% by weight ethylene, propylene, or other olefin-based polymers; polypropylene most common	Lightest man-made fiber, low moisture absorption, high strength, chemical and soil resistance
Polyester—at least 85% by weight of an ester based on dihydric alcohol and terephthalic acid*	High strength, resistance to stretching and shrinking, dyes well, resists chemicals, crisp and resilient both wet and dry, fabric can be heat-set
Saran—at least 80% by weight vinylidene-chloride-based polymer	Resistance to chemicals, sunlight, and outdoor exposure; nonflammable
Spandex—at least 85% by weight polyurethane*	Easily dyed; strength and durability; high elastic recovery, resistance to chemicals

Table 14.2 Classification and properties of principal man-made fibers† [2]
(*Continued*)

Vinal—a manufactured fiber in which the fiber-forming substance is any long-chain synthetic polymer composed of at least 50% by weight of vinyl alcohol units and in which the total of the vinyl alcohol units and any one or more of the various acetal units is at least 85% by weight of the fiber	High resistance to chemicals
Vinyon—at least 85% by weight vinyl-chloride-based polymer	Resistance to chemicals; softens at low temperatures
Fibers from nonfibrous natural substances	
Glass—composed of silicon dioxide and other components drawn from molten state	Great strength and resistance to heat, flame and most chemicals; will not shrink, stretch, or absorb water
Metal—various metals and alloys drawn to fine-diameter fiber	Properties similar to metals from which they drawn
Rubber—natural or synthetic rubber drawn into fiber form	Elastic, often is a core around which other fibers are wrapped
Azlon—manufactured from naturally occurring proteins in corn, peanuts, and milk*	Soft feel, blends with other fibers, used like wool

*Heterochain polymers.

†This tabulation is based on the Federal Trade Commission's Textile Fiber Products Identification Act, which assigns generic names to the various types of man-made fibers.

Within each of the two major classes the polymers have been arranged in families. Although some properties and uses are mentioned, more details on individual polymers are in App. 3.

14.2 THE POLYOLEFINS

Ethylene and propylene from petroleum-refining operations or from natural gas are the cheapest raw materials for polymer production. Aside from being the starting materials for a variety of other monomers, they are combined in an impressive array of products (Table 14.3). Branched polyethylene, its chlorosulfonated, rubbery derivative, and butyl rubber date back to the 1930s and 1940s. The 1950s saw the introduction of linear polyethylene and isotactic polypropylene. Copolymers of ethylene with propylene and with polar monomers came along in the 1960s, as did the isotactic polymer of 4-methylpentene-1.

The range of properties that can be achieved within the olefin family goes from highly crystalline, linear polymers of high T_m to completely amorphous, rubbery materials that are branched or cross-linked (in the case of cross-linked butyl, ethylene-propylene copolymer (EPM) and ethylene-propylene-diene terpolymer (EPDM) rubbers). Even in the homopolymers of ethylene, mechanical behavior

Table 14.3 The polyolefins

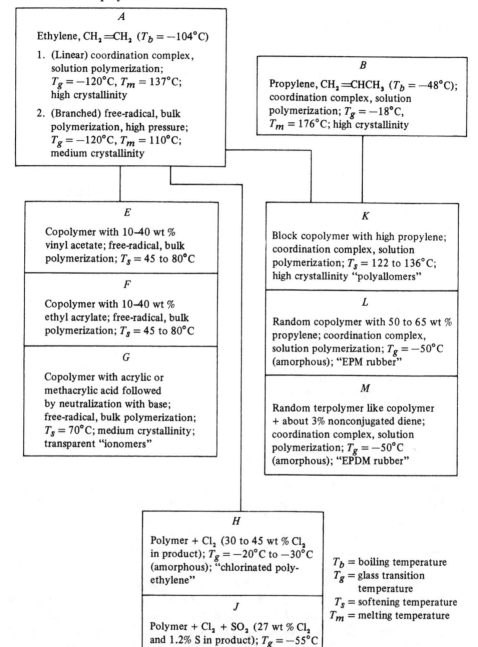

A

Ethylene, $CH_2=CH_2$ ($T_b = -104°C$)

1. (Linear) coordination complex, solution polymerization; $T_g = -120°C$, $T_m = 137°C$; high crystallinity

2. (Branched) free-radical, bulk polymerization, high pressure; $T_g = -120°C$, $T_m = 110°C$; medium crystallinity

B

Propylene, $CH_2=CHCH_3$ ($T_b = -48°C$); coordination complex, solution polymerization; $T_g = -18°C$, $T_m = 176°C$; high crystallinity

E

Copolymer with 10–40 wt % vinyl acetate; free-radical, bulk polymerization; $T_s = 45$ to $80°C$

F

Copolymer with 10–40 wt % ethyl acrylate; free-radical, bulk polymerization; $T_s = 45$ to $80°C$

G

Copolymer with acrylic or methacrylic acid followed by neutralization with base; free-radical, bulk polymerization; $T_s = 70°C$; medium crystallinity; transparent "ionomers"

K

Block copolymer with high propylene; coordination complex, solution polymerization; $T_s = 122$ to $136°C$; high crystallinity "polyallomers"

L

Random copolymer with 50 to 65 wt % propylene; coordination complex, solution polymerization; $T_g = -50°C$ (amorphous); "EPM rubber"

M

Random terpolymer like copolymer + about 3% nonconjugated diene; coordination complex, solution polymerization; $T_g = -50°C$ (amorphous); "EPDM rubber"

H

Polymer + Cl_2 (30 to 45 wt % Cl_2 in product); $T_g = -20°C$ to $-30°C$ (amorphous); "chlorinated polyethylene"

J

Polymer + Cl_2 + SO_2 (27 wt % Cl_2 and 1.2% S in product); $T_g = -55°C$ (amorphous); "chlorosulfonated polyethylene," "CSM rubber"

T_b = boiling temperature
T_g = glass transition temperature
T_s = softening temperature
T_m = melting temperature

Table 14.3 The polyolefins (*Continued*)

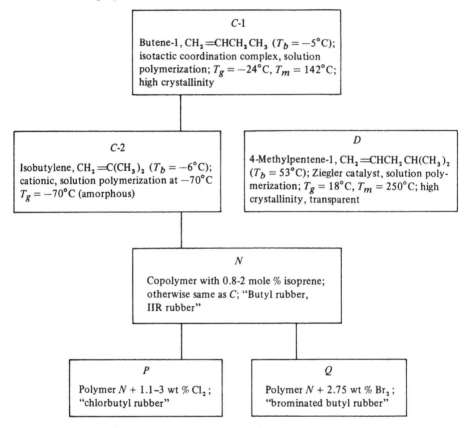

C-1

Butene-1, CH_2=$CHCH_2CH_3$ ($T_b = -5°C$); isotactic coordination complex, solution polymerization; $T_g = -24°C$, $T_m = 142°C$; high crystallinity

C-2

Isobutylene, CH_2=$C(CH_3)_2$ ($T_b = -6°C$); cationic, solution polymerization at $-70°C$ $T_g = -70°C$ (amorphous)

D

4-Methylpentene-1, CH_2=$CHCH_2CH(CH_3)_2$ ($T_b = 53°C$); Ziegler catalyst, solution polymerization; $T_g = 18°C$, $T_m = 250°C$; high crystallinity, transparent

N

Copolymer with 0.8-2 mole % isoprene; otherwise same as C; "Butyl rubber, IIR rubber"

P

Polymer N + 1.1-3 wt % Cl_2; "chlorbutyl rubber"

Q

Polymer N + 2.75 wt % Br_2; "brominated butyl rubber"

will be affected by molecular weight and branching (Fig. 14.1). A high-molecular-weight, linear polyethylene is dense, reflecting its high crystallinity. Also, it is highly viscous, almost rubbery, in the melt at, say 200°C. Branching decreases crystallinity, as does copolymerization. Molecular weights of a few hundred to 10,000 are typical of the naturally occurring waxes recovered from petroleum.

The free-radical polymerization of ethylene requires extremely high pressures. In a typical scheme (Fig. 14.2), 10 to 30% of the monomer might be converted in a single pass at 1000 to 3000 atm and 100 to 200°C. While a peroxide might be introduced, a small amount (0.05%) of oxygen is an effective initiator. Under these conditions the ethylene is well above its critical pressure and temperature and the polymer is dissolved. Stirred reactors have been used, but there are obvious advantages in using a tubular reactor at high pressures. A peristaltic flow in which ethylene is continuously pumped into the tube and product periodically released from

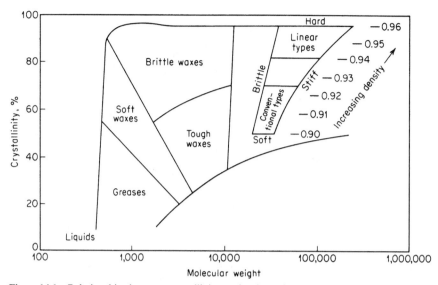

Figure 14.1 Relationships between crystallinity, molecular weight, and mechanical properties of polyethylene [3]. (After Kratz and Lyle, *Encyclopedia of Chemical Technology,* First Suppl., Interscience, 1957.)

the other end is used to approximate a plug flow through the tube and decrease the buildup of polymer on the walls. In a single-train system with a capacity of 100×10^6 kg/yr, the tube might be over 1 km long with an inside diameter of 5 cm and an outside diameter of 15 cm.

In any kind of reactor, the chain transfer between growing polymer and "dead" chains occurs at high pressures, resulting in the branched structure discussed in Sec. 4.4. The same apparatus can be used to produce copolymers of ethylene with polar monomers such as vinyl acetate and ethyl acrylate. As the ethylene content goes down, the crystallinity decreases and room-temperature flexibility increases. Some items in which this controlled degree of flexibility is useful are tubing, protective coatings (waxes and polishes), adhesives, and shoe soles. Copolymerization with carboxylic acids followed by neutralization with a base (usually NaOH) gives the class of *ionomers.* Although still crystalline, the spherulite size is decreased, making films of ionomers much more transparent than those of polyethylene. As might be expected, the ionic groups facilitate adhesion and printing. The ionic groups act as somewhat labile cross-links in the melt, so that ionomer sheets are rubbery and can be vacuum-formed more easily than polyethylene. It is possible also to graft an ionic monomer such as acrylic acid on a "dead" polyethylene or polypropylene. Under these circumstances the ionic groups will probably be clustered at the grafting points rather than randomly distributed as they would be in a

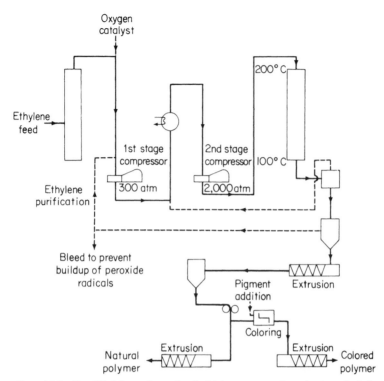

Figure 14.2 Simplified flow scheme for the high-pressure polymerization of ethylene [4]

random copolymerization. Thus, while both products would combine the main features of a polyolefin with ionic character, they may well differ in their applicability for a given end use.

Ziegler-Natta catalysts, the metallocenes, and supported metal oxides (Phillips process) all are capable of producing linear polyethylene (see Sec. 4.6). These systems are sensitive to water and polar compounds, operate at moderate pressures, and usually are somewhat slower than most ionic systems. Reactor residence times of several hours are typical. In order to obtain the highest electrical resistivity, catalyst residues may have to be removed. As indicated in the flow sheet (Fig. 14.3), the metal oxide might be CrO_3 supported on Al_2O_3 (or a silica-alumina base), and polymerization takes place at a pressure of a few hundred psi and a temperature high enough that the polymer dissolves to form a dope up to a concentration of about 10%. The Ziegler process (Fig. 14.4) may be carried out at a lower temperature and, consequently, a lower pressure. Depending on the solvent, the molecular weight, and the temperature, the polymer may stay in solution or precipitate to form a slurry. In the Ziegler process hydrogen is an effective chain transfer agent and can be added to regulate the molecular weight. In the Phillips process the

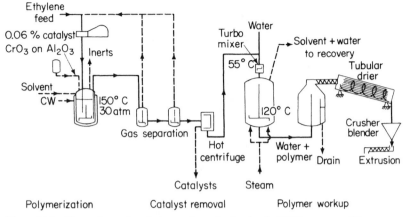

Figure 14.3 Flow scheme for ethylene polymerization by the Phillips process [4]

molecular weight is strongly temperature-dependent, increasing about threefold as the temperature is decreased from 170 to 140°C. Generalizations about these two major processes are dangerous, since hundreds of significant variations in catalyst composition are recorded in the literature and no two companies use exactly the same recipe. The newer metallocene (single-site) catalysts resemble the conven-

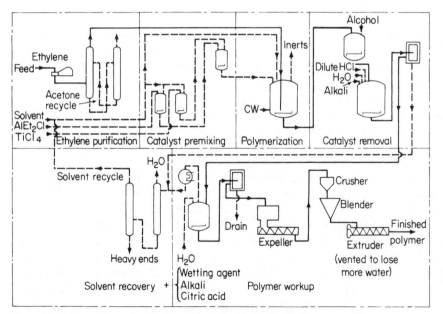

Figure 14.4 Flow scheme for ethylene polymerization by the Ziegler process [4]

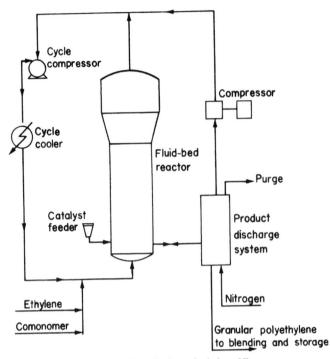

Figure 14.5   Gas-phase polymerization of ethylene [6]

tional Ziegler catalysts in that they can be used in many of the same polymerization processes that have been described. However, the metallocene catalysts have allowed a control over processing properties not previously attained. Very narrow molecular weight distributions coupled with long chain branching make a polyethylene which is more shear-thinning than older materials (see Sec. 4.6). Copolymerization with other α-olefins also can be coupled with control of molecular weight distribution so that materials ranging from highly crystalline to highly elastic have been brought to market.

Gas-phase polymerization for ethylene was commercialized in 1968. Elimination of solvent recovery and catalyst removal as process steps give significant economies. In one installation [5] with a capacity of 100×10^6 kg/yr, the fluid bed reactor (Fig. 14.5) has a diameter of 4.3 m (14 ft). Purified ethylene and powdered catalyst (modified chromia on silica) are fed continuously, and polyethylene fluff is removed intermittently from the reactor through a gas-lock chamber. Reactor pressure is about 20 atm and the temperature is 85 to 100°C. Heat of polymerization is removed as the gas is circulated through external coolers. Only about 2 or 3% of the monomer is converted per pass, but overall monomer loss is kept below 1% of the feed by recycling. The growing particles remain in the reactor 3 to 5 h and attain a final diameter of about 500 μm. Since as much as 600,000 kg of

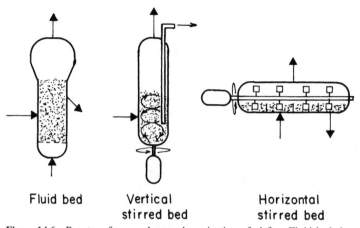

| Fluid bed | Vertical
stirred bed | Horizontal
stirred bed |

Figure 14.6 Reactors for gas-phase polymerization of olefins. Fluid beds have been used by Union Carbide, Shell, BP Chemicals, Himont, and Mitsui Petrochemical, among others. BASF developed the vertical stirred bed. The horizontal stirred bed has been used by El Paso and Amoco [7]. (Courtesy of Gulf Publishing Co.)

polymer can be formed per kilogram of chromium, catalyst removal is not necessary for many applications. Molecular weight can be decreased using hydrogen as a chain transfer agent, and short chain branching can be introduced with 1-butene to decrease crystallinity.

In the 1980s, the process was modified to produce a polyethylene similar in density to the older, high-pressure-process material [6]. In the trade the terms HDPE (high-density, or linear, polyethylene) and LDPE (low-density, or branched, polyethylene) are now supplemented by LLDPE (linear, low-density polyethylene). While all the properties of the high-pressure polymer are not achieved by the newer material, the economics of the low-pressure, solvent-free process are such that about half the volume in markets once dominated by branched LDPE will be in competition with LLDPE. At least in the immediate future few new high-pressure plants are likely to be built.

The fluid bed is only one of several gas-phase polyolefin processes. Stirred beds, arranged either horizontally or vertically (Fig. 14.6), do not depend on gas velocity and are less sensitive to uniformity of gas flow. The stirred beds have been used primarily for polypropylene [7]. The fluid bed has also been adapted for production of polypropylene and propylene copolymers. A 90×10^6 kg/yr plant uses two reactors so that one can be dedicated to homopolymer and the other to modified polymers (Fig. 14.7).

Ziegler-type systems have been used to produce a number of stereoregular polymers. Isotactic and syndiotactic polymers of many olefins have been made in the laboratory. Two of these, isotactic polypropylene and poly(4-methylpentene-1), are available commercially. Production parallels the scheme of Fig. 14.3. To obtain

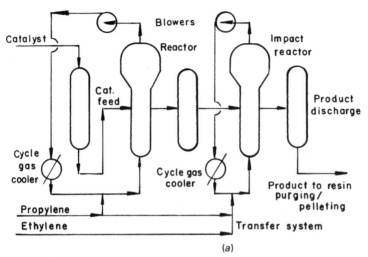

Figure 14.7 Two fluid-bed reactors (one for propylene homopolymer and the other for random and impact copolymers) are used at the Union Carbide Seadrift, Texas, plant. (*a*) The flow sheet [8] is a modification of the ethylene process of Fig. 14.5. (Courtesy of Gulf Publishing Co.)

a polymer of maximum isotactic content, the dope may be precipitated in such a way as to leave the atactic portion behind in solution. However, advances in catalyst preparation have increased the isotactic portion in most processes so that removal of the atactic fraction is not needed. Also, productivity (kilogram of polymer per gram of catalyst metal) usually is high enough so that removal of catalyst residues is seldom necessary. Without the high productivity and high-stereo-selectivity catalysts, the gas-phase processes would certainly lose their economic advantage.

The metallocene "single-site" catalysts (see Sec. 4.6) have been used in gas-phase and slurry reactors of various descriptions. The fact that metallocenes have been easily adapted to existing production lines has sped the introduction of the new polyethylenes into commercial markets. Other monomers that can be polymerized with metallocene catalysts include propylene, other α-olefins, and styrene.

Polyallomers are block copolymers of propylene with small amounts of ethylene. The lifetime of a single growing chain apparently is long enough that blocks can be achieved by a rhythmic alternation of ethylene and propylene in the monomer feed. The materials still are highly crystalline. However, when a random copolymer is made with almost equal amounts of ethylene and propylene, the product is amorphous and rubbery (EPM rubber). Peroxides can be used to cross-link either polyethylene or EPM by hydrogen atom abstraction followed by radical combination. The mechanism was outlined in Sec. 11.3.

The cross-linking is done in a separate operation after the polymer has been made. The rubber is compounded and molded or extruded into its final shape

Figure 14.7 *(Continued)* Two fluid reactors (one for propylene homopolymer and the other for random and impact copolymers) are used at the Union Carbide Seadrift, Texas, plant. (*b*) The reactor towers are the prominent feature. (Photo courtesy of Union Carbide Corp.)

before being cured. By incorporating a third monomer, which contributes a favored cross-linking site, a wider variety of cross-linking agents can be used. Some non-conjugated dienes are especially useful, since only one double bond polymerizes, leaving a pendant unsaturation that is reactive with the sulfur and sulfur-bearing vulcanizing systems long used in the rubber industry. Some dienes used to the extent of about 3% in EPDM (ethylene-propylene-diene monomer) rubber are:

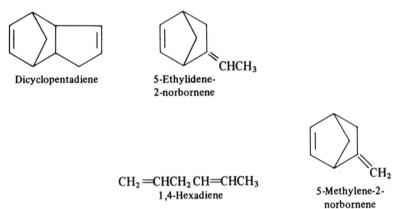

Dicyclopentadiene

5-Ethylidene-
2-norbornene

CH_2=CHCH$_2$ CH=CHCH$_3$
1,4-Hexadiene

5-Methylene-2-
norbornene

A rubbery material can be made directly from branched polyethylene by postchlorination. A major use of this material is as a polymeric plasticizer for poly(vinyl chloride). If SO_2 is added during chlorination, chlorosulfonic groups are added. These act as preferred cross-linking sites when the polymer is heated with a metal oxide and a source of water.

$$-CH_2- + Cl_2 + SO_2 \xrightarrow{\text{boiling CCl}_4} -\underset{\underset{Cl}{\overset{|}{\underset{SO_2}{|}}}}{CH}- + HCl \uparrow$$

In a typical commercial polymer there is one chlorosulfonyl group for each 200 backbone carbon atoms. Magnesia (MgO) and a hydrogenated wood resin (as a source of H_2O) can be used as a cross-linking system.

A perfectly alternating, linear copolymer of ethylene with carbon monoxide was scheduled for commercial production in 1996 [9]. A slurry-type polypropylene plant was converted to produce the new polymer, which is labeled as an aliphatic polyketone, since the oxygen is not part of the chain. The polymer can be modified by the random addition of propylene as a third monomer. Some properties of the semicrystalline copolymer include a density of 1.24 g/cm^3, elongation at yield and break of 25% and 350%, respectively, and a heat deflection temperature of 100°C under load compared to a T_m of 220°C. However, in a composite with 30 wt% glass fiber, the heat deflection temperature rises to 215°C. Suggested uses include gears and bearings as well as other electrical and mechanical components.

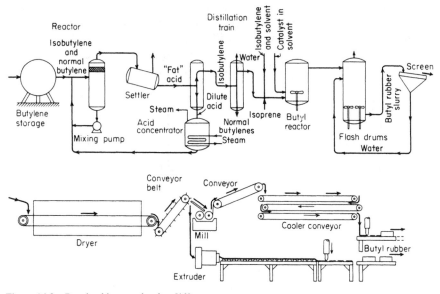

Figure 14.8   Butyl rubber production [10]

$$-CH_2-CH_2-\overset{\overset{\textstyle O}{\textstyle \|}}{C}-$$

Aliphatic polyketone

There are some dramatic contrasts between the cationic polymerization of iso-butylene and the free-radical and coordination complex (both Ziegler and Phillips) methods. Few free-radical processes can be operated below room temperature. Even when the radicals can be generated, the rate of propagation is low. On the other hand, the cationic growth of isobutylene proceeds rapidly at −65°C. Like the coordination complex systems, most cationic catalysts are inactivated by water. Unlike them, however, the radical life is short, so long reactor residence times are not necessary. The major engineering problem becomes that of heat removal at a low temperature. Cooling coils containing liquid ethylene are used inside the reactor (Fig. 14.8). The continuous feed to the reactor consists of two solutions, the monomers (25% isobutylene, 2 to 3% isoprene, balance methyl chloride) and the catalyst (0.4% AlCl$_3$ in methyl chloride). About 3 parts of the latter are sprayed into 10 parts of the former solution. The exothermic reaction is almost instantaneous, and a slurry of fine polymer particles overflows into an agitated flash tank with an excess of hot water. The methyl chloride and any unreacted hydrocarbons are flashed off and sent to a recovery system. An antioxidant and a lubricant (zinc stearate) are added to the slurry in the flash tank also. Screening, filtering, and drying in a tunnel dryer are followed by a forming operation and packaging. The isoprene is used to provide a preferred cross-linking site. The copolymer has been chlorinated and brominated to provide a different reactive group for cross-linking. Metal oxide cross-linking with a halogenated group gives a structure which is more stable at high temperatures than the sulfur cross-link usually employed with residual unsaturation.

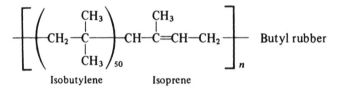

Use Pattern

Blow-molded bottles from polyethylene (squeeze bottles and detergent and bleach bottles) and packaging film are everyday examples of large-volume uses (Table 14.4). Gasoline tanks for many automobiles are blow-molded from HDPE (See Sec. 12.6). Except for butyl rubber and chlorosulfonated polyethylene, the polyolefins are not easily dissolved at room temperature, so they have not been widely used in solvent coatings or adhesives. Melt coating or laminating of polyethylene on paper or cardboard is important for making containers for milk and other liquids. Butyl rubber had a large market in inner tubes before the advent of the tubeless tire. Most tubeless tires today have inner liners of chlorobutyl or bromobutyl rubber to enhance air-pressure retention. An all-butyl tire was offered for a time. Tires still make use of nonstaining polyolefin rubbers for white sidewalls.

14.3 THE ABS GROUP

Homopolymers and interpolymers from acrylonitrile, butadiene, and styrene offer a wide range of properties (Table 14.5). It has been pointed out (Sec. 1.3) that the synthetic rubber program during World War II established almost overnight a tremendous capacity for styrene and butadiene. Butadiene is made by the dehydrogenation of butane or butene from petroleum. Since benzene can be made from petroleum by catalytic reforming, much styrene owes its origin to petroleum too. Benzene is alkylated with ethylene ($AlCl_3$ catalyst) and dehydrogenated to make styrene. The third member of the ABS triumvirate, acrylonitrile, can be made by several routes. A single-stage catalyzed reaction of propylene with ammonia and oxygen is popular:

$$CH_2{=}CH{-}CH_3 + NH_3 + \tfrac{3}{2}O_2 \xrightarrow[\text{400-500°C}]{\substack{\text{molybdenum-}\\\text{bismuth-based catalyst}}} CH_2{=}CH{-}CN + 3H_2O$$

Table 14.4 End use pattern for polyolefins

Polymer*	Molded plastic	Extruded plastic	Film	Laminates	Tires	Rubber specialties	Fibers	Adhesives and coatings	Polymer additives
A, C−1	• •	• •	• •	•			•	•	
B, D, K	• •	• •	• •				• •		
C−2								•	•
E, F	• •	• •	•	•				•	
G	•	•	• •						
H								•	•
J					•	• •		•	
L						• •			
M					•	• •			
N, P, Q			•		•	• •		•	•

A discussion of the stereoregular homopolymers of butadiene is delayed until Sec. 14.4. Although the monomer also is homopolymerized by one producer by a free-radical, emulsion technique, the styrene copolymer (SBR) affords a more complete example (Fig. 14.9). A continuous emulsion polymerization makes use of a series of twelve 7000-gal reactors, each 20 ft tall and 90 in (inside) in diameter. Each reactor is compartmented and stirred, so that the emulsion passes from one end of the train to the other in a kind of plug flow. Residence time in each reactor is 1.0 to 1.25 h, and the heat removed in each may be as high as 132,000 Btu/h. Either "hot" rubber is made with a peroxide initiator at 50°C or "cold" rubber is made with a redox couple at 5°C. A chain transfer agent, dodecyl mercaptan, is added to regulate molecular weight. After reaching a conversion of about 75%, the reaction is short-stopped by the addition of hydroquinone, and unreacted monomers are removed without coagulation in vacuum flash tanks and falling-film strippers. The relative reactivities are such that butadiene is consumed more rapidly than styrene ($r_1 = 1.4$, $r_2 = 0.8$). Consequently, a feed concentration of about 30% styrene is used to achieve a 25% styrene polymer. About 70% of the SBR rubber goes to make tires in which carbon black will be used as a reinforcement and oil as an extender-plasticizer. Most producers sell SBR as master batches that have been compounded before coagulation. Common loadings are 37.5 or 50 parts of oil per 100 of rubber and 40 to 75 parts of carbon black per 100 of rubber.

Latex paints are based on the stripped latex together with suitable pigments, stabilizers, and other additives. Higher styrene contents give somewhat better abrasion resistance (from a higher T_g), and conversions may be run almost to completion (with high molecular weights and branching), since film formation does not involve much polymer flow. Some latexes are intentionally cross-linked by copolymerizing a tetrafunctional monomer, divinyl benzene. These gelled latex particles act like organic fillers and are used to smooth out extrusion of linear rubber. The random copolymer produced by a coordination complex catalyst in solution,

Table 14.5 The ABS polymers

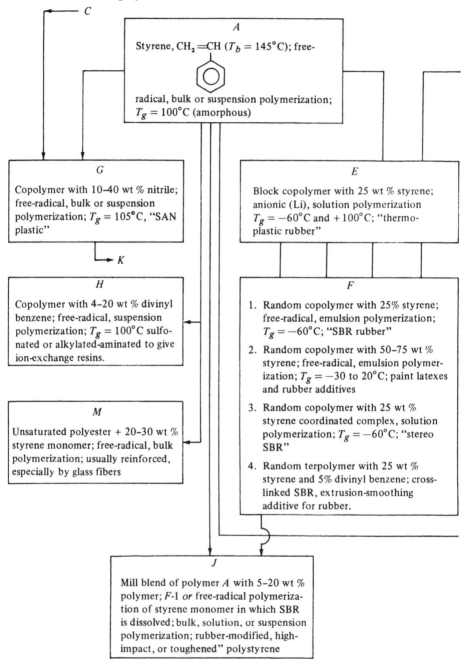

C

A

Styrene, CH_2=CH ($T_b = 145°C$); free-

radical, bulk or suspension polymerization; $T_g = 100°C$ (amorphous)

G

Copolymer with 10–40 wt % nitrile; free-radical, bulk or suspension polymerization; $T_g = 105°C$, "SAN plastic"

E

Block copolymer with 25 wt % styrene; anionic (Li), solution polymerization $T_g = -60°C$ and $+100°C$; "thermo-plastic rubber"

K

H

Copolymer with 4–20 wt % divinyl benzene; free-radical, suspension polymerization; $T_g = 100°C$ sulfo-nated or alkylated-aminated to give ion-exchange resins.

F

1. Random copolymer with 25% styrene; free-radical, emulsion polymerization; $T_g = -60°C$; "SBR rubber"

2. Random copolymer with 50–75 wt % styrene; free-radical, emulsion polymer-ization; $T_g = -30$ to $20°C$; paint latexes and rubber additives

3. Random copolymer with 25 wt % styrene coordinated complex, solution polymerization; $T_g = -60°C$; "stereo SBR"

4. Random terpolymer with 25 wt % styrene and 5% divinyl benzene; cross-linked SBR, extrusion-smoothing additive for rubber.

M

Unsaturated polyester + 20–30 wt % styrene monomer; free-radical, bulk polymerization; usually reinforced, especially by glass fibers

J

Mill blend of polymer A with 5–20 wt % polymer; F-1 *or* free-radical polymeriza-tion of styrene monomer in which SBR is dissolved; bulk, solution, or suspension polymerization; rubber-modified, high-impact, or toughened" polystyrene

Table 14.5 (*Continued*)

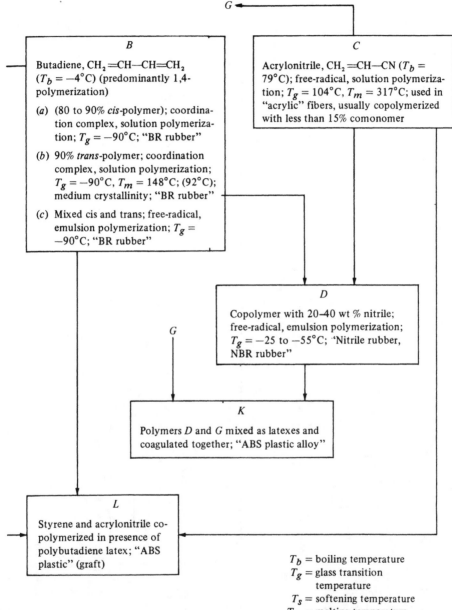

G

B	C
Butadiene, $CH_2 = CH - CH = CH_2$ ($T_b = -4°C$) (predominantly 1,4-polymerization)	Acrylonitrile, $CH_2 = CH - CN$ ($T_b = 79°C$); free-radical, solution polymerization; $T_g = 104°C$, $T_m = 317°C$; used in "acrylic" fibers, usually copolymerized with less than 15% comonomer

(*a*) (80 to 90% *cis*-polymer); coordination complex, solution polymerization; $T_g = -90°C$; "BR rubber"

(*b*) 90% *trans*-polymer; coordination complex, solution polymerization; $T_g = -90°C$, $T_m = 148°C$ (92°C); medium crystallinity; "BR rubber"

(*c*) Mixed cis and trans; free-radical, emulsion polymerization; $T_g = -90°C$; "BR rubber"

D

Copolymer with 20–40 wt % nitrile; free-radical, emulsion polymerization; $T_g = -25$ to $-55°C$; "Nitrile rubber, NBR rubber"

G

K

Polymers *D* and *G* mixed as latexes and coagulated together; "ABS plastic alloy"

L

Styrene and acrylonitrile co-polymerized in presence of polybutadiene latex; "ABS plastic" (graft)

T_b = boiling temperature
T_g = glass transition temperature
T_s = softening temperature
T_m = melting temperature

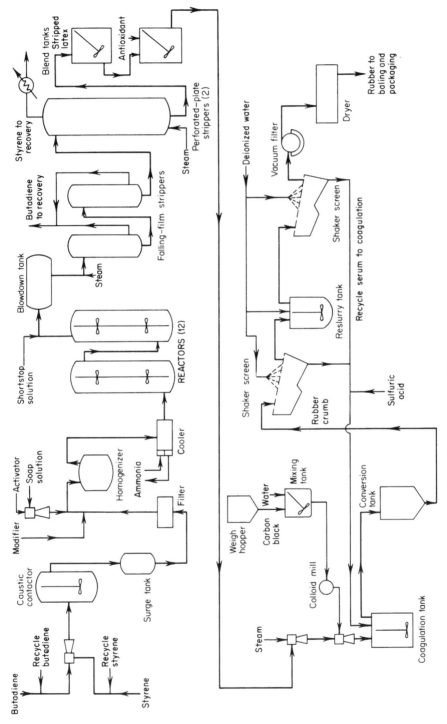

Figure 14.9 Continuous process for styrene-butadiene rubber production [10]

568

Table 14.6 Formulation and properties for vinyl-terminated butadiene-acrylonitrile casting [11]

Component	Parts by weight
Hycar VTBN (B.F. Goodrich Chemical Co.)	50
Styrene	50
Dimethylaniline	0.2
Benzoyl peroxide	2.5
TiO$_2$	20

Stir together at 27°C, cast and hold at 27°C overnight followed by 10 min at 120°C.

Property	Value
Tensile strength	8.2 MPa (1210 psi)
Elongation	320%
Stress at 100% elongation	· 1.2 MPa (170 psi)
Hardness (Shore A)	39

"stereo SBR," has many of the same properties as the free-radical material. It probably contains less branching and less 1,2-polymerization of the butadiene. On the other hand, block copolymers produced by rhythmic alternation of styrene and butadiene in the feed of a Ziegler polymerization are quite different in mechanical behavior from conventional SBR (see Sec. 4.6, Insertion Polymerization).

Copolymerization of acrylonitrile with butadiene is usually carried out in free-radical emulsion polymerization with high conversion. Because so many modifications are marketed (for example, low, medium, and high nitrile content) and the total volume is smaller than that of SBR, production is most often batchwise rather than continuous. The outstanding property of NBR, or *nitrile* rubber, as it is called, is resistance to swelling in nonpolar solvents. With increasing acrylonitrile content, the cohesive energy density increases, so swelling in nonpolar, low-cohesive-energy-density solvents decreases (Sec. 2.5). The term "solvent-resistant rubber" is used, but it should be regarded cautiously because a polar rubber actually swells more than a nonpolar one in a polar solvent.

In the 1970s a family of polymers became available that is based on a 1:5 mole ratio of acrylonitrile to butadiene [11]. These liquid polymers (mol wt of 3000 to 4000) are made with a variety of end groups. Carboxyl, amine, and vinyl groups allow a choice of subsequent reactions. The carboxyl and amine groups react with epoxies and can be part of castings or adhesives. A vinyl group at each end makes the "prepolymer" capable of chain polymerization with a functionality of four (or more, if the original polymer is branched). A casting compound is illustrated in Table 14.6. The classification is complex from the standpoint that the butadiene

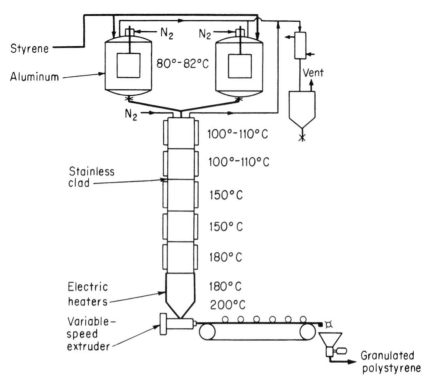

Figure 14.10 Continuous polymerization of styrene [12]

and acrylonitrile presumably form a random copolymer, which is then block-copolymerized with the styrene. Thus the result is part random, part block, an interesting hybrid.

The atactic homopolymer of styrene is produced by free-radical initiation in bulk or suspension polymerization. Few details of commercial bulk polymerization have been made public. A two-stage scheme is shown in Fig. 14.10. A peroxide (or a controlled amount of oxygen) might be added to obtain the low conversion in stirred stainless-steel reactors over a period of 10 to 30 h. Final polymerization to high conversion is obtained in a tower with a temperature that rises from 100 to 200°C. A scraper can be used to help move the polymer. Finally, a vented extruder simultaneously removes unreacted monomer and forms the product into molding pellets. Suspension polymerization in a batch process might require 8 to 12 h at 70 to 100°C. Coagulation is not necessary, since the particles are large enough to be filtered out directly after steam treatment to remove unreacted monomer. Suspension polymerization also is ideal for producing the cross-linked beads required for ion-exchange resins. In the presence of about 10% divinyl benzene, the monomer beads are gelled at a low conversion, triggering the Trommsdorff effect (Sec. 5.1).

This cuts the time for almost 100% conversion down to about 2½ h. From the 1000-gal, glass-lined reactors (Fig. 14.11), the beads (0.30 to 1 mm in diameter) are transferred after washing and drying to separate kettles, where they are reacted with sulfuric acid or chloromethyl ether followed by trimethylamine, making a quaternary ammonium group. A weak anion-exchange group results if dimethyl amine is used instead.

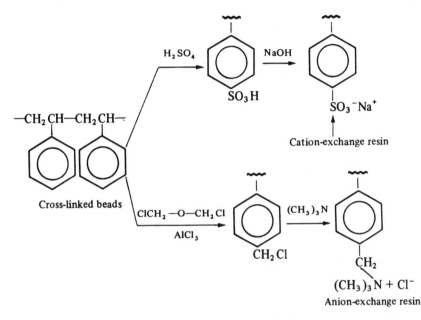

For applications that require the crystal clarity of polystyrene, but where the chemical resistance and tensile strength are marginal, a copolymer of styrene with acrylonitrile is available (SAN). Some grades will bear more than twice the stress that conventional polystyrene does before crazing or cracking. Where clarity is not needed, polystyrene can be toughened by mechanically blending it with a rubbery polymer. Blends can be made by milling polystyrene with SBR rubber on a hot mill or by coagulating latexes of the two ingredients together. However, it is more common to dissolve SBR or polybutadiene rubber in styrene monomer and then to polymerize in bulk. This results in some grafting along with homopolymerization. In these rubber-modified resins, maximum toughness is achieved by selecting the two components so that they are compatible enough to wet each other well but incompatible to the extent that the rubber will be present as discrete droplets about 0.005 cm in diameter within a continuous polystyrene matrix. The impact strength can be raised to over 1 ft-lb/in (53 J/m) of notch from the unmodified value of 0.3 by a rubber content of 10%.

The combination of all three monomers is different enough to result in a separate class of materials, the ABS resins. Simultaneous terpolymerization does not

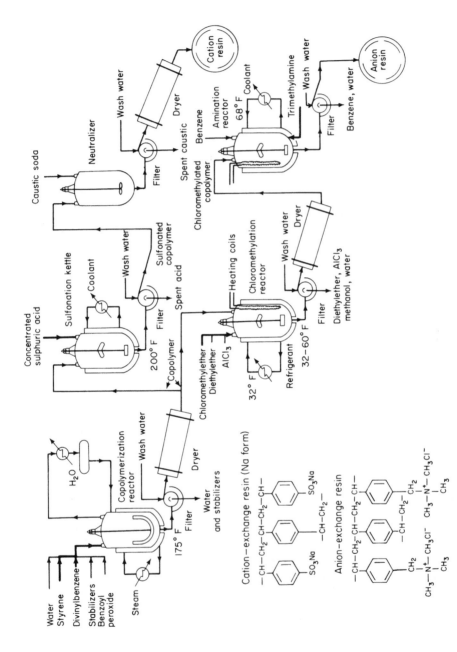

Figure 14.11 Ion-exchange resins based on polystyrene beads [13]

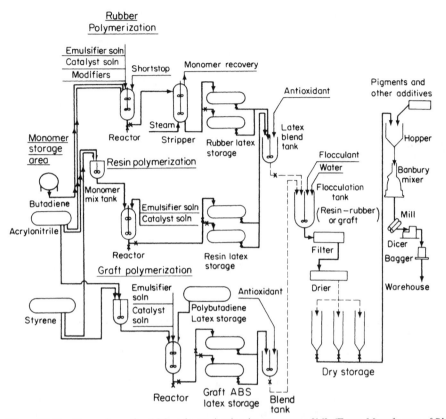

Figure 14.12 Flow scheme for ABS resin production by two routes [14]. (From *Manufacture of Plastics,* by W. Mayo Smith, copyright 1964 by Reinhold Publishing Corp., by permission of Van Nostrand Reinhold Co.)

give the desired heterogeneity that goes along with high impact strength and general toughness. The two processes favored are a blending of latexes of styrene-acrylonitrile and acrylonitrile-butadiene copolymers or a grafting of styrene and acrylonitrile on a polybutadiene backbone in latex form. In the blending process (Fig. 14.12), the butadiene-acrylonitrile copolymerization usually is short-stopped at about 75% conversion and residual monomer is stripped before blending. The styrene-acrylonitrile copolymerization can be carried substantially to completion, as can the grafting of the two monomers on a polybutadiene latex. Typical recipes include, in each case, surfactants, a water-soluble peroxide, and a chain transfer agent (mercaptan) (Table 14.7).

The three monomers representing glassiness (styrene), rubberiness (butadiene), and polarity (acrylonitrile), can be placed at the corners of a triangular composition diagram (Fig. 14.13). Even the triangular diagram is not enough to cata-

Table 14.7 Typical polymerization recipes for ABS components (parts by weight) [14]

	Rubber	Resin	Graft
Water	180	200	200
Butadiene	65		
Styrene	–	70	42
Acrylonitrile	35	30	24
Rubber latex (solids)	–	–	34
Sodium cetyl sulfate	3.0		–
Sodium alkyl aryl sulfonate	–	2.0	
Sodium disproportionated rosin	–	–	2.0
Cumene hydroperoxide	0.2	–	–
Potassium persulfate	–	0.30	0.2
Sodium bisulfite	–	0.01	
Mercaptans (C_{12})	1.0	0.35	1.0
Time	17 h	4 h	Not given
Temperature	41°C	50°C	50°C
Conversion	75%	100%	100%

log all the significant variations that are possible. Neither steric variations (tacticity and cis-trans isomerism) nor cross-linking are shown. As indicated previously, both random and block copolymers of styrene and butadiene have the same ratio of monomers, but they differ in properties. On the diagram the "barrier" resins, which appear at high acrylonitrile content, are so called because of their low permeability to oxygen and carbon dioxide. Functional groups affect permeability strongly (Table 14.8). This is reflected in the series of styrene-acrylonitrile copolymers (Table 14.9). Such polymers are used as margarine tubs and vegetable oil bottles.

Even more complex are compositions that are physical mixtures of polymers. One important outlet for ABS resins is in blends (*alloys*) with poly(vinyl chloride). In Europe, rigid poly(vinyl chloride) is used for many sheet-forming applications such as door liners and crash pads. In the United States, ABS has been used. It

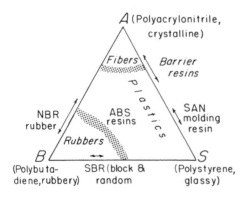

Figure 14.13 ABS compositions. Besides the homopolymers, interpolymers of the three building blocks give rise to numerous modifications. Locations indicated are only approximate

Table 14.8 Effect of functional groups on permeability* [15]

R group in +CH$_2$CHR+	Permeability to oxygen
—OH (dry)	0.002
—CN	0.035
—Cl	8.02
—F	15.0
—CH$_3$	150
—C$_6$H$_5$	416
—H·	501

*Permeability is in units of cm^3/day for an area of 100 in^2 and a driving force of 1 atm over a thickness of 0.001 in.

would appear that alloys may be developed that combine the flame retardancy inherent in poly(vinyl chloride) together with the toughness and ease of fabrication of ABS.

Fibers bearing the generic name *acrylic fiber* contain at least 85% acrylonitrile. *Modacrylics* contain 35 to 85% acrylonitrile. Most familiar to the consumer by trade-marked names such as Orlon, Acrilan, Creslan, and Dynel, these fibers appear in rugs, upholstery, and apparel. Acrylonitrile may be solution-polymerized in water from which the polymer precipitates or in a polar solvent such as acetonitrile. Varying amounts of comonomers are used to modify mechanical properties or to facilitate the dying of products. In spinning fibers from solution, a wet technique may be used with a water-miscible solvent spun into a water bath. Dry spinning with evaporation of solvent is another possibility.

In 1995, the commercial availability of syndiotactic polystryene was announced in Japan and in the United States [16]. The polymer is produced using metallocene catalysis and, unlike the atactic polymer, is highly crystalline, with a T_m of 270°C. Most grades offered have been glass-reinforced. Alloying with nylon

Table 14.9 Gas permeability* of acrylonitrile-styrene co-polymers [15]

Weight % acrylonitrile in copolymer	Permeability to oxygen	Permeability to carbon dioxide
0	416	1250
25	66.8	217
60	4.5	7.5
67	2.3	5.3
82	0.25	0.83
100	0.035	0.15

*Permeability is in units of cm^3/day for an area of 100 in^2 and a driving force of 1 atm over a thickness of 0.001 in.

Table 14.10 End-use pattern for the ABS group of polymers

Polymer*	Molded plastic	Extruded plastic	Film	Laminates	Tires	Rubber specialties	Fibers	Adhesives and coatings	Polymer additives
A, G J	• •	•	•					•	
B					• •	•		•	•
C							• •		
D						• •		•	•
E						• •			
F-1, F-3					• •	• •		•	•
F-2								• •	•
F-4						•			• •
H (ion-exchange)									
K, L	• •	• •	•	• •				•	•
M	•			• •				•	

and other polymers is said to improve processing. The heat deflection temperature under load is 256°C for an alloy with nylon which is reinforced with 30 wt% glass fiber. Some applications for the homopolymer and its alloys include heat-, steam-, and oil-resistant food containers, molded circuit boards, and radiator tanks for automobiles.

Use Pattern

Each of the members of the ABS group can be regraded as primarily a thermoplastic molding resin, as a rubber, or as a fiber (Table 14.10). An exception is the divinylbenzene-styrene copolymer, which is the basis for ion-exchange resins. This copolymer does not fit the three aforementioned end use categories because it is cross-linked and glassy. The *cast polyesters,* in which styrene monomer is polymerized in the presence of an unsaturated polymer, are discussed later in the section on polyesters. Styrene also finds use in the coatings industry as a modifier for other resins. Styrenated alkyds and oils are uses that undoubtedly make use of grafting to some extent. However, the heavyweight of the ABS group in the coatings business is the high-styrene SBR latex. The interior water-based paints introduced right after World War II were made from SBR latexes. The vinyl and acrylic latexes now dominate the market.

14.4 THE DIENE POLYMERS

Although conjugated diene monomers are capable of exhibiting a functionality of four in polymerization reactions, the important examples are linear or branched polymers with residual unsaturation (Table 14.11). Natural rubber (*cis*-1,4-

Table 14.11 The dienes (butadiene included in table 14.5)

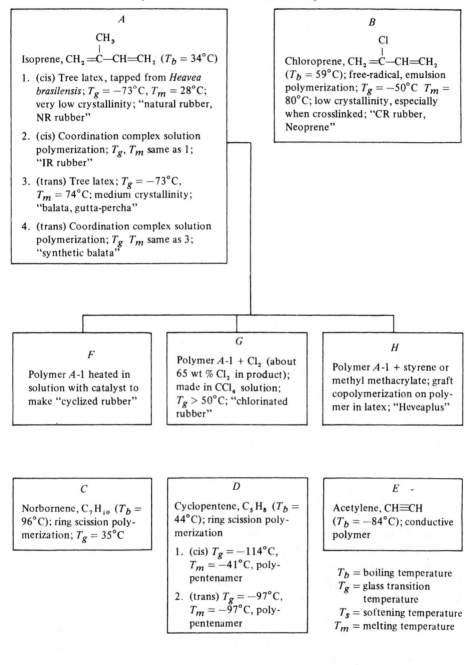

A

Isoprene, $CH_2\!=\!\underset{\underset{CH_3}{|}}{C}\!-\!CH\!=\!CH_2$ ($T_b = 34°C$)

1. (cis) Tree latex, tapped from *Heavea brasilensis*; $T_g = -73°C$, $T_m = 28°C$; very low crystallinity; "natural rubber, NR rubber"

2. (cis) Coordination complex solution polymerization; T_g, T_m same as 1; "IR rubber"

3. (trans) Tree latex; $T_g = -73°C$, $T_m = 74°C$; medium crystallinity; "balata, gutta-percha"

4. (trans) Coordination complex solution polymerization; T_g T_m same as 3; "synthetic balata"

B

Chloroprene, $CH_2\!=\!\underset{\underset{Cl}{|}}{C}\!-\!CH\!=\!CH_2$ ($T_b = 59°C$); free-radical, emulsion polymerization; $T_g = -50°C$ $T_m = 80°C$; low crystallinity, especially when crosslinked; "CR rubber, Neoprene"

F

Polymer *A*-1 heated in solution with catalyst to make "cyclized rubber"

G

Polymer *A*-1 + Cl_2 (about 65 wt % Cl_2 in product); made in CCl_4 solution; $T_g > 50°C$; "chlorinated rubber"

H

Polymer *A*-1 + styrene or methyl methacrylate; graft copolymerization on polymer in latex; "Heveaplus"

C

Norbornene, C_7H_{10} ($T_b = 96°C$); ring scission polymerization; $T_g = 35°C$

D

Cyclopentene, C_5H_8 ($T_b = 44°C$); ring scission polymerization

1. (cis) $T_g = -114°C$, $T_m = -41°C$, polypentenamer

2. (trans) $T_g = -97°C$, $T_m = -97°C$, polypentenamer

E

Acetylene, $CH\!\equiv\!CH$ ($T_b = -84°C$); conductive polymer

T_b = boiling temperature
T_g = glass transition temperature
T_s = softening temperature
T_m = melting temperature

polyisoprene) was harvested and used in the Western Hemisphere for centuries before it was introduced to Europe over 200 years ago. Because of the remoteness of the tropical climates where rubber could be grown, Germany in World War I and Germany, Russia, and the United States in World War II intensified research on duplicating the natural product on which so much of their transportation depended. Substitutes such as SBR rubber were found, but synthesis of the stereoregular polymer was not accomplished until the 1950s and 1960s. There was no rush to put synthetic "natural" rubber into production, however, because by this time the "substitutes" had come to be appreciated for their own properties.

Natural rubber is recovered as a latex by tapping trees. Although the process resembles maple sugar tapping, the latex is carried in a separate layer from the sap of the rubber tree. The latex contains about 35% rubber and another 5% of solids comprised of proteins, sugars, resins, and salts. Most tree latex is coagulated and formed into sheets 3 or 4 mm thick. At one time, smoking (exposure to fumes from burning wood) was favored as a method of preserving against mold and bacterial attack. The same process is used for curing hams and fish. Nowadays, other types of fungicides and bactericides can be used with air-dried sheets. While there are many standard grades of natural rubber, names frequently encountered are "ribbed smoked sheets," and "pale crepes." Major producers include Malaysia, Indonesia, Thailand, and Sri Lanka. Almost half the world's supply today is sold in the form of *technically specified rubber* (TSR), which establishes standards for preparation, properties, and packaging. In 1992, worldwide production of natural rubber was 5.6×10^9 kg, compared to 9.1×10^9 kg for all synthetic rubber.

Since the middle 1970s, attempts to recover *cis*-polyisoprene from guayule have received the support of U.S. and Mexican governments [17]. Guayule is a shrub native to the southwestern United States and northern Mexico. Unlike the rubber tree, which is tapped again and again using a sizable input of hand labor, the guayule shrub is harvested in its entirety. The process can be automated. Besides the 20% by weight of rubber that can be extracted with solvents, dried guayule has another 20% of resin and wax that may have commercial value. Two major incentives for development of guayule are the goal of independence from an overseas source of raw materials as far as the Americas are concerned, and the productive use of arid and semiarid lands in the United States and Mexico.

The naturally occurring *trans*-1,4-polyisoprene is recovered as a latex from trees also. Gutta-percha and balata are obtained from trees in southeast Asia and South America, respectively. The unvulcanized material crystallizes enough to be quite dimensionally stable at room temperature. It may have been the first thermoplastic resin, since it was used as an electrical insulation for submarine cable in England and Germany [18] as long ago as 1848. One application for the vulcanized gutta-percha, golf-ball covers, indicates the typical properties of the polymer. Another novel application is as a splint in orthopedics. The synthetic *trans*-1,4-polyisoprene made with Ziegler catalysts is preferred. Direct molding is used to customize the splint. When heated to 70 or 80°C, the polymer becomes amorphous

Table 14.12 Structure of polymers from conjugated dienes [19]

Monomer	Initiator	Polymer structure, %			
		cis-1,4	*trans*-1,4	1–2	3–4
Butadiene	Free radical, 50°C	19	60	21	
	Butyl lithium–hexane, 20°C	33	55	12	
	Cobalt chelate–AlEt$_2$ Cl,				
	Al/Co = 5 : 20	99	1		
	TiCl$_4$-AlR$_3$, Al/Ti < 1	6	91	3	
Isoprene	Free radical, 50°C	18	75	5	5
	Butyl lithium–pentane	93	–	–	7
	TiCl$_4$-AlEt$_3$, Al/Ti > 1	96	–	–	4
	TiCl$_4$-AlEt$_3$, Al/Ti < 1	–	95	–	5

and a sheet can be handled and shaped quite easily. It can be held against the skin, where it becomes dimensionally stable after a few minutes due to crystallization. It can replace plaster of Paris, which is the older material used for this purpose. The polymer has the advantages of being light, durable, attractive, and unaffected by water.

The remarkable degree of control over structure in diene polymers is illustrated in Table 14.12. These values are selected from among hundreds found in the literature. A commercially successful catalyst system may use similar systems, but economic considerations will modify the requirements. Predominantly cis polymers have the largest market, although the trans polymers are available. The major differences between isoprene or butadiene polymerization by the Ziegler process and the process described for ethylene and propylene (Sec. 14.2) are that the reaction may be short-stopped at 80 to 90% conversion, that monomers are less volatile and somewhat more difficult to recover, and that the polymer remains soluble in the medium and must be precipitated by adding a diluent. In one process (Fig. 14.14), a modified Ziegler catalyst (aluminum, cobalt) in benzene is charged together with butadiene to the first in a line of cooled, agitated reactors. In 4 to 6 h total residence time at 20 to 50°C, conversion reaches 80 to 90% and is short-stopped, presumably by an alcohol or some other reactive material. Just before short-stopping, an additional catalyst may be added to make a molecular weight "jump," probably involving a dimerization. This is a useful device for achieving the high molecular weight needed for oil-extended rubber. Since the jump occurs at high conversion, only the final cement or dope has a high viscosity. An efficient solvent-recovery system is essential for economic operation.

The emulsion polymerization of chloroprene by free-radical initiation dates back to the early 1930s. While the *trans*-1,4 structure predominates, it is not all head-to-tail, and there is some branching. Most chloroprene is made by chlorina-

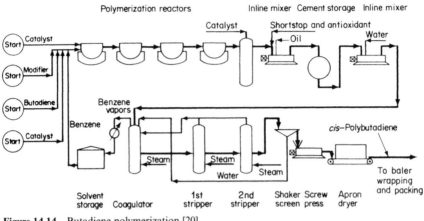

Polymerization reactors   Inline mixer Cement storage Inline mixer

Figure 14.14   Butadiene polymerization [20]

tion of butadiene followed by isomerization to give 3,4-dichlorobutene-1. Then hydrogen chloride is removed by aqueous sodium hydroxide [21].

$$\underset{\text{3,4-Dichlorobutene-1}}{CH_2=CH-\overset{\overset{\displaystyle Cl}{|}}{CH}-\overset{\overset{\displaystyle Cl}{|}}{CH_2}} \xrightarrow[\text{+ NaCH}]{-HCl} \underset{\text{Chloroprene}}{CH_2=CH-\overset{\overset{\displaystyle Cl}{|}}{C}=CH_2}$$

Some monomer may still be made by an older process in which hydrogen chloride is added to vinylacetylene. The continuous emulsion polymerization uses a series of stirred reactors similar to SBR production, although only one is shown in Fig. 14.15. A typical residence time is 2 h for 70% conversion at 40°C. Some grades are produced in batch reactors, and conversions can range up to 90% [22]. Sulfur or sulfur compounds are included, possibly as antioxidants or as chain transfer agents. Several of the commercial rubbers probably have small amounts of comonomer such as styrene, acrylonitrile, or 2,3-dichloro-1,3-butadiene. Some latex is used directly for dipped goods and paints, but most of it is coagulated. The coagulation differs from the method used for SBR in that after neutralizing the emulsifier with acetic acid, the unstable latex is coagulated on the surface of a chilled, rotating drum. The strong film of polymer which forms rapidly can be pulled off the drum continuously, and washed and dried in film form before being gathered into a rope which is chopped into short pieces for packaging. The unvulcanized rubber crystallizes sufficiently to appear hard and horny, with little tendency to flow. Cross-linking can be carried out by heating with 5 parts ZnO and 4 parts MgO per 100 parts of rubber. As with any rubber, many other cross-linking combinations are used.

Two derivatives of natural rubber that have been produced for many years are cyclized rubber and chlorinated rubber. The first of these may be produced in solu-

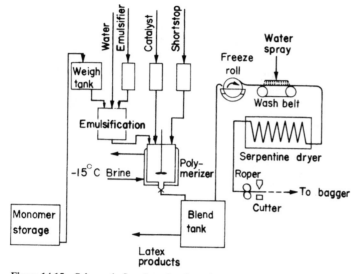

Figure 14.15 Schematic flowsheet for the polymerization and isolation of polychloroprene [22]

tion by heating with a strong acid such as H$_2$SnCl$_6$. The structure that results resembles a ladder to some extent (Sec. 11.3). It is still quite soluble and finds use in adhesives and coatings. Chlorinated rubber with about 65% chlorine in the product has been used as a coating for many years. One particular application is in traffic paints (actually lacquers). The chlorinated rubber is a good binder for the pigments and has excellent abrasion resistance.

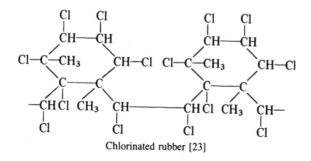

Chlorinated rubber [23]

Two other derivatives that have received attention are thermoplastic natural rubber blends (TPNR), and epoxidized natural rubber (ENR). TPNR often are blends of natural rubber with polypropylene [24]. Applications include automotive components and floor tiles. ENR-50, with 50 mol% epoxidized double bonds, can be used where oil resistance or air impermeability are needed together with high strength [25]. Some applications include hoses and seals.

The ring-scission polymerization of cyclopentene gives a product very similar to the diene polymers, in that one unsaturation remains in the main chain for each monomer unit introduced (see Sec. 4.8). Although a considerable literature [26–28] exists on polypentenamers, neither the cis nor the trans polymer has achieved widespread commercial use. On the other hand, a related rubber was introduced in 1976 in France and has since been offered in other countries. The monomer is norbornene (1,3-cyclopentylenevinylene) made by the Diels-Alder condensation of ethylene with cyclopentadiene:

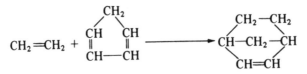

The ring-opening polymerization can be carried out by a variety of catalysts, for example, WCl_6 with aluminum triethyl [29].

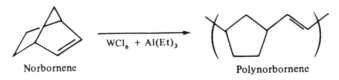

| Norbornene | Polynorbornene |

There can be a considerable amount of cis configuration as well as the trans shown here. The commercial product is a high-molecular-weight (2×10^6) plastic with a T_g of 35°C. To be used as a rubber, plasticizing oils must be added to lower T_g. A fascinating feature of the material is its ability to absorb large amounts of oil and still remain a useful rubber. One mixing technique which can be used is dry-blending of polymer and plasticizer, a method widely used with poly(vinyl chloride). If the oil is added while the powdered polymer is tumbled, the final mixture remains a free-flowing powder which can be fed directly to an extruder. In a typical recipe (Table 14.13), the polymer is less than 20% of the final weight of the product. Some applications have been engine mounts, roll coverings, sealing materials, and tail-light gaskets for automobiles [30].

Polyacetylene is another polymer with residual chain unsaturation:

$$CH{\equiv}CH \longrightarrow +CH{=}CH+$$

Catalytic polymerization of acetylene gas on a glass surface gives a lustrous, silvery film with a fair amount of mechanical strength [31]. The polycrystalline films (mainly trans isomer) can be "doped" with small amounts of AsF_5 to give conductivities of over 100 (ohm·cm)$^{-1}$. Control of electrical and magnetic properties could lead to applications in solar-cell technology [32].

Solid-state polymerization of some diacetylenes proceeds via a biradical to give a backbone in which a triple and a double bond remain for each 4-carbon repeat unit in the chain [33, 34].

Table 14.13 Compound based on polynorbornene [30]

Recipe	Parts by weight
Polynorbornene (Norsorex, American Cyanamid Company)	100
Zinc oxide	5
Stearic acid	1
Carbon black (HAF)	200
Aromatic oil	180
Paraffinic oil	20
Accelerator (*N*-cyclohexylbenzothiazole-2-sulfenamide)	5
Sulfur	1

Properties after molding 30 min at 150°C	
Tensile strength	20 MPa (2900 psi)
Elongation at break	320%
Hardness (Shore *A*)	55
Brittle temperature	−40°C

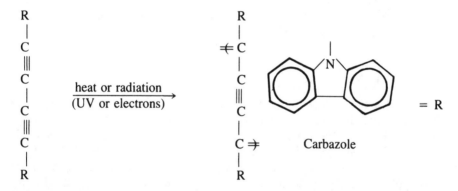

Carbazole

When R is the carbazole unit, large crystals of the monomer can be polymerized in the solid state to give macroscopic single crystals of the polymer with dimensions up to several centimeters. Often the colors that result in the highly conjugated unsaturated polymers are very striking. The heat of polymerization is high, about 37 kcal/mol. Some proposed uses for various diacetylenes are as time-temperature indicators, radiation indicators, nonlinear optical components, and holographic imaging media.

Use Pattern

The dienes enter into all the typical applications of natural rubber (Table 14.14). Tires consume about 70% of the *cis*-polyisoprene and butadiene. Polychloroprene,

Table 14.14 End-use pattern for the dienes

Polymer*	Molded plastic	Extruded plastic	Film	Laminates	Tires	Rubber specialties	Fibers	Adhesives and coatings	Polymer additives
A-1, A-2					• •	• •	•	•	
A-3, A-4						• •		•	
B						• •		•	
C (Foam)						•			
D								• •	
E								• •	
F						•			•

while not suited for tires because of heat buildup, is the most widely used oil-resistant rubber. It is much more resistant to oxidation at high temperatures than the hydrocarbon dienes or SBR. More than any other polymer, it merits the paradoxical ecomium of being the general-purpose specialty rubber. The grafting of methyl methacrylate on natural rubber has been carried out on a commercial scale by polymerizing the monomer in a rubber latex. The primary use of the grafted material is as an additive to improve flow of unmodified rubber.

14.5 THE VINYLS

The group designated as vinyl polymers excludes vinyl benzene (styrene), vinyl cyanide (acrylonitrile), and vinyl fluoride (Table 14.15). The observation recorded by Regnault in 1838, that vinylidene chloride is converted by exposure to sunlight in a sealed glass tube to an insoluble polymer, often is cited as the beginning of polymer science. However, 100 years passed before commercial use began. In the 1930s copolymers of vinyl acetate with a vinyl chloride were introduced. They are more soluble than the homopolymers of vinyl chloride and are used in lacquers. The unique affinity of vinyl polymers for plasticizers was discovered at about the same time. No other thermoplastic matches the ability of poly(vinyl chloride) and its copolymers to form stable, dry, flexible solutions in nonvolatile liquids. Although triaryl phosphates and dialkyl phthalates dominate the plasticizer market today, thousands of liquids are marketed as plasticizers. As a generalization, one could say that every liquid with a volatility less than that of water at room temperature has been suggested as a vinyl plasticizer at one time or another. In Germany and the United States in the 1930s and 1940s, homopolymers were developed using suspension and emulsion techniques that could yield useful plasticized items.

Until the introduction of exterior latex paints, most poly(vinyl acetate) was hydrolyzed to alcohol and reacted to form the butyral adhesive layer for laminating

Table 14.15 The vinyls

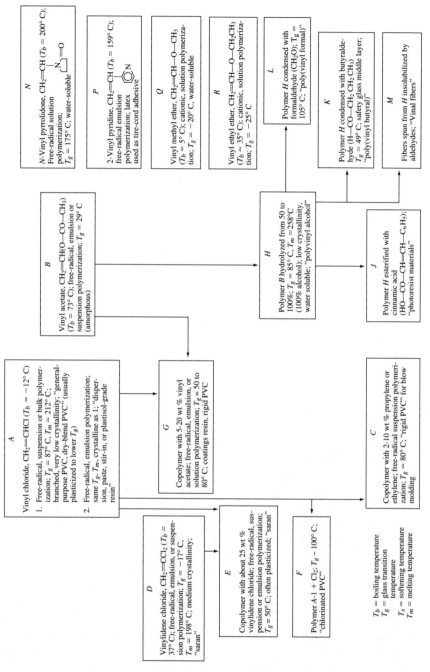

T_b = boiling temperature
T_g = glass transition temperature
T_s = softening temperature
T_m = melting temperature

A
Vinyl chloride, CH₂=CHCl ($T_b = -12°$ C)

1. Free-radical, suspension or bulk polymerization; $T_g = 87°$ C, $T_m = 212°$; branched, very low crystallinity; "general-purpose PVC, dry-blend PVC" (usually plasticized to lower T_g)

2. Free-radical, emulsion polymerization; same T_g, T_m; crystalline as 1; "dispersion, paste, stir-in, or plastisol-grade resin"

D
Vinylidene chloride, CH₂=CCl₂ ($T_b = 37°$ C); free-radical, emulsion, or suspension polymerization; $T_g = -17°$ C, $T_m = 198°$ C; medium crystallinity; "saran"

B
Vinyl acetate, CH₂=CH(O—CO—CH₃) ($T_b = 73°$ C); free-radical, emulsion or suspension polymerization; $T_g = 29°$ C (amorphous)

E
Copolymer with about 25 wt % vinylidene chloride; free-radical, suspension or emulsion polymerization; $T_g = 50°$ C; often plasticized; "saran"

F
Polymer A-1 + Cl₂; $T_g \sim 100°$ C; "chlorinated PVC"

G
Copolymer with 5-20 wt % vinyl acetate; free-radical, emulsion, or solution polymerization; $T_g = 50$ to 80° C; coatings resin, rigid PVC

C
Copolymer with 2-10 wt % propylene or ethylene; free-radical suspension polymerization; $T_g = 80°$ C; "rigid PVC" for blow molding

H
Polymer B hydrolyzed from 50 to 100%; $T_g = 85°$ C, $T_m = 258°$C (100% alcohol); low crystallinity, water soluble; "polyvinyl alcohol"

J
Polymer H esterified with cinnamic acid (HO—CO—CH=CH—C₆H₅); "photoresist materials"

N
N-Vinyl pyrrolidone, CH₂=CH ($T_b = 200°$ C); Free-radical solution polymerization; $T_g = 175°$ C; water-soluble

P
2-Vinyl pyridine, CH₂=CH ($T_b = 159°$ C); free-radical emulsion polymerization; latex used as tire-cord adhesive

Q
Vinyl methyl ether, CH₂=CH—O—CH₃ ($T_b = 5°$ C); cationic, solution polymerization; $T_g = -20°$ C, water-soluble

R
Vinyl ethyl ether, CH₂=CH—O—CH₂CH₃ ($T_b = 35°$ C); cationic, solution polymerization; $T_g = -25°$ C

L
Polymer H condensed with formaldehyde (CH₂O); $T_g = 105°$ C; "poly(vinyl formal)"

K
Polymer H condensed with butyraldehyde (H—CO—CH₂ CH₂ CH₃); $T_g = 49°$ C; safety glass middle layer; "poly(vinyl butyral)"

M
Fibers spun from H insolubilized by aldehydes; "Vinal fibers"

585

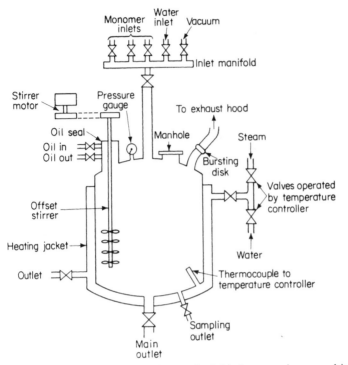

Figure 14.16 Typical polymerization vessel suitable for suspension or emulsion polymerization of vinyl chloride [35]

safety glass. The intermediate alcohol still has no single major market in the United States, but it is made into an important fiber in Japan.

Most vinyl chloride is polymerized by a suspension process. Jacketed, stirred reactors like the one in Fig. 14.16 vary in size from 2000 to 10,000 gal (8 to 40 m³). Individual reactors with volumes of 200 m³ have been made with bottom-entering agitators [36]. The charge to one of these "super-sized" reactors includes 100 × 10³ kg of monomer. Reactors are stainless steel or glass-lined and must handle a pressure of about 1.5 MPa (220 psi). An initiator soluble in the monomer, such as lauroyl peroxide, is added to monomer, which is dispersed in about twice its weight of water containing 0.01 to 1% of a stabilizer such as poly(vinyl alcohol). Over a period of 8 to 15 h at 50 to 80°C, conversion may reach 80 to 90%. Polymer precipitates within the monomer droplets to form a reticulated structure. If the unreacted monomer is blown off at the end of the period by reducing the pressure, porous particles are obtained which can adsorb large volumes of plasticizer. *Dry-blend* resins are those which can adsorb 20 to 40 parts of plasticizer per 100 parts of polymer and still remain as free-flowing powders. Mixing and handling such systems is easier than operating with sticky, semifluid mixtures. The suspension polymer can be dried in rotary dryers, in spray dryers, or in combination grinding mill-classifier-dryers. Special suspension recipes are used with high conversion to give solid, nonporous resins. These are harder to dissolve in plasticizers. However, they are useful as additives in plastisols, since they do not increase the viscosity very much.

Dispersion (also called plastisol or stir-in) resins are made by emulsion polymerization in vessels similar to those used for suspension polymerization. The ease with which these polymers can be suspended in the plasticizer to give a free-flowing liquid is a result of the small particle size of the emulsion and the residual emulsifier. When the latex is spray-dried, most of the emulsifier remains on the surface of the particles (actually aggregates of latex particles) and helps the plasticizer to wet them. Plastisol molding was described in Sec. 12.6, and some effects of plasticization were described in Sec. 3.3.

Bulk polymerization has the advantage that no suspending or emulsifying agent is left in the resin to detract from optical clarity. A two-stage, batch process (Fig. 14.17) operates on a principle similar to that described for continuous polystyrene (Sec. 14.2). "Prepolymerization" in a stirred tank (0.016% azodiisobutyronitrile, 130 rpm, 60°C, 1 MPa) is carried to about 10% conversion. Further reaction to about 75% conversion occurs in a scraped-surface autoclave with slow agitation over a period of 10 to 12 h or more. Porous particles with diameters of 100 to 200 μm result after monomer is blown off and the polymer is recovered.

Copolymerization of vinyl chloride with vinyl acetate or vinylidene chloride gives polymers of lower T_g and better solubility. The main advantage of propylene copolymers is said to be stability and clarity in unplasticized, blow-molded bottles. Postchlorination of poly(vinyl chloride) yields a material of higher T_g that has been recommended for pipes that can carry hot water without deforming.

The latexes that result from vinyl acetate polymerization have been used for exterior water-based paints. Emulsion polymerization can be carried to high conversion fairly rapidly. The rate is controlled by continuous addition of monomer. In one process, 1 m³ of water containing 3% poly(vinyl alcohol) and 1% surfactant is heated up to 60°C [38]. Two streams, monomer and an aqueous persulfate solution, are fed in over a period of 4 or 5 h while the temperature rises to 70 or 80°C. Rate of addition is limited primarily by rate of heat removal. A final heating to 90°C may be used to react the last bit of monomer. When the latex is to be used as an adhesive, a plasticizer may be added. The popular white glues for paper and wood are poly(vinyl acetate) suspensions plasticized with liquids such as castor oil.

Suspension polymerization is favored for poly(vinyl acetate) that is to be hydrolyzed to alcohol. A charge of 2000 kg of vinyl acetate containing 1.5 kg of benzoyl peroxide can be dispersed in 1 m³ of water containing 1 kg of poly(vinyl alcohol) [38]. In 6 to 8 h at 65 to 75°C (reflux), conversion is high enough that monomer can be steam-stripped out, the suspended beads can be hardened by cooling to 10°C, and the product can be filtered and dried.

Hydrolysis of the acetate can be carried out in concentrated methanol solution

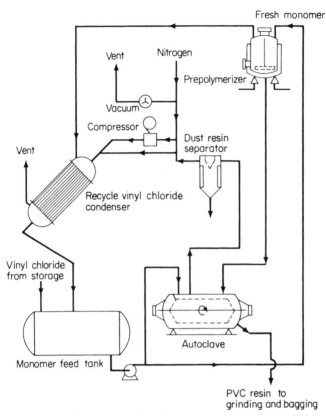

Figure 14.17 Continuous bulk polymerization of vinyl chloride [37]

with a sodium methoxide catalyst. Various degrees of hydrolysis are marketed. The gelatinous poly(vinyl alcohol) precipitates as a slurry in the methanol. Some liquid is removed by expression (between steel rolls, for example), and the balance is taken off in a dryer. The product is ground to a fine powder for most applications. Poly(vinyl alcohol) is water-soluble and has been mentioned frequently as the stabilizing agent in many suspension polymerizations. All the other customary uses (thickening agent, gelling agent, cosmetic ingredient) of water-soluble polymers are conceivable for poly(vinyl alcohol). Fibers are produced in Japan by extruding a concentrated aqueous solution of poly(vinyl alcohol) into a sodium sulfate solution containing sulfuric acid and formaldehyde, which acts as a cross-linking agent.

When butyraldehyde is reacted with poly(vinyl alcohol) suspended in ethanol with sulfuric acid as a catalyst (5 to 6 h, 80°C), an intramolecular condensation is favored. If the reaction were irreversible, one would expect a great many isolated unreacted hydroxyl groups in the product. However, rearrangements do occur and

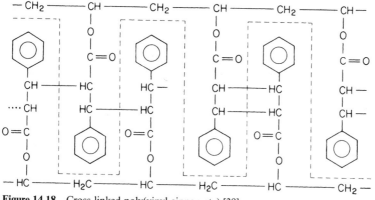

Figure 14.18   Cross-linked poly(vinyl cinnamate) [39]

the product may contain fewer than 10% unreacted OH groups rather than the theoretical limit of 13.5%.

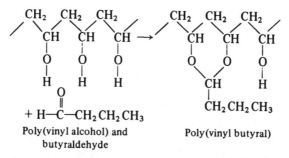

Poly(vinyl alcohol) and butyraldehyde

Poly(vinyl butyral)

Poly(vinyl formal) can be made by direct reaction of formaldehyde with poly(vinyl acetate) in aqueous sulfuric acid. The cyclic intramolecular structure is formed again.

Like poly(vinyl alcohol), an ester with pendant unsaturation cannot be made directly from the appropriate monomer. One such commercial polymer is produced by the reaction of cinnamoyl chloride with poly(vinyl alcohol) dispersed in pyridine [39]. A dry film of the product can be deposited from a solution and easily redissolved. However, when it is exposed to light, extensive cross-linking insolubilizes the film (Fig. 14.18). Lithographic plates are made by this photosensitive insolubilization. Poly(vinyl alcohol) itself also can be photoinsolublized in the presence of chromate ions and by exposure to much higher intensities of ultraviolet light than the cinnamate [39]. Photoresists are discussed also in Sec. 12.4, Coatings.

Use Pattern

Some familiar items made from plasticized poly(vinyl chloride) include garden hoses, laboratory tubing, and numerous film products such as shower curtains,

Table 14.16 End-use pattern for the vinyls

Polymer*	Molded plastic	Extruded plastic	Film	Laminates	Tires	Rubber specialties	Fibers	Adhesives and coatings	Polymer additives
A	• •	• •	• •	•				•	
B								• •	
C	• •								
D	•		• •				• •		
E								• •	
F		• •							
G	• •	•		•				• •	
H			•					• •	• •
J								• •	
K				•				• •	
L								• •	
M							• •		
N								•	•
P					•	•			
Q, R								•	

diaper covers, and raincoats (Table 14.16). Blow-molded bottles from unplasticized vinyls have been commercially available for some time in Europe and the United States. Pipe and glazing are more common outlets for rigid vinyls. Although far outweighed by the chloride and acetate, several other vinyl monomers are homopolymerized industrially. Poly(vinyl ethyl ether) finds use in pressure-sensitive adhesives. Poly(vinyl pyrrolidone) (PVP) probably has its major market as a home-permanent wave-setting resin. It has unique complexing properties which make it valuable in some pharmaceutical and physiological applications. The iodine complex of PVP is an effective antiseptic that is much lower in toxicity than iodine itself. Also, about 7% PVP added to whole blood allows it to be frozen, stored at liquid-nitrogen temperatures for years, and thawed out without destroying blood cells. Poly(vinylidene chloride) is molded into pipe and fittings for corrosion-resistant applications. However, it is probably most familiar in the form of a household plastic film, Saran Wrap (Dow), which is plasticized by acetyl tributyl citrate. Because it is a safe and stable barrier material with low energy absorption, it is widely used as a cover for foods that are heated in microwave ovens.

14.6 THE ACRYLICS

Acrylic and methacrylic acid and their esters are included in the acrylic group (Table 14.17). Otto Röhm published his doctoral thesis dealing with acrylic compounds in 1901 in Germany. First in Germany (1927) and then in the United States (1931), he was associated with the commercialization of acrylic polymers as coating

Table 14.17 The acrylics

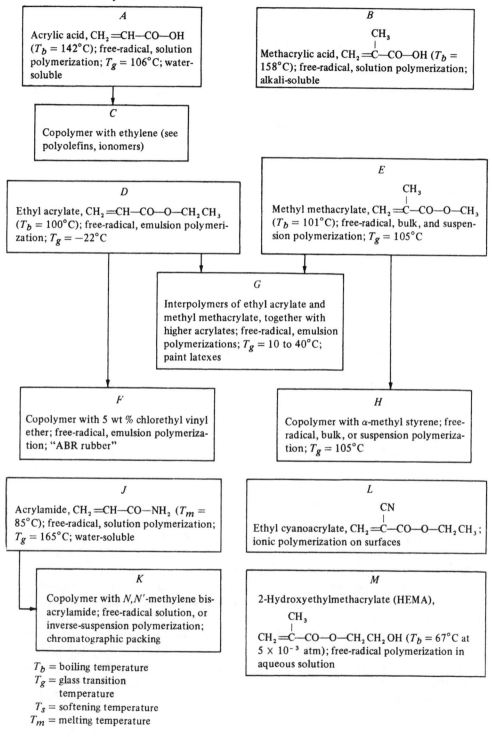

A

Acrylic acid, $CH_2\!=\!CH\!-\!CO\!-\!OH$ ($T_b = 142°C$); free-radical, solution polymerization; $T_g = 106°C$; water-soluble

B

Methacrylic acid, $CH_2\!=\!\overset{\overset{\displaystyle CH_3}{|}}{C}\!-\!CO\!-\!OH$ ($T_b = 158°C$); free-radical, solution polymerization; alkali-soluble

C

Copolymer with ethylene (see polyolefins, ionomers)

D

Ethyl acrylate, $CH_2\!=\!CH\!-\!CO\!-\!O\!-\!CH_2CH_3$ ($T_b = 100°C$); free-radical, emulsion polymerization; $T_g = -22°C$

E

Methyl methacrylate, $CH_2\!=\!\overset{\overset{\displaystyle CH_3}{|}}{C}\!-\!CO\!-\!O\!-\!CH_3$ ($T_b = 101°C$); free-radical, bulk, and suspension polymerization; $T_g = 105°C$

G

Interpolymers of ethyl acrylate and methyl methacrylate, together with higher acrylates; free-radical, emulsion polymerizations; $T_g = 10$ to $40°C$; paint latexes

F

Copolymer with 5 wt % chlorethyl vinyl ether; free-radical, emulsion polymerization; "ABR rubber"

H

Copolymer with α-methyl styrene; free-radical, bulk, or suspension polymerization; $T_g = 105°C$

J

Acrylamide, $CH_2\!=\!CH\!-\!CO\!-\!NH_2$ ($T_m = 85°C$); free-radical, solution polymerization; $T_g = 165°C$; water-soluble

L

Ethyl cyanoacrylate, $CH_2\!=\!\overset{\overset{\displaystyle CN}{|}}{C}\!-\!CO\!-\!O\!-\!CH_2CH_3$; ionic polymerization on surfaces

K

Copolymer with N,N'-methylene bis-acrylamide; free-radical solution, or inverse-suspension polymerization; chromatographic packing

M

2-Hydroxyethylmethacrylate (HEMA),

$CH_2\!=\!\overset{\overset{\displaystyle CH_3}{|}}{C}\!-\!CO\!-\!O\!-\!CH_2CH_2OH$ ($T_b = 67°C$ at 5×10^{-3} atm); free-radical polymerization in aqueous solution

T_b = boiling temperature
T_g = glass transition temperature
T_s = softening temperature
T_m = melting temperature

resins. Later in the 1930s, poly(methyl methacrylate) sheets and molding compounds became available. Latex paints based on ethyl acrylate for exterior application achieved importance in the 1950s and 1960s. Acrylonitrile has been mentioned previously (Sec. 14.3) as a fiber former and as a comonomer in oil-resistant rubber.

Acrylates can be made by a single-step process from acetylene or by a two-step process from ethylene oxide. Methacrylates can be made by selective oxidation of isobutylene followed by esterification or by a two-step process starting with acetone.

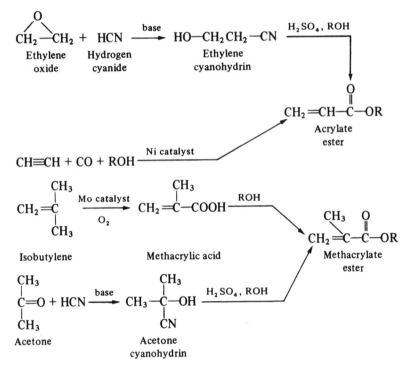

The quiescent bulk polymerization of methyl methacrylate was used as an example in Sec. 5.2. A stirred bulk polymerization like that used for styrene (Sec. 14.3) could be adapted for methyl methacrylate. A modification which bears a little more resemblance to the vinyl chloride bulk polymerization (Sec. 14.5) is *granulation polymerization* [40]. A small amount of water, some lubricant, and either a water- or oil-soluble catalyst is agitated with monomer (or a monomer-polymer syrup) in a heavy-duty dough mixer. Fluffy, granular particles result and can be washed, dried, and extruded into molding pellets. The absence of emulsifier or suspending agent gives a clear product, and the granular nature of the agitated mass gives better temperature control than a viscous, homogeneous mass would afford.

A suspension process of poly(methyl methacrylate) has been described in enough detail to point up some interesting features [41]. The chemicals in the aqueous phase include sodium polyacrylate (suspending agent) and buffers. Polymerization takes only about an hour at 95 to 110°C, 40 psig (0.3 MPa), which means that a 10-m^3 reactor can produce more polymer than a 40-m^3 reactor for styrene or vinyl chloride with their typical reaction times of 8 to 15 h. The advantage is cut somewhat by charging and emptying times. A saving is realized by using the back pressure to push the warm suspension into a separate slurry cooling tank and immediately recharging the reactor. Also, a cyclic fluidized-bed dryer is used rather than the continuous rotary dryer shown in Fig. 14.11. The end product can be molding powder (pellets) or extruded sheet.

The polymerization of ethyl acrylate most often is carried out in emulsion. A process such as that used for vinyl acetate is suitable (Sec. 14.4). Like vinyl acetate, the monomer is slightly water-soluble, so "true" emulsion polymerization kinetics are not followed. That is, there is initiation of monomer dissolved in water in addition to that dissolved in growing polymer particles. Ethyl acrylate is distinguished by its rapid rate of propagation. Initiation of a 20% monomer emulsion at room temperature by the redox couple persulfate-metabisulfite can result in over 95% conversion in less than a minute. To control the temperature, a continuous addition of monomer at a rate commensurate with the heat transfer capacity of the reactor is necessary.

A copolymer of ethyl acrylate with 5% chloroethyl vinyl ether has been marketed as an oil-resistant, high-temperature-stable, specialty elastomer for some years. The pendent chlorine group gives a preferred cross-linking site with polyfunctional amines. The relatively high T_g of the rubber (*ca.* − 28°C) can be lowered by plasticization, but this can be temporary if oils or solvents are in contact with the product and leach out the plasticizer. Interpolymers of ethyl acrylate with methyl methacrylate and with higher acrylates and methacrylates are widely used as latex paints and as additives for paper and textiles. An important parameter of such a system is the minimum film-forming temperature, which one might expect to be related to the T_g. It can be seen (Fig. 14.19) that an almost linear relationship exists between this temperature and composition for the simple binary system. The film hardness measured at room temperature also is expected to increase as the interval between room temperature and T_g increases.

The water-soluble polymers of acrylamide, acrylic acid, and methacrylic acid probably are produced by solution polymerization. Interpolymers of these monomers are used in sewage coagulation, drag reduction, adhesives, coatings, and textile and paper sizes.

In the last 40 years, contact lenses have become an important method of correcting eye defects. While there are other reasons for choosing "contacts" over conventional glasses, the major reason is appearance. There are two basic types, both based on acrylic polymers. The hard lenses generally are made from poly(methyl methacrylate). Most soft lenses are made from poly(2-hydroxyethyl methacrylate), also called poly(HEMA).

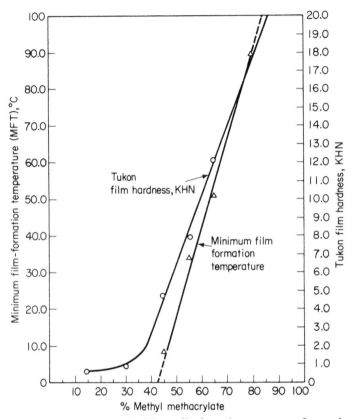

Figure 14.19 Hardness and minimum film formation temperature for copolymers of ethyl acrylate and methyl methacrylate [42]. The Knoop hardness number (KHN) is obtained in a Tukon Indentation Tester (ASTM D1474) and is inversely proportional to the length of an indentation made by a diamond-shaped tool

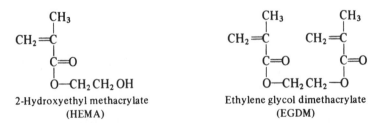

While the soft lenses require more care and may be of somewhat lower optical quality, they are usually more comfortable and easier to get accustomed to. As it is used, poly(HEMA) is slightly cross-linked because of the dimethacrylate (EGDM) present. It is a swollen hydrogel in contact with tears, but does not dissolve. Copolymers of HEMA with vinyl pyrrolidone have been used for lenses also.

Table 14.18 End-use pattern for the acrylics

Polymer*	Molded plastic	Extruded plastic	Film	Laminates	Tires	Rubber specialties	Fibers	Adhesives and coatings	Polymer additives
A								• •	
B								• •	
C	• •	•							
D								• •	
E, H	• •	• •						•	
F						• •			
G, L								• •	
J								• •	•
K (packing)									
M	•								

*See Table 14.17 for code

The polymerization process for HEMA lenses is a casting operation. A spinning mold is used so that the outside diameter of the lens is determined by the mold surface and the inside diameter depends on the rotational speed of spinning which alters the free surface. Free-radical polymerization of the HEMA-water mixture takes place in the mold. This particular hydrogel is only one of many that have been studied. While poly(HEMA) dominates the soft contact lens market, many other hydrogels have been tested as soft tissue replacements in the human body [43].

Use Pattern

To the layman, the term acrylic plastic has become synonymous with poly(methyl methacrylate), although the trade names such as Plexiglas, Lucite, Perspex, and Acrylite might be even better known (Table 14.18). The high light transmittance of poly(methyl methacrylate) has led to its use in a variety of decorative forms. Industrially, one single large market for molded, dyed polymer has been automotive exterior signal lenses. The toughness and stability of the polymer are important. In the automobile interior and in appliance panels, "internally lighted" numerals and letters can be molded in, since the transparent plastic transmits light so well that it is scattered only at interruptions such as raised or roughened surfaces. Exterior glazing in large structures such as the United States Pavilion at Expo '67 (Montreal) and the Astrodome (Houston) was made from white or gray translucent panels. Aircraft glazing has been formed from stretched sheets for many years. Acrylics can be toughened by rubber addition in the manner of polystyrene, but with a sacrifice in light transmission.

Some other uses of acrylate and methacrylate interpolymers are as viscosity index improvers in lubricating oil, ion-exchange resins, textile additives, paper coating, thermosetting (baking) coatings, and polymeric plasticizers in poly(vinyl chloride). Cyanoacrylate adhesives were described in Sec. 12.4, Adhesives.

14.7 THE FLUOROCARBON POLYMERS

Although many fluorocarbon polymers are commercially available, poly(tetrafluoroethylene) (Table 14.19) is estimated to command about 90% of the market. This polymer, under DuPont's trade name, Teflon, has been made since the early 1940s. The combination of high temperature stability, chemical resistance, and surface lubricity has made it useful in many industrial applications and, more recently, in coated kitchenware such as frying pans, cookie sheets, pots, and pans. In order to make the monomer, chloroform is treated with hydrofluoric acid to give an intermediate which is subjected to pyrolysis.

$$2HF + \underset{\text{Chloroform}}{CHCl_3} \rightarrow \underset{\substack{\text{Monochloro-} \\ \text{difluoromethane}}}{CHClF_2} \xrightarrow{\text{Pt, 700°C}} \underset{\substack{\text{Tetrafluoro-} \\ \text{ethylene}}}{\tfrac{1}{2}CF_2{=}CF_2} + HCl$$

Emulsion polymerization of the monomer gives a latex that is used directly in some applications, notably in thin coatings. One solid form sold is the coagulated latex with aggregates about 0.5 mm in diameter made up from the latex particles, which are 0.1 μm in diameter. Another solid form also has granules about 0.5 mm in diameter apparently made by a suspension polymerization, since no finer ultimate particle is discernible [44].

Special techniques have been developed to process the polymer into useful items. The polymer is dense (2.2 g/cm³), insoluble at temperatures below its T_m (327°C), and highly crystalline (see Fig. 8.19 for thermal transitions). The favored method for forming resembles techniques developed for powdered metals. Particles are brought into intimate contact by high-pressure preforming or by suspension in an oil which is removed before fusing. The particles are sintered together by heating to 370°C followed by cooling. Thin coatings adhere well to metals, but the polymer is very hard to print on or to fasten to another surface by ordinary adhesives. Good adhesion usually requires a chemical alteration of the surface. A dispersion of sodium metal in kerosene will attack poly(tetrafluoroethylene) sufficiently to allow adhesion to metals and plastics. It is one of the few reagents that have any effect on this polymer. Semifinished forms that are available include tape, film, tubing, and fibers. The tape may be skived, that is, shaved from a molded cylinder. Extruded tape, made by sintering an extruded paste of powder in oil, is much smoother but somewhat more expensive. Spun fibers of poly(tetrafluoroethylene) are quite expensive. However, a less costly process has been developed which produces fibrous-porous shapes by a replicating technique.

Poly(chlorotrifluoroethylene) and the copolymer of fluorinated ethylene and propylene represent polymers that can be molded by conventional techniques while

Table 14.19 The fluorocarbon polymers

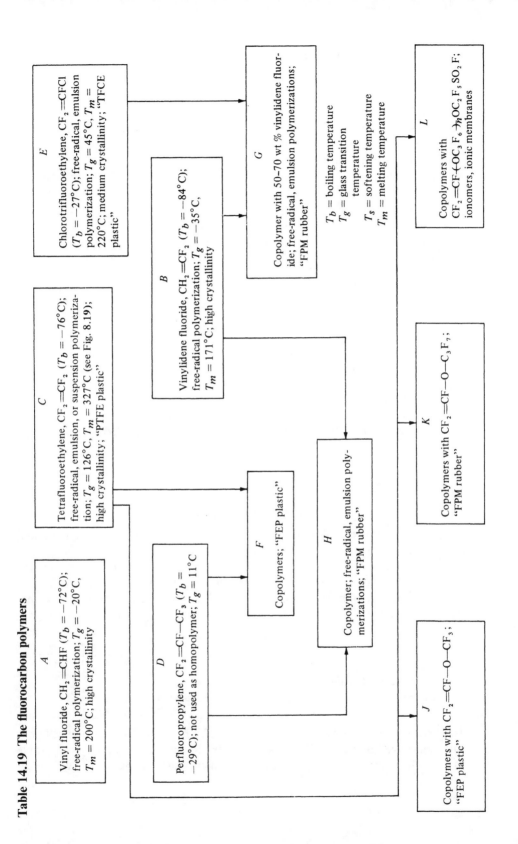

preserving a large measure of the chemical resistance and high-temperature stability of poly(tetrafluoroethylene). There is a sacrifice in these properties, however. Films based on vinyl fluoride and vinylidene fluoride are much more stable than the corresponding chloride products. Their use has been recommended in outdoor applications where weather resistance and moisture containment are important. While the fluorine-carbon bond is very strong in these polymers, the carbon-hydrogen bond is susceptible to oxidation, making them less chemically resistant than a fully fluorinated material. On the other hand, they are much more dimensionally stable, not having the multiple transitions near room temperature that are associated with poly(tetrafluoroethylene)'s tendency to cold-flow.

Rubbers based on fluoro or chloro-fluoro polymers make use of carbon-hydrogen bonds in a comonomer to provide cross-linking sites. Peroxides can be used to abstract hydrogen atoms, leaving chain radicals to join in carbon-carbon links. However, amines such as hexamethylenediamine are used more often. One widely used FPM rubber (Table 14.19) can be cross-linked by heating with an alicyclic amine salt and magnesium oxide at 150°C for 30 min followed by a "post-cure" of 24 h at 200°C. The vulcanizate can be used for continuous service as high as 200°C and yet can be useful down to its brittle temperature of −40°C. Solvent and chemical resistance are excellent except for amines, ketones, and a few other polar chemicals.

In 1970, DuPont introduced a perfluoroelastomer under the trade name of Kalrez which is based on tetrafluoroethylene and perfluoromethylvinylether [45].

$$CF_2 = \underset{\underset{\underset{CF_3}{|}}{\overset{|}{O}}}{CF}$$

Perfluoromethylvinylether

A third perfluorinated monomer (perhaps 1-hydropentafluoropropylene) provides a cross-linking site. The random copolymer has a molar ratio of 3:2 for the two major monomers, C_2F_4 and C_3F_6O, respectively, and a T_g of −12°C. The rubbery material and its cross-linked derivative are quite stable for months at 288°C.

Two other monomers that have been used with C_2F_4 are a vinyl propyl ether and an ionic ether:

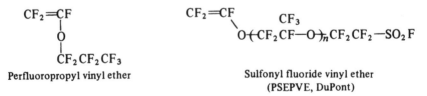

Perfluoropropyl vinyl ether

Sulfonyl fluoride vinyl ether
(PSEPVE, DuPont)

The first monomer has been used to modify PTFE to make it thermoplastic but with somewhat better stability than FEP based on perfluoropropylene (Table 14.15). The second monomer makes copolymers that are ionomers, swelling in

Table 14.20 End-use pattern for fluorocarbon polymers

Polymer*	Molded plastic	Extruded plastic	Film	Laminates	Tires	Rubber specialties	Fibers	Adhesives and coatings	Polyme: additive
A			• •						
B, L			• •						
C	• •	• •	•	•			•	• •	•
E	• •	•							
F, J	• •	• •						•	
G, K						• •		•	
H						• •		•	

water. Films of copolymers (Nafion, DuPont) have been used as membranes in electrolytic cells and in ion recovery units [46].

Use Pattern

The workhorse of the fluorocarbon family, poly(tetrafluoroethylene), is molded and extruded only by the special techniques mentioned previously (Table 14.20). A familiar form of film is the tape used as a pipe thread "dope." The cold flow of the polymer to seal the threaded pipe joint is a real advantage here. Fibers have been made from the polymer and woven into textile for special applications such as filter fabrics, protective clothing, and surgical thread. The rubbery copolymers are most often seen as O-rings and gaskets, but tubing and sheets are also used.

KEYWORDS FOR CHAPTER 14

Section

2. Polyolefins
   Ionomer
   LDPE
   LLDPE
   HDPE
   Polyallomer
3. ABS
   Nitrile rubber
   Barrier resin
   Cast polyester
   Ion-exchange resin

4. Norbornene
   Cyclized rubber
   Polyacetylene
5. Dry-blend resin
   Poly(vinyl butyral)
6. Granulation polymerization
   HEMA
7. Poly(tetrafluoroethylene)

PROBLEMS

14.1 In Fig. 14.9, what is the purpose of sending the monomers through a caustic contactor?

14.2 Describe two methods for coagulating a latex (see Figs. 14.9, 14.14, and 14.15).

14.3 Assuming 15% conversion per pass, calculate the energy needed to compress ethylene to 2000 atm and 100°C for the production of 1 lb of polymer. At $0.03/kWh, is this a significant production cost?

14.4 Explain the origin of short- and long-chain branches in polyethylene produced at high pressures.

14.5 Why are so many synthetic rubbers copolymers?

14.6 What difference in structure might be expected between cold- (5°C) and hot-process (50°C) styrene-butadiene rubber?

14.7 Why are some ethylene-propylene copolymers stiff and crystalline (at room temperature), while others are soft and amorphous?

14.8 Vinyl acetate is polymerized, hydrolyzed, and reacted with butyraldehyde to give a soluble product.

(a) Why is it not hydrolyzed first, then polymerized?
(b) What is a major use of the product from each of the three steps?

REFERENCES

1. ASTM Standard D-1418-77.
2. "Guide to Man-Made Fibers," Man-Made Fiber Producers Association, Washington, D.C., 1977.
3. Winding, C. C., and G. D. Hiatt: "Polymeric Materials," p. 280, McGraw-Hill, New York, 1961.
4. Miles, D. C., and J. H. Briston: "Polymer Technology," pp. 134 to 136, Chemical Publishing, New York, 1965.
5. *Chem. Eng.,* 80:72 (November 26, 1973).
6. *Chem. Eng.,* 86:83 (December 3, 1979).
7. Sinclair, K. B.: *Hydrocarbon Proc.,* 64(7):81 (1985).
8. Burdett, I. D.: *Hydrocarbon Proc.,* 65(11):75 (1986).
9. *Mod. Plastics,* 72(3):19 (1995).
10. Winding, C. C., and G. D. Hiatt: "Polymeric Materials," pp. 383–387, McGraw-Hill, New York, 1961.

11. "Hycar Vinyl-Terminated Liquid Polymers," B. F. Goodrich Chemical Co., Cleveland, 1977.
12. Winding, C. C., and G. D. Hiatt, "Polymeric Materials," p. 317, McGraw-Hill, New York, 1961.
13. Guccione, E.: *Chem. Eng.,* 70:138 (Apr. 15, 1963).
14. Nelb, R. G., in W. M. Smith (ed.): "Manufacture of Plastics," vol. 1, pp. 437, 451, Reinhold, New York, 1964.
15. "Environmental Impact of Nitrile Barrier Materials," pp. 24, 25, Monsanto Company, St. Louis, July 19, 1973.
16. *Mod. Plastics,* 72(3):21 (1995).
17. Hager, T., A. MacArthur, D. McIntyre, and R. Seeger: *Rubber Chem. Technol.,* 52:693 (1979).
18. Stern, H. J.: "Rubber: Natural and Synthetic," 2d ed., p. 12, Palmerton, New York, 1967.
19. Cooper, W., and G. Vaughan, in A. D. Jenkins (ed.): "Progress in Polymer Science," vol. 1, pp. 100, 107, 109, 128, 130, Pergamon, New York, 1967.
20. *Hydrocarbon Processing,* 54:176 (November 1975).
21. Prescott, J. H.: *Chem. Eng.,* 78:47 (Feb. 8, 1971).
22. Johnson, P. R.: *Rubber Chem. Technol.,* 49:650 (1976).
23. Stern, H. J.: "Rubber, Natural and Synthetic," p. 53, Palmerton, New York, 1967.
24. Tinker, A. J.: *NR Technol.,* 18:30 (1987).
25. Gelling, I. R.: *NR Technol.,* 18:21 (1987).
26. Calderon, N.: *Revs. Macromol. Chem.,* C7:105 (1972).
27. Calderon, N., and R. L. Hinrichs: *Chem. Tech.,* 4:627 (1974).
28. Dall'Asta, G.: *Rubber Chem. Technol.,* 47:511 (1974).
29. Matsumoto, S. K. Komatsu, and K. Igarashi, in T. Saegusa and E. Goethals (eds.): "Ring-Opening Polymerization," pp. 303–317, ACS, Washington, D.C., 1977.
30. Walker, D.: *Rubber and Plastics News,* p. 1, Oct. 3, 1977.
31. Shirakawa, H., and S. Ikeda: *Polymer J.,* 2:231 (1971).
32. *Physics Today,* 32:19 (September 1979).
33. Chance, R. R.: "Diacetylene Polymers," in *Enc. Polym. Sci. Eng., 2d ed.,* 4:767 (1987).
34. Bloor, D., and R. R. Chance (eds.): "Polydiacetylenes: Synthesis, Structure and Electronic Properties," M. Nijhoff (Netherlands), 1985.
35. Brydson, J. A.: "Plastics Materials," p. 160, Van Nostrand, Princeton, N.J., 1966.
36. Terwiesch, B.: *Hydrocarbon Processing,* 55:117 (November 1976).
37. Krause, A.: *Chem. Eng.,* 72:72 (Dec. 20, 1965).
38. Vona, J. A., J. R. Costanza, H. A. Cantor, and W. J. Roberts, in W. M. Smith (ed.): "Manufacture of Plastics," vol. 1, pp. 230, 234, Reinhold, New York, 1964.
39. Kosar, J.: "Light Sensitive Sytems," pp. 66, 141–142, Wiley, New York, 1965.
40. Miles, D. C., and J. H. Briston: "Polymer Technology," p. 220, Chemical Publishing, New York, 1965.
41. Guccione, E.: *Chem. Eng.,* 73:138 (June 6, 1966).
42. "Emulsion Polymerization of Monomeric Acrylic Esters, SP-154," Rohm and Haas Co., Philadelphia, 1960.
43. Andrade, J. D. (ed.): "Hydrogels for Medical and Related Applications," ACS, Washington, D.C., 1976.
44. Brydson, J. A.: "Plastics Materials," p. 204, Van Nostrand, Princeton, N.J., 1966.
45. Arnold, R. G., A. L. Barney, and D. C. Thompson: *Rubber Chem. Technol.,* 46:619 (1973).
46. Vaughan, D. J.: *DuPont Innovation,* 4:10 (Spring 1973).

General References

Albright, L. F.: "Processes for Major Addition-Type Plastics and Their Monomers," McGraw-Hill, New York, 1974.
Arthur, J. C., Jr. (ed.): "Polymers for Fibers and Elastomers," ACS, Washington, D.C., 1984.

Barrett, K. E. J. (ed.): "Dispersion Polymerization in Organic Media," Wiley-Interscience, New York, 1975.

Bhowmick, A. K., and H. L. Stephens (eds.): "Handbook of Elastomers," Dekker, New York, 1988.

Bishop, R. B.: "Practical Polymerization for Polystyrene," Cahners, Boston, 1972.

Blackley, D. C.: "Synthetic Rubbers," Elsevier Applied Science, New York, 1983.

Bloor, D., and R. R. Chance (eds.): "Polydiacetylenes: Synthesis, Structure, and Electronic Properties," M. Nijhoff (Netherlands), 1985.

Brighton, C. A., G. Pritchard, and G. A. Skinner: "Styrene Polymers: Technology and Environmental Aspects," International Ideas, Philadelphia, 1979.

Brydson, J. A.: "Plastics Materials," 4th ed., Butterworth, Woburn, Mass., 1982.

Burgess, R. H. (ed.): "Manufacturing and Processing of PVC," Elsevier Applied Science, New York, 1981.

Butters, G. (ed.): "Particulate Nature of PVC," Elsevier Applied Science, New York, 1982.

Chien, J. C. W.: "Polyacetylene," Academic Press, New York, 1984.

Chung, T. C. (ed.): "New Advances in Polyolefins," Plenum, New York, 1993.

Cowie, J. M. G. (ed.): "Alternating Copolymers," Plenum, New York, 1985.

Culbertson, B. M., and J. E. McGrath (eds.): "Advances in Polymer Synthesis," Plenum, New York, 1985.

Davidson, R. L. (ed.): "Handbook of Water-Soluble Gums," McGraw-Hill, New York, 1980.

De, S. K., and A. K. Bhowmick (eds.): "Thermoplastic Elastomers from Rubber-Plastic Blends," Prentice-Hall, Englewood Cliffs, N.J., 1991.

Dragutan, V., A. T. Balaban, and M. Dimonie: "Olefin Metathesis and Ring-Opening Polymerization of Cyclo-Olefins," Wiley, New York, 1986.

Driver, W. E.: "Plastics Chemistry and Technology," Van Nostrand Reinhold, New York, 1979.

Dubois, J. H., and F. W. John: "Plastics," 5th ed., Van Nostrand Reinhold, New York, 1974.

Dyson, R. W. (ed.): "Specialty Polymers," Chapman & Hall (Methuen), New York, 1987.

Eirich, F. R. (ed.): "Science and Technology of Rubber," Academic Press, New York, 1978.

El-Aasser, M. S., and J. W. Vanderhoff (eds.): "Emulsion Polymerisation of Vinyl Acetate," Elsevier Applied Science, New York, 1981.

Elias, H.-G.: "Macromolecules, Synthesis, Materials, and Technology," 2d ed., Plenum, New York, 1984.

Elias, H.-G., and F. Vohwinkel: "New Commercial Polymers—2," Gordon & Breach, New York, 1986.

Erusalimskii, B. L.: "Mechanisms of Ionic Polymerization," Plenum, New York, 1986.

Finch, C. A. (ed.): "Chemistry and Technology of Water-Soluble Polymers," Plenum, New York, 1983.

Flick, E. W. (ed.): "Industrial Synthetic Resins Handbook," 2d ed., Noyes, Park Ridge, N.J., 1991.

Flick, E. W.: "Water-Soluble Resins: An Industrial Guide," 2d ed., Noyes, Park Ridge, N.J., 1991.

Fontanille, M., and A. Guyot (eds.): "Recent Advances in Mechanistic and Synthetic Aspects of Polymerization," (Reidel Holland), Kluwer Academic, Norwell, Mass., 1987.

Frank, H. D.: "Polypropylene," Gordon & Breach, New York, 1968.

Glass, J. E. (ed.): "Water-Soluble Polymers," ACS, Washington, D.C., 1986.

Glass, J. E. (ed.): "Polymers in Aqueous Media," ACS, Washington, D.C., 1989.

Harper, C. A. (ed.): "Handbook of Plastics, Elastomers, and Composites," 2d ed., McGraw-Hill, New York, 1992.

Henderson, J. N., and T. C. Bouton (eds.): "Polymerization Reactors and Processes," ACS, Washington, D.C., 1979.

"High Performance Polymers," Springer-Verlag, Berlin, 1994.

Holliday, L. (ed.): "Ionic Polymers," Applied Science, London, 1975.

Kaminsky, W., and H. Sinn (eds.): "Transition Metals and Organometallics as Catalysts for Olefin Polymerization," Springer, New York, 1988.

Karger-Kocsis, J. (ed.): "Polypropylene: Structure, Blends, and Composites," Chapman & Hall, New York, 1994.

Kniel, L., O. Winter, and K. Stork: "Ethylene: Keystone to the Petrochemical Industry," Dekker, New York, 1980.

Koleske, J. V., and L. H. Wartman, "Poly(vinyl chloride)," Gordon & Breach, New York, 1969.

Krivoshei, I. V., and V. M. Skorobogatov: "Polyacetylene and Polyarylenes: Synthesis and Conductive Properties," Gordon & Breach, New York, 1991.

Legge, N. R., G. Holden, and H. E. Schroeder (eds.): "Thermoplastic Elastomers," Macmillan, New York, 1987.

Margolis, J. M. (ed.): "Engineering Thermoplastics," Dekker, New York, 1985.

Martuscelli, E., and C. Marchetta (eds.): "New Polymeric Materials: Reactive Processes and Physical Properties," VNU Science, Utrecht (Netherlands), 1987.

Meier, D. J. (ed.): "Block Copolymers," Gordon & Breach, New York, 1983.

Meltzer, Y. L.: "Water-Soluble Polymers," Noyes Data, Park Ridge, N.J., 1979.

Meltzer, Y. L.: "Water-Soluble Polymers (Developments since 1978)," Noyes Data, Park Ridge, N.J., 1981.

Miles, D. C., and J. H. Briston: "Polymer Technology," Chemical Publishing, New York, 1979.

Molyneux, P.: "Water Soluble Synthetic Polymers: Properties and Behavior," 2 vols., CRC Press, Boca Raton, Fla., 1984.

Morton, M. (ed.): "Rubber Technology," 3d ed., Van Nostrand Reinhold, New York, 1987.

Nass, L. I. (ed.): "Encyclopedia of PVC," vol. 3, 2d ed., Dekker, New York, 1992.

Nass, L. I., and C. A. Heiberger (eds.): "Encyclopedia of PVC: Resin Manufacture and Properties," vol. 1, 2d ed., Dekker, New York, 1986.

Nass, L. I., and C. A. Heiberger (eds.): "Encyclopedia of PVC, Compound Design and Additives," vol. 2, 2d ed., Dekker, New York, 1987.

Pritchard, J. G.: "Poly(vinyl alcohol)," Gordon & Breach, New York, 1969.

Roberts, A. D. (ed.): "Natural Rubber Science and Technology," Oxford Univ. Press, New York, 1988.

Roesky, H. W. (ed.): "Rings, Clusters, and Polymers of Main Group and Transition Elements," Elsevier, Amsterdam, 1989.

Rubin, I. D.: "Poly-1-butene," Gordon & Breach, New York, 1986.

Sakurada, I.: "Polyvinyl Alcohol Fibers," Dekker, New York, 1985.

Schildknecht, C. E.: "Polymer Processes," Wiley-Interscience, New York, 1956.

Schildknecht, C. E., and I. Skeist (eds.): "Polymerization Processes," Wiley-Interscience, New York, 1977.

Seymour, R. B., and T. Cheng (eds.): "Advances in Polyolefins," Plenum, New York, 1988.

Seymour, R. B., and G. S. Kirshenbaum (eds.): "High-Performance Polymers: Their Origin and Development," Elsevier, New York, 1986.

Seymour, R. B., and H. F. Mark (eds.): "Applications of Polymers," Plenum, New York, 1988.

Sittig, M.: "Polyolefin Production Processes," Noyes Data, Park Ridge, N.J., 1976.

Sittig, M.: "Vinyl Chloride and PVC Manufacture," Noyes Data, Park Ridge, N.J., 1978.

Smith, W. M. (ed.): "Manufacture of Plastics," Van Nostrand-Reinhold, New York, 1964.

Svec, P., L. Rosik, Z. Horak, and F. Vecerka: "Styrene-based Plastics and Their Modification," Prentice-Hall, Englewood Cliffs, N.J., 1989.

Tess, R. W., and G. W. Poehlein (eds.): "Applied Polymer Science," 2d ed., ACS, Washington, D.C., 1985.

Titow, W. V.: "PVC Plastics: Properties, Processing, and Applications," Elsevier, New York, 1990.

Tötsch, W., and H. Gaensslen: "Polyvinylchloride: Environmental Aspects of a Common Plastic," Elsevier, New York, 1992.

Van der Ven, S.: "Polypropylene and Other Polyolefins: Polymerization and Characterization," Elsevier, New York, 1990.

Walker, B. M. (ed.): "Handbook of Thermoplastic Elastomers," Van Nostrand Reinhold, New York, 1979.

Walker, B. M., and C. P. Rader (eds.): "Handbook of Thermoplastic Elastomers," 2d ed., Van Nostrand Reinhold, New York, 1988.

Wessling, R. A. (ed.): "Poly(vinylidene chloride)," Gordon & Breach, New York, 1977.

Whelan, A., and J. L. Craft (eds.): "Developments in PVC Production and Processing—1," Applied Science, London, 1977.

Wickson, E. J. (ed.): "Handbook of PVC Formulating," Wiley, New York, 1993.

Wilson, A. D., and H. J. Prosser (eds.): "Developments in Ionic Polymers—1," Elsevier Applied Science, New York, 1982.

Wilson, A. D., and H. J. Prosser (eds.): "Developments in Ionic Polymers—2," Elsevier Applied Science, New York, 1986.

FIFTEEN

HETEROCHAIN POLYMERS

15.1 POLYESTERS

The alkyd resins (described in Sec. 12.4) were developed in the 1920s and 1930s as an outgrowth of the earlier cooked varnishes, which combined unsaturated oils with natural resins. Similar unsaturated polyesters were dissolved in styrene monomer to make casting and laminating resins in the 1940s. Films and fibers based on poly(ethylene terephthalate) were developed in England in the 1940s. During the 1950s, polycarbonates of bisphenol A were offered in Germany and in the United States.

The condensation of an alcohol with an acid to give an ester with water as a by-product is one way to make polymers (Table 15.1). In alkyd resin cooking (Sec. 12.4) it was mentioned that additional reactions are those between anhydride and alcohol or epoxide and between free acid and alcohol and esters (transesterification). The alkyd resin kettle also could be used to make a poly(ethylene maleate) for use as a casting or laminating resin. It is more likely, however, that other monomers would be included in the resin to give branching, higher reactivity, or resistance to hydrolysis. The simple polyester with its repeating dyadic structure can be dissolved in styrene (about 70 parts polyester to 30 parts styrene) and polymerized at room temperature by a free-radical route using a redox couple. (Methyl ethyl ketone peroxide plus cobalt naphthenate would be typical.) Glass-reinforced boat hulls and automobile bodies are made from such systems. In some cases, monomers more expensive than styrene can be justified. Methyl methacrylate, triallyl cyanurate, and diallyl phthalate have been suggested by various sources.

Table 15.1 Polyesters

I. Unsaturated polyesters
   A. Branched alkyd resins for coating

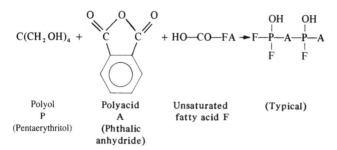

         Polyol   Polyacid   Unsaturated   (Typical)
            P   A   fatty acid F
   (Pentaerythritol) (Phthalic
                  anhydride)

   B. Linear or long-branched resins for casting with styrene monomer

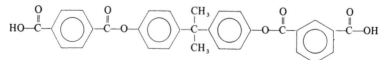

$$16\ HO-CH_2CH_2-OH + 15 \quad \text{Maleic anhydride} \xrightarrow[\text{5 h, } N_2 \text{ atm}]{180-200°C}$$

Ethylene glycol   Maleic anhydride

$$HO-CH_2CH_2 \leftarrow O-CO-CH=CH-CO-O$$
$$-CH_2CH_2 \rightarrow_n OH + H_2O$$

$$n\ (\text{average}) = 15$$

II. Saturated polyesters
   A. Terephthalates and related esters

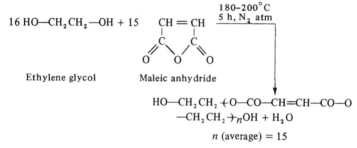

$$T_g = 80°C,\ T_m = 265°C,\ \text{medium cryst.}$$

Same ester with $HO-CH_2CH_2CH_2CH_2-OH$ in place of glycol, $T_m = 228°C$.

Aromatic polyester based on terephthalic and isophthalic acids with bisphenol A.
T_g about 175°C. Typical units shown.

Table 15.1 (*Continued*)

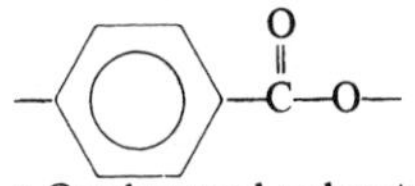

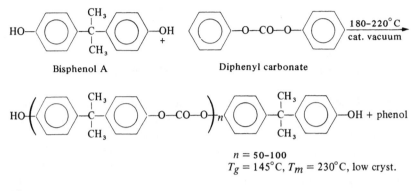

p-Oxybenzoyl polyester

B. Polycarbonates

Bisphenol A + Diphenyl carbonate

$n = 50–100$
$T_g = 145°C$, $T_m = 230°C$, low cryst.

Also

$$HO—R—OH + Cl—CO—Cl \xrightarrow[\text{pyridine}]{25–35°C} \text{polymer} + HCl$$

C. Polylactone

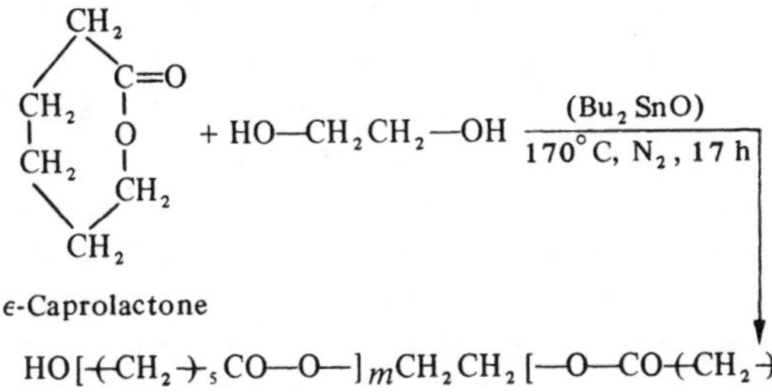

ε-Caprolactone + $HO—CH_2CH_2—OH \xrightarrow[170°C, N_2, 17 h]{(Bu_2 SnO)}$

$$HO[(CH_2)_5 CO—O—]_m CH_2 CH_2 [—O—CO(CH_2)_5]_n OH$$

$T_g = -75°C$
$m + n = 10$

Diol for extension by diisocyanate

The free-radical polymerization of diallyl diglycol carbonate ester proceeds more slowly than with corresponding divinyl monomers.

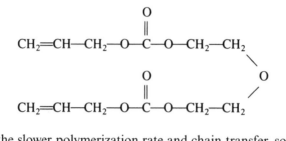

As a result of the slower polymerization rate and chain transfer, soluble branched polymers often persist up to 25% conversion. The network polymer obtained on complete conversion contains ester and carbon-chain linkages. Most popularly known as CR-39 (PPG Ind.) since its introduction in the 1930s, this monomer (together with analogous materials and comonomers) still finds specialty uses. The bulk polymerization proceeds in two steps, so that the actual casting of the final product starts with a prepolymerized syrup much as in the methyl methacrylate sheet process (see Sec. 5.2). The hard, abrasion-resistant, transparent character of the polymer makes it suitable for optical applications such as cast spectacle lenses [1].

Ester interchange is favored for terephthalates because the free acid is very insoluble and difficult to incorporate into a reaction system. Two stages are used. In the first, methanol is displaced from the terephthalic acid ester by the diol. In the second, the excess diol is driven off at high temperatures and low pressures. A continuous process (Fig. 15.1) starts with a molar ratio of dimethyl terephthalate (DMT) to glycol of 1:1.7. Methanol is removed from the horizontally sectioned vessel over a period of 4 h with a rise in temperature to 245°C at a pressure of 1 atm. In two polycondensation reactors the pressures are about 0.020 and 0.001 atm, successively, and the exit temperatures are 270 and 280°C. The resultant polymer melt is low enough in viscosity that pumps can be used to extrude fibers directly or to make chips as an intermediate form for storage.

In the mid-1970s, beverage bottles made of poly(ethylene terephthalate) (PET) became common. The process of stretch-blow molding (see Sec. 12.6, Molding) made possible strong bottles of high clarity. The most popular sizes are 2 liter and 32 oz. PET has the unfavorable property in thick sections of crystallizing slowly over a long period of time after molding. This results in some dimensional changes due to shrinkage and warping. Also in the 1970s, poly(butylene terephthalate) (PBT) became popular. Although the T_m is 228°C compared to 265°C for PET, PBT is much easier to work with. Reinforced with glass fibers, the polymer competes with nylon and metal for such applications as gears, machine parts, small pump housings, and insulators. When the terephthalic acid is replaced by 2,6-naphthalene dicarboxylic acid, the T_g is raised about 50°C. Oxygen permeability

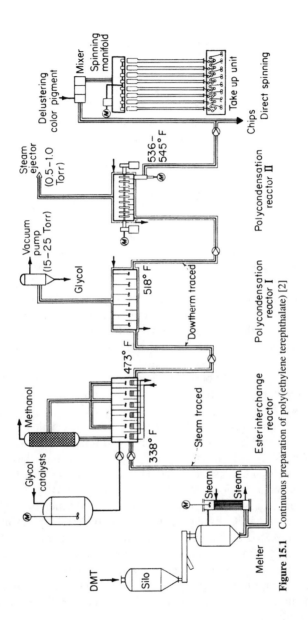

Figure 15.1 Continuous preparation of poly(ethylene terephthalate) [2]

609

and UV resistance are also improved. However, the polymer, abbreviated as PEN, crystallizes more slowly than PET. The polymer is more expensive than PET or PBT.

Another related material is a thermoplastic elastomer made by using two diols differing in molecular flexibility:

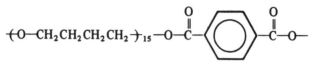

Repeat unit in hard segment

Soft segment

The final polyester [3] with a molecular weight of about 30,000 is made up of soft, flexible segments (mol wt about 1500) and an equal number of hard segments (mol wt about 700). At a molding temperature above 200°C the materials are thermoplastic. Below 150°C the "hard segments" crystallize to form massive cross-links similar to the glassy domains in the styrene-butadiene or styrene-isoprene block copolymers (see Sec. 4.5, Ionic Polymerization). The lower temperature limit for rubbery behavior is set by glass formation at about −50°C. Between −50°C and 150°C, the materials are rubbery and can be used for molded tires, snowmobile tracks, wire and cable insulation, gaskets, seals, and other mechanical goods (see also Sec. 15.10).

Several aromatic polyesters (polyarylates) are made which have high T_g's because of the stiff backbones containing many aromatic rings. They are part of a class of polymers termed "liquid crystal polymers" because they can remain ordered in both the liquid and solid states. The nematic structure is typical, especially when the polymers are molded or extruded (see Sec. 3.9). Poly (p-oxybenzoyl ester) is a compact, linear, crystalline polymer which does not melt below its decomposition temperature of 550°C. Like poly(tetrafluoroethylene) (abbreviated PTFE), the ester can be fabricated by compression sintering at temperatures above a crystal-crystal transition (330 to 360°C). The polymer can be combined with PTFE (see Sec. 14.7) in a plasma-spray coating. Typically a 1:3 ratio of polyester to PTFE is applied using an argon-hydrogen plasma. When powdered plastics are deposited and sintered on a metal surface, wear-resistant coatings with low coefficients of friction are produced.

An important ingredient for a number of heterochain polymers is bisphenol

A, which is produced by the condensation of acetone with phenol. Bisphenol A can be combined directly with phosgene in the presence of an HCl acceptor such as pyridine. In analogy with the terephthalate condensation, phenol can be condensed with phosgene in a prior step to make diphenyl carbonate. This material can then be purified by distillation before the reaction with bisphenol A in which phenol is eliminated.

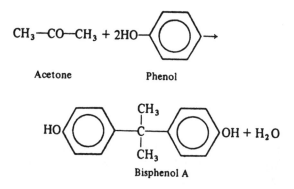

The polymer of isophthalic acid with bisphenol A (Table 15.1) is amorphorus with a heat deflection temperature (near T_g) of 173°C. Melt temperatures for molding can be as high as 350 to 400°C. Copolymers of p-hydroxybenzoic acid with the bisphenol A, terephthalic acid ester are also made. Other commercial examples of liquid crystal polymers are the copolymers of p-hydroxybenzoic acid with 6,2-hydroxynaphthoic acid.

A variety of aliphatic polyesters of molecular weight near 1000 are made with terminal hydroxyl groups to form the basis for urethane polymers. A ring scission polymerization has been used for one such material (Table 15.1). A product which is less likely to crystallize on aging can be made if a copolymer of α-methyl caprolactone and caprolactone is made.

Use Pattern

Aliphatic polyesters have been used as lubricants and as vinyl plasticizers (Table 15.2). Hydrolytic stability is a factor to be considered in many applications. As a rule, aryl acids and branched-chain diols resist hydrolysis. Fibers and films of terephthalic acid esters and polycarbonates are not particularly sensitive to moisture. Alkyd resins are so highly cross-linked in the final coating that water is not a major problem, although alkaline solutions of soaps and detergents can cause film failure. It is in the urethane foams that hydrolytic stability has been a significant factor. Flammability is a parameter that is important for all materials. Polyesters for casting resins can be modified by copolymerization with a chlorine-rich diacid

Table 15.2 End-use pattern for polyesters

Polymer	Molded plastic	Extruded plastic	Film	Lam-inates	Tires	Rubber specialties	Fibers	Adhesives and coatings	Polymer additives
Alkyd resins								• •	
Linear, unsaturated polymers	•			• •					
Poly(ethylene terephthalate)			• •				• •		
Polycarbonate	• •	•							
Diol-terminated polymers	• •					•			

to decrease flammability. The Diels-Alder condensation product of hexachloro-cyclopentadiene with maleic anhydride has a chlorine content of 55% and is a popular monomer for this purpose.

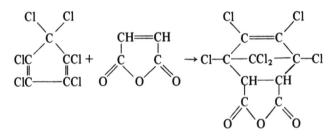

Polycarbonate has been used for some years as "safety" glazing. In thin sheets the plastic has extremely high impact strength. Both in the form of sheets and as molded lamp housings, it has been a favored material for vandal-resistant, rugged applications. A silicone treatment for polycarbonate surfaces is claimed to make the abrasion resistance of the polymer approach that of ordinary inorganic window glass.

15.2 POLYETHERS

Polymers of formaldehyde form spontaneously in aqueous solutions of the monomer, but a useful molding material was not marketed until the late 1950s. The three-membered ring of ethylene and propylene oxide opens readily and was used for low-molecular-weight surfactants in the 1940s. Urethane prepolymers became important in the 1950s; high-molecular-weight poly(ethylene oxide) became available in 1958; and rubbers based on epichlorohydrin came on the market in 1966.

The epoxy resins from bisphenol A and epichlorohydrin were produced in the 1940s, and the high-molecular-weight phenoxy resins appeared in the 1960s. Polysulfide rubber was the first successful synthetic rubber in the United States back in the 1920s. The aromatic polyethers poly(phenylene oxide) and polysulfone appeared in the 1960s. Although polysaccharides and silicones could be regarded as polyethers, they will be treated separately.

Formaldehyde forms low-molecular-weight polymers and ring compounds so easily that it must usually be prepared as an integral part of the polymerization process. Thermal decomposition of a waxy, anhydrous polymeric formaldehyde yields a dry, monomeric vapor that can be dissolved in an inert hydrocarbon such as heptane along with an initiator. Initiators include Lewis acids such as BF_3 (Table 15.3, *IA*) as well as amines, phosphines, arsines, stibenes, organometallic compounds, and transition metal carbonyls. Esterification of the end-group hydroxyls may take place in a separate, subsequent operation. The polymer is often referred to as acetal homopolymer, polyacetal, or poly(oxymethylene). All of these terms distinguish the useful, stable polymer from polyformaldehyde, the thermally unstable, waxy material.

Some processes for making high-molecular-weight polymers from formaldehyde start with trioxane, so that an authentic ring scission is involved. The basic problem with making useful molding materials based on formaldehyde is the instability of the chain structure and its tendency to "unzip."

Trioxane

End blocking with acetic acid is one answer (Table 15.3). Another seems to be the copolymerization of formaldehyde with monomers that will interrupt the chain sequence. Ethylene oxide is one such monomer.

Polymers containing ethylene oxide have one prominent property, an affinity for water. Homopolymers, even with molecular weights in excess of a million, are soluble in water. Segmented, block copolymers with propylene oxide provide a wide range of materials with variable hydrophilic-lipophilic balance (HLB) because the propylene oxide polymers are not water-soluble above a molecular weight of about 500. Random copolymers of ethylene oxide and propylene oxide, homopolymers of propylene oxide, and homopolymers of tetrahydrofuran in the molecular weight range of 1000 to 3000 are often extended by reaction with diisocyanates in making foams, rubbers, and spandex-type fibers (Table 15.3).

Epoxy resins make use of several different reactions. To make the simplest member of the series, 2 mol of epichlorohydrin are reacted with 1 mol of bisphenol A at 60°C with flake NaOH added to neutralize the HCl that is formed. Actually, more than 2 mol of epichlorohydrin is charged, and the excess is removed after the

Table 15.3 Polyethers

I. Ring-scission polymerization

A.

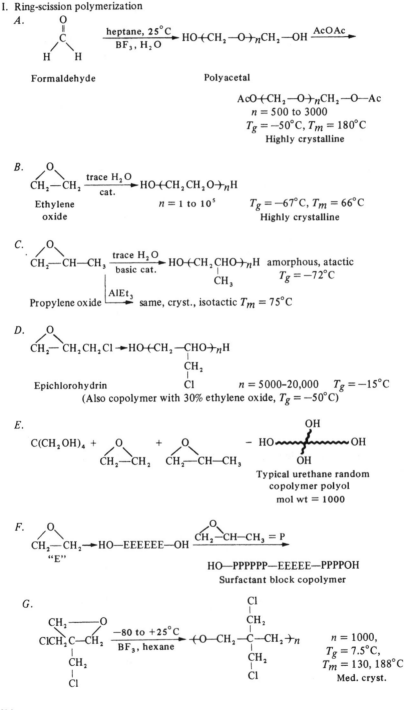

Formaldehyde → Polyacetal

$$HO(CH_2-O)_nCH_2-OH \xrightarrow{AcOAc}$$

$$AcO(CH_2-O)_nCH_2-O-Ac$$
$n = 500$ to 3000
$T_g = -50°C$, $T_m = 180°C$
Highly crystalline

B. Ethylene oxide

$$HO(CH_2CH_2O)_nH$$
$n = 1$ to 10^5
$T_g = -67°C$, $T_m = 66°C$
Highly crystalline

C. Propylene oxide

$$HO(CH_2CHO)_nH \quad \text{amorphous, atactic}$$
CH_3 $\quad T_g = -72°C$

$\xrightarrow{AlEt_3}$ same, cryst., isotactic $T_m = 75°C$

D. Epichlorohydrin

$$CH_2-CH_2CH_2Cl \rightarrow HO(CH_2-CHO)_nH$$
CH_2
Cl $\quad n = 5000-20,000 \quad T_g = -15°C$
(Also copolymer with 30% ethylene oxide, $T_g = -50°C$)

E.

$$C(CH_2OH)_4 + \quad + \quad - \quad HO\sim\sim\sim OH$$

Typical urethane random
copolymer polyol
mol wt = 1000

F.

$$CH_2-CH_2 \rightarrow HO-EEEEEE-OH \quad CH_2-CH-CH_3 = P$$
"E"

$$HO-PPPPPP-EEEEE-PPPPOH$$
Surfactant block copolymer

G.

$$ClCH_2C-CH_2 \xrightarrow[BF_3, hexane]{-80 \text{ to } +25°C} (O-CH_2-C-CH_2)_n$$

$n = 1000$,
$T_g = 7.5°C$,
$T_m = 130, 188°C$
Med. cryst.

614

Table 15.3 (*Continued*)

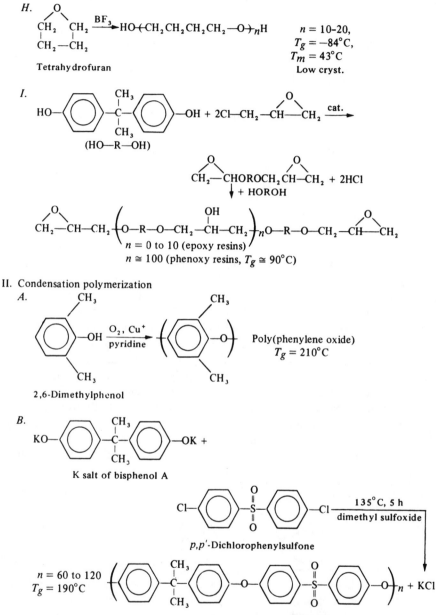

H.

Tetrahydrofuran

$n = 10\text{-}20,$
$T_g = -84°C,$
$T_m = 43°C$
Low cryst.

I.

(HO—R—OH)

$n = 0$ to 10 (epoxy resins)
$n \cong 100$ (phenoxy resins, $T_g \cong 90°C$)

II. Condensation polymerization

A.

2,6-Dimethylphenol

Poly(phenylene oxide)
$T_g = 210°C$

B.

K salt of bisphenol A

p,p'-Dichlorophenylsulfone

135°C, 5 h
dimethyl sulfoxide

$n = 60$ to 120
$T_g = 190°C$

Polysulfone

Table 15.3 Polyethers (*Continued*)

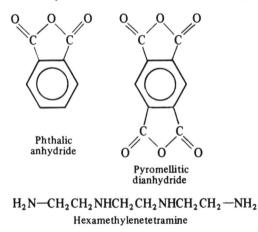

$C.$ Cl—CH_2CH_2—Cl + Cl—$CH_2CH_2OCH_2OCH_2CH_2Cl$ + $Na_2S \cdot S_x$ $\xrightarrow[\text{emulsion}]{H_2O}$

$HS(\!-\!R\!-\!S\!-\!S\!-\!)\!R\!-\!SH$ $\xleftarrow[\substack{20 \text{ to } 50 \text{ h} \\ \text{(chain extension)}}]{PbO_2, \, 25°C}$

$T_g = -20$ to $-60°C$

$HS(\!-\!CH_2CH_2\!-\!S_yCH_2CH_2OCH_2OCH_2CH_2\!-\!S\!-\!)\!H$
Typical SH-terminated polymer
mol wt = 1000 to 20,000
(HS—R—SH)

p-Dichlorobenzene   Poly(phenylene sulfide), $T_m = 288°C$

reaction. Higher-molecular-weight resins call for sturdy equipment to agitate the molten polymer. In the "taffy" process (Fig. 15.2), epichlorohydrin and bisphenol A are charged to the reactor with flake NaOH and a small amount of inert solvent. The temperature is allowed to rise to 90 or 95°C. The taffy, which forms rapidly, is an emulsion of about 30% concentrated brine in molten polymer (perhaps containing some solvent). Several cycles of washing with water are followed by stripping of solvent and water under a vacuum at 150°C. Very high molecular weights can be made by reacting low-molecular-weight polymers with additional bisphenol A with the aid of a catalyst. It can be seen that both condensation and ring scission are used for these materials. The largest use of epoxy resins (adhesives, coatings, laminates) calls for a "curing" through the oxirane rings. Popular "curing agents" include acid anhydrides and polyamines, which form ester and imine bonds, respectively. Polyfunctionality leads to network structures of high thermal stability.

Phthalic
anhydride

Pyromellitic
dianhydride

H_2N—$CH_2CH_2NHCH_2CH_2NHCH_2CH_2$—$NH_2$
Hexamethylenetetramine

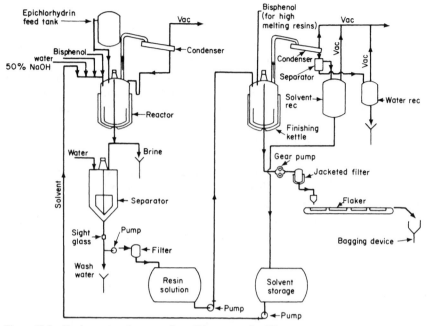

Figure 15.2   Equipment and process for solid epoxy resins [4]

A number of other epoxy compounds have been made by treating an unsaturated structure with peracetic acid. Some are complex, such as epoxidized soybean oil, which is widely used as a plasticizer-stabilizer for poly(vinyl chloride). Some other structures are

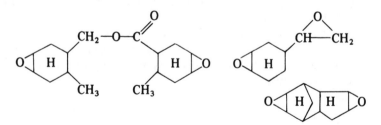

Specialty rubbers are made from epichlorohydrin (Table 15.3, I*D*).

To be useful at high temperatures, thermoplastic polymers must combine chain stiffness (for a high T_g) with thermal stability to enable molding at temperatures higher than the projected use temperature. Phenylene groups in the polymer backbone contribute both of these properties. In poly(phenylene oxide), a novel oxidative-coupling technique is applicable which gives a chemically stable thermoplastic resin (Table 15.3, II*A*). The polymer is difficult to process. A number of alloys (polymer blends) with polystyrene are marketed (Noryl, General Electric

Co.) which combine dimensional stability and processability. The polysulfones also combine phenylene groups in a backbone to yield a thermoplastic molding material useful up to temperatures of 140 to 170°C (Table 15.3, II*B*). Alloys of polysulfones with ABS resin form a separate family of materials.

The aliphatic thioethers (polysulfides) are unique among condensation polymers in being produced in an emulsion polymerization. The rank x of the inorganic sulfide used in the condensation (for example, $Na_2S \cdot S_x$) has an influence on the number of sulfur atoms in each polymer linkage. The slow oxidation of the end groups by lead peroxide at room temperature to give a high-molecular-weight (somewhat cross-linked) polymer is a very desirable change when the tacky or liquid polymers are cast or applied as pastes in the case of sealants. The sulfur analog of poly(phenylene oxide) is poly(phenylene sulfide), which can be made by a condensation reaction between *p*-dichlorobenzene and sodium sulfide. The highly crystalline polymer is solvent resistant and has a low coefficient of friction. It can be injection-molded at about 300°C and also can be used as a powder spray coating.

Use Pattern

The importance of regularity, polarity, and chain stiffness becomes apparent in the uses of polyethers (Table 15.4). Only polyformaldehyde, poly(ethylene oxide), and the polymer based on 2,2-dichloromethylpropylene-1,3-oxide are crystalline. Since poly(ethylene oxide) is water-soluble, only the remaining two polyethers are useful structural materials. Phenoxy, phenylene oxide, and sulfone polymers are useful thermoplastics because the chain stiffness contributed by aromatic rings in the backbone raises T_g well above room temperature. The aliphatic ether backbone is more conducive to low T_g's, which are suitable for rubbery applications. Fully cured epoxy resins are dimensionally stable network polymers.

A chlorinated polyether (Table 15.3, I*G*) was produced until 1972 by Hercules under the trade name Penton. Because of its resistance to inorganic and organic reagents, it was used in injection-molded valves and pump heads and as a lining for pipe, valves, and pumps.

15.3 POLYSACCHARIDES

Of the polysaccharides, cellulose receives the most attention from polymer chemists, since it is the basic polymer contained in cotton and wood. Cotton may have a cellulose content as high as 90% when dry, whereas wood contains about 50% of other compounds, notably lignin (about 30%), a natural phenolic polymer that is the binder in wood, as well as sugars and salts. What appear to be minor structural changes in cellulose-related polymers lead to major differences in properties. The water solubility of some starch components and the well-known gelling action of

Table 15.4 End-use pattern for polyethers

Polymer	Molded plastic	Extruded plastic	Film	Lam-inates	Tires	Rubber specialties	Fibers	Adhesives and coatings	Polymer additives
Polyacetal	• •	•							
Poly(ethylene oxide)			•					•	•
Diol-terminated polymers and copolymers (in urethanes)	•					•		•	•
Block copolymers									•
Poly(2,2-dichloromethyl propylene-1,3-oxide)	• •	• •							
Epoxy resins	•			• •				• •	•
Phenoxy resins	• •	•						• •	
Poly(phenylene oxide)	• •	• •					•	•	
Polysulfone	• •	• •						•	
Polysulfide						• •		• •	
Poly(phenylene sulfide)	• •	•						•	

pectin and agar in water are due primarily to differences in the stereoconfiguration. In addition to the hydroxyl groups of cellulose, polysaccharides may contain carboxyl groups (pectin), sulfate groups (carrageenan from Irish Moss and agar), or amide groups (chitin) [5, 6]. Typical structures are shown in Fig. 15.3. The half-ester sulfate group in agar is not shown, since it occurs only once every tenth repeat unit on the average. All of the polysaccharides result from biosynthesis in plants and animals. Some attempt has been made to carry out the production of polysaccharides by industrial fermentation. One such polymer contains mannose and mannose 6-phosphate and is used as a water-soluble emulsifier, thickening agent, or gelling agent for foods and other products. The flow sheet (Fig. 15.4) makes provision for refining the crude product. Corn sugar, yeast extracts, and salts are charged to a fermenter as a dilute solution (2.25% dextrose, 0.4% yeast extracts, 0.5% dibasic phosphate, and 0.05% $MgSO_4$). The batch is inoculated with a yeast, *Xanthomonas campestris,* and the reactor is stirred and aerated with 0.5 vol of air per volume of solution per minute. After 4 days the viscosity has increased to 70 poise and the solution can be spray-dried to recover a crude polymer which is about 60% polysaccharide. Dilution and precipitation by a quaternary ammonium compound can be used to produce about 68 lb of refined polymer per 100 lb of sugar. The disadvantages of the process are the dilute solutions and long holdup

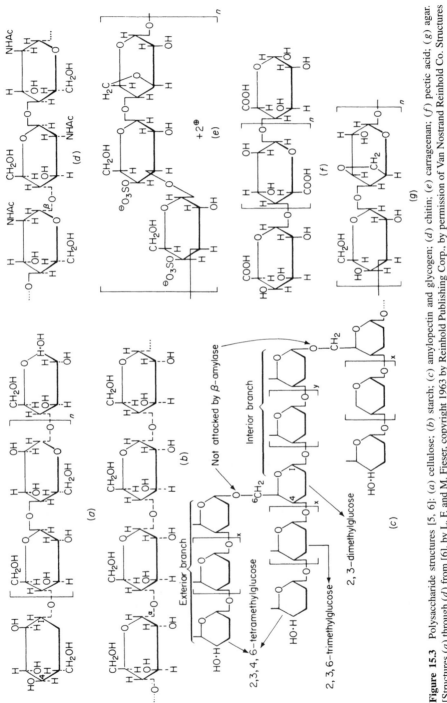

Figure 15.3 Polysaccharide structures [5, 6]: (*a*) cellulose; (*b*) starch; (*c*) amylopectin and glycogen; (*d*) chitin; (*e*) carrageenan; (*f*) pectic acid; (*g*) agar. [Structures (*a*) through (*d*) from [6], by L. F. and M. Fieser, copyright 1963 by Reinhold Publishing Corp., by permission of Van Nostrand Reinhold Co. Structures (*e*), (*f*), and (*g*) from [5] by permission of Academic Press, New York.]

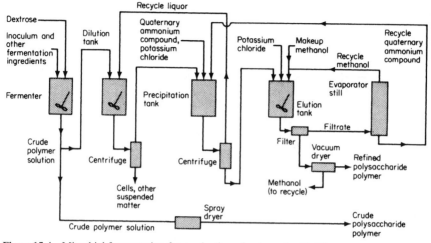

Figure 15.4 Microbial fermentation for production of polysaccharide [7]

times. On the other hand, a stereoregular structure of this complexity would be very expensive to produce by any present-day synthetic method.

Historically, the plastics industry often dates its birth to about 1865 (Parkes, England) or 1868 (Hyatt, U.S.) when the plasticization of cellulose nitrate by camphor permitted the molding of simple objects. Regenerated cellulose in the 1900s and cellulose acetate in the 1920s pretty much displaced the nitrate from all markets except lacquers and explosives.

At first glance there would seem to be little incentive for regenerating cellulose. However, wood pulp or the short fibers left from cotton recovery (linters) cannot be molded or spun into useful forms unless they are changed chemically. In the viscose process (Table 15.5), cellulose in sheet form resembling blotter paper is steeped in aqueous alkali for 2 to 4 h in a rack in which it can then be pressed to squeeze out excess liquid. After the alkali cellulose is shredded, it is aged 2 or 3 days, during which the crystalline structure is disrupted and the molecular weight may be decreased. With carbon disulfide added in a rotating drum called a barratte, the xanthate is formed, which dissolves in alkali. The orange, unpleasant-smelling solution (7 to 8% cellulose, 6.5 to 7% NaOH) is ripened for 4 to 5 days, during which time the distribution of xanthate groups presumably becomes more uniform. For rayon fibers the solution is spun through spinnerets with many small holes into a salt-acid bath. A number of operations may be carried out on the fibers, including washing, bleaching, drawing, and twisting and crimping. For cellophane a slot orifice is substituted for the spinneret, and aftertreatment may include plasticization by glycerol or coating by a moisture barrier such as cellulose nitrate. In the United States alone, rayon fiber production was around 130×10^6 kg in 1994 and cellophane production was much less. Sausage casing for millions of "skinless" hot dogs is made by the same process using a tubular die.

Table 15.5 Cellulose* derivatives

I. Esters
    A. Nitrate

$$R(OH)_3 + \xrightarrow[\text{H}_2\text{O}]{\text{HNO}_3,\ \text{H}_2\text{SO}_4} R(ONO_2)_3 \qquad \text{Trinitrate, } T_g \cong 53°C$$

    B. Acetate (also propionate, butyrate)

$$R(OH)_3 + \xrightarrow[\text{H}_2\text{SO}_4,\ \text{CH}_2\text{Cl}_2]{\text{HOAc, AcOAc}} R(OAc)_3 \xrightarrow[\text{heat}]{\text{H}_2\text{O}} R(OAc)_2 OH$$

$$\begin{array}{cc}\text{Triacetate} & \text{Diacetate} \\ T_g = 105°C & T_g = 120°C\end{array}$$

II. Ethers
    A. Reactions with alkali cellulose $(R(OH_3) + NaOH, H_2O)$

Alkali cellulose

$$+\ Cl-CH_3 \text{ or } (CH_3)_2SO_4 \xrightarrow{50-100°C} R(OCH_3)_m(OH)_{3-m}$$
$$m = 1.7\text{-}1.9,$$
methyl cellulose

$$+\ ClCH_2CH_3 \text{ or } (CH_3CH_2)_2SO_4 \xrightarrow[\text{6-12 h}]{90-150°C} R(OCH_2CH_3)_m(OH)_{3-m}$$
$$m = 2.4\text{-}2.5,\ T_g = 43°C,$$
ethyl cellulose

$$+\ Cl-CH_2-CO-ONa \xrightarrow[\text{2-4 h}]{40-50°C} R(OCH_2CO-ONa)_m(OH)_{3-m}$$
$$m = 0.65\text{-}1.4,$$
carboxymethyl cellulose

    B. Reaction with slurried cellulose

$$R(OH)_3 + CH_2\overset{O}{-}CH_2 \xrightarrow[\text{diluent (alcohol or ketone)}]{30°C,\ 15\ h} RO(CH_2CH_2O)_nH_m(OH)_{3-m}$$
$$n = 1 \text{ (some 2,3 etc.)}$$
$$m = 2\text{-}2.5$$
Hydroxyethyl cellulose

III. Regenerated cellulose
    A. Viscose

$$\text{Alkali cellulose} + CS_2 \longrightarrow R(O-CS-S^+Na^-)_m(OH)_{3-m} \xrightarrow{\text{H}_2\text{SO}_4} R(OH)_3$$
$$m \cong 0.5$$

    B. Vulcanized fiber

$$R(OH)_3 + ZnCl_2 + 2H_2O \longrightarrow \text{complex} \xrightarrow{\text{wash}} R(OH)_3$$

*Cellulose $= R(OH)_3$.

An important reason for decreased interest in rayon is the problem of waste disposal from the production process. In the past, sulfur compounds were not recovered, and were the cause of some air and water pollution. The search for an alternative process which would not involve disposal of by-products has not yet yielded a large-scale solution. However, one plant has been built that produces regenerated cellulose fiber by wet or dry spinning from a solution of cellulose in *n*-methyl morpholine (or similar solvents) [8].

In the production of vulcanized fiber, the cellulose is not dissolved but is highly plasticized with a concentrated zinc chloride solution so that it can be formed into sheets as thick as 2 in. Washing out the salt solution is a slow process. The product retains some of the fibrous appearance of the starting material but is very dense and strong. It is oil-resistant but does absorb water. Sheets can be punched or machined for use as gaskets, bobbins, shuttles, and printed-circuit boards.

The flammability of cellulose nitrate and the necessity of having a solvent present during forming operations have pretty much eliminated it as a molding plastic. Popular household cements for paper and wood may be solutions of the nitrate in ketones and esters. Lacquers for wood and metal also do not suffer from the limitations mentioned. One molded product that is slow to yield to the use of other polymers is the Ping-Pong ball with its peculiar requirements of resilience and toughness. The preferred plasticizer is still camphor,

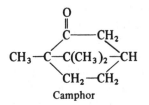

Camphor

Acetylation of cellulose requires three steps. A preliminary treatment of cellulose with glacial acetic acid for 1 or 2 h is followed by reaction with acetic anhydride with a trace of sulfuric acid as a catalyst. In order to keep the temperature of the reaction at about 50°C, methylene chloride is refluxed. A typical charge would be [9]:

Cellulose	100 parts	pretreatment
Glacial acetic acid	35 parts	
Acetic anhydride	300 parts	
Methylene chloride	400 parts	acetylation
Sulfuric acid	1 part	

The reaction product is the triacetate, from which some film and fiber are produced. Most of the triacetate is hydrolyzed to the diacetate by adding water to the reaction product and ripening the mixture for several days. The diacetate is soluble in acetone or halogenated solvents, whereas the traicetate is soluble in fewer, more expensive solvents such as chloroform. The diacetate is used in films, fibers, and coatings. Like vinyl chloride polymers, the acetate usually is plasticized, dimethyl phthalate being more suitable than the higher esters favored for vinyls. Cellulose propionate, butyrate, and a mixed acetate-butyrate are marketed for film, molding, and coating applications.

Ethyl cellulose resembles cellulose acetate in many of its qualities (toughness, transparency) and applications (films, coatings). The other cellulose ethers are water-soluble and have the usual string of applications such as thickening agents, textile and paper sizes, cosmetic and pharmaceutical suspending agents, and water-based paint additives. Except for the hydroxyethyl cellulose, the ethers are made by condensation of alkali cellulose with a chlorinated or sulfated compound. An acid, HCl or H_2SO_4, is eliminated in the reaction. Where low degrees of substitution are encountered, the ether groups may not be evenly distributed but may be clustered toward the outside of the original cellulose fiber structure. In this respect they differ from viscose rayon and cellulose acetate, where ripening or aging steps are included to increase uniformity.

Use Pattern

As befits a senior citizen of the polymer industry, the polysaccharide group is stable but rather sedate (Table 15.6). The imminent demise of cotton, rayon, and cellophane at the hands of new synthetics has been predicted often but has not materialized. When cotton was attacked by nylon and polyesters for use in apparel, wash-and-wear finishes based on melamine and urea resins gave it new life. Polyethylene film was supposed to take over cellophane markets but instead found new markets and left some old ones to cellophane. Almost all cellophane is coated to lower the moisture sensitivity and permeability and to provide a plastic surface that can be heat sealed. Saran (vinylidene chloride-vinyl chloride copolymer), cellulose nitrate, poly(vinyl acetate), and other ethenic polymers are used as coatings. A significant application for cellulose acetate film has been found in seawater desalination by reverse osmosis. The film is strong enough to resist the high pressures involved, and it has a good permeability to water and a low permeability to salt. Most cigarette filters are made of cellulose acetate fibers.

15.4 POLYAMIDES AND RELATED POLYMERS

Naturally occurring proteins and synthetic nylons are polyamides, as familiar to most of us as are our own bodies and clothing (Table 15.7). Proteins are found in all living cells. Simple proteins yield only α-amino acids upon hydrolysis. Proteins can be divided into two major classes: the fibrous and the globular proteins. The fibrous proteins act as structural materials in animals in much the same way that cellulose acts for plants. Keratin, the protein of hair, horn, feathers, and fingernails;

Table 15.6 End-use pattern for polysaccharides

Polymer	Molded plastic	Extruded plastic	Film	Lam-inates	Tires	Rubber specialties	Fibers	Adhesives and coatings	Polymer additives
Regenerated cellulose (cellophane, rayon)			• •				• •		
Cellulose nitrate	•							• •	
Cellulose acetate	• •	• •	• •				• •	• •	
Ethyl cellulose	• •	• •	• •					• •	
Methyl cellulose								•	• •
Water-soluble polysaccharides								•	• •

fibroin, the protein of silk; and collagen, the protein of connective tissue, are all fibrous proteins with the common property of water insolubility. Collagen yields gelatin when boiled in water. *Globular* proteins are soluble in water or aqueous solutions of acids or bases. Albumin (eggs), casein (milk), and zein (corn) are examples of globular proteins. The common amino acids and representative analyses of some proteins are shown in Tables 15.8 and 15.9. Although great progress has been made in analyzing proteins, their synthesis has proceeded much more slowly. The first total synthesis of an enzyme, ribonuclease A, was reported in 1969 [11]. In this work, accomplished almost simultaneously by research groups at Rockefeller University and at Merck and Company, 19 amino acids were assembled into the protein which has 124 units in a definite sequence (Fig. 15.5). In one technique, each amino acid was added in sequence after the first, valine, was fully bound to an insoluble substrate. To do this, 369 chemical reactions requiring 11,931 steps were carried out in an automated apparatus. Much can be learned about protein behavior and conformation by studying simpler molecules, such as polypeptides, which are polymers of a single α-amino acid. Predictions of helix formation and of solution properties are possible in many cases [12].

The poly(α-amino acid)s are sometimes referred to as nylon 2. The term *nylon* originally was a trademark for the polyamide based on hexamethylene diamine and adipic acid. Later on it became a generic term. The numerals following the name designate the number of carbon atoms in the chain between successive amide groups. The dyadic nylons have two numbers, the first for the diamine and the second for the diacid. Monadic nylons such as polycaprolactam require only one number. Although they have been studied as protein models, the synthetic nylon 2 polymers have not been commercialized as fibers or plastics. While some new polymers have been introduced recently (nylon 4, nylon 11, nylon 12), nylon 66 and

Table 15.7 Polyamides and polyimides

I. Dyadic polyamides

A. $H_2N+CH_2+_6NH_2$ + $HO-CO+CH_2+_4CO-OH$ $\xrightarrow{\text{methanol}}$

   Hexamethylene   Adipic acid
   diamine

$$NH_2-R_1NH_3^{+-}O-CO-R_2-CO-OH$$
Nylon salt

Nylon salt + acetic acid (HO—Ac) $\xrightarrow[\text{3 h}]{220-280°C}$

$$Ac+NH+CH_2+_6NH-CO+CH_2+_4CO+_nNH-R_1-NH-Ac$$
Nylon 66, $T_g = 50°C$, $T_m = 265°C$, $n = 35-45$

B. $H_2N+CH_2+_6NH_2$ + R_3 ... $CH=CH-R_3 \longrightarrow +NH+CH_2+_6NH-CO-X-CO+_n$

$n = 5-10$
Amorphous

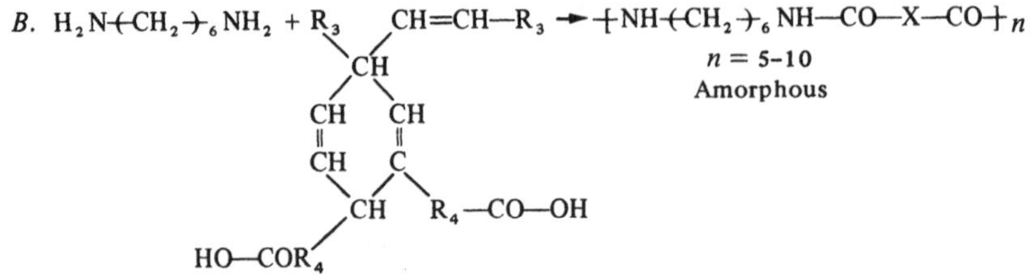

"Dimer acid" = $HO-CO-X-CO-OH$

C.

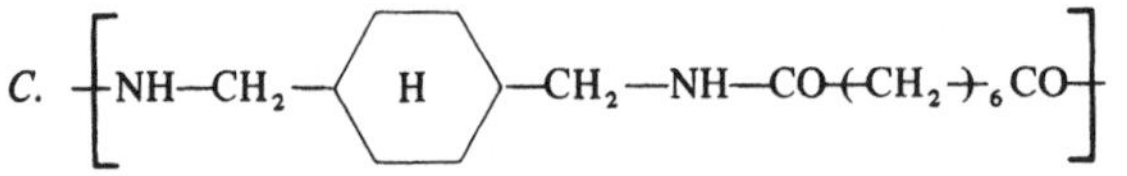

Poly(1,4-cyclohexylenedimethylene suberamide),
$T_g = 86°C$, $T_m = 296°C$

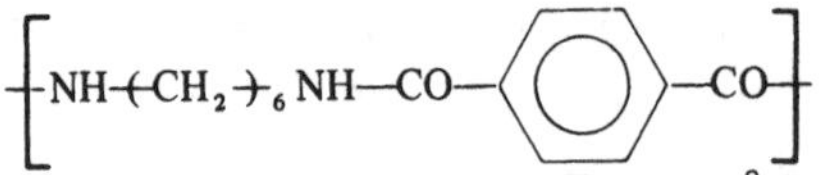

Nylon 6T(terephthalic acid), $T_m = 370°C$

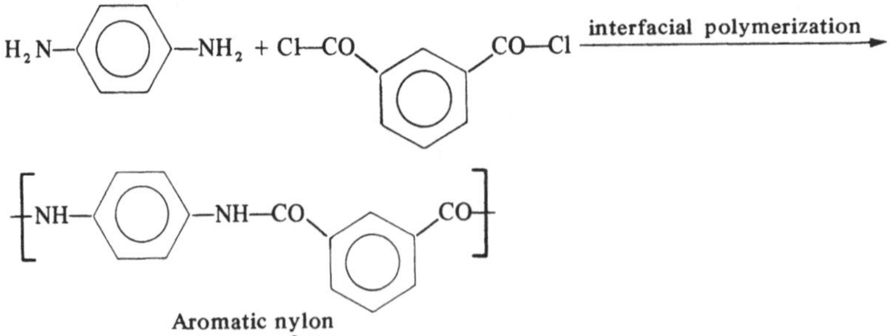

$\xrightarrow{\text{interfacial polymerization}}$

Aromatic nylon
$T_m = 375°C$

Table 15.7 (*Continued*)

II. Monadic polyamides

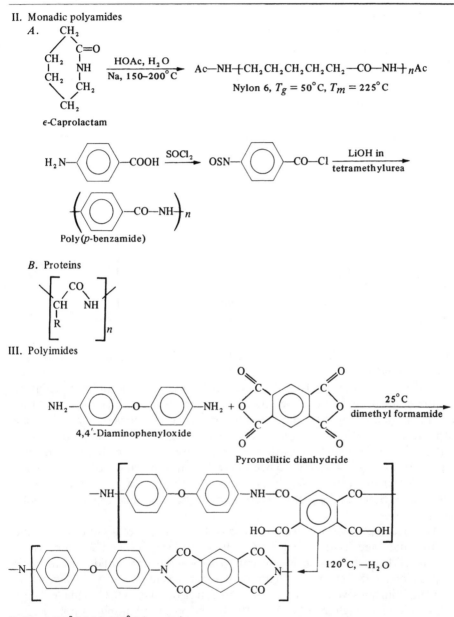

A. ε-Caprolactam

$$\text{Ac—NH} \{ \text{CH}_2 \text{CH}_2 \text{CH}_2 \text{CH}_2 \text{CH}_2 \text{—CO—NH} \}_n \text{Ac}$$
Nylon 6, $T_g = 50°C$, $T_m = 225°C$

Poly(p-benzamide)

B. Proteins

III. Polyimides

4,4′-Diaminophenyloxide

Pyromellitic dianhydride

Stable at 420°C (air), 500°C (vacuum).
Tensile strength at 260°C is two-thirds that at 25°C

Table 15.7 Polyamides and polyimides (*Continued*)

IV. Polybenzimidazoles

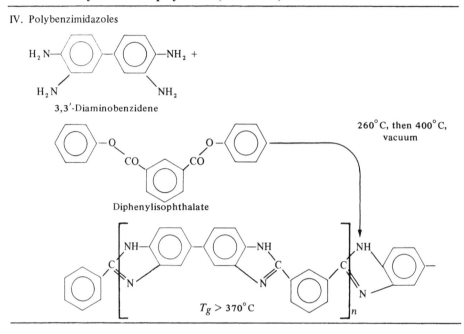

3,3'-Diaminobenzidene

Diphenylisophthalate

260°C, then 400°C, vacuum

$T_g > 370°C$

nylon 6 have been produced for a longer time and dominate the markets for synthetic polyamides.

In the 1930s a team of chemists at DuPont led by W. H. Carothers investigated the properties of aliphatic polyesters at some length before discovering that an aliphatic polyamide of the same molecular weight as the polyester would have a much higher T_m and could be spun into a useful fiber. Nylon 66, made from a dyad of a diamine and a diacid each with six carbons, was commercialized before 1940 and continues to grow in importance today. Nylon 6, based on the ring scission polymerization of ε-caprolactam, was produced in Europe before it was introduced to the United States in the 1950s. The 1960s brought the aromatic nylons and polyimides.

The first step in making dyadic nylons is the formation of a salt of amine and acid (without condensation) (Fig. 15.6). The second step involves actual polycondensation. The salt solution is concentrated to about 75% solids before being charged along with a chain terminator such as acetic acid into an autoclave, where a residence time of several hours and a temperature of up to 280°C yield a polymer with a molecular weight of 12 to 15 × 10³. In one process, continuous polymerization takes place in a coiled-tube reactor. Water vapor escapes as a fast-moving, turbulent core while the nylon melt moves along the wall in laminar flow (Fig. 15.7). In this process, the incoming salt solution is 47% solids and the total resi-

Table 15.8 Some amino acids isolated from proteins [10]

Common name	Systematic name	Structural formula	Discoverer and date
Tryptophan	α-Amino-β-indol-propionic acid	(indole ring)—$CH_2 \cdot CH \cdot COOH$ with NH_2 on the CH; ring NH	Hopkins and Cole 1901
Proline	Pyrrolidine-α-carboxylic acid	H_2C——CH_2 / H_2C $CH \cdot COOH$ \ NH (pyrrolidine ring)	E. Fischer 1901
Hydroxyproline	γ-Hydroxypyrroli-dine-α-carboxylic acid	HO—CH——CH_2 / H_2C $CH \cdot COOH$ \ NH (ring)	E. Fischer 1901
Cystine	Di-(α-amino-β-thiopropionic acid	H_2C—S—S—CH_2 / HC—NH_2 HC—NH_2 / COOH COOH	Wollaston 1810
Methionine	α-Amino-γ-methyl-thiol-n-butyric acid	H_3C—S—$CH_2 \cdot CH_2 \cdot CH \cdot COOH$ with NH_2	Mueller 1922
Aspartic acid	Aminosuccinic acid	$COOH \cdot CH_2 \cdot CH \cdot COOH$ with NH_2	Plisson 1827
Glutamic acid	α-Aminoglutaric acid	$COOH \cdot (CH_2)_2 \cdot CH \cdot COOH$ with NH_2	Ritthausen 1866
Hydroxy-glutamic acid	α-Amino-β-hydroxy-glutaric acid	$COOH \cdot CH_2 \cdot CH \cdot CH \cdot COOH$ with OH NH_2	Dakin 1918
Lysine	α-ε-Diaminocaproic acid	$CH_2 \cdot (CH_2)_3 \cdot CH \cdot COOH$ with NH_2 NH_2	Drechsel 1889
Hydroxylysine	α-ε-Diamino-β-hydroxy-n-caproic acid	$CH_2 \cdot (CH_2)_2 \cdot CH \cdot CH \cdot COOH$ with NH_2 OH NH_2	Schryver et al. 1925
Arginine	α-Amino-δ-guanidine-n-valeric acid	HN—$CH_2 \cdot (CH_2)_2 \cdot CH \cdot COOH$; $C=NH$ with NH_2; NH_2	Schulze and Steiger 1886
Histidine	α-Amino-β-imidazol-propionic acid	(imidazole ring) N——CH / HC C—$CH_2 \cdot CH \cdot COOH$ \ N ; with NH_2	Kossel 1896 Hedin 1896

Table 15.8 Some amino acids isolated from proteins [10] (*Continued*)

Common name	Systematic name	Structural formula	Discoverer and date
Glycine (glycocoll)	Aminoacetic acid	$\overset{\displaystyle NH_2}{\underset{\displaystyle}{\mid}}$ HCH·COOH	Braconnot 1820
Alanine	α-Aminopropionic acid	$\overset{\displaystyle NH_2}{\mid}$ CH₃·CH·COOH	Strecker 1850
Valine	α-Aminoisovaleric acid	H₂C NH₂ CH₃·CH·CH·COOH	Gorup-Besanez 1856
Leucine	α-Aminoisocaproic acid	H₃C NH₂ CH₃·CH·CH₂·CH·COOH	Proust 1819
Isoleucine	α-Amino-β-ethyl-β-methyl-propionic acid	H₅C₂ NH₂ CH·CH·COOH H₃C	F. Ehrlich 1903
Serine	α-Amino-β-hydroxy-propionic acid	OH NH₂ CH₂·CH·COOH	Cramer 1865
Threonine	α-Amino-β-hydroxy-*n*-butyric acid	NH₂ CH₃·CH·CH·COOH OH	Schryver and Buston 1925; Rose et al., 1935
Phenylalanine	α-Amino-β-phenyl-propionic acid	⬡—CH₂·CH·COOH, NH₂	Schulze and Barbieri 1879
Tyrosine	α-Amino-β-(*p*-hydroxyphenyl) propionic acid	HO—⬡—CH₂·CH·COOH, NH₂	Liebig 1846
Iodogorgoic acid	3,5-Diiodotyrosine	I, HO—⬡—CH₂·CH·COOH, NH₂, I	Drechsel 1896
Thyroxine	β-3,5-Diiodo-4-(3′,5′-diiodo-4-hydroxy) phenyl-α-aminopropionic acid	I, I HO—⬡—O—⬡—CH₂·CH·COOH, NH₂, I, I	Kendall 1915

Table 15.9 Partial analysis of some proteins [10]

	Egg albumin	Lact-albumin	Serum albumin	Serum globulin	Edestin (hemp)	Glutenin (wheat)	Gliadin (wheat)	Zein (maize)	Keratin (wool)	Fibroin (silk)	Gelatin	Salmine (salmon sperm)	Casein (cow's milk)	Vitellin (egg yolk)	Insulin (cryst.)	Pepsin (cryst.)
Glycine	0.0	0.4	0.0	3.5	3.8	0.9	0.0	0.0	0.6	40.5	25.5	–	0.5	1.1		
Alanine	2.2	2.4	2.7	2.2	3.6	4.7	2.0	9.8	4.4	25.0	8.7	–	1.9	0.2		
Valine	2.5	3.3	–	–	6.3	0.2	3.4	1.9	2.8	–	0.0	4.3	7.9	2.4		
Leucine and isoleucine	10.7	14.0	20.0	18.7	14.5	6.0	6.6	25.0	11.5	2.5	7.1	–	9.7	11.0	30.0	
Phenyl-alanine	5.1	1.3	3.1	3.8	3.1	2.0	2.4	7.6	–	11.5	1.4	–	3.9	2.8	+	
Tyrosine	4.2	2.0	4.7	6.7	4.6	4.5	3.4	5.9	4.8	11.0	0.0	–	6.6	5.0	12.2	10.3
Tryptophan	1.3	2.7	0.5	2.3	2.5	1.7	1.1	0.2	1.8	–	0.0	–	2.2	2.5	–	2.2
Threonine	–	–	–	–	–	–	–	–	–	1.5	–	–	3.6	–	2.7	
Glutamic acid	14.0	12.9	7.7	8.2	19.2	25.7	43.7	31.3	12.9	–	5.8	–	21.8	12.2	30.0	18.6
Hydroxy-glutamic acid	1.4	10.0	–	–	–	1.8	2.4	2.5	–	–	0.0	–	10.5	–		
Aspartic acid	6.1	9.3	3.1	2.5	10.2	2.0	0.8	1.8	2.3	–	3.4	–	4.1	0.5	–	6.8
Proline	4.2	3.8	1.0	2.8	4.1	4.2	13.2	9.0	4.4	1.0	9.5	11.0	9.0	4.0	+	
Hydroxy-proline	–	–	–	–	–	–	–	–	–	–	14.1	–	0.2	–		
Serine	–	1.8	0.6	–	0.3	0.7	–	1.0	2.9	13.6	0.4	7.8	5.0	–	3.6	
Cystine	1.3	4.3	6.1	1.0	1.0	1.8	–	0.8	13.1	–	0.2	–	0.3	1.2	12.2	1.4
Methionine	4.6	2.6	–	–	2.1	–	2.1	2.3	–	–	–	–	3.4	–	0.0(?)	
Arginine	5.2	3.0	4.8	5.2	15.8	4.7	2.9	1.8	7.8	0.7	8.2	87.4	3.8	7.8	3.2	2.7
Histidine	1.4	2.1	1.2	0.9	2.2	1.8	1.5	0.8	0.7	0.1	1.0	–	2.5	1.2	8.0	0.1
Lysine	6.4	8.8	6.9	6.2	2.2	1.9	0.6	0.0	2.3	0.3	5.9	–	6.0	5.4	2.2	2.1
Ammonia	1.4	1.3	–	–	2.3	4.0	5.2	3.6	–	–	0.4	–	2.3	–	1.7	8.8

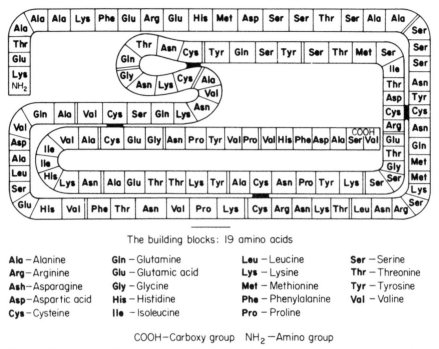

The building blocks: 19 amino acids

Ala – Alanine	Gln – Glutamine	Leu – Leucine	Ser – Serine
Arg – Arginine	Glu – Glutamic acid	Lys – Lysine	Thr – Threonine
Ash – Asparagine	Gly – Glycine	Met – Methionine	Tyr – Tyrosine
Asp – Aspartic acid	His – Histidine	Phe – Phenylalanine	Val – Valine
Cys – Cysteine	Ile – Isoleucine	Pro – Proline	

COOH – Carboxy group   NH₂ – Amino group

Figure 15.5 Structure of ribonuclease A. Double lines indicate where small units were fused together in one method of synthesis. The second method involved adding monomers one at a time starting from the carboxy end [11]

dence time is about 1 h at 390 psig (2.7 MPa), 290°C. In a pilot-plant unit the coil consisted of 300 ft of ⅜-in-diameter stainless-steel tubing followed by 120 ft of 1¼-in-diameter tubing. A reservoir at the end of the reactor allowed steam to separate and escape. A production rate of 40 lb/h (18 kg/h) is claimed for a unit of this size.

Not all nylons are highly crystalline. The polyamide based on dimerized fatty acids (Table 15.7, I*B*) is soluble in common solvents and contains residual unsaturation. It is used as a coating material and as a reactive curing agent for epoxy resins in coatings and adhesives. Ordinary nylon 66 is soluble only in highly polar solvents such as 90% formic acid or metacresol. Treatment of nylon with formaldehyde and an alcohol while the polymer is dissolved in 90% formic acid (phosphoric acid is added as a catalyst) results in the *N*-alkylated derivative, which is soluble in methanol and ethanol and can be used in coatings [15]. The coatings can be crosslinked by baking in the presence of an acid (20 min at 120°C with 2% citric acid is typical).

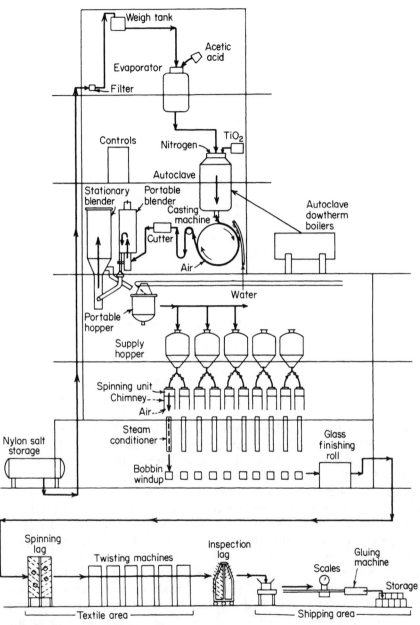

Figure 15.6 Nylon 66 production [13]

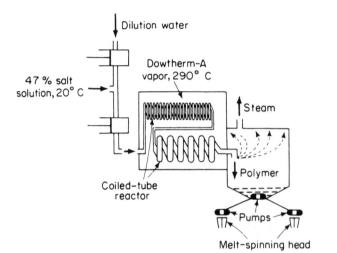

Figure 15.7   Coiled-tube reactor for continuous polymerization of nylon 66 [14]

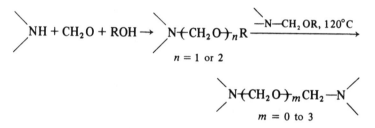

Other commercially available nylons such as nylon 610 (using sebacic acid, $HO-CO-(CH_2)_8CO-OH$) are based on aliphatic dyads. The T_g and T_m can be increased by incorporating cyclic or aromatic structures into the backbone. The nylons in Table 15.7, IC, can be made by interfacial polymerization (see Sec. 5.3), since the amide formation is rapid between aroyl chlorides and aromatic amines.

Nylon 6 is made by a ring-scission polymerization which results in an equilibrium mixture of high-molecular-weight, linear polymer and cyclic monomer. High-strength fibers require removal of the unreacted monomer. This can be done in a continuous process (Fig. 15.8) that starts with molten caprolactam. Titanium dioxide can be added at the very start, since it is inert during the reaction and serves only as a delusterant in the final fiber. After a residence time in the reactor of about 18 h at 260°C, the melt viscosity reaches about 100 Pa·s. The unreacted monomer (about 10% of the batch) is removed and recovered in a falling-film evaporator at 260 to 280°C at an absolute pressure of about 0.001 atm.

Large cast pieces of nylon 6 can be produced directly from monomer by using a catalyst such as sodium metal or a Grignard reagent. Much higher molecular weights can be tolerated and less unreacted monomer is left in the molded piece

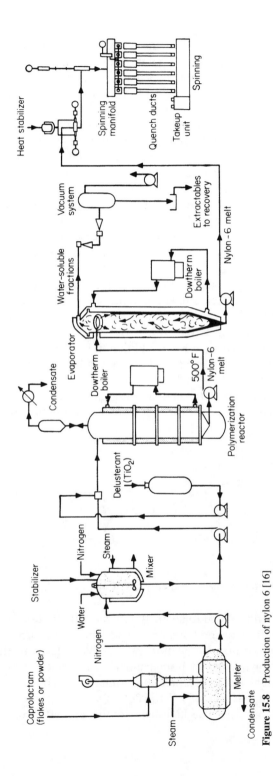

Figure 15.8 Production of nylon 6 [16]

635

when the temperature is limited to the range between the melting point of ε-caprolactam, 72°C, and the melting point of the polymer, 225°C. In one example, N-acetyl caprolactam (0.005 mol) and methyl magnesium bromide (0.005 mol) were added to 1 mol of caprolactam at 130°C. The mixture became solid in 4 min. The liquid system can be poured into a mold and converted to the solid *in situ* [17].

The polyimides and polybenzimidazoles (Table 15.7, II and IV) represent attempts to combine chain stiffness with thermal stability. A two-step synthesis is usually required. In the first step a linear, soluble, and fusible polymer is made by forming amide links. At this stage the polymer can be cooled and stored and then later heated and fabricated into a final shape such as a film, fiber, or molded object. In the second step, at a higher temperature than that used in the first step, heterocyclic ring structures are made by intramolecular condensation. While the structures shown in Table 15.7 predominate, there is some intermolecular cross-linking also. At any rate, the materials are nominally thermosetting, since they cannot be formed again by reheating after extensive ring formation. They have the disadvantage of not being easily formed compared to the aromatic polyamides. However, in general, they preserve their strength and flexibility to higher temperatures than can be attained by the true thermoplastics. Commercial polyimide film maintains its flexibility and strength at extremely low as well as high temperatures. It is useful at the boiling point of helium, 4 K. In addition, it has been used as long as 15 min at 530°C. Its major use has been as an electrical insulation in motors, transformers, and capacitors.

The reaction of an aromatic diamine with maleic anhydride results in a *bismaleimide* that combines the reactivity of the double bonds with the chemical stability of the aromatic imide structure.

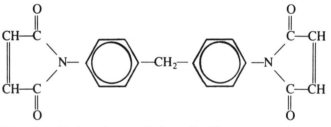

Bismaleimide based on methylene dianiline

Since the homopolymers tend to be brittle, coreactants are used with the bismaleimides to improve strength and toughness. Also, coreactants such as amines or allyl phenols make the starting material more processable by lowering the melting point of the mixture. A major use of bismaleimides is in printed-circuit boards. With glass fiber reinforcement, the compression molding of the board typically proceeds in two stages: an initial stage of 175 to 230°C for 1 to 4 h followed by a postcure at 230°C for another 4 h. The high-temperature properties of such composites

Table 15.10 End-use pattern for polyamides and related polymers

Polymer	Molded plastic	Extruded plastic	Film	Lam-inates	Tires	Rubber specialties	Fibers	Adhesives and coatings	Polymer additives
Nylon 66	•	•	•	•	•		•	•	
Nylon 6	•	•	•	•	•		•	•	
Aromatic nylons	•		•				•	•	
Amorphous nylons								•	•
Polyimides	•		•				•	•	
Poly(benzimidazole)	•		•				•	•	
N-Alkylated nylons								•	•

generally are superior to those of most epoxy-based composites. Aircraft structural components are contemplated as future uses [18].

Use Pattern

Aromatic polyamides (Table 15.7, IC) have been used for flame-resistant garments for firefighters and astronauts and as light-weight body armor for military and law-enforcement agents. In some tests, seven layers of fabric made from an aromatic polyamide were sufficient to protect against the effect of most common handguns. Such a "bullet-proof vest" weighs less than 2 kg. Fiber type, fabric weave, and garment design are all factors in making such a product.

A liquid crystal polymer such as poly(p-benzamide) can be wet-spun from concentrated sulfuric acid into dilute acid. The rigid molecules undergo a great deal of flow orientation during spinning. High-temperature stability and high strength have led to a variety of applications. Aramid fibers compete with steel and glass as the belt material in radial tires.

Nylon 66 and nylon 6 accounted for most of the 1.25×10^9 kg of nylon fibers as well as the 0.42×10^9 kg of nylon plastics used in 1994. Some of the large-volume fiber products include carpets, apparel, and tire cord. Smaller markets include strings for guitars, tennis racquets, and fishing lines. Amorphous nylons are used in adhesives and coatings (Table 15.10).

Although food processing is a large field of study in itself, the development of fat substitutes is a particular example of a useful polymer modification [20]. In one process, [21], nonagregated microparticles (ca. 1 μm in diameter) are produced by the high-speed shearing of milk whey. The whey is the watery part of milk that is left when the casein-rich portion (curd) has been coagulated and removed as in cheese making. The microparticles, mainly protein in composition, can be used to impart a texture to food products similar to that obtained with fat emulsions. The advantage of the nonfat microparticle is the reduced calorie content in a food such as ice cream.

Table 15.11 Aldehyde condensation products

I. General reaction

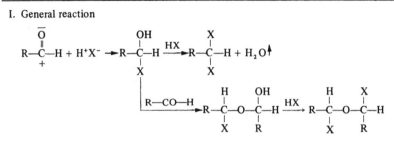

II. Reactants

R = H—   Formaldehyde

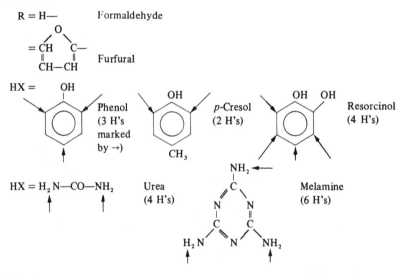

= CH   C—   Furfural

HX =   OH   Phenol (3 H's marked by →)

OH   p-Cresol (2 H's)   CH₃

OH   OH   Resorcinol (4 H's)

HX = H₂N—CO—NH₂   Urea (4 H's)

Melamine (6 H's)

III. Curative

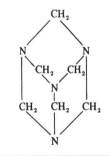

$C_4H_8N_6$, Hexamethylenetetramine

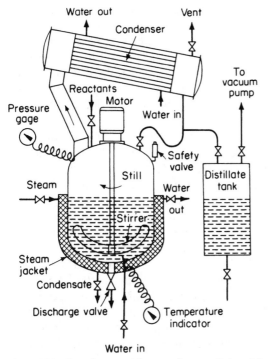

Figure 15.9 Reaction vessel for manufacture of phenol-formaldehyde resins [19]

15.5 ALDEHYDE CONDENSATION POLYMERS

Historians disagree about the birthday of plastics. Some contend that cellulose nitrate in the 1860s started it all, but purists hold out for 1907, when Dr. Leo H. Baekeland patented the resins based on phenol and formaldehyde, an entirely synthetic polymer (Table 15.11). The urea-formaldehyde resins were introduced in the 1930s, opening up the possibility of light colors that were impractical in phenolics. Sterilizable, rugged plastic dinnerware based on melamine resins made a great impact on the market after World War II. The exciting growth of thermoplastics in the 1950s tended to overshadow the thermosets. But in the 1960s it was found that urea and melamine resins could be used to impart permanent press to cottons and other fabrics. The phenolic resins were found to perform well as ablative materials in reentering space vehicles. Over the intervening years so many markets became important for the aldehyde condensation polymers that no one use can be said to dominate the market for any one polymer. Coatings and adhesives, laminating adhesives for plywood, bonding agents for grinding wheels and fiberglass filters, and paper-treating agents all claim significant portions of the production of aldehyde condensation polymers.

A typical *one-stage* phenolic resin might be made in a vessel of the type shown in Fig. 15.9. An excess of formaldehyde (as a 40% solution in water) with the phenol and an alkaline catalyst (NH_3 or Na_2CO_3) is refluxed for an hour or so. A charge of 1.5 mol aldehyde per mole of phenol is typical. A branched, water-soluble, hydroxyl-bearing polymer of low molecular weight called a *resole* is formed. Dehydration by heating under a vacuum may take 3 or 4 h. The polymer is removed and cooled. Casting resins, bonding resins, and resins for laminating paper and wood are made this way. The product, when heated with further dehydration, will cross-link completely to give a hard, infusible network like that pictured in Table 4.7.

For a *two-stage* resin, about 0.8 mol formaldehyde per mole of phenol might be used with an acid catalyst (oxalic acid or sulfuric acid is typical). Over a heating period of 2 to 4 h at reflux, the soluble, fusible *novolak* resin forms (Fig. 15.10). Water is removed at temperatures as high as 160°C (higher temperatures than can be tolerated with resoles). The molten, low-molecular-weight polymer is cooled, and the glassy product is crushed, blended with a lubricant and an "activator" and filler, then rapidly mixed on differential rolls. The resultant blend is quickly cooled and made into a molding powder by cutting and blending. The activator is usually hexamethylenetetramine (structure in Table 15.11, III). This compound is the reaction product of 6 mol of formaldehyde and 4 mol of ammonia. When heated in the final molding operation, excess formaldehyde for polymerization is generated together with an alkaline catalyst. In compression molding, a cycle of 10 min at 165°C is usually sufficient to yield a hard, dense, thermoset network. Wood flour (ground wood) is a popular filler, but numerous other fillers, including cellulose fibers and glass fibers, are used.

In considering melamine molding resin production it should be noted that the filler (alpha cellulose in this case) is added to the wet syrup from the reaction kettle without removing water, rather than being mixed with dry resin as in the case of the phenolic resin. The mixture is then dehydrated, ground, mixed, and finally converted into a molding powder. A molar ratio of 2:1 (formaldehyde to melamine) and a pH of 8.6 are maintained until condensation has proceeded to the desired point as evidenced by insolubility of a sample in cold water. To make an unfilled resin for laminating paper, the syrup is spray-dried after reaching the required viscosity. The dried powder is redissolved by the consumer in order to steep his layers of paper.

As with the polyamides, the melamine or urea condensation products can be *N*-alkylated with formaldehyde and an alcohol such as methanol or butanol. The resulting compositions have good solubility in common solvents and are used in textile treatment and in protective coatings.

Use Pattern

As mentioned above, no single use dominates the aldehyde resin market (Table 15.12). Over a period of many years, dozens of minor variations in the resins have been produced. Substitution of furfural for formaldehyde adds ethenic functional-

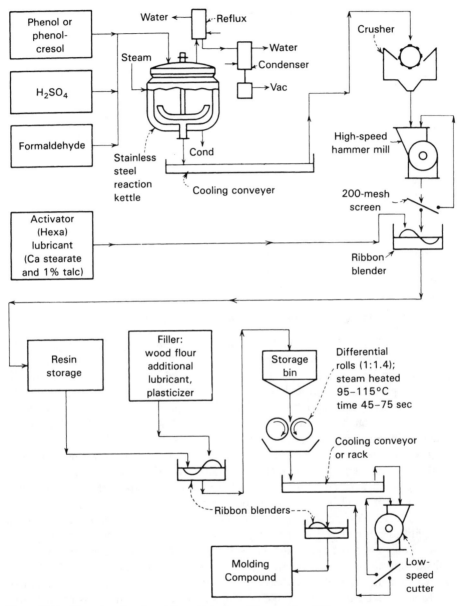

Figure 15.10 Flow sheet for phenol-formaldehyde molding powders [22]

Table 15.12 Amino and phenolic resins—applications

	Phenolics	Aminos
Abrasives	x	
Fibrous and granulated wood binder	x	x
Foundry and shell moldings	x	
Brake linings, etc.	x	
Laminating resins	x	x
Molding materials	x	x
Plywood	x	x
Protective coatings	x	x
Thermal insulation	x	
Paper treating	–	x
Textile treating	–	x

ity. Phenolic resins have been made into ion-exchange resins by sulfonation and other reactions used with polystyrene resins (see Sec. 14.3).

15.6 POLYMERS BASED ON ISOCYANATE REACTIONS

During World War II, products in all the major categories of Table 15.13 were investigated in Germany. It was not until the 1950s that "urethane" foam was produced in the United States on a large scale. Flexible foam continues to be the dominant product, even though many other items are made on a commercial scale.

Foams were discussed in Sec. 12.6. The diols and polyols used usually are polyethers, although polyesters are preferred for some applications such as textile backings because they adhere more readily to the substrates.

Cast urethane elastomers can be made by a variety of routes similar to those used for foams. A popular scheme with polyester diols is to make a prepolymer with excess diisocyanate and then to chain-extend or cross-link the prepolymer with a small molecule such as a diamine or a polyol. Two such reactants are shown in Table 15.13, IV. Thermoplastic urethane elastomers have been called "virtually cross-linked" polymers because they appear at room temperature to be cross-linked and thus resist creep and have rather low hysteresis. However, the cross-links disappear as the temperature is raised, so that the polymer can be molded or extruded like an ordinary thermoplastic resin. Thus, these are "thermoplastic elastomers" (see Sec. 15.10). The cross-links may be hydrogen bonds between urethane groups as well as allophanate bonds. They re-form on cooling. A typical elastomer is made by combining a polyester diol (adipic acid with 1,4-butanediol) with diisocyanate (MDI). The segmented nature of the polymer, polyester sections separating the urethane links, allows the cross-links to form at isolated points with flexible chains leading from one cross-link to the next. Covalent cross-links can be

Table 15.13 Isocyanate reaction polymers

I. Products
  A. Foams
  Flexible or rigid depending mainly on branched structure of diol
  "Prepolymer" or "one-shot"
  Polyester or polyether diol
  B. Elastomers
  Cast
  Thermoplastic
  Millable, vulcanizable
  C. Fibers
  Nylon-like
  Spandex, elastic
  D. Adhesives and coatings

II. Isocyanate reactions

$$
\text{I—N=C=O} + \quad
\begin{cases}
\text{H}_2\text{O} & \longrightarrow \text{I—NH}_2 + \text{CO}_2 \\
\text{HO—R (alcohol)} & \longrightarrow \text{I—NH—CO—O—R (urethane)} \\
\text{H}_2\text{N—R (amine)} & \longrightarrow \text{I—NH—CO—NH—R} \\
& \qquad \text{(disubstituted urea)} \\
\text{HO—CO—R (acid)} & \longrightarrow \text{I—NH—CO—R (amide)} + \text{CO}_2 \\
\text{R}_1\text{NH—CO—NH—R}_2\ \text{(urea)} & \longrightarrow \text{I—NH—CO—NR}_1\text{—CO—NH—R}_2 \\
& \qquad \text{(biuret)} \\
\text{R}_1\text{NH—CO—O—R}_2\ \text{(urethane)} & \longrightarrow \text{I—NH—CO—NR}_1\text{—CO—O—R}_2 \\
& \qquad \text{(allophanate)}
\end{cases}
$$

I = organic group

III. Commercial diisocyanates

Tolylene-2,4-diisocyanate (TDI), often
 used as mixture with 2,6 isomer

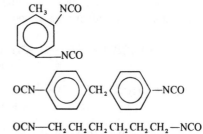

Methylene-bis(4)phenyl isocyanate (MDI)

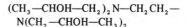

Hexamethylene diisocyanate (HDI)   OCN—CH$_2$CH$_2$CH$_2$CH$_2$CH$_2$CH$_2$—NCO

IV. Chain-extending agents

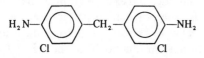

4,4'-methylene-bis-(2-chloro aniline)

(CH$_3$—CHOH—CH$_2$)$_2$N—CH$_2$CH$_2$—
N(CH$_2$—CHOH—CH$_3$)$_2$

N,N,N',N'-Tetrakis(2-hydroxypropyl)
ethylene diamine

Table 15.13 (*Continued*)

V. Preparation of urethane elastomer by amine extension of prepolymer

$$HO-polyester-OH + I(NCO)_2 \rightarrow$$

    Diol   Diisocyanate

$$OCN(I-NHCO-O-polyester-O-CONH)_n I-NCO$$

    Isocyanate-terminated prepolymer

        $n = 2$ to 5

$$OCN-prepolymer-NCO + H_2N-amine-NH_2 \rightarrow$$

$$H_2N-amine-NH-CO-NH-prepolymer-NH$$

                               $\overset{|}{C}=O$

                               $\overset{|}{}$

$$H_2N-amine-NH-CO-NH-prepolymer-NH-CO-NH-amine-N-CO-NH-etc.$$

    Portion of "cured" elastomer with biuret link

introduced into such an elastomer by heating the elastomer with a peroxide. Several "millable, vulcanizable" gums are now available that incorporate in the diol (polyester or polyether) some unsaturation allowing conventional sulfur-type curing as is used in vulcanizing natural rubber.

One nylonlike fiber has been produced in Germany for some time based on 1,4-butanediol and hexamethylene diisocyanate. The structure resembles nylon 66 except for two additional oxygens per repeating dyad. Their properties are similar except that the urethane has lower water adsorption and somewhat better electrical and mechanical stability on aging. However, the urethane apparently has not shown sufficient advantage to warrant introduction in the United States.

$$-(O-CH_2CH_2CH_2CH_2-O-CO-NH-CH_2CH_2CH_2CH_2CH_2CH_2-NH-CO-)$$

    Polyurethane

$$-(CH_2CH_2CH_2CH_2-CO-NH-CH_2CH_2CH_2CH_2CH_2CH_2-NH-CO-)$$

    Nylon 66

Elastic fibers that are at least 85% by weight polyurethanes have the generic name spandex. As a primitive example, one could take the thermoplastic elastomer mentioned above, dissolve it in dimethyl formamide, and extrude it through a spinneret into an aqueous bath, which will cause the polymer to precipitate, the dimethyl formamide dissolving in the water. The fibers may be dried, drawn, and put through all the processes associated with fiber production (see Sec. 12.5). One commercial fiber appears to be the result of chain-extending with a diamine, a prepolymer based on MDI and poly(tetrahydrofuran). The fibers find use in foundation garments, surgical hose, and swimsuits. They compete here with natural rubber thread made by slitting extruded, vulcanized rubber sheets.

In coatings and adhesives, isocyanates may be used as reactive species at room temperature, or by modification, at high temperatures. Most "urethane" varnishes, however, are probably solutions of alkyd resins modified by reaction of free hydroxyl groups with diisocyanates.

Use Pattern

Except for the nylonlike urethane fiber, the isocyanate polymers are characterized by a segmented structure in which the rather polar urethane groups act as crosslinks and the connecting chains are varied in stiffness and polarity depending on the application. In the elastomeric and coatings applications, the one property most often cited as being superior for polyurethanes over other materials is abrasion resistance. Pneumatic tires said to outlast conventional rubber tires have been constructed from unfilled cast elastomers. Other properties such as traction and cost presumably have kept them from becoming commercially available. Solid tires for fork-lift trucks are made from cast polyurethanes, especially for locations where oil resistance is a prerequisite. By far the major product group for the isocyanate-based polymers has been flexible foams for bedding, furniture, and rug underlays. Reaction injection molding (RIM, see Sec. 12.6, Molding), which produces parts with densities about 80 to 90% of the corresponding solid materials, is widely used for automobile parts such as bumpers and grills.

15.7 SILICONES

The need during World War II for a heat-resistant, flexible electrical insulating material accelerated development of the silicones on the basis of research previously done by Kipping (England) and Hyde (U.S.). The Dow Corning Corporation (1943) and General Electric Company (1946) were the first producers of fluids and rubber based on silicones. In addition to the heat resistance associated with silicones, the low-temperature flexibility inherent in the Si—O bond (Sec. 2.2) and the low surface tension of silicones suit them for many applications despite their relatively high cost. Most silicone monomers start with reactions of silicon metal or $SiCl_4$ derived from Cl_2 + SiC. Originally, the Grignard synthesis was widely used commercially (Table 15.14, IC). The "direct" reaction of alkyl and aryl chlorides with silicon metal is favored today (Table 15.14, IA, B). In either case, monomers of variable functionality are formed and must be separated prior to hydrolysis (Table 15.14, IF). Since a monochlorotrialkyl silane, $ClSiR_3$, has only one hydrolyzable group, we can abbreviate it in the hydrolyzed form as M—. Because the free hydroxyls are not stable in general, the product of hydrolysis is $R_3Si—O—SiR_3$, abbreviated M—M. The difunctional group, R_2SiO, is abbreviated D, $RSiO_{1.5}$ by T, and SiO_2 by Q.

A peculiarity of the hydrolysis of dichlorosilanes is the formation of many ring structures. In addition to the cyclic tetramer, D_4, simple mixing of $(CH_3)_2SiCl_2$

Table 15.14 Silicones

I. Monomer production

A. $CH_3Cl + Si \xrightarrow{Cu,\ 300°C} (CH_3)_mSiCl_{m-4}$

B. $3HCl + Si \xrightarrow{300°C} HSiCl_3 + H_2$

C. $CH_3MgCl + SiCl_4 \xrightarrow{ethyl\ ether} (CH_3)_mSiCl_{m-4} + MgCl_2$

D. $CH{=}CH + HSiCl_3 \xrightarrow{peroxide} CH_2{=}CH{-}SiCl_3$

E.
$$R_1{-}CH{=}CH_2 + \underset{\underset{R_2}{|}}{\overset{\overset{H}{|}}{Si}}{-}O{-} \xrightarrow{Pt} R_1{-}CH_2CH_2{-}\underset{\underset{R_2}{|}}{Si}{-}O{-}$$

F.
$$(R)_2SiCl_2 + H_2O \rightarrow {-}\underset{\underset{R}{|}}{\overset{\overset{R}{|}}{Si}}{-}O{-}$$

II. Product classes
   A. Fluids
   B. Rubber
      Dimethyl siloxane
      Methyl, phenyl copolymer
      Fluoromethyl siloxane
   C. Resins
   D. Surfactants
   E. Coupling agents

with alkaline water leads to appreciable quantities of D_3, D_5, D_6, etc. Vapor-liquid chromatography is able to confirm the presence in hydrolyzate of rings up to D_{40}. Production of many silicone compounds starts with redistilled hydrolyzate, which is predominantly D_4.

By heating D_4 together with the appropriate amount of M—M in the presence of a strong base (KOH) or acid (H_2SO_4), an equilibrium composition may be reached in which the molecular weight of the linear portion is regulated by the amount of end groups (M) added.

$$M{-}M + xD_4 \xrightarrow{KOH,\ 170°C} M{-}(D_4)_x{-}M$$

Neutralization of catalyst and vacuum distillation can give a linear fluid with most of the cyclic species removed. The relationship between melt viscosity and molecular size is shown in Fig. 7.10. Dimethyl siloxanes with viscosities at room temperature of up to 100,000 centistoke are used as lubricants, antifoaming agents, and additives for cosmetics, pharmaceutical preparations, and hundreds of other con-

Table 15.15 Temperature dependence of viscosity for polysiloxanes [23]

	Viscosity at 25°C, cs*	VTC [Eq. (15.1)]
Substituents:		
Dimethyl (M—M)	0.65	0.31
Dimethyl (MD₃M)	2.0	0.48
Dimethyl	10	0.57
Dimethyl	10^2	0.60
Dimethyl	10^3	0.62
Dimethyl	10^4	0.61
Dimethyl	10^5	0.61
Methylphenyl	482	0.88
Methylethyl	1300	0.71
Diethyl	800	0.90
Methylhydrogen	25	0.50
Mineral oil (typical)	110	0.91

*1 cs = 1 mPa·s divided by density (g/cm³).

sumer products. An interesting feature of dimethyl siloxanes is their low viscosity-temperature coefficient [VTC, Eq. (15.1)] compared with that of hydrocarbon lubricants.

$$VTC = 1 - \frac{\text{viscosity at } 210°F}{\text{viscosity at } 100°F} \qquad (15.1)$$

The effect is lost, however, when phenyl, ethyl, or other groups are substituted for methyl (Table 15.15). Only the substitution of hydrogen for methyl lowers the VTC.

When D_4 is heated to 170°C with a small amount of KOH and no added end blocker, high-molecular-weight gums are produced. The molecular weight distribution corresponds roughly to the "most probable" distribution (Sec. 6.2) for the linear portion. The cyclic species form a separate distribution. Gel permeation chromatography (Sec. 6.4) shows the initial distribution and the distribution after neutralization of catalyst and vacuum stripping (Fig. 15.11). No redistribution seems to occur in the linear portion, and most of the cyclic species are removed. Most commercial gums are copolymers with 0.2 to 0.6% vinylmethylsiloxane. The vinyl group provides a preferential cross-linking site with peroxides. The tensile strength of unfilled, cross-linked silicone rubber is only a few hundred pounds per square inch at best. Fillers, especially finely divided silica, raise this into the range of 1200 to 2000 psi (8 to 14 MPa). Even with the vinyl comonomer and cross-linking, dimethyl siloxanes become stiff at −54°C because of crystallization. The stiffening temperature is brought close to the glass-transition temperature (−110°C) by copolymerization with phenylmethyl or diphenyl siloxane (see Sec. 3.5). Some silicone fluids with hydrolyzable or reactive end groups have been com-

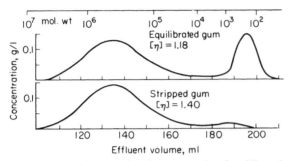

Figure 15.11 Gel permeation chromatographs of equilibrated and vacuum-stripped dimethyl siloxane gums [24]

pounded with fillers and catalysts to make RTV (room-temperature vulcanizing) rubber. An example of a one-package system that converts to a rubber on exposure to moist air was given in Sec. 12.4.

Silicone resins for coatings resemble phenol-formaldehyde polymers in that polymerization takes place in two stages. A trifunctional monomer such as phenyltrichlorosilane may be hydrolyzed to the relatively stable hexamer (and related structures modified by comonomers for better solubility).

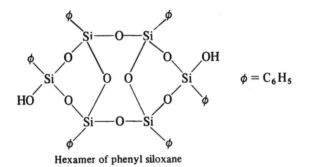

$\phi = C_6H_5$

Hexamer of phenyl siloxane

This compound when heated alone or with carboxylic acids and esters will undergo further condensation and rearrangement to give heat-stable, network polymers.

Surfactants based on block copolymers of dimethylsiloxane with poly(ethylene oxide) are unique in regulating the cell size in polyurethane foams. One route to such polymers uses reaction IE, Table 15.14, between a polysiloxane and an allyl ether of polyethylene oxide [25]. Increasing the silicone content makes the surfactant more lipophilic, while increasing the poly(ethylene oxide) content makes it more hydrophilic.

The coupling agents used to change the surfaces of silica and glass fillers (Sec. 9.8) are substituted alkoxy silanes in monomeric form.

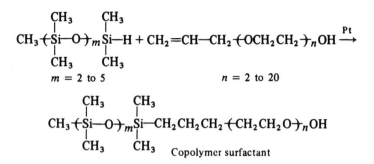

$$m = 2 \text{ to } 5 \qquad\qquad n = 2 \text{ to } 20$$

Copolymer surfactant

They may polymerize on the surface of the filler, or they can be copolymerized with other organic monomers.

The incorporation of bulky groups into the silicone chain would be expected to raise T_m and T_g. Attempts have been made to do this without sacrificing the high-temperature stability of the silicones. One such material is *p*-bis(dimethylhydroxysilyl) benzene, made from *p*-dibromobenzene and $(CH_3)_2SiCl_2$ by the Grignard reaction. Polymerization is carried out by refluxing the monomer in benzene with an amine salt acting as a catalyst (*n*-hexylamine 2-ethylhexoate) [26].

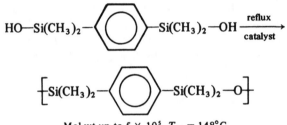

Mol wt up to 5×10^5, $T_m = 148°C$

Copolymers with dimethylsiloxane are also possible.

Fluorosilicone elastomers are based on trifluoropropylmethyl siloxane:

$$\begin{array}{c} CH_3 \\ | \\ +Si-O+ \\ | \\ CH_2 \\ | \\ CH_2 \\ | \\ CF_3 \end{array}$$

Such elastomers have some of the solvent resistance associated with fluorocarbon polymers. It would seem that a more highly fluorinated monomer would be better.

Unfortunately, fluorine on carbons next to or one carbon away from silicon generally lead to thermal or hydrolytic instability [27].

Chains of silicon atoms (polysilanes) differ in many properties from those with alternating silicon and oxygen (polysiloxanes). On exposure to UV or ionizing radiation, for example, most polysiloxanes undergo cross-linking whereas the polysilanes usually undergo chain scission. Stable organopolysilanes can be prepared by the reaction of a silyl dihalide with sodium [28, 29]:

$$\text{Cl}-\underset{\underset{\text{R}''}{|}}{\overset{\overset{\text{R}'}{|}}{\text{Si}}}-\text{Cl} \; + \; 2\,\text{Na} \; \longrightarrow \; +\underset{\underset{\text{R}''}{|}}{\overset{\overset{\text{R}'}{|}}{\text{Si}}}+ \; + \; 2\,\text{NaCl}$$

where R′ and R″ can be various alkyl or aryl groups. In one example, when R′ is *p-tert*-butylphenyl and R″ is methyl, a high-molecular-weight, soluble polymer can be produced with some interesting properties. Although it is stable in air, a film of the polymer adsorbs UV light strongly. In fact, so much energy can be adsorbed that the polymer will undergo extensive random chain scission. An image can be recorded in such a film by the evaporation of the small fragments that result from irradiation [29]. Some other uses proposed for polysilanes include electrically conducting materials (when doped with arsenic pentafluoride) and impregnating agents for ceramic materials. It has also been shown that poly(dimethylsilane), R′ = R″ = CH_3 can be converted to poly(carbosilane) and eventually to β-silicon carbide:

$$\text{Me}_2\text{SiCl}_2 \xrightarrow{\text{Na}} +\text{Me}_2\text{Si}+ \xrightarrow{320°\text{C,Ar}} +\underset{\underset{CH_3}{|}}{\overset{\overset{H}{|}}{\text{Si}}}-CH_2-\underset{\underset{CH_3}{|}}{\overset{\overset{H}{|}}{\text{Si}}}-CH_2+ \xrightarrow{200°\text{C,air};1300°\text{C}} \beta\text{-SiC}$$

(two-step process)

Use Pattern

It was pointed out above that silicone fluids are used as additives in many products. Their use alone is limited; high-temperature laboratory baths and special lubricants are examples. Silicone rubber and surfactants also find large markets.

15.8 POLYPHOSPHAZENES

Polymers with the—P=N—backbone were produced as long ago as the 1890s. Cyclic phosphonitrilic chloride (which is made from PCl_5 and NH_4Cl) is opened by ring-scission polymerization.

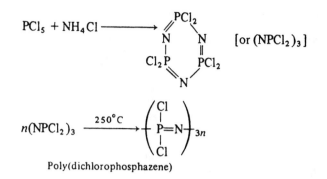

Poly(dichlorophosphazene)

A cross-linked version of this "inorganic rubber" was described by Stokes in 1895 [30, 31]. However, moisture, even atmospheric moisture, soon hydrolyzes this polymer to a brittle material. In the 1960s and 1970s, preparation of the dichloropolymer was made reproducible so that it could be used as a precursor for various derivatives. The polymerization of the cyclic trimer proceeds at about 250 to 270°C in a vacuum over a period of 20 to 40 h with reflux. Precipitation of the polymer in hexane or pentane leaves unreacted trimer and some other low-molecular-weight species in solution. Yields may range from 15 to 75%.

The chlorine atoms on the backbone of the polymer may be replaced by organic groups such as alkoxy, aryloxy, or amino groups. Most of these derivative polymers do not exhibit the hydrolytic instability of the parent polymer. In order to carry out the substitution, the chlorine-bearing polymer is added dropwise to a solution containing sodium alkoxide, or aryloxide in a solvent such as tetrahydrofuran or benzene. A commercial rubber (PNF, Firestone Tire and Rubber Co.) is based on alkoxides of trifluoroethyl alcohol or heptafluoroisobutyl alcohol [32].

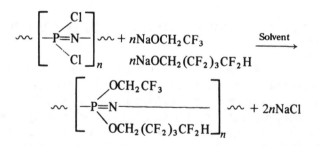

the product is referred to as a phosphonitrilic fluoroelastomer, a "semiorganic" rubber. The useful range is from the T_g of about $-68°C$ up to about 175°C. PNF can be cross-linked by peroxides, sulfur, or high-energy electrons. The outstanding properties of the rubber are excellent oil and fuel resistance, good dynamic properties, and good abrasion resistance. Disadvantages are water resistance that is only fairly good, and a price higher than that of fluorosilicones.

Table 15.16 Some polyorganophosphazenes [31]

Repeat group in compound	T_g, °C	Solvent
—N=PCl$_2$—	−63	Benzene
—N=P(OCH$_3$)$_2$—	−76	Methanol
—N=P(OCH$_2$CH$_3$)$_2$—	−84	Alcohols
—N=P(OCH$_2$CF$_3$)$_2$—	−66	Ketones
—N=P(OC$_6$H$_5$)$_2$—	−8	Benzene
—N=P(NH—C$_6$H$_5$)$_2$—	91	Benzene
—N=P(N(CH$_3$)$_2$)$_2$—	−4	Aqueous acid
—N=P(NC$_5$H$_{10}$)$_2$—	19	Benzene

Numerous other derivatives have been reported in the literature. A few examples are tabulated (Table 15.16). In addition to those with phosphazene in the backbone, some contain—P=N—in side groups. The reaction of styrene with a vinyl-substituted trimer gives a copolymer that is flame-retardant in a standard test [33].

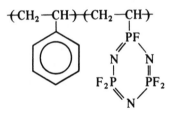

15.9 ENGINEERING RESINS AND HIGH-PERFORMANCE POLYMERS

In industrial practice, polymer categories may be based on a variety of considerations. Producers of polymers may well use a compositional basis like the one used in the various sections of the previous chapter and this one. Raw materials and techniques for many products may be common to one such group. However, users of polymers often create categories by performance properties such as those summarized in Sec. 10.1. Within a major industry, polymers compete on a price and performance basis. As an example, in the rubber industry one can differentiate among general-purpose rubbers (for tires, wire and cable insulation), specialty rubbers (oil-resistant, ozone-resistant, heat-resistant), and thermoplastic elastomers (which need no covalent cross-linking step).

The term *engineering resin* is used to designate a group of mainly heterochain thermoplastics that compete with some die-cast metals such as zinc, aluminum,

Table 15.17 Selected engineering thermoplastics [34]

Formula index (Fig. 15–12)	Polymer (abbreviation)	Trade name(s)	Distortion temperature under load, °C Neat (reinforced)
1	Polyetheretherketone (PEEK)	Victrex	165 (282)
2	Polyamideimide (PAI)	Torlon	274 (278)
3	Polyarylsulfone	Udel	174
4	Polyarylsulfone	Victrex	203
5	Polyarylsulfone	Astrel	274
6	Polyimide (not melt processible)	Kapton	245+ (350)
7	Polyetherimide (PEI)	Ultem	200 (223)
8	Polyphenylenesulfide (PPS)	Ryton	111 (241)
9	Poly(ethylene terephthalate) (PET)	Valox, Rynite	150 (224)
10	Poly(butylene terephthalate) (PBT)	Valox	130 (160)
11	Polyarylate, aromatic polyester	Ardel, Durel	170
12	Polycarbonate (PC)	Lexan, Merlon	130 (142)
13	Polyacetal	Delrin	136 (155)
14	Polyacetal	Celcon	110 (165)
15	Nylon 66	Zytel	100 (250)
16	Nylon 6	Zytel	80 (205)
17	Modified poly(phenyleneoxide) (PPO)	Noryl	90 (160)

and magnesium in plumbing parts, hardware, and automotive parts [34]. Generally speaking, the engineering resins command a higher price than the "commodity plastics." Most people would agree on the selection shown in Table 15.17 and Fig. 15.12. The polyimide is included by some even though it is not melt-processible. All of the polymers included here can be used without reinforcement, but glass fibers or mineral reinforcement generally improve dimensional stability at elevated temperatures (See Table 9.1). The flexural modulus and tensile strength as a function of temperature are included for some of these materials in Figs. 15.13 and 15.14.

A list of materials regarded as *high-performance polymers* (Table 15.18) overlaps the list of Table 15.17 to some extent, but also includes thermosets such as epoxy, phenolic, and silicone molding resins. As with most molded plastics, high-temperature performance is enhanced by reinforcement with glass, boron, or carbon fibers. The thermosets are seldom used without reinforcement. Several are included in Figs. 15.13 and 15.14.

15.10 THERMOPLASTIC ELASTOMERS

It has been noted in several places (Secs. 4.6, 15.1, and 15.6) that certain thermoplastic polymers can be quite elastic (rubbery) at room temperature. They form a

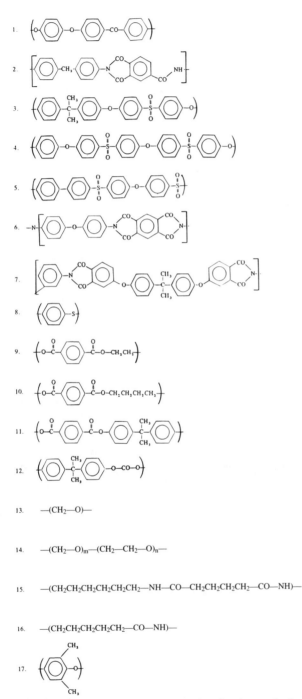

13. —(CH₂—O)—

14. —(CH₂—O)ₘ—(CH₂—CH₂—O)ₙ—

15. —(CH₂CH₂CH₂CH₂CH₂CH₂—NH—CO—CH₂CH₂CH₂CH₂—CO—NH)—

16. —(CH₂CH₂CH₂CH₂CH₂—CO—NH)—

17.

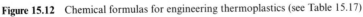

Figure 15.12 Chemical formulas for engineering thermoplastics (see Table 15.17)

654

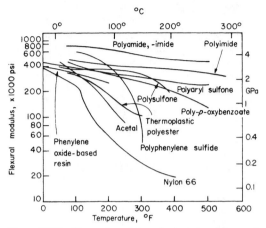

Figure 15.13 Flexural modulus of selected molding resins (unreinforced) [35]

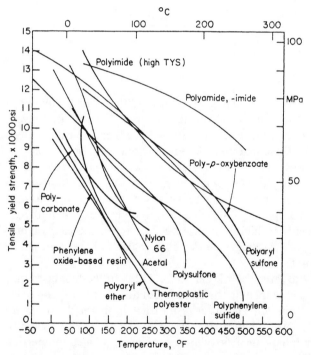

Figure 15.14 Tensile yield strength of selected molding resins (unreinforced) [35]

655

Table 15.18 Properties of some high-performance plastics* [36]

	Tensile strength, MPa	Flexural modulus, GPa	Heat deflection temperature, °C (at 1.8 MPa)	Continuous use temperature, °C
Polyimide				
Thermoplastic	118	3.3	132 to 138	290
Thermoset	62 to 90	–	over 245	–
Thermoset, filled	19 to 39	22	290 to 350	260
Poly(amide -imide)				
Unfilled	92	5	280	290
Filled	78 to 90	5 to 8	285 to 295	290
Polyarylsulfone	90	2.7	–	260
Aromatic polyester	17	7.1	–	–
Aromatic copolyester	70	3.1	over 315	300
Aromatic polyamide	115	4.4	–	–
Polyphenylene sulfide				
Unfilled	70	4.1	137	205 to 260
Filled	145	15	over 215	205 to 260
Silicone resin, filled	25 to 45	7 to 17	over 480	over 315
Poly(tetrafluoroethylene)	15 to 35	–	–	260
Epoxy (mineral-filled)	70 to 140	17 to 30	120 to 260	–
Phenolic (glass-fiber-filled)	35 to 125	14 to 23	150 to 315	–

*Unfilled unless designated otherwise.

distinct class of materials, the *thermoplastic elastomers* (TPE), which have been defined (ASTM D 1566) as "a family of rubber-like materials that, unlike conventional vulcanized rubber, can be processed and recycled like thermoplastic materials." The typical structure of these materials is the coexistence of "soft" domains that give rubbery behavior with "hard" domains that act as heat-labile cross-links. Although the thermoplastic elastomers compete in many of the applications traditionally assigned to conventional rubbers, the sensitivity of the cross-links to heat makes them unsuitable for an application such as automobile tires, where the repeated flexing invariably leads to high temperatures. However, the ease of processing and the possibility of recycling material both make TPE very attractive from an economic standpoint.

Payne and Rader [37] cite three generic classes of TPEs. In the first class they put the block copolymers. These include the styrene-butadiene-styrene, copolyesters, polyurethanes, and polyamides. The latter result when aromatic diamines are used in the "hard" segments and aliphatic diamines are used in the "soft" segments, both in combination with aliphatic dicarboxylic acids. The mechanical properties, processing conditions, and solvent resistance differ for the various copolymers.

Within each type, a spectrum of hardness values is obtainable by variation of the size and chemical composition of the segments.

The second class is made up of elastomer/thermoplastic blends, abbreviated TEO. Typical of this class is a dispersion of 20 to 30 parts of rubber based on ethylene-propylene-diene monomer (EPDM) in a continuous phase of 70 to 80 parts of a plastic such as isotactic polypropylene. Partial covalent cross-linking of the EPDM particles, which may be less than 10 μm in diameter, results in improved elastic recovery and solvent resistance. Another example of the second class is a dispersion of particles of nitrile rubber in a continuous phase of a PVC compound.

The third class consists of elastomeric "alloys" (EA), which are defined as [37] "highly vulcanized rubber systems with the vulcanization having been done dynamically in the melted plastic phase." The difference between the highly cross-linked EA and the uncross-linked or loosely cross-linked TEO can appear as a fivefold improvement in ultimate tensile strength and much improved elastic recovery. Evidence has been presented to show that bonding occurs between the phases when the cross-linking is done in the melted plastic [38]. Some other EAs are made with isotactic polypropylene as the continuous phase by dispersing and dynamically cross-linking nitrile rubber, natural rubber, or butyl rubber.

KEYWORDS FOR CHAPTER 15

Section

1. Alkyd
   Ester interchange
   PET, PBT, PEN
2. Polyacetal
3. Viscose process
   Vulcanized fiber
4. Nylon
   Poly(benzimidazole)
5. One-stage resin
   Two-stage resin
6. Spandex
7. Viscosity temperature coefficient
8. Polyphosphazene
9. Engineering resin
   High-performance plastics
10. Thermoplastic elastomer (TPE)
    TEO
    EA

PROBLEMS

15.1 How many grams of I and II are needed to make 100 g of a poly(dimethyl siloxane) oil with $M_n = 10,000$?

$$(CH_3)_3SiCl \qquad (CH_3)_2SiCl_2$$
$$\text{I} \qquad\qquad \text{II}$$

15.2 Explain why 2,5-dimethyl phenol gives a high-molecular-weight polymer when oxidized but only a low-molecular-weight material with formaldehyde.

15.3 In the "equilibrium" polymerization of tetrahydrofuran, how can residual monomer be removed without depolymerizing the linear polyether?

15.4 What polymers make use of bisphenol A? What properties does it contribute to the polymers?

15.5 A fabricated sample of "cast unsaturated polyester" has the following properties:

(*a*) It becomes somewhat less stiff at 100°C.
(*b*) It does not flow, even at 300°C.
(*c*) It becomes somewhat weaker on soaking in hot, dilute sodium hydroxide.

Explain how each of these properties is related to the chemical structure of the sample.

15.6 Both propylene oxide and styrene oxide can be polymerized to give isotactic polymers that can crystallize.

(*a*) Write formulas for the two polymers.
(*b*) Identify the asymmetric carbons.
(*c*) Which of the two polymers would you expect to have the highest T_g? Why?

15.7 Why are most of the new thermoplastics with high heat distortion temperatures heterochain polymers rather than carbon chain polymers? Why is not ring-scission polymerization used more often for them?

15.8 In a certain country with an "emerging" economy, a governmental decision has been made to establish a polymer industry based on native resources. No oil refineries or coal tar operations exist. Cotton (pure cellulose) and sugar are the major raw materials. In addition, electrolytic processes can be used, since hydroelectric power abounds. Your first step is to set up plants to produce from sugar, sequentially: ethyl alcohol, acetic acid, and butadiene. From electrolysis you now have NaOH, Cl_2, HCN, and HCl. Sulfuric and nitric acids are also produced locally.

Outline the steps necessary to establish consumer-product plants for (include reactions where possible):

Cellulose acetate film
cis-Polybutadiene rubber (to be used in tires)
Nylon 66 (fibers and molding compound)

(Where you have to import materials, say so.)
Outline briefly the feasibility of producing at least two other polymers and give some idea of where they might be useful. Remember that an intermediate in butadiene production can be

$$\underset{\displaystyle \text{CH}_3}{\text{HO—CH—CH}_2 \text{—CH}_2 \text{—OH}}$$

$\text{CH}_3\text{CH}_2\,\text{Cl}$ is possible, too.

REFERENCES

1. Schildknecht, C. E.: "Diallyl and Related Polymers," in *Enc. Polym. Sci. Eng.,* 2d ed., 4:779 (1986).
2. Ellwood, P.: *Chem. Eng.,* 74:98 (Nov. 20, 1967).
3. Witsiepe, W. K.: *Polymer Preprints,* 13:588 (April 1972).
4. Hutz, C. E., in W. M. Smith (ed.): "Manufacture of Plastics," p. 507, Reinhold, New York, 1964.
5. Whistler, R. L., and J. N. BeMiller (eds.): "Industrial Gums," Academic Press, New York, 1959.
6. Fieser, L. F., and M. Fieser: "Organic Chemistry," 3d ed., Heath, Boston, 1963.
7. *Chem. Eng.,* 73:116 (Oct. 11, 1966).
8. Warner, S. B.: "Fiber Science," p. 105, Prentice-Hall, Englewood Cliffs, N.J., 1995.
9. Brydson, J. A.: "Plastics Materials," p. 366, Van Nostrand, Princeton, N.J., 1966.
10. Mitchell, P. H.: "A Textbook of Biochemistry," 2d ed., pp. 97, 98, and 106, McGraw-Hill, New York, 1950.
11. Gutte, B., and R. B. Merrifield: *J. ACS,* 91:501 (1969); W. Sullivan, *New York Times,* p. 1, Jan. 17, 1969; *Chem. Eng. News,* p. 15 (Jan. 20, 1969).
12. Fasman, G. D. (ed.): "Poly-α-amino Acids," Dekker, New York, 1967.
13. Winding, C. C., and G. D. Hiatt: "Polymeric Materials," p. 258, McGraw-Hill, New York, 1961.
14. *Chem. Eng. News,* 43:49 (June 28, 1965).
15. Brydson, J. A.: "Plastics Materials," p. 311, Van Nostrand, Princeton, N.J., 1966.
16. *Chem. Eng.,* 73:178 (Nov. 7, 1966).
17. Butler, J. M., R. M. Hedrick, and E. H. Mottus (Monsanto): U.S. Patent 3,018,273, 1962.
18. King, J. J.: "Bismaleimide," in *Mod. Plastics Enc. Issue,* p. 18, 1986.
19. Miles, D. C., and J. H. Briston: "Polymer Technology," p. 44, Chemical Publishing, New York, 1961.
20. Thayer, A. M.: *Chem. Eng. News,* 70(25):26 (1992).
21. Bringe, N. A., and D. R. Clark, in M. Yalpani (ed.): "Science for the Food Industry of the 21st Century," chap. 5, ATL Press, Mount Prospect, Ill., 1993.
22. Winding, C. C., and G. D. Hiatt: "Polymeric Materials," p. 225, McGraw-Hill, New York, 1961.
23. Barry, A. J., and H. N. Beck, in F. G. A. Stone and W. A. G. Graham (eds.): "Inorganic Polymers," Academic Press, New York, 1962.
24. Rodriguez, F., R. A. Kulakowski, and O. K. Clark: *Ind. Eng. Chem. Prod. Res. Develop.,* 5:121 (1966).
25. Kanner, B., W. G. Reid, and I. H. Petersen: *Ind. Eng. Chem. Prod. Res. Develop.,* 6:88 (1967).

26. Merker, R. L., and M. J. Scott: *J. Polymer Sci.,* A2:15 (1964).
27. Warrick, E. L., O. R. Pierce, K. E. Polmanteer, and J. C. Saam: *Rubber Chem. Technol.,* 52:437 (1979).
28. Miller, R. D., D. Hofer, D. R. McKean, C. G. Willson, R. West, and P. T. Trefonas, III: "Soluble polysilane derivatives," in L. F. Thompson, C. G. Willson, and J. M. J. Frechet (eds.): *Materials for Microlithography,* ACS Symp. Ser. 266, p. 292, ACS, Washington, D.C., 1984.
29. Miller, R. D., D. Hofer, J. Rabolt, R. Sooriyakumaran, C. G. Willson, G. N. Fickes, J. E. Guillet, and J. Moore: "Soluble polysilanes in photolithography," in M. J. Bowden and S. R. Turner (eds.): *Polymers for High Technology,* ACS Symp. Ser. 346, p. 170, ACS, Washington, D.C., 1987.
30. Allcock, H. R.: "Phosphorus-Nitrogen Compounds," Academic Press, New York, 1972.
31. Allcock, H. R.: *Angew. Chem. Internat. Ed.* (English), 16:147 (1977).
32. Kyker, G. S., and T. A. Antkowiak: *Rubber Chem. Technol.,* 47:32 (1974).
33. DuPont, J. G., and C. W. Allen: *Macromolecules,* 12:169 (1979).
34. Clagett, D. C.: "Engineering Plastics," in *Enc. Polym. Sci. Eng.,* 2d ed., 6:94 (1986).
35. Raia, D. D.: *Plastics World,* 31:75 (Sept. 17, 1973).
36. Markle, R. A., in J. K. Craver and R. W. Tess (eds.): "Applied Polymer Science," chap. 39, ACS, Washington, D.C., 1975.
37. Payne, M. T., and C. P. Rader: "Elastomer Technology Handbook," chap. 14, CRC Press, Boca Raton, Fla., 1993.
38. Coran, A. Y., R. P. Patel, and D. Williams: *Rubber Chem. Technol.,* 55:116 (1982).

General References

Abadie, M. J. M., and B. Sillion (eds.): "Polyimides and Other High-Temperature Polymers," Elsevier, Amsterdam, 1991.

Allcock, H. R.: "Phosphorus-Nitrogen Compounds," Academic Press, New York, 1972.

Arthur, J. C., Jr. (ed.): "Polymers for Fibers and Elastomers," ACS, Washington, D.C., 1984.

Ashida, K. and K. C. Frisch (eds.): "International Progress in Urethanes," vol. 6, Technomic, Lancaster, Pa., 1993.

Baer, E., and A. Moet (eds.): "High Performance Polymers," Hanser-Gardner, Cincinnati, Ohio, 1991.

Bailey, F. E., Jr., and J. V. Koleske (eds.): "Alkylene Oxides and Their Polymers," Dekker, New York, 1991.

Bauer, R. S. (ed.): "Epoxy Resin Chemistry," ACS, Washington, D.C., 1979.

Bauer, R. S. (ed.): "Epoxy Resin Chemistry II," ACS, Washington, D.C., 1983.

Bessonov, M. I., M. M. Koton, V. V. Kudryavtsev, and L. A. Laius: "Polyimides," Plenum, New York, 1987.

Bessonov, M. I., and V. A. Zubkov (eds.): "Polyamic Acids and Polyimides: Synthesis, Transformations, and Structure," CRC Press, Boca Raton, Fla., 1993.

Bhowmick, A. K., and H. L. Stephens (eds.): "Handbook of Elastomers," Dekker, New York, 1988.

Blackley, D. C.: "Synthetic Rubbers," Elsevier Applied Science, New York, 1983.

Bruins, P. F. (ed.): "Unsaturated Polyester Technology," Gordon & Breach, New York, 1976.

Brydson, J. A.: "Plastics Materials," 4th ed., Butterworth, Woburn, Mass., 1982.

Buist, J. M. (ed.): "Developments in Polyurethanes—1," Elsevier Applied Science, New York, 1978.

Carfagna, C. (ed.): "Liquid Crystalline Polymers," Pergamon, New York, 1994.

Carraher, C. E., Jr., and L. H. Sperling (eds.): "Renewable-Resource Materials," Plenum, New York, 1986.

Cassidy, P. E.: "Thermally Stable Polymers: Syntheses and Properties," Dekker, New York, 1980.

Chum, H. L. (ed.): "Polymers from Biobased Materials," Noyes Data, Park Ridge, N.J., 1991.

Ciferri, A., and I. M. Ward (eds.): "Ultra-High Modulus Polymers," International Ideas, Philadelphia, 1979.

Culbertson, B. M., and J. E. McGrath (eds.): "Advances in Polymer Synthesis," Plenum, New York, 1985.

Davidson, R. L. (ed.): "Handbook of Water-Soluble Gums," McGraw-Hill, New York, 1980.

De, S. K., and A. K. Bhowmick (eds.): "Thermoplastic Elastomers from Rubber-Plastic Blends," Prentice-Hall, Englewood Cliffs, N.J., 1991.

DiStasio, J. I. (ed.): "Epoxy Resin Technology (Developments Since 1979)," Noyes Data, Park Ridge, N.J., 1982.

Driver, W. E.: "Plastics Chemistry and Technology," Van Nostrand Reinhold, New York, 1979.

DuBois, J. H., and F. W. John: "Plastics," 5th ed., Van Nostrand Reinhold, New York, 1974.

Dumitriu, S. (ed.): "Polymeric Biomaterials," Dekker, New York, 1993.

Dusek, K. (ed.): "Epoxy Resins and Composites, 1–4," Springer-Verlag, New York, 1985–1986.

Dusek, K., et al. (eds.): "Key Polymers," Springer-Verlag, New York, 1985.

Dyson, R. W. (ed.): "Specialty Polymers," Chapman & Hall (Methuen), New York, 1987.

Dyson, R. W. (ed.): "Engineering Polymers," Chapman & Hall, New York, 1990.

Elias, H.-G.: "Macromolecules, Synthesis, Materials, and Technology," 2d ed., Plenum, New York, 1984.

Elias, H.-G., and F. Vohwinkel: "New Commercial Polymers—2," Gordon & Breach, New York, 1986.

Ellis, B. (ed.): "Chemistry and Technology of Epoxy Resins," Chapman & Hall, New York, 1993.

Feger, C., M. M. Khojasteh, and M. S. Htoo (eds.): "Advances in Polyimide Science and Technology," Technomic, Lancaster, Pa., 1993.

Feger, C., M. M. Khojasteh, and J. E. McGrath (eds.): "Polyimides: Materials, Chemistry and Characterization," Elsevier, Amsterdam, 1989.

Finch, C. A. (ed.): "Chemistry and Technology of Water-Soluble Polymers," Plenum, New York, 1983.

Fishman, M. L., R. B. Friedman, and S. J. Huang (eds.): "Polymers from Agricultural Coproducts," ACS, Washington, D.C., 1994.

Flick, E. W.: "Epoxy Resins Curing Agents, Compounds, and Modifiers: An Industrial Guide," Noyes Data, Park Ridge, N.J., 1987.

Flick, E. W.: "Water-Soluble Resins: An Industrial Guide," 2d ed., Noyes Data, Park Ridge, N.J., 1991.

Flick, E. W. (ed.): "Industrial Synthetic Resins Handbook," 2d ed., Noyes Data, Park Ridge, N.J., 1991.

Fontanille, M., and A. Guyot (eds.): "Recent Advances in Mechanistic and Synthetic Aspects of Polymerization," (Reidel Holland), Kluwer Academic, Norwell, Mass., 1987.

French, A. D., and J. W. Brady (eds.): "Computer Modeling of Carbohydrate Molecules," ACS, Washington, D.C., 1990.

Frisch, K. C., and D. Klempner (eds.): "Advances in Urethane Science and Technology," 12 vols., Technomic, Lancaster, Pa., 1971–1992.

Gilbert, R.: "Cellulosic Polymers," Hanser-Gardner, Cincinnati, Ohio, 1994.

Glass, J. E. (ed.): "Water-Soluble Polymers," ACS, Washington, D.C., 1986.

Glass, J. E. (ed.): "Polymers in Aqueous Media," ACS, Washington, D.C., 1989.

Gum, W. F., Jr., W. Reise, and H. Ulrich (eds.): "Reaction Polymers," Hanser-Gardner, Cincinnati, Ohio, 1992.

Gupta, S. K., and A. Kumar: "Reaction Engineering of Step Growth Polymerization," Plenum, New York, 1987.

Hamerton, I. (ed.): "Chemistry and Technology of Cyanate Ester Resins," Chapman & Hall, New York, 1994.

Harper, C. A. (ed.): "Handbook of Plastics, Elastomers, and Composites," 2d ed., McGraw-Hill, New York, 1992.

Hepburn, C.: "Polyurethane Elastomers," Elsevier, New York, 1992.

"High Performance Polymers," Springer-Verlag, Berlin, 1994.

Inagaki, H., and G. O. Phillips (eds.): "Cellulosics Utilization," Elsevier, New York, 1989.

Knop, A., and L. A. Pilato: "Phenolic Resins," Springer-Verlag, New York, 1985.

Knop, A., and W. Scheib: "Chemistry and Application of Phenolic Resins," Springer-Verlag, New York, 1979.

Koerner, G., M. Schulze, and J. Weis: "Silicones: Chemistry and Technology," CRC Press, Boca Raton, Fla., 1992.

Kohan, M. I. (ed.): "Nylon Plastics," Wiley-Interscience, New York, 1973.

Lee, J. A., and D. L. Mykkanen: "Metal and Polymer Matrix Composites," Noyes Data, Park Ridge, N.J., 1987.

Legge, N. R., G. Holden, and H. E. Schroeder (eds.): "Thermoplastic Elastomers," Macmillan, New York, 1987.

Lin, S.-C.: "High-Performance Thermosets," Hanser-Gardner, Cincinnati, Ohio, 1993.

Lucke, H.: "Aliphatische Polysulfide," Hüthig & Wepf, Basel, Switzerland, 1992.

Margolis, J. M. (ed.): "Engineering Thermoplastics," Dekker, New York, 1985.

Margolis, J. M. (ed.): "Advanced Thermoset Composites," Van Nostrand Reinhold, New York, 1986.

Mark, J. E., H. R. Allcock, and R. C. West: "Inorganic Polymers: An Introduction," Prentice-Hall, Englewood Cliffs, N.J., 1992.

Martuscelli, E., and C. Marchetta (eds.): "New Polymeric Materials: Reactive Processes and Physical Properties," VNU Science, Utrecht, Netherlands, 1987.

May, C. A. (ed.): "Epoxy Resins," 2d ed., Dekker, New York, 1988.

Meier, D. J. (ed.): "Block Copolymers," Gordon & Breach, New York, 1983.

Meltzer, Y. L.: "Water-Soluble Polymers," Noyes Data, Park Ridge, N.J., 1979.

Meltzer, Y. L.: "Water-Soluble Polymers (Developments Since 1978)," Noyes Data, Park Ridge, N.J., 1981.

Meyer, B.: "Urea-Formaldehyde Resins," Addison-Wesley, Reading, Mass., 1979.

Miles, D. C., and J. H. Briston: "Polymer Technology," Chemical Publishing, New York, 1979.

Mittal, K. L. (ed.): "Polyimides," 2 vols., Plenum, New York, 1984.

Molyneux, P.: "Water Soluble Synthetic Polymers: Properties and Behavior," 2 vols., CRC Press, Boca Raton, Fla., 1984.

Moncrieff, R. W. (ed.): "Man-Made Fibres," Halsted-Wiley, New York, 1975.

Nevell, T. P., and S. H. Zeronian (eds.): "Cellulose Chemistry and Its Applications," Wiley, New York, 1985.

Oertel, G. (ed.): "Polyurethane Handbook," 2d ed., Hanser-Gardner, Cincinnati, Ohio, 1993.

Ranney, M. W.: "Epoxy Resins and Products," Noyes Data, Park Ridge, N.J., 1977.

Ranney, M. W.: "Silicones," 2 vols., Noyes Data, Park Ridge, N.J., 1977.

Ray, N. H.: "Inorganic Polymers," Academic Press, New York, 1978.

Riew, C. K., and J. K. Gillham (eds.): "Rubber-Modified Thermoset Resins," ACS, Washington, D.C., 1984.

Robinson, J. S. (ed.): "Fiber-Forming Polymers," Noyes Data, Park Ridge, N.J., 1980.

Rochow, E. G.: "Silicon and Silicones," Springer-Verlag, New York, 1987.

Roesky, H. W. (ed.): "Rings, Clusters, and Polymers of Main Group and Transition Elements," Elsevier, Amsterdam, 1989.

Ropp, R. C.: "Inorganic Polymeric Glasses," Elsevier, Amsterdam, 1992.

Schildknecht, C. E.: "Polymer Processes," Wiley-Interscience, New York, 1956.

Schildknecht, C. E., and I. Skeist (eds.): "Polymerization Processes," Wiley-Interscience, New York, 1977.

Schultz, J., and S. Fakirov: "Solid State Behavior of Linear Polyesters and Polyamides," Prentice-Hall, Englewood Cliffs, N.J., 1990.

Serafini, T. T. (ed.): "High Temperature Polymer Matrix Composites," Noyes Data, Park Ridge, N.J., 1987.

Seymour, R. B., and G. S. Kirshenbaum (eds.): "High-Performance Polymers: Their Origin and Development," Elsevier, New York, 1986.

Seymour, R. B., and H. F. Mark (eds.): "Applications of Polymers," Plenum, New York, 1988.

Sheats, J. F., C. E. Carraher, Jr., and C. U. Pittman, Jr. (eds.): "Metal-Containing Polymeric Systems," Plenum, New York, 1985.

Smith, W. M. (ed.): "Manufacture of Plastics," Van Nostrand Reinhold, New York, 1964.

Stille, J. K., and T. W. Campbell (eds.): "Condensation Monomers," Wiley, New York, 1972.

Tess, R. W., and G. W. Poehlein (eds.): "Applied Polymer Science," 2d ed., ACS, Washington, D.C., 1985.

Tomanek, A.: "Silicones & Industry," Hanser-Gardner, Cincinnati, Ohio, 1992.

Walker, B. M., and C. P. Rader (eds.): "Handbook of Thermoplastic Elastomers," 2d ed., Van Nostrand Reinhold, New York, 1988.

Walsh, K. A. (ed.): "Methods in Protein Sequence Analysis—1986," Humana Press, Clifton, N.J., 1987.

Whistler, R. L., and J. N. BeMiller (eds.): "Industrial Gums: Polysaccharides and Their Derivatives," 3d ed., Academic Press, San Diego, Calif., 1993.

Wilson, D., H. D. Stenzenberger, and P. M. Hergenrother (eds.): "Polyimides," Chapman & Hall, New York, 1990.

Wisian-Neilson, P., H. R. Allcock, and K. J. Wynne (eds.): "Inorganic and Organometallic Polymers II: Advanced Materials and Intermediates," ACS, Washington, D.C., 1995.

Woods, G.: "The ICI Polyurethanes Handbook," Wiley, New York, 1990.

Young, R. A., and R. M. Rowell: "Cellulose Structure, Modification, and Hydrolysis," Wiley, New York, 1986.

Zeldin, M., K. J. Wynne, and H. R. Allcock (eds.): "Inorganic and Organometallic Polymers," ACS, Washington, D.C., 1988.

Ziegler, J. M., and F. W. G. Fearon: "Silicon-Based Polymer Science," ACS, Washington, D.C., 1990.

CHAPTER
SIXTEEN

ANALYSIS AND IDENTIFICATION OF POLYMERS

16.1 THE PURPOSE OF POLYMER ANALYSIS

A variety of situations call for the identification of polymers. A customer wishes to paint or glue an item or to use an object in a novel environment. It may be awkward or impossible to obtain the composition from the supplier. In another common case, it may simply be detective work to find a chemical reason why a competitor's product behaves differently from your own. Contaminants in fabrication plants can be very hard to trace. For example, a small amount of a highly unsaturated polymer, such as natural rubber, can prevent the proper cross-linking of a large amount of a more saturated material, such as butyl or EPDM rubber. Of course, many of the tests described as useful in identification are also run routinely as quality controls on known polymers. While infrared spectroscopy is extremely helpful as an identification tool, it is also applied to the measurement of crystallinity and branching in some polymers (notably polyethylene). Physical testing, molecular weight determination, and electrical testing have been discussed previously. The larger polymer producers perform almost all of these tests routinely. The smaller, more specialized laboratories of many fabricators and consumers may not possess all of the instruments needed for these tests. Independent laboratories will perform them on a consulting basis. Where the instruments are expensive or the test method is quite involved, recourse to independent laboratories may be wise.

16.2 INFRARED AND ULTRAVIOLET SPECTROSCOPY

The atoms constituting a molecule are in constant vibration. When the number of atoms in a molecule exceeds 10 or so, the number of possible modes of vibration becomes very large. Fortunately, many frequencies are characteristic of localized bonds. Thus the absorption of light by the stretching of the C—H bond almost always occurs at frequencies between 2880 and 2900 cm^{-1}. Most infrared (IR) spectrophotometers use a glowing light source to provide light with wavelengths from 2.5 to about 15 μm. An alternative system of nomenclature that is popular refers to the reciprocal of wavelength as *wave number* or frequency. A wavelength of 2.5 μm corresponds to a wave number or frequency of 4000 cm^{-1}, 10 μm is equivalent to 1000 cm^{-1}, and so on. A rotating prism or diffraction grating breaks up the light from the source into a spectrum from which various wavelengths are progressively isolated with the aid of filters. A programmed slit compensates for the varying intensity of the source at different wavelengths.

The sample may be a thin unsupported film (usually about 0.001 in thick); a film evaporated on a nonabsorbing substrate such as NaCl; a solution held in a nonabsorbing cell (often made of NaCl); a pressed, clear wafer of the material to be analyzed mixed with KBr powder; or a mull of the material with a heavy paraffin oil held between NaCl flats. Other alkali and alkaline-earth halides are used as substrates. Water and alcohols are not used as solvents generally, because they absorb strongly in the infrared region and may corrode the substrates.

In a double-beam instrument the absorption by H_2O and CO_2 vapors is canceled out. By using matched cells in such a machine, one can also compensate for the solvent. It must be remembered that even with compensation, high absorption by the solvent in both cells still reduces the sensitivity of the method at that particular frequency.

The positions of some characteristic absorption bands are summarized in Table 16.1. Although these are valuable in analyzing spectra, we can often learn much by regarding the entire spectrum as a kind of fingerprint. In this way we can identify the common homopolymers rapidly by simple comparison with standard spectra. Copolymers, terpolymers, and mixtures introduce complications, although the spectra may be simply additive. Some bands are shifted by changes in environment. The OH and NH bands are moved by hydrogen bonding. In fact, the shift of the OD band in deuterated methanol when it is mixed with another liquid is used as a measure of hydrogen bonding in classifying solubility and swelling data (Sec. 2.5).

In Figs. 16.1 through 16.3, spectra for some common polymers are reproduced. Since these are transmission spectra, absorption bands appear as valleys. The characteristic band frequencies of Table 16.1 can be compared with specific polymers. The OH band appears in poly(vinyl alcohol) and not in polyethylene, for example. Spectra for the same polymer may appear differently depending on several factors. The thickness or concentration will affect the absolute value of absorption. Since

Table 16.1 Positions of characteristic infrared bands [1]

Group	Frequency range, cm^{-1}
OH stretching vibrations	
Free OH	3610–3645 (sharp)
Intramolecular hydrogen bonds	3450–3600 (sharp)
Intermolecular hydrogen bonds	3200–3550 (broad)
Chelate compounds	2500–3200 (very broad)
NH stretching vibrations	
Free NH	3300–3500
Hydrogen-bonded NH	3070–3350
CH stretching vibrations	
$\equiv$C—H	3280–3340
$=$C—H	3000–3100
C—CH$_3$	2872 ± 10, 2962 ± 10
O—CH$_3$	2815–2832
N—CH$_3$ (aromatic)	2810–2820
N—CH$_3$ (aliphatic)	2780–2805
CH$_2$	2853 ± 10, 2926 ± 10
CH	2880–2900
SH stretching vibrations	
Free SH	2550–2600
C$\equiv$N stretching vibrations	
Nonconjugated	2240–2260
Conjugated	2215–2240
C$\equiv$C stretching vibrations	
C$\equiv$CH (terminal)	2100–2140
C—C$\equiv$C—C	2190–2260
C—C$\equiv$C--C$\equiv$CH	2040; 2200
C$=$O stretching vibrations	
Nonconjugated	1700–1900
Conjugated	1590–1750
Amides	~1650
C$=$C stretching vibrations	
Nonconjugated	1620–1680
Conjugated	1585–1625
CH bending vibrations	
CH$_2$	1405–1465
CH$_3$	1355–1395, 1430–1470
C—O—C vibrations in esters	
Formates	~1175
Acetates	~1240, 1010–1040
Benzoates	~1275
C—OH stretching vibrations	
Secondary cyclic alcohols	990–1060
CH out-of-plane bending vibrations in substituted ethylenic systems	
—CH$=$CH$_2$	905–915, 985–995
—CH$=$CH— (cis)	650–750
—CH$=$CH— (trans)	960–970
C$=$CH$_2$	885–895
	790–840

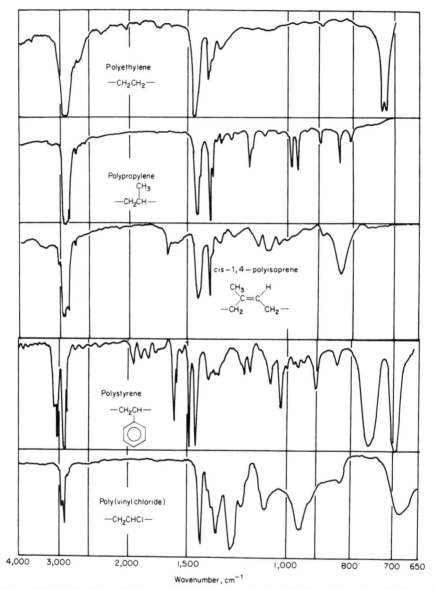

Figure 16.1 Typical transmission spectra in the infrared region [2]. The ordinate in each diagram is 0 to 100% transmission

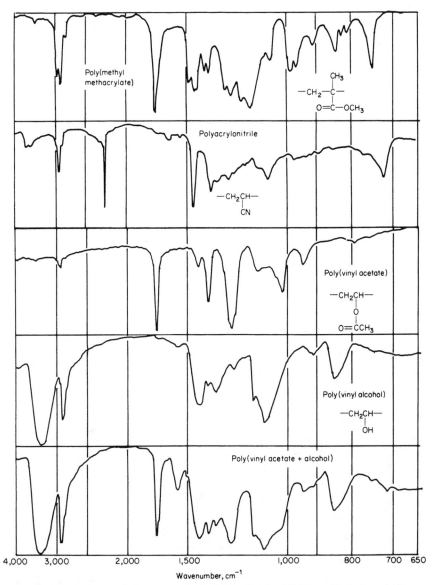

Figure 16.2 Typical transmission spectra in the infrared region [2]. The ordinate in each diagram is 0 to 100% transmission. The incompletely hydrolyzed sample of poly(vinyl acetate) illustrates the additivity of spectra

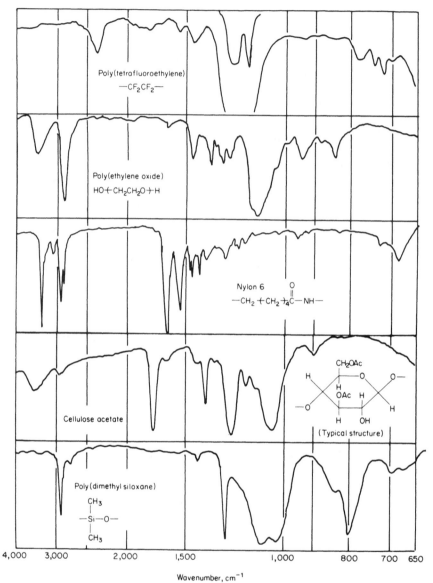

Figure 16.3 Typical transmission spectra in the infrared region [2]. The ordinate in each diagram is 0 to 100% transmission. Since the cellulose acetate is made by hydrolysis of the triacetate, the structure shown is only one of many that may exist in the polymer. The average composition does correspond to the diacetate

the hydroxyl group in the poly(ethylene oxide) shown in Fig. 16.3 occurs as an end group, the molecular weight will affect the intensity of that band relative to the bands for the ether and hydrocarbon bonds. In poly(vinyl alcohol), Fig. 16.2, an incompletely hydrolyzed sample will show vestiges of the acetate group remaining.

Infrared analysis can be applied to surfaces [3] and to pyrolyzed fragments of polymeric materials [4] for purposes of identification.

In addition to the obvious use in identifying unknown polymers and the estimation of amounts of comonomers, impurities, or end groups in polymers, infrared analysis, with some modification, has been used for measuring crystallinity and orientation of specific groups in crystalline polymers. Chemical reactions of polymers can be followed by the appearance or disappearance of functional groups such as oxygen groups in oxidation or free hydroxyl groups from hydrolysis.

The absorption of light in the near ultraviolet (wavelengths of 0.2 to 0.4 μm, 2000 to 4000 Å) arises from electronic excitation of the molecules. Two important features are that the ultraviolet absorption is generally indicative of multiple unsaturation and the absorption bands are often rather broad. One does not usually obtain a "fingerprint" with many individual bands as in the infrared region. On the other hand, absorption in the ultraviolet region can be quite strong and can be carried out in aqueous solutions. The column effluent from chromatographic separations of water-soluble polymers can be monitored by examining the absorption at a single wavelength.

16.3 DIFFERENTIAL THERMAL METHODS

Very often the limiting parameter for the useful application of a polymer is dimensional or chemical stability. The temperatures at which physical transitions occur are important to the first criterion, and the temperature of the onset of chemical reaction is important to the second. Some subtle changes in structure can be observed when the polymer sample is compared to another material undergoing a similar heating process but not undergoing any transitions or reactions.

Differential scanning calorimetry (DSC) uses a servo system to supply energy to a sample and a reference material in such a way that both undergo a linearly rising temperature. The difference in energy supplied between sample and reference is a measure of various thermal events, the most common of which are the glass transition, crystallization, and melting. Other processes can take place such as evaporation and chemical reactions. Although the most common test uses an increasing temperature, some processes such as crystallization and chemical reactions can be characterized by the energy supplied under isothermal conditions.

In one commercial instrument, the polymer sample is held in an aluminum pan with a crimped-on lid (Fig. 16.4). The reference most often is simply an empty pan. Both sample and reference are heated to give a constant rate of temperature change by separate platinum resistance heaters, which are controlled by the signals

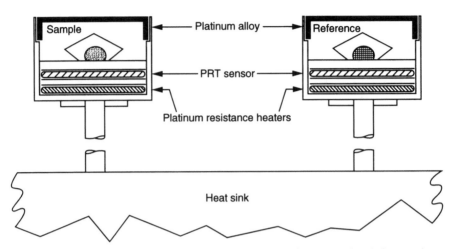

Figure 16.4 Differential scanning calorimeter (power compensation type). The platinum resistance thermometer (PRT sensor) detects a small error signal between the programmed temperature and the sample temperature and calls for more or less power to each heater to keep both holders on program. The difference in power required between sample and reference is amplified and recorded. (Courtesy of Perkin-Elmer Corp.)

from platinum resistance thermometers. The difference in power needed to keep both at the same temperature is amplified and provides the information on thermal events. For a properly calibrated DSC, the level of the power flux is a direct measure of the specific heat of the sample. Since the specific heat of a polymer melt is greater than that of the corresponding glass, a discontinuity is expected at T_g. When crystallization or melting occurs, the area under a peak represents the enthalpy change.

A sample mass of 5 to 15 mg and a heating rate of 10 to 40°C/min are typical conditions. As indicated earlier in Sec. 3.8, DSC has become accepted as a standard measure of T_g (ASTM D3418). In an illustrative example (Fig. 16.5), mixtures of poly(vinyl chloride) with a plasticizer, di-2-ethylhexylphthalate, are sealed in aluminum pans. At the rapid rate of temperature increase used, each test requires only 5 or 10 min. The transitions seen for mixtures like these are not as clear-cut as those for the pure components. As indicated in Sec. 3.3, it is not unusual for the transition zone for mixtures to be broader than it is for individual components. The "T_g" measured by DSC at a fast heating rate (Fig. 16.5) is substantially higher than the flex temperature of Fig. 3.4, which corresponds to a much longer time scale of testing. Some other typical effects can be seen when a partly crystalline polymer is examined (Fig. 16.6). A "thermoplastic polyester" exhibits a melting transition at about 250°C. The sample cooled at 20°C/min shows very little change at the expected glass transition point (77°C). The sample quenched from the melt, having a greater amorphous fraction, shows a substantial shift in heat flow at 77°C

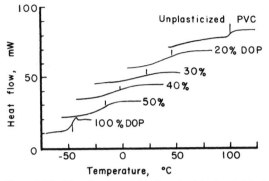

Figure 16.5 The glass transition for plasticized poly(vinyl chloride) as determined by DSC [5]. The heating rate was 40°C/min. (Reprinted from *American Laboratory,* 20(1):36 (1988); copyright by International Scientific Communications, Inc.)

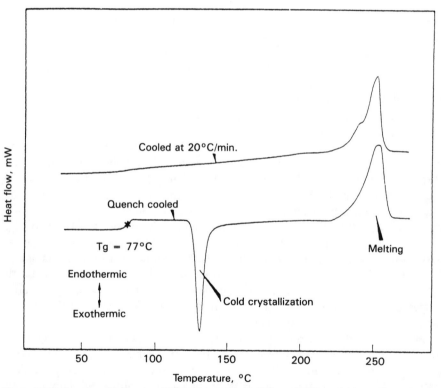

Figure 16.6 Samples of a thermoplastic polyester show different DSC traces depending on preparation kinetics [6]. Rapid quenching yields a sample that, on heating, shows a distinct glass transition as well as crystallization and melting. The heating rate was 20°C/min. (Courtesy of Perkin-Elmer Corp.)

and crystallization at 135°C followed by melting at 250°C (refer also to Figs. 3.17 and 3.18 in Sec. 3.8).

In conventional *differential thermal analysis* (DTA), the temperature at the center of the sample is compared with a reference material (often powdered alumina) as both are heated at a uniform rate. Any change in the sample's specific heat as at T_g, any structural change that is endothermic or exothermic as at T_m, or chemical reactions will show as changes in the temperature difference between sample T_s and reference T_r. A primitive instrument consists of thermocouples inserted in the sample and the reference. The sample and the reference material are held in a heated metal block at a controlled temperature T_0. As T_0 is raised (a rate of 10°C/min is common), T_s and T_r follow perhaps by as little as 0.1°C. If an endothermic reaction takes place, T_s will lag behind T_r temporarily. If an exothermic reaction takes place, T_s will exceed T_r temporarily. Superficially, DTA resembles DSC in that temperatures of a sample and a reference material are increased at almost the same rate. However, in DTA the difference in temperature, ΔT, is the measured quantity rather than the heat. If an endothermic event such as crystallization takes place, the sample temperature will lag slightly behind the reference temperature, giving a peak in the ΔT vs. T(average) plot similar in appearance to that in Fig. 16.6. The method is a sensitive one for detecting T_m, but the area under the peak is not easily related quantitatively to the enthalpy change.

Some workers have superimposed a sinusoidal pattern on the heating rate in order to derive more detailed information from the heat flow pattern [7]. Many other applications have been found for DSC, including the study of kinetics of polymerization or cross-linking, absorption, adsorption and desorption, and almost every other possible thermal event.

In *thermogravimetric analysis* (TGA), a sample is weighed continuously as the temperature is raised. Volatilization, chemical reaction, and dehydration are some of the processes that affect the sample weight. When many compounds have to be screened for their applicability in a high-temperature environment, TGA may be the only test needed. For example, if a rubbery gasket material is needed for short-term service at 300°C, the limitation on most materials will not involve T_g or T_m. Moreover, the chemical stability may be affected by fillers, cross-linking agents, antixoidants, plasticizers, and lubricants. Whether the test is run in a vacuum or in an atmosphere of air or oxygen will be important. When TGA is combined with DTA or when the gases evolved are examined by gas chromatography, infrared spectroscopy, or mass spectroscopy, an even more complete picture of the mechanism of reaction can be put together.

Somewhat related is *torsional braid analysis* [8]. In this method the polymer is applied to a glass braid, which becomes the restoring element in a torsion pendulum. A programmed temperature rise is imposed and the dynamic mechanical properties of the system are followed. Both changes in rigidity and damping can be correlated with physical and chemical changes in the polymer structure.

Commercial equipment is available for measuring mechanical properties (modulus, hysteresis) and electrical properties (dielectric constant, loss tangent) as a function of temperature. Although some use a continuous temperature scan, others raise the temperature in steps of 5°C or so, allowing the samples, which are larger than those for DSC, to come to a more even temperature. The continuous scan will usually reflect rate-dependent behavior so that the loss peaks will occur at higher temperatures at higher rates. *Frequency multiplexing* increases the amount of information from a single scan. At each temperature, the mechanical or electrical test is run at a number of frequencies covering two or more decades. Mechanical frequencies up to 200 Hz are common. Electrical tests can cover greater ranges, say 10 to 10,000 Hz. Time-temperature superposition allows a further generalization of the results obtained over a wide frequency range.

Thermally stimulated current (TSC) has been used to characterize various thermal processes [9]. In this method, a sample is subjected to a high voltage which orients dipoles in the material. The temperature is lowered with the voltage applied, trapping the polarized dipoles. When the sample is reheated, small currents are detected as the material relaxes. TSC can be operated in such a way as to obtain a complete map of relaxation times vs. temperature. Even thin films of coatings and adhesives are conveniently studied by this method.

16.4 NUCLEAR MAGNETIC RESONANCE SPECTROSCOPY

When a compound containing hydrogen is placed in a strong magnetic field and irradiated with a radio-frequency signal, it is found that energy absorption occurs at discrete frequencies. Transitions between different spin orientation levels are taking place in the hydrogen atom. Fortunately, few of the other nuclei found in organic compounds have a net magnetic moment and spin, so that except for fluorine, chlorine, and certain isotopes of carbon, nitrogen, and oxygen, the only effects seen are those of the protons. It is fortunate, also, that the exact frequency at which energy is absorbed is very sensitive to the atomic environment of the proton. The scan of absorption vs. frequency for ethanol, for example, gives three widely separated peaks with areas in the ratio 1:2:3. Tetramethyl silane, in which all the protons are equivalent, is often used as an internal standard and assigned a frequency value τ of 10. Most other protons have values of τ between 0 and 10, representing various degrees of *chemical shift* of the fundamental frequency by environment.

In practice, about 0.5 ml of polymer solution (say, 10% in an aprotic solvent such as carbon tetrachloride) is placed in a thin glass tube between the poles of a strong magnet having a field strength of about 10,000 gauss. A radio-frequency field is imposed on the sample at right angles to the magnetic field. Various devices are used to detect the absorption of energy at resonance as the protons undergo changes in spin orientation level.

Some typical *nuclear magnetic resonance* (NMR) spectra (Fig. 16.7) show the broadening of the peaks in the polymer compared with small molecules because of the lower mobility in the polymer. Of course, the hydroxyl peak at $\tau = 4.2$

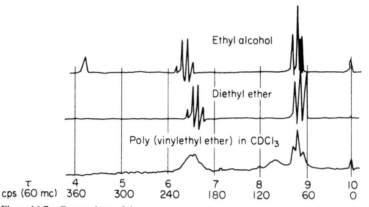

Figure 16.7 Comparison of the proton magnetic resonance spectra of ethanol, diethyl ether, and poly-(vinyl ethyl ether) [10]

appears only in the alcohol in this series. The chain CH peak in the polymer under-lies the CH_2 multiplet at $\tau = 6.3$ to 6.6 and is not seen separately.

The applications of NMR spectroscopy, in addition to compositional analysis, include the detection of isotactic-atactic ratios, monomer sequence distributions in copolymers, and other configurational variations. A narrowing of the resonance peak is typical at phase transitions, making it possible to use NMR of solid poly-mers to measure T_g and T_m.

A related technique using microwave frequencies is capable of detecting the changes in energy levels due to the magnetic moment of electrons. Such *electron paramagnetic resonance* (EPR), also called *electron spin resonance* (ESR), can be studied when there is a net unpaired electronic magnetic moment in the molecule being investigated. Free radicals from polymerization or oxidation, especially long-lived ones trapped in the glassy solid state, can be measured. In most free-radical polymerizations, the radical concentration is down around 10^{-8} mol/liter and is difficult to detect by EPR. The bulk copolymerization of norbornene and sulfur dioxide, however, appears to proceed with radical concentrations of around 10^{-5} mol/liter, so that detection of radicals at room temperature over a period of hours and weeks is feasible [11].

The structure of these "living" polymers is postulated as

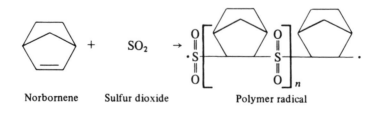

Norbornene   Sulfur dioxide   Polymer radical

16.5 MISCELLANEOUS SPECTROSCOPIC METHODS

Many new methods of analysis have been extended to polymer systems. Some techniques include the following [12–14].

1. Fourier-transform infrared analysis, by which the entire spectrum can be scanned in about 20 s.
2. Raman spectroscopy for analysis of chain-stretching motions in lamellar regions.
3. Carbon-13 NMR using computer-enhancement of weak signals; "magic-angle" (54.7°) spinning with cross-polarization to reduce broadening of lines [15].
4. Neutron scattering, which extends dynamic light-scattering techniques to movements of very small chain segments.
5. Analysis of electrons ejected when a polymer surface is bombarded with x-rays (ESCA, electron spectroscopy for chemical analysis). The energy spectrum can be used to identify the chemical species present in the surface layers.

16.6 CHEMICAL TESTS AND IDENTIFICATION SCHEMES

There are many ways by which polymers may be identified. In textile fibers, microscopic examination of the cross section and texture may be sufficient for the more complex, naturally occurring polymers. Solubility and staining tests can also be used. These have been summarized by Morrison [16]. When a polymer or compound ingredient can be extracted and made into a film or pressed wafer, the infrared spectrum may be the most direct and most powerful tool available. An analytical scheme for identifying many common plastics is given by Gruben and Leiner [17]. Other "spot" tests have been suggested by Feigl [18, 19]. As a general rule, it is best to try to separate a sample suspected of being a mixture by extraction before attempting to carry out chemical tests. Plasticizers, antioxidants, surfactants, and other small molecules are usually not bound chemically to the polymer and should be removed by preliminary treatment. Thin-layer chromatography (TLC) has been used mainly for identification of compounding ingredients that have been extracted from a product, but it can be used for low-molecular-weight polymers also [20, 21].

KEYWORDS FOR CHAPTER 16

Section

2. Infrared spectra (IR)
3. Differential scanning calorimetry (DSC)
   Thermogravimetric analysis (TGA)
   Torsional braid analysis

4. Nuclear magnetic resonance (NMR)
   Electron paramagnetic resonance (EPR)
6. Spot tests

PROBLEMS

16.1   Three samples of flexible tubing are subjected to the following tests:

(*a*) Held in burner flame. Tubes A and C burn.

(*b*) Soaked overnight in chloroform. Tube A swells, B, becomes very stiff and somewhat smaller on drying out, C is unaffected.

(*c*) Immersed in a lead nitrate solution, A turns black; B and C are unaffected.

Which of the three is vulcanized rubber, polyethylene, or plasticized poly(vinyl chloride)?

16.2   Try to identify the group associated with the four most prominent absorption bands in poly(vinyl acetate). Use Table 16.1 as a guide.

16.3   One characteristic property by which unmodified polystyrene can often be identified is the clattering sound made by dropping a molded piece on a hard surface. An inexpensive protractor or comb tossed on a table top can be identified this way. What factors might be involved in causing this property?

16.4   How would microscopic observation of the cross section of a fiber sample differentiate between nylon 6 and rayon?

16.5   A sample of rubber shows no flow in a creep experiment, indicating that it is cross-linked. It contains no fillers. What additional simple tests might be run to decide whether the material is natural rubber, styrene-butadiene rubber, or acrylonitrile-butadiene rubber?

16.6   You are given a material to identify. It is in the form of a stiff, transparent sheet about 0.01 in thick. It burns slowly with a smell of vinegar. It dissolves completely in acetone. Is it polyethylene, polypropylene, poly(vinyl chloride), poly(vinyl acetate), nylon 6, or cellulose diacetate?

REFERENCES

1. Cole, A. R. H., in K. W. Bentley (ed.): "Elucidation of Structures by Physical and Chemical Methods," part 1, (Technique of Organic Chemistry, vol. XI), pp. 33ff, Interscience, New York, 1963.
2. Nyquist, R. A.: "Infrared Spectra of Plastics and Resins," 2d ed., Dow Chemical Co., Midland, 1961.
3. Willis, H. A., and V. J. I. Zichy, in D. T. Clark and W. J. Feast (eds.): "Polymer Surfaces," Wiley, New York, 1978.
4. Gross, D.: *Rubber Chem. Technol.*, 48:289 (1975).
5. Brennan, W. P., R. B. Cassel, and M. P. DiVito: *American Laboratory*, 20(1):32 (1988).

6. Brennan, W. P.: "Characterization and Quality Control of Engineering Thermoplastics by Thermal Analysis," Instrument Div., Perkin-Elmer Corp., 1977.
7. Reading, M.: *Trends Polym. Sci.,* 1:248 (1993).
8. Gilham, J. K.: *Polym. Eng. Sci.,* 19:749 (1979).
9. Matthiesen, A., F. McIntyre, J. P. Ibar, and J. Saffell: *Amer. Lab.,* 23(1):44 (1991).
10. Slichter, W. P.: *J. Chem. Educ.,* 45:10 (1968).
11. Zutty, N. L., and C. W. Wilson III: *Tetrahedron Lett.,* 1963(30):2181 (1963).
12. Stinson, S.: *Chem. & Eng. News,* 86:20 (Jan. 8, 1979).
13. Collins, E. A., and C. A. Daniels, in J. K. Craver and R. W. Tess (eds.): "Applied Polymer Science," chap. 7, ACS, Washington, D.C., 1975.
14. Craver, C. D., in J. K. Craver and R. W. Tess (eds.): "Applied Polymer Science," chap. 8, ACS, Washington, D.C., 1975.
15. Miknis, F. P., V. J. Bartuska, and G. E. Maciel: *American Laboratory,* 11:19 (Nov. 1979).
16. Morrison, R. D.: *Am Dyestuff Reptr.,* 52:28 (Oct. 28, 1963); also, see ASTM D276–62T in the "ASTM Standards on Textile Materials."
17. Gruben, A., and H. H. Leiner: *Mod. Plastics,* 37:88 (July 1960).
18. Feigl, F., and V. Anger: *Mod. Plastics,* 37:151 (May 1960).
19. Feigl, F., V. Gentil, and E. Jungreis: *Textile Res.,* 28:891 (1958).
20. Kreiner, J. G.: *Rubber Chem. Technol.,* 44:381 (1971).
21. Kotaka, T., and J. L. White: *Rubber Chem. Technol.,* 48:310 (1975).

General References

Bark, L. S., and N. S. Allen (eds.): "Analysis of Polymer Systems," Elsevier Applied Science, New York, 1982.
Bässler, H. (ed.): "Optical Techniques to Characterize Polymer Systems," Elsevier, Amsterdam, 1989.
Berezkin, V. G., V. R. Alishoyev, and I. B. Nemirovskaya: "Gas Chromatography of Polymers," Elsevier, New York, 1977.
Braun, D.: "Simple Methods for Identification of Plastics," 2nd ed., Hanser-Gardner, Cincinnati, Ohio, 1986.
Cheremisinoff, N. P., and P. N. Cheremisinoff (eds.): "Handbook of Advanced Materials Testing," Dekker, New York, 1995.
Clark, D. T., and W. J. Feast (eds.): "Polymer Surfaces," Wiley, New York, 1978.
Craver, C. D. (ed.): "Polymer Characterization: Spectroscopic, Chromatographic, and Physical Instrumental Methods," ACS, Washington, D.C., 1983.
Craver, C. D., and T. Provder (eds.): "Polymer Characterization: Physical Property, Spectroscopic, and Chromatographic Methods," ACS, Washington, D.C., 1990.
Crompton, T. R.: "The Analysis of Plastics," Pergamon, New York, 1984.
Crompton, T. R.: "Analysis of Polymers," Pergamon, New York, 1989.
Crompton, T. R.: "Practical Polymer Analysis," Plenum, New York, 1993.
Crompton, T. R. (ed.): "Chemical Analysis of Additives in Plastics," 2d ed., Pergamon, New York, 1977.
Dawkins, J. V. (ed.): "Developments in Polymer Characterization—1," Elsevier Applied Science, New York, 1978.
Dawkins, J. V. (ed.): "Developments in Polymer Characterization—2," Elsevier Applied Science, New York, 1980.
Dawkins, J. V. (ed.): "Developments in Polymer Characterization—3," Elsevier Applied Science, New York, 1982.
Dawkins, J. V. (ed.): "Developments in Polymer Characterization—4," Elsevier Applied Science, New York, 1983.

Dawkins, J. V. (ed.): "Developments in Polymer Characterization—5," Elsevier Applied Science, New York, 1986.

Garton, A.: "Infrared Specroscopy of Polymer Blends," Hanser-Gardner, Cincinnati, Ohio, 1992.

Hemsley, D. A.: "The Light Microscopy of Synthetic Polymers," Oxford Univ. Press, New York, 1984.

Hummel, D. O. (ed.): "Polymer Spectroscopy," Verlag Chemie, Deerfield Beach, Fla., 1974.

Hummel, D. O., and F. Scholl: "Infrared Analysis of Polymers, Resins, and Additives, an Atlas," vol. 1, Halsted (Wiley), New York, 1969.

Hummel, D. O., and F. Scholl: "Atlas of Polymer and Plastics Analysis," 2d ed., Verlag Chemie, Deerfield Beach, Fla., vols. 1 and 2, 1978; vol. 3, 1980.

Hunt, B. J., and M. I. James (eds.): "Polymer Characterization," Chapman & Hall, New York, 1993.

Ishida, H. (ed.): "Fourier Transform Infrared Characterization of Polymers," Plenum, New York, 1987.

Kämpf, G.: "Characterization of Plastics Using Physical Methods," Macmillan, New York, 1986.

Kline, G. M. (ed.): "Analytical Chemistry of Polymers," parts 1–3, Interscience, New York, 1959, 1962.

Koenig, J. L.: "Spectroscopy of Polymers," ACS, Washington, D.C., 1991.

Koenig, J. L. (ed.): "Probing Polymer Structures," ACS, Washington, D.C., 1979.

Krause, A., A. Lange, and M. Ezrin: "Plastics Analysis Guide," Macmillan, New York, 1983.

Lever, A. E., and J. A. Rhys: "The Properties and Testing of Plastics Materials," Temple Press Books, London, 1968.

Liebman, S. A., and E. J. Levy (eds.): "Pyrolysis and GC in Polymer Analysis," Dekker, New York, 1985.

Mark, J. E., A. Eisenberg, W. W. Graessley, L. Mandelkern, and J. L. Koenig: "Physical Properties of Polymers," ACS, Washington, D.C., 1984.

Mathot, V. B. F. (ed.): "Calorimetry and Thermal Analysis of Polymers," Hanser-Gardner, Cincinnati, Ohio, 1994.

Mitchell, J., Jr. (ed.): "Applied Polymer Analysis and Characterization," Hanser-Gardner, Cincinnati, Ohio, 1992.

Pham, Q. T., R. Pétiaud, and H. Waton: "Proton and Carbon NMR Spectra of Polymers," 3 vols., Wiley, New York, 1984.

Pham, Q. T., R. Pétiaud, H. Waton, and M.-F. Llauro-Darricades: "Proton and Carbon NMR Spectra of Polymers," CRC Press, Boca Raton, Fla., 1991.

Porter, R. S., and J. F. Johnson (eds.): "Analytical Calorimetry," Plenum, New York, 1976.

Provder, T., M. W. Urban, and H. G. Barth (eds.): "Hyphenated Techniques in Polymer Characterization: Thermal-Spectroscopic and Other Methods," ACS, Washington, D.C., 1995.

Randall, J. C., Jr. (ed.): "NMR and Macromolecules Sequence, Dynamic, and Domain Structure," ACS, Washington, D.C., 1984.

Rochow, T. G., and E. G. Rochow: "Resinography," Plenum, New York, 1976.

Sadler Standard Spectra, "Monomers and Polymers," vols. I–VII, Sadler Research Lab, Philadelphia, 1966.

Sawyer, L., and D. T. Grubb: "Polymer Microscopy," Chapman & Hall (Methuen), New York, 1987.

Schmidt-Rohr, K., and H. W. Spiess: "Multidimensional Solid-State NMR and Polymers," Academic, New York, 1994.

Schmitz, J. V., and W. E. Brown (eds.): "Testing of Polymers," 4 vols., Interscience, New York, 1965–1969.

Schröder, E., G. Müller, and K.-F. Arndt: "Polymer Characterization," Hanser-Gardner, Cincinnati, Ohio, 1989.

Siesler, H. W., and K. Holland-Moritz: "Infrared and Raman Spectroscopy of Polymers," Dekker, New York, 1980.

Spells, S. J. (ed.): "Characterization of Solid Polymers: New Techniques and Developments," Chapman & Hall, New York, 1994.

Turi, E. A. (ed.): "Thermal Characterization of Polymeric Materials," Technomic, Lancaster, Pa., 1981.

Urbanski, J., W. Czerwinski, K. Janicka, F. Majewska, and H. Zowall: "Handbook of Analysis of Synthetic Polymers and Plastics," Wiley, New York, 1977.

Vîlcu, R., and M. Leca: "Polymer Thermodynamics by Gas Chromatography," Elsevier, Amsterdam, 1989.

Wake, W. C., B. K. Tidd, and M. J. R. Loadman: "Analysis of Rubber and Rubber-like Polymers," Elsevier Applied Science, New York, 1983.

Walsh, K. A. (ed.): "Methods in Protein Sequence Analysis—1986," Humana Press, Clifton, N.J., 1987.

White, J. R., and D. Campbell: "Polymer Characterization," Chapman & Hall, New York, 1989.

Wu, T. K., and J. Mitchell, Jr. (eds.): "Polymer Analysis," Applied Polym. Symp. 34, Wiley, New York, 1978.

Zbinden, R.: "Infrared Spectroscopy of High Polymers," Academic Press, New York, 1964.

LIST OF SYMBOLS

Symbol	Common unit	Definition or label	Section where first used
A	Variable	Used as an arbitrary constant	
A	cm^2, in^2	Area	7.1
A_t	m^2	Total surface area of particles	5.5
A_u	cm^2	Area, original unstretched	8.2
A, A_p, A_i, A_t	s^{-1}	Collision frequency factors	4.4
A_2	dl·mol/g^2	Second virial coefficient	6.6
A_3	dl^2·mol/g^3	Third virial coefficient	6.6
Å		Angstrom unit, 10^{-10} m	
B	Variable	Used as an arbitrary constant	
B	· · ·	Factor in modulus equation	8.7
C_2	Pa	Constant in Eq. (8.24)	8.2
C_f	farad	Capacitance	10.6
C_i, C_s	· · ·	Chain transfer constant	5.5
C_t	· · ·	Couette term	7.8
C_z	Variable	Lumped constant	7.7
C'	(mol/g)$^{1/2}$	Constant in Eq. (7.11)	7.3
CED	cal/cm^3	Cohesive energy density	2.5
D	· · ·	Abbreviation for bifunctional unit	15.7
D, D_i, $\overline{D}_L$, $\overline{D}_A$, $\overline{D}_V$	cm, in	Diameter or average diameter	6.1
D_b	cm	Barrel diameter inside extruder	12.5
D_S	cm	Diameter of extruder die	7.8
ΔE_v	cal	Internal energy of vaporization	2.5

Symbol	Common unit	Definition or label	Section where first used
E	Pa	Young's modulus	8.1
$E(\theta)$	Pa	Distribution of relaxation times	8.4
$E(t)$	Pa	Time-dependent modulus	8.3
E_i	Pa	Modulus of individual element	8.5
E_r	Pa	Modulus measured in stress-relaxation mode	11.3
E_t	Pa	Modulus measured intermittently	11.3
E_T	Pa	Modulus of an array of elements	8.5
E^*	Pa	Complex modulus	8.3
E'	Pa	Real part of complex modulus	8.3
E''	Pa	Imaginary part of complex modulus	8.3
ε	V	Voltage	10.6
F_1, F_2	$\cdots$	Mole fractions of monomers 1 and 2 in polymer	4.8
ΔG_m	J	Free energy of mixing	2.5
G	Pa	Shear modulus	8.1
G_N^0	Pa	Rubbery plateau shear modulus	8.7
$\mathfrak{g}$	J/m^2	Strain energy release rate	9.7
ΔH_f	J/mol	Enthalpy of fusion, pure solvent	6.6
ΔH_m	J	Enthalpy of mixing	2.6
H_p	hp	Horsepower	12.5
ΔH_p	J/mol	Enthalpy of polymerization	5.1
ΔH_u	J/mol	Enthalpy of fusion for polymer repeat unit	3.4
ΔH_v	J/mol	Enthalpy of vaporization	2.5
H'	Variable	Lumped constant in Debye equation	6.8
$\bar{H}(log\ \theta)$	Pa·s	Distribution of relaxation times	8.4
[I]	mol/liter	Initiator concentration	4.4
I_g	lumen	Light scattered	6.7
I_{gs}	lumen	Light scattered by solvent	6.7
I_m, I_r	g·cm²	Moment of inertia	8.3
J	Pa^{-1}	Compliance	8.1
K, K', K''	Variable	Constants in Eqs. (7.16), (7.25), and (7.34)	7.3
K_{lc}	MPa·m$^{1/2}$	Critical stress intensity factory *or* fracture toughness	9.7
K_0	s^{-1}	Constant rate of strain	8.3
K_c	$\cdots$	Dielectric constant	10.6
L	cm	Length	7.7, 8.1
L_u	cm	Unstressed length	8.1
M	g	Mass	7.1
M	$\cdots$	Used as an abbreviation for monofunctional unit	15.7
M	g/mol	Molecular weight	6.1
[M]	mol/liter	Monomer concentration	4.4
[M·]	mol/liter	Concentration of radical chains	4.4
M_n	g/mol	Number-average molecular weight	6.1

Symbol	Common unit	Definition or label	Section where first used
M_v	g/mol	Viscosity average molecular weight	7.3
M_w	g/mol	Weight-average molecular weight	6.1
M_0	g/mol	Molecular weight of repeat unit	6.2
M_i, M_1, M_2	g/mol	Molecular weight of specific components	6.1, 6.6
N	$\cdots$	Number of items	7.6, 9.2
N_i'	$\cdots$	Number of molecules of i	2.6
N_i	$\cdots$	Mole fraction of i	2.6
$\left.\begin{array}{l} N, N_i, N_1, N_2 \\ N_E, N_A, N_D, \\ N_o, N_c, N_S \end{array}\right\}$	mol	Number of moles of specific components	4.7, 4.9 6.3, 11.3
N	mol/cm^3	Moles of polymer chains per unit volume	8.2
N	mol/liter	Concentration of monomer units per unit volume	11.3
N_p	mol/liter	Concentration of particles per unit volume	5.5
$P, P(0, N_2)$, etc.	dyn/cm^2, Pa	Pressure	6.6
$P(\theta)$	$\cdots$	Complex function of rotation angle	6.8
PDI	$\cdots$	Polydisperity index	6.1
P_n	Pa	Normal stress coefficient	7.8
$(P_t)_x$	$\cdots$	Probability of mole fraction of x-mer	6.2
$\Delta P/L$	Pa/m	Pressure drop per unit length	7.7
Q	$\cdots$	Used as an abbreviation for tetrafunctional monomer unit	15.7
Q	$\cdots$	Constant in Price-Alfrey scheme	4.9
Q	cm^3/s	Volumetric rate of flow	7.6
Q_e	kg/h	Extruder capacity	12.5
$\dot{Q}$	J/s	Rate of energy dissipation in viscous flow	7.2
R	cal/mol·deg	Gas constant	
R_p	mol/liter·s	Rate of polymerization	4.4
R_3, R_4	ohm	Electrical resistance	10.6
$[R\cdot]$	mol/liter	Concentration of initiator radicals	4.4
$[R_S]$	mol/liter	Concentration of initiator seeds	4.5
S	J/K	Entropy	8.2
ΔS_m	J/K	Entropy of mixing	2.6
$[S]$	mol/liter	Surfactant concentration	5.5
T	$\cdots$	Used as an abbreviation for a trifunctional monomer unit	15.7
T	K	Temperature	
T_f	K	Freezing temperature of pure solvent	6.6
T_g	K	Glass transition temperature	3.5
T_m	K	Melting temperature	3.4
$(T_m)_0$	K	Melting temperature, pure polymer	3.4

Symbol	Common unit	Definition or label	Section where first used
T_R	K	Reference temperature	7.5
$\mathscr{T}$	N·m	Torque	7.7
U_p	W	Power used in dielectric heating	10.6
V	cm^3	Volume	8.1
V	cm^3/g	Specific volume	7.5
V_f	cm^3/g	Free volume	7.5
V_g	cm^3/mol	Molar volume of solvent as vapor	2.5
V_0	cm^3/g	Specific volume extrapolated to 0 K	7.5
V_r	cm^3	Effluent volume	6.4
$V_r(M)$	cm^3	Effluent volume for molecular weight M	6.4
V_u	cm^3	Original volume	3.4
V_1	cm^3/mol	Molar volume of solvent	2.5
W	J	Work	8.2
W_i	g	Weight of component i	6.1
W_x	g	Weight of x-mer	6.2
X	$\cdots$	The ratio of f_1/f_2	4.8
Y	$\cdots$	The ratio of F_1/F_2	4.8
Z_w	$\cdots$	Weight-average number of chain atoms, $M_w/$(mol wt per chain atom)	7.4
a	$\cdots$	Constant in Eqs. (7.16), (7.33)	7.3, 7.5
a	cm	Bond length	7.2
a	$\cdots$	$x_n/(x_w - x_n)$	6.3
a	cm	Thickness	8.1
a	m^2	Surface area of a particle	5.5
a_s	m^2	Area occupied by a mole of surfactant	5.5
a_T	$\cdots$	Shift factor	8.3
b	cm	Width	8.1
c	gm/dl, gm/cm^3	Concentration	6.6
c	cm/s	Velocity of light in vacuum, 3.00×10^8 cm/s	11.1
c_p	cal/g·°C	Specific heat	12.5
c_r	cm	Crack length	9.8
d	cm	Plate spacing, sample thickness	10.6
e	$\cdots$	Polarity parameter in Price-Alfrey scheme	4.9
e	$\cdots$	End correction	7.8
f	$\cdots$	Frequency factor	4.4
f	$\cdots$	Functionality of monomer	4.2, 4.7
f	N	Force	
f_1, f_2	$\cdots$	Mole fraction in monomer feed	4.9
f', f'_g	$\cdots$	Fractional free volume at temperature T or T_g	7.5
h	cm	Height	6.6
i	$\cdots$	Subscript referring to component i	

Symbol	Common unit	Definition or label	Section where first used
i	· · ·	Denoting imaginary part of complex variable, square root of -1	
k	Variable	Arbitrary factor in non-Newtonian equations	7.6
k	J/K	Boltzmann constant	2.6
$k, k_i, k_t, k_p,$ $k_2, k_3, k_4,$ $k_{12}, k_{22}, k_{21},$ k_{11}	Variable	Rate constants	4.4, 4.9
k_c	W/m·°C)	Thermal conductivity	10.6
k', k'', k''', k_L	Variable	Constants in equations for concentration dependence of viscosity	7.3
m	mol/liter	Concentration of polymer molecules	11.3
m	g	Masses used in torsion pendulum	8.3
n	· · ·	Exponent in power law	7.6
n	· · ·	Sample index number	9.2
n_C, n_D, n_F	· · ·	Refractive index	10.7
n	· · ·	Number of links in freely oriented chain	7.2
p	Pa	Vapor pressure	2.5
p	· · ·	Fraction of monomer converted to polymer	6.2
p_B, p_F	· · ·	Fraction of monomer converted to polymer at gelling point for bifunctional B monomer and polyfunctional F monomer	4.7
p_c	nm	Packing length	8.7
r	cm	Radius	7.7
r	· · ·	Ratio of hydroxyl to carboxyl groups	4.7
r_1, r_2	· · ·	Relative reactivity ratios	4.9
$(\overline{r^2})^{1/2}$	cm	Root-mean-square end-to-end distance	7.2
$(\overline{r_0^2})^{1/2}$	cm	Root-mean-square end-to-end distance in a theta solvent	7.2
s	· · ·	Standard deviation	9.2
$(\overline{s^2})^{1/2}$	cm	Radius of gyration	7.2
t	s	Time	
u	cm/s	Velocity	7.2
u_e	J	Energy loss per cycle	8.3
u_t	m/s	Terminal velocity	Prob. 5.13
v	m³	Particle volume	5.5
v_1, v_2	· · ·	Volume fractions of components 1 and 2	2.5, 3.4
w_i	· · ·	Weight fraction of component i	3.5
w_x	· · ·	Weight fraction of x-mer	6.2

Symbol	Common unit	Definition or label	Section where first used
x	$\cdots$	Degree of polymerization (also x_w, x_n, etc.)	4.4
x	$\cdots$	Mole fraction of one component in liquid phase	4.8
y	cm	Distance	7.1
z	$\cdots$	Exponent in power law	7.6
z	$\cdots$	$d(\ln \dot{\gamma})/d(\ln \tau)$	7.6
$\Gamma(a + 1)$	$\cdots$	Gamma function of $a + 1$	6.3
Δ	$\cdots$	Signifies difference	
Δ	$\cdots$	Logarithmic decrement	8.3
Λ	Pa	Factor in Eq. (8-25)	8.3
Σ	$\cdots$	Summation sign	
Υ	$\cdots$	Coupling-disproportionation factor	4.4
Φ	$\cdots$	Constant in Eq. (7-21), $2.1 \pm 0.2 \times 10^{21}$ dl/mol·cm^3	7.3
$\Omega, \Omega_b, \Omega_c$	rad/s	Rate of rotation, also referred to bob or cup	7.3, 7.7
Ω_1, Ω_2	$\cdots$	Probabilities of states 1 and 2	8.2
α	$\cdots$	In Eq. (4.45), $\alpha = r_2/(1 - r_2)$	4.9
α	$\cdots$	Expansion factor, $(\overline{r^2/r_0^2})^{1/2}$	7.2
α	K^{-1}	Volumetric coefficient of thermal expansion for melt minus that of glass	7.5
$\alpha, \alpha_x, \alpha_y, \alpha_z$	$\cdots$	Extension ratio, also in x, y, and z directions	8.1, 8.2
α_e	K^{-1}	Linear coefficient of thermal expansion	10.5
α_r	$\cdots$	Relative volatility	4.9
β	$\cdots$	In Eq. (4.45), $\beta = r_1/(1 - r_1)$	4.9
β	g/mol	Constant in Eq. (7-33)	7.5
β	$\cdots$	Die swell	7.8
β	$\cdots$	Normalizing factor	8.2
γ	$\cdots$	Shear strain	7.1, 8.1
γ	$\cdots$	A constant, function of r_1 and r_2 in Eq. (4-45)	4.9
γ_c	dynes/cm	Critical surface tension	12.4
γ_c	$\cdots$	Hydrogen bonding index	2.5
γ_R	$\cdots$	Recoverable shear strain	7.8
γ_e	mN/m, J/m^2	Specific surface energy	9.7
$\dot{\gamma}$	s^{-1}	Rate of shear, $d\gamma/dt$	7.1
$\dot{\gamma}_w$	s^{-1}	Rate of shear at wall	7.6
$(\dot{\gamma}_w)_a$	s^{-1}	Apparent rate of shear at wall	7.6
$\dot{\gamma}_{gm}$	s^{-1}	Rate of shear at geometric mean radius	7.7
$\dot{\gamma}^+$	s^{-1}	Rate of shear at reference radius	7.7
δ	$\cdots$	A constant, function of r_1 and r_2 in Eq. (4-45)	4.9

Symbol	Common unit	Definition or label	Section where first used
δ	$\cdots$	Loss angle	8.3
δ, δ_1, δ_2	$(cal/cm^3)^{1/2}$ (Hildebrand)	Solubility parameter, also referred to components 1, solvent, and 2, polymer	2.5
δ_T, δ_d, δ_p, δ_h	$(cal/cm^3)^{1/2}$	Total, nonpolar, polar, and hydrogen-bonding solubility parameters, respectively	2.5
$\sin \delta$	$\cdots$	Power factor	10.6
$\tan \delta$	$\cdots$	Dissipation factor	8.3, 10.6
ε	$\cdots$	Elongation, $\varepsilon = \gamma - 1$	8.1
ε_b	$\cdots$	Elongation at break	9.6
ε_1, ε_2	$\cdots$	Elongation of components 1 and 2 in Maxwell model	8.3
$\dot{\varepsilon}$	s^{-1}	Rate of elongation, $d\varepsilon/dt$	8.3
$\dot{\varepsilon}_{ij}$	s^{-1}	Rate of strain tensor	7.8
η	Pa·s	Viscosity	7.1
η_0, η_∞	Pa·s	Viscosity at low and high shear rates	7.6
η_g, η_R	Pa·s	Viscosity at T_g and at T_r	7.5
η_{sp}	$\cdots$	Specific viscosity, $\eta_r - 1$	7.3
η_{sp}/c	dl/g	Reduced viscosity	7.3
η_r	$\cdots$	Relative viscosity, η/η_s	7.3
η_s	$\cdots$	Solvent viscosity	7.3
$[\eta]$	dl/g	Intrinsic viscosity	7.3
$[\eta]_\theta$	dl/g	Intrinsic viscosity at theta temperature	7.3
η'	Pa·s	Dynamic viscosity	8.3
$(\ln \eta_r)/c$	dl/g	Inherent viscosity	7.3
θ	degree	Supplement of bond angle	7.2
θ	s	Relaxation time	8.2
θ_c	degree	Contact angle	12.4
θ_f	K	Flory temperature (theta temperature)	7.2
θ_t	s	Characteristic relaxation time	7.6
θ'	degree	Bond angle	7.2
λ	m	Wavelength	11.1
λ	$\cdots$	Coefficient in Eq. (8.25)	8.3
μ	m^3/s	Particle growth rate	5.5
μ	$\cdots$	Geometric factor in Eq. (8-4)	8.1
ν	$\cdots$	Poisson's ratio	8.1
ν_n	Variable	Kinetic chain length	4.4
π	Pa	Osmotic pressure	6.6
π		$3.14159 \ldots$	
ρ	g/cm^3	Density	6.6
ρ_i	$\cdots$	Ratio of reactive groups on reactant i to all groups that react the same way in the system	4.7
ρ_p	g/cm^3	Particle density	Prob. 5.13
ρ_r	ohm·cm	Volume resistivity	10.6

Symbol	Common unit	Definition or label	Section where first used
ρ_r	radicals/s	Rate of radical generation	5.5
σ	Pa	Tensile stress	8.1
σ	$\cdots$	Steric factor	7.2
σ_b	Pa	Stress at break	9.1
$\sigma_b(n)$	Pa	Stress at break for nth sample	9.1
$\overline{\sigma_b}$	Pa	Modal strength for sample population	9.1
σ_i	Pa	Stress in element i	8.4
σ_0	Pa	Fixed stress	8.3
σ_T	Pa	Total stress of an array of elements	8.4
τ	Pa	Shear stress	7.1
τ	s	Time of particle formation	5.5
τ_{gm}	Pa	Shear stress at geometric mean radius	7.7
τ_w	Pa	Shear stress at wall	7.7
$(\tau_w)_t$	Pa	"True" shear stress at wall	7.8
τ_{ij}	Pa	Stress tensor	7.8
ϕ	degree	Bond rotation angle	7.2
ϕ_i	s	Relaxation time	7.6
ϕ_0	s	Maximum relaxation time	7.6, 7.8
$[\phi]$	mol/liter	Concentration of ester bonds	11.3
χ	$\cdots$	Polymer-solvent interaction parameter	2.5
ψ	degree	Cone angle	7.7
ψ'	degree	Angle of twist	8.1
ω	rad/s	Frequency	7.3
ω_c	Hz (cycle/s)	Frequency	10.6

HARMONIC MOTION OF A MAXWELL MODEL

A2.1 TRIGONOMETRIC NOTATION

Starting with a sinusoidal input of strain in a Maxwell element (Sec. 8.3), we derive the resulting sinusoidal stress. First we let the strain ε be a function of a maximum strain ε_m and time t with a frequency ω.

$$\varepsilon = \varepsilon_m \sin \omega t \tag{A2.1}$$

For the Maxwell element:

$$\frac{d\varepsilon}{dt} = \frac{1}{E}\frac{d\sigma}{dt} + \frac{\sigma}{E\theta} \tag{A2.2}$$

Differentiating Eq. (A2.1):

$$\frac{d\varepsilon}{dt} = \varepsilon_m \cos \omega t \tag{A2.3}$$

Rearranging Eq. (A2.2):

$$\frac{d\sigma}{dt} + \frac{\sigma}{\theta} = \omega\varepsilon_m E \cos \omega t \tag{A2.4}$$

This is a simple linear differential equation of the form

$$\frac{dy}{dx} + Py = Q$$

The general solution for such an equation, when P and Q are functions of x only, is

$$y \exp(\psi) = \int \exp(\psi)Q \, dx + C \qquad \psi = \int P \, dx \qquad (A2.5)$$

For Eq. (A2.4), the analogy is

$$\psi = \frac{t}{\theta} \qquad (A2.6)$$

$$\sigma \exp\left(\frac{t}{\theta}\right) = \omega \varepsilon_m E \int \exp\left(\frac{t}{\theta}\right) \cos \omega t \, dt + C \qquad (A2.7)$$

$$\sigma \exp\left(\frac{t}{\theta}\right) = \frac{\omega \varepsilon_m E \theta}{1 + \omega^2 \theta^2} (\cos \omega t + \omega\theta \sin \omega t) \exp\left(\frac{t}{\theta}\right) + C \qquad (A2.8)$$

or

$$\sigma = \frac{\omega\theta}{1 + \omega^2\theta^2} \varepsilon_m E(\cos \omega t + \omega\theta \sin \omega t) + C \exp\left(\frac{-t}{\theta}\right) \qquad (A2.9)$$

The second term on the right is a transient one that drops out in the desired steady-state solution for $t/\theta \gg 1$.

We define an angle δ by

$$\tan \delta = \frac{1}{\omega\theta} = \frac{\sin \delta}{\cos \delta} \qquad \text{and} \qquad \sin \delta = \frac{1}{(1 + \omega^2\theta^2)^{1/2}} \qquad (A2.10)$$

Then, making use of trigonometric identities:

$$\cos \omega t + \omega\theta \sin \omega t = \frac{\cos \omega t(\sin \delta)}{\sin \delta} + \frac{\sin \omega t(\cos \delta)}{\sin \delta} \qquad (A2.11)$$

$$= \frac{\sin (\omega t + \delta)}{\sin \delta} \qquad (A2.12)$$

$$= (1 + \omega^2\theta^2)^{1/2} \sin (\omega t + \delta) \qquad (A2.13)$$

Finally, combining Eqs. (A2.13) and (A2.9) with the transient term dropped, we arrive at

$$\sigma = \frac{\omega\theta}{(1 + \omega^2\theta^2)^{1/2}} \varepsilon_m E \sin(\omega t + \delta) \qquad (A2.14)$$

A2.2 COMPLEX NOTATION

In this case, we start with a complex strain, the real part of which is the actual strain:

$$\varepsilon^* = \varepsilon_m \exp(i\omega t) \tag{A2.15}$$

The motion of the Maxwell element, in terms of a complex strain, is

$$\frac{d\sigma^*}{dt} = \frac{1}{E}\frac{d\sigma^*}{dt} + \frac{\sigma^*}{E\theta} \tag{A2.16}$$

Differentiating Eq. (A2.15):

$$\frac{d\varepsilon^*}{dt} = i\omega a_m \exp(i\omega t) \tag{A2.17}$$

Rearranging Eqs. (A2.16) and (A2.17):

$$\frac{d\sigma^*}{dt} + \frac{\sigma^*}{\theta} = E\frac{d\varepsilon^*}{dt} = i\omega\varepsilon_m E \exp(i\omega t) = Q \tag{A2.18}$$

As in App. A2.1, the general solution is

$$\sigma^* \exp\!\left(\frac{t}{\theta}\right) = \int \exp\!\left(\frac{t}{\theta}\right) Q\, dt + C \tag{A2.19}$$

$$\int \exp\!\left(\frac{t}{\theta}\right) Q\, dt = i\omega\varepsilon_m E \int \exp\!\left(i\omega t + \frac{t}{\theta}\right) dt = \frac{i\omega\varepsilon_m E \exp(i\omega t + t/\theta)}{i\omega + 1/\theta} \tag{A2.20}$$

Substitution and rearrangement with Eq. (A2.19) yields

$$\sigma^* = \frac{i\omega\varepsilon_m \theta E \exp(i\omega t)}{i\omega\theta + 1} + C \exp\!\left(-\frac{t}{\theta}\right) \tag{A2.21}$$

Once again, the second term on the right-hand side is a transient term which drops out at $t/\theta \gg 1$. Multiplying both numerator and denominator by $1 - i\omega\theta$ and substituting ε^* for its equivalent $\varepsilon_m \exp(i\omega t)$ gives

$$\sigma^* = \frac{\omega^2\theta^2\varepsilon^* E + i\omega\theta\varepsilon^* E}{1 + \omega^2\theta^2} \tag{A2.22}$$

The definition of complex modulus E^* is

$$E^* = E' + iE'' = \frac{\sigma^*}{\varepsilon^*} \tag{A2.23}$$

But Eq. (A2.22) gives

$$\frac{\sigma^*}{\varepsilon^*} = \frac{E\omega^2\theta^2}{1 + \omega^2\theta^2} + \frac{i(E\omega\theta)}{1 + \omega^2\theta^2} \tag{A2.24}$$

Identifying Eqs. (A2.23) and (A2.24) together, we concluded that

$$E' = \frac{E\omega^2\theta^2}{1 + \omega^2\theta^2} \quad \text{and} \quad E'' = \frac{E\omega\theta}{1 + \omega^2\theta^2} \tag{A2.25}$$

By the conventions of complex notation we find a loss angle δ, where

$$E^* = [(E')^2 + (E'')^2]^{1/2} \exp(i\delta) \quad \text{and} \quad \tan\delta = \frac{E''}{E'} \tag{A2.26}$$

From Eqs. (A2.25):

$$\tan\delta = \frac{1}{\omega\theta} \tag{A2.27}$$

$$[(E')^2 + (E'')^2]^{1/2} = \frac{E\omega\theta}{(1 + \omega^2\theta^2)^{1/2}} \tag{A2.28}$$

Also

$$\sigma^* = E^*\varepsilon^* = \frac{E\omega\theta\varepsilon^*}{(1 + \omega^2\theta^2)^{1/2}} \exp(i\delta) \tag{A2.29}$$

The actual stress σ is the real part of σ^*. Equation (A2.29) shows that σ^* depends on E and ε^* but leads by an angle δ. The maximum value of stress σ_m occurs at $\omega t + \delta = 0$.

$$\sigma_m = \frac{E\omega\theta\varepsilon_m}{(1 + \omega^2\theta^2)^{1/2}} \quad \text{and} \quad \sigma^* = \sigma_m \exp(i\omega t + i\delta) \tag{A2.30}$$

Also we can write

$$E' = \frac{\sigma_m\omega\theta}{\varepsilon_m(1 + \omega^2\theta^2)^{1/2}} \tag{A2.31}$$

Thus we can evaluate E, θ, E', or E'' from experimental measurement of ε_m, σ_m, and δ at a known value of ω.

SELECTED PROPERTIES OF POLYMER SYSTEMS

Table A3.1 Typical properties* of polymers used for molding and extrusion†

	ASTM test method	ABS medium impact	Acetal	Cellulose acetate
Specific gravity	D792	1.03 to 1.06	1.42	1.22 to 1.34
Refractive index	D542	–	1.48	1.46 to 1.50
Tensile strength, psi $\times$ 10^{-3} *	D638, D651	6 to 7.5	9.5 to 12	1.9 to 9.0
Elongation, %	D638	5 to 25	25 to 75	6 to 70
Tensile modulus, psi $\times$ 10^{-5}	D638	3 to 4	5.2	0.65 to 4.0
Compressive strength, psi $\times$ 10^{-3}	D695	10.5 to 12.5	18	3 to 8
Flexural yield strength, psi $\times$ 10^{-3}	D790	11 to 13	14	2 to 16
Impact strength, notched Izod, ft·lb/in	D256	3 to 6	1.3 to 2.3	1 to 7.8
Hardness, Rockwell	D785	R107 to 115	M94 to R120	R34 to 125
Flexural modulus, psi $\times$ 10^{-5}	D790	3.7 to 4.0	3.8 to 4.3	–
Compressive modulus, psi $\times$ 10^{-5}	D695	2.0 to 4.5	6.7	–
Thermal conduct., (cal/s·cm·K) $\times$ 10^4	C177	4.5 to 8.0	5.5	4 to 8
Specific heat, cal/g·K	· · ·	0.3 to 0.4	0.35	0.3 to 0.4
Linear therm. exp. coeff., K^{-1} $\times$ 10^5	D696	8 to 10	10	8 to 18
Continuous use temperature, °C	· · ·	71 to 93	90	60 to 105
Deflection temp., °C at 0.45 MPa	D648	102 to 107	124	50 to 100
Volume resistivity, ohm·cm	D257	2.7 $\times$ 10^{16}	1.0 $\times$ 10^{15}	10^{10} to 10^{14}
Dielectric strength, kV/in	D149	350 to 500	500	250 to 500
Dielectric constant at 1 kHz	D150	2.4 to 4.5	3.7	3.4 to 7.0
Dissipation factor at 1 kHz	D150	0.004 to 0.007	0.004	0.01 to 0.07
Deleterious media	D543	Conc. oxidizing acids, organic solvents	Strong acids, some other acids and bases	Strong acids and bases
Solvents (room temperature) (Cl.H. = chlorinated hydrocarbons)	· · ·	Ketones, esters, some Cl.H.	None	Ketones, esters, Cl.H.

*Conversion factors: 1000 psi = 6.895 MPa; 1 ft·lb/in = 53.4 J/m; 1 cal = 4.187 J; 1 kV/in = 0.0394 MV/m.
†Based on data from *Modern Plastics Encyclopedia*.

Table A3.1 (*Continued*)

Cellulose acetate-butyrate	Cast epoxy glass fiber fill	Fluoro polymers		Ionomers
		$-CF_2CFCl-$	$-CF_2CF_2-$	
1.15 to 1.22	1.6 to 2.0	2.1 to 2.2	2.14 to 2.20	0.93 to 0.96
1.46 to 1.49	–	1.425	1.35	1.51
2.6 to 6.9	5 to 20	4.5 to 6	2 to 5	3.5 to 5.0
40 to 88	4	80 to 250	200 to 400	350 to 450
0.5 to 2.0	30	1.5 to 3	0.58	0.2 to 0.6
2.1 to 7.5	18 to 40	4.6 to 7.4	1.7	–
1.8 to 9.3	8 to 30	7.4 to 9.3	–	–
1 to 11	0.3 to 10	2.5 to 2.7	3.0	6.0 to 15
R31 to 116	M100 to 112	R75 to 95	D50 to 55 (Shore)	D50 to 65 (Shore)
–	20 to 45	–	–	–
–	–	–	–	–
4 to 8	4 to 10	4.7 to 5.3	6.0	5.8
0.3 to 0.4	0.19	0.22	0.25	0.55
11 to 17	1 to 5	4.5 to 7	10	12
60 to 105	150 to 260	175 to 200	290	70 to 95
54 to 108	–	126	121	38
10^{11} to 10^{15}	Over 10^{14}	1.2×10^{18}	Over 10^{18}	Over 10^{16}
250 to 400	300 to 400	500 to 600	480	900 to 1100
3.4 to 6.4	3.5 to 5.0	2.3 to 2.7	Under 2.1	2.4
0.01 to 0.04	0.01	0.023 to 0.027	Under 0.0002	0.0015
Strong acids and bases	None	None	None	Acids, esp. strong, oxidizing acids
Ketones, esters, Cl.H.	None	Swells in Cl.H.	None	None

Table A3.1 Typical properties* of polymers used for molding and extrusion†
(*Continued*)

	ASTM test method	Melamine-formaldehyde (cellulose fill)	Nylon 66 (moisture-conditioned)	Nylon 6 (moisture-conditioned)
Specific gravity	D792	1.47 to 1.52	1.13 to 1.15	1.12 to 1.14
Refractive index	D542	–	1.53	–
Tensile strength, psi $\times$ 10^{-3}	D638, D651	5 to 13	11	10
Elongation, %	D638	0.6 to 1.0	300	300
Tensile modulus, psi $\times$ 10^{-5}	D638	11 to 14	–	1.0
Compressive strength, psi $\times$ 10^{-3}	D695	33 to 45	–	–
Flexural yield strength, psi $\times$ 10^{-3}	D790	9 to 16	6.1	5.0
Impact strength, notched Izod, ft·lb/in	D256	0.2 to 0.4	2.1	3.0
Hardness, Rockwell	D785	M115 to 125	R120	R119
Flexural modulus, psi $\times$ 10^{-5}	D790	–	1.75 to 4.1	1.4
Compressive modulus, psi $\times$ 10^{-5}	D695	–	–	2.5
Thermal conduct., (cal/s·cm·K) $\times$ 10^4	C177	6.5 to 10	5.8	5.8
Specific heat, cal/g·K	· · ·	0.4	0.4	0.38
Linear therm. exp. coeff., K^{-1} $\times$ 10^5	D696	4 to 4.5	8.0	8.0 to 8.3
Continuous use temperature, °C	· · ·	99	80 to 150	80 to 120
Deflection temp., °C at 0.45 MPa	D648	43	180 to 240	150 to 185
Volume resistivity, ohm·cm	D257	0.8 to 2 $\times$ 10^{12}	10^{14} to 10^{15}	10^{12} to 10^{15}
Dielectric strength, kV/in	D149	270 to 300	385 to 470	440 to 510
Dielectric constant at 1 kHz	D150	7.8 to 9.2	3.9 to 4.5	4.0 to 4.9
Dissipation factor at 1 kHz	D150	0.015 to 0.036	0.02 to 0.04	0.011 to 0.06
Deleterious media	D543	Strong acids and bases	Strong acids	Strong acids
Solvents (room temperature) (Cl.H = chlorinated hydrocarbons)	· · ·	None	Phenol and formic acid	Phenol and formic acid

*Conversion factors: 1000 psi = 6.895 MPa; 1 ft·lb/in = 53.4 J/m; 1 cal = 4.187 J; 1 kV/in = 0.0394 MV/m.
†Based on data from *Modern Plastics Encyclopedia*.

Table A3.1 (*Continued*)

Phenol-Formaldehyde (cellulose fill)	Polycarbonate	Phenoxy	Poly(phenylene oxide)	Polypropylene
1.37 to 1.46	1.2	1.17 to 1.18	1.06	0.900 to 0.910
–	1.586	1.598	–	1.49
5 to 9	9.5	8 to 9	9.6	4.5 to 6.0
0.4 to 0.8	110	50 to 100	60	100 to 600
8 to 17	3.5	3.5 to 3.9	3.55	1.6 to 2.25
25 to 31	12.5	10.4 to 12.0	16.4	5.5 to 8.0
7 to 14	13.5	12 to 13	13.5	6 to 8
0.2 to 0.6	16	1.5 to 2.0	5.0	0.4 to 1.0
E64 to 95	M70	R115 to 123	R119	R80 to 102
10 to 12	3.4	3.75 to 4.0	3.6 to 4.0	1.7 to 2.5
–	3.5	3.2 to 3.4	–	1.5 to 3.0
4 to 8	4.7	4.2	5.2	2.8
0.35 to 0.40	0.3	0.4	–	0.46
3.0 to 4.5	6.8	5.7 to 6.1	3.3 to 5.9	8.1 to 10.0
150 to 175	121	77	–	120 to 160
–	138	83 to 88	–	225 to 250
10^9 to 10^{13}	2.1×10^{16}	10^{10} to 10^{13}	10^{18}	Over 10^{16}
200 to 400	400	400 to 520	400 to 500	500 to 660
4.4 to 9.0	3.02	4.1	2.6	2.2 to 2.6
0.04 to 0.20	0.0021	0.002	0.00035	Under 0.0018
Strong bases and oxidizing acids	Bases and strong acids	Strong oxidizing acids	None	Strong oxidizing acids
None	Aromatic and Cl.H.	Aromatic and Cl.H.	Aromatic and Cl.H.	None

Table A3.1 Typical properties* of polymers used for molding and extrusion†
(*Continued*)

	ASTM test method	Polyethylene		Poly(methyl methacrylate)
		Low density	High density	
Specific gravity	D792	0.91 to 0.925	0.94 to 0.965	1.17 to 1.20
Refractive index	D542	1.51	1.54	1.49
Tensile strength, psi × 10^{-3}	D638, D651	0.6 to 2.3	3.1 to 5.5	7 to 11
Elongation, %	D638	90 to 800	20 to 130	2 to 10
Tensile modulus, psi × 10^{-5}	D638	0.14 to 0.38	0.6 to 1.8	3.8
Compressive strength, psi × 10^{-3}	D695	–	2.7 to 3.6	12 to 18
Flexural yield strength, psi × 10^{-3}	D790	–	1.0	13 to 19
Impact strength, notched Izod, ft·lb/in	D256	No break	0.5 to 20	0.3 to 0.5
Hardness, Rockwell	D785	D40 to 51 (Shore)	D60 to 70 (Shore)	M85 to 105
Flexural modulus, psi × 10^{-5}	D790	0.08 to 0.6	1.0 to 2.6	4.2 to 4.6
Compressive modulus, psi × 10^{-5}	D695	–	–	3.7 to 4.6
Thermal conduct., (cal/s·cm·K) × 10^4	C177	8.0	11 to 12	4 to 6
Specific heat, cal/g·K	· · ·	0.55	0.55	0.35
Linear therm. exp. coeff., K^{-1} × 10^5	D696	10 to 22	11 to 13	5 to 9
Continuous use temperature, °C	· · ·	80 to 100	120	60 to 88
Deflection temp., °C at 0.45 MPa	D648	38 to 49	60 to 88	80 to 107
Volume resistivity, ohm·cm	D257	Over 10^{16}	Over 10^{16}	Over 10^{14}
Dielectric strength, kV/in	D149	450 to 1000	450 to 500	400
Dielectric constant at 1 kHz	D150	2.25 to 2.35	2.30 to 2.35	3.0 to 3.6
Dissipation factor at 1 kHz	D150	Under 0.0005	Under 0.0005	0.03 to 0.05
Deleterious media	D543	Oxidizing acids	Oxidizing acids	Strong bases and strong, oxidizing acids
Solvents (room temperature) (Cl.H. = chlorinated hydrocarbons)	· · ·	None	None	Ketones, esters, aromatic and Cl.H.

*Conversion factors: 1000 psi = 6.895 MPa; 1 ft·lb/in = 53.4 J/m; 1 cal = 4.187 J; 1 kV/in = 0.0394 MV/m.

†Based on data from *Modern Plastics Encyclopedia*.

Table A3.1 (*Continued*)

Polysulfone	Polystyrene		Poly(vinyl chloride)	
	Gen. purpose	Impact-resistant	Rigid	Plasticized
1.24	1.04 to 1.05	1.03 to 1.06	1.30 to 1.58	1.16 to 1.35
1.633	1.59 to 1.60	–	1.52 to 1.55	–
10.2 (yield)	5.3 to 7.9	3.2 to 4.9	6 to 7.5	1.5 to 3.5
50 to 100	1 to 2	13 to 50	2 to 80	200 to 450
3.6	3.5 to 4.85	2.6 to 4.65	3.5 to 6	–
13.9 (yield)	11.5 to 16	4 to 9	8 to 13	0.9 to 1.7
15.4 (yield)	8.7 to 14	5 to 12	10 to 16	–
1.2	0.25 to 0.40	0.5 to 11	0.4 to 20	–
M69, R120	M65 to 80	M20 to 80	D65 to 85 (Shore)	A40 to 100 (Shore)
3.9	4.3 to 4.7	3.3 to 4.0	3 to 5	–
3.7	–	–	–	–
2.8	2.4 to 3.3	1.0 to 3.0	3.5 to 5.0	3.0 to 4.0
0.31	0.32	0.32 to 0.35	0.2 to 0.28	0.3 to 0.5
5.2 to 5.6	6 to 8	3.4 to 21	5 to 10	7 to 25
150 to 175	66 to 77	60 to 79	65 to 80	65 to 80
180	75 to 100	75 to 95	57 to 82	–
5×10^{16}	Over 10^{16}	Over 10^{16}	Over 10^{16}	10^{11} to 10^{15}
425	500 to 700	300 to 600	425 to 1300	300 to 1000
3.13	2.4 to 2.65	2.4 to 4.5	3.0 to 3.3	4 to 8
0.001	0.0001 to 0.0003	0.0004 to 0.002	0.009 to 0.017	0.07 to 0.16
None	Strong, oxidizing acids	Strong, oxidizing acids	None	None
Aromatic hydrocarbons	Aromatic and Cl.H.	Aromatic and Cl.H.	Ketones, esters, swelling in aromatic and Cl.H.	Plasticizer may be extracted. Otherwise like rigid PVC

Table A3.2 Typical properties of representative textile fibers*

Fiber identification		Breaking tenacity, g/denier		Elongation at break, %	
Generic name	Chemical name	Standard	Wet	Standard	Wet
1. Rayon (viscose)	Regenerated cellulose				
a. Regular		0.7–3.2	0.7–1.8	15–30	20–40
b. High tenacity		3.0–5.7	1.9–4.3	9–26	14–34
2. Acetate	Cellulose acetate				
a. Diacetate		1.2–1.4	0.8–1.0	25–45	35–50
b. Triacetate		1.1–1.3	0.8–1.0	26–40	30–40
3. Spandex	Segmented polyurethane	0.7–0.9	· · ·	400–625	· · ·
4. Fluorocarbon	Poly(tetrafluoroethylene)	0.9–2.0	0.9–2.0	19–140	19–140
5. Glass	(Silica, silicates)	9.6–19.9	6.7–19.9	3.1–5.3	2.2–5.3
6. Polyester	Poly(ethylene terephthalate)	2.2–9.5	2.2–9.5	12–55	12–55
7. Acrylic	Polyacrylonitrile	2.0–2.7	1.6–2.2	34–50	34–60
8. Nylon					
a. Nylon 6		4.0–9.0	3.7–8.2	16–50	19–47
b. Nylon 66		3.0–9.5	2.6–8.0	16–66	18–70
9. Aramid					
a. Kevlar (DuPont)		21.7	21.7	2.5–4	2.5–4
b. Nomex (DuPont)		4.0–5.3	3.0–4.1	22–32	20–30
10. Olefin					
a. Polyethylene (branched)		1.0–3.0	1.0–3.0	20–80	20–80
b. Polyethylene (linear)		3.5–7.0	3.5–7.0	10–45	10–45
c. Polypropylene		3.0–8.0	3.0–8.0	14–80	14–80
11. Cotton	α-Cellulose	3.0–4.9	3.0–5.4	3–10	
12. Wool	Protein	1.0–2.0	0.8–1.8	20–40	

*Based on information in *Textile World*, 128(8):57 (1978). Tensile strength (MPa) = tenacity (g/denier) $\times$ density (g/cm^3) $\times$ 88.3.

Table A3.2 (*Continued*)

Specific gravity	Water absorbed at 70°F, 65% rel. humidity, %	Thermal stability
1.46–1.54	11–13	*a.* Loses strength at 150°C *b.* Decomposes at 175–240°C
1.32 1.3	6.4 3.2	*a.* Sticks at 175–205°C; softens, 205–230°C; melts, 260°C *b.* Melts at 300°C
1.21	1.3	Sticks at 215°C
2.1	Nil	Melts at about 288°C
2.49–2.55	Nil	Softens at 730–850°C; does not burn
1.38	0.4–0.8	Sticks at 230°C; melts, 250°C
1.17	1.5	Shrinks 5% at 253°C
1.14 1.14	2.8–5.0 4.2–4.5	*a.* Melts at 216°C; decomposes, 315°C *b.* Sticks at 230°C; melts, 250–260°C
1.44 1.38	4.5–7 6.5	*a.* Decomposes at 500°C *b.* Decomposes at 370°C
0.92 0.95 0.90	Nil Nil 0.01–0.1	*a.* Softens at 105–115°C; melts, 110–120°C; shrinks 5% at 75°C *b.* Softens at 115–125°C; melts, 125–138°C; shrinks 5% at 75–80°C *c.* Softens at 140–175°C; melts, 160–177°C; shrinks 5% at 100–130°C
1.54	7–8.5	Decomposes at 150°C
1.32	11–17	Decomposes at 130°C

Table A3.3 Typical properties of cross-linked rubber compounds

a. Diene-based polymers and copolymers

	cis-1,4-Polyisoprene (natural rubber, also made synthetically) NR (nat.);* IR (syn.)	cis-1,4-Polybutadiene BR	Styrene-butadiene random copolymer, 25 wt % styrene SBR	Styrene-butadiene block copolymer, about 25% styrene (YSBR)	Polychloroprene (neoprene) CR	Butadiene-acrylonitrile random copolymer, variable % acrylonitrile, NBR	Reclaimed rubber (whole tires) (mainly NR and SBR)
Gum stock (cross-linked, unfilled):							
Density, g/cm³	0.93	0.93	0.94	0.94-1.03	1.23	1.00	1.2 (compd'd)
Tensile strength, psi	2500-3000	200-1000	200-400	1700-3700	3000-4000	500-1000	500-1000
Resistivity, ohm·cm, log	15-17	...	15	13	11	10	
Dielec. const. at 1 kHz	2.3-3.0	2.3-3.0	3.0	3.4	9.0	13	
Diss. factor, at 1 kHz	0.002-0.003	0.002-0.003	0.003	0.01	0.03	0.055	
Dielectric str., kV/in	...	...	...	485	150-600		
Reinforced stock:							
Tensile strength, psi	3000-4000	2000-3500	2000-3500	1000-3000	3000-4000	3000-4000	500-1000
Elong. at break, %	300-700	300-700	300-700	500-1000	300-700	300-700	300-400
Hardness, Shore A	20-100	30-100	40-100	40-85	20-100	30-100	50-100
Resilience	Excellent	Excellent	Good	Excellent	Good	Fair	Good
Stiffening temp., °C	-30 to -45	-35 to -50	-20 to -45	-50 to -60	-10 to -30	0 to -30	-20 to -45
Brittle temp., °C	-60	-70	-60	-70	-40 to -55	-15 to -55	-60
Cont. high temp. limit, °C	100	100	110	65	120	120	100

Resistance to:

	1	2	3	4	5	6	7
Acid	Good	Good	Good	Good	Good	Good	Good
Alkali	Good	Good	Good	Good	Good	Good	Good
Gasoline and oil	Poor	Poor	Poor	Poor	Good	Excellent	Poor
Aromatic hydrocarbons	Poor	Poor	Poor	Poor	Fair	Good	Poor
Ketones	Good	Good	Good	Poor	Poor	Poor	Good
Chlorinated solvents	Poor	Poor	Poor	Poor	Poor	Poor	Poor
Oxidation	Good	Good	Good	Good	Excellent	Fair	Good
Ozone	Poor	Poor	Poor	Poor	Excellent	Poor	Poor
Gamma radiation	Good	Good	Good	Good	Poor	Fair	Good
Advantages	High resilience and strength; abrasion-resistant	Low heat buildup in flexing; good resilience; abrasion-resistant	Low price; good wearing; bonds easily	Easily injection-molded; not cross-linked	Flame-resistant; good resistance to oxygen, oil, and gasoline	Excellent resistance to oil and gasoline (non-swelling)	Low price; easy processing; good weathering
Typical applications	Tires, tubes, belts, bumpers, tubing, gaskets, seals, foamed matresses and padding	Tire treads, mechanical goods	Tires, mechanical goods	Toys, rubber bands, mechanical goods	Gaskets, tubing, O-rings, seals, gasoline hose	Gaskets, tubing, O-rings, gasoline hose	Tire carcasses, battery boxes, friction tape, soles and heels

Table A3.3 Typical properties of cross-linked rubber compounds (*Continued*)

b. Saturated, carbon-chain polymers

	Polyisobutylene (Butyl rubber, copolymer with 0.5–2% isoprene) IIR	Ethylene-propylene random copolymer, 50% ethylene EPM	Ethylene-propylene random terpolymer, 50% ethylene EPDM	Chlorosulfonated polyethylene CSM	Poly(ethyl acrylate), usually a copolymer ACM	Vinylidene fluoride-chloro-trifluoro ethylene random copolymer FKM	Vinylidene fluoride-hexa-fluoro propylene random copolymer FKM
Gum stock (cross-linked, unfilled):							
Density, g/cm³	0.92	0.86	0.86	1.12–1.28	1.10	1.85	1.85
Tensile str., psi†	2500–3000	500	200	2500	200–400	200–2500	2000
Resistivity, ohm·cm, log	17	16	16	14	...	14	13
Dielect. const., 1 kHz	2.1–2.4	3.0–3.5	3.0–3.5	7–10	...	6	...
Diss. factor, at 1 kHz	0.003	0.004–0.008	0.004–0.008	0.03–0.07	...	0.05	0.03–0.04
Dielectric str., kV/in	600	900	900	500	...	600	250–750
Reinforced stock:							
Tensile str., psi	2000–3000	1000–3000	1000–3500	3000	1500–2500	1500–2500	1500–2500
Total elong., %	300–700	200–300	200–300	300–500	250–350	300–400	300–400
Hardness, Shore A	30–100	30–100	30–100	50–100	40–100	50–90	50–90
Stiff'g. temp., °C	−25 to −45	−40	−40	−10 to −30	...	−35	...
Brittle temp., °C	−60	−50 to −75	−50 to −75	−40 to −55	−30	−50	−45
Cont. high temp. limit, °C	120	150	150	160	175	200	250
Resilience	Fair	Good	Good	Good	Fair	Fair	Fair

Resistance to:					
Acid	Excellent	Excellent	Good	Fair	Excellent
Alkali	Excellent	Excellent	Good	Poor	Good
Gasoline and oil	Poor	Poor	Good	Good	Excellent
Aromatic hydrocarbons	Poor	Fair	Fair	Good	Excellent
Ketones	Excellent	Good	Poor	Poor	Poor
Chlorinated solvents	Poor	Poor	Poor	Poor	Good
Oxidation	Excellent	Excellent	Excellent	Excellent	Excellent
Ozone	Excellent	Excellent	Excellent	Excellent	Excellent
Gamma radiation	Poor	Poor	Good	Good	Good
Advantages	Low gas permeability; outstanding resistance to oxygen and ozone	Outstanding resistance to oxygen and ozone; good color stability, good electrical stability	Flame-resistant; outstanding resistance to oxygen and ozone; good color stability	Oil- and heat-resistant; flame-resistant	High resistance to most liquids; good retention of strength at high temperatures; good weathering
Typical applications	Inner tubes, weather stripping, shock absorbers	Wire and cable installation, weather stripping, hose, belts, footwear	Gaskets, weather stripping, wire and cable insulation, gasoline hose	Automotive transmission seals, oil hose, O-rings	O-rings, tubing, valve seats, shaft seals, gaskets, diaphragms

Table A3.3 Typical properties of cross-linked rubber compounds (*Continued*)

c. Heterochain polymers

	Poly(dimethyl siloxane) (silicone rubber), usually copolymer with vinyl groups VMQ	Poly(dimethyl siloxane) copolymer with phenyl-bearing siloxane and vinyl groups PVMQ	Room-temp. vulcanizing silicone	Polysulfide ET and EOT	Polyurethane AU and EU	Poly(epichlorohydrin) CO	Epichlorohydrin-ethylene oxide random copolymer, 32% eth. oxide ECO
Gum stock (cross-linked, unfilled):							
Density, g/cm³	0.98	0.98	1.0–1.3 (cmpd'd)	1.35	1.25	1.36	1.27
Tensile strength, psi[†]	50–100	50–100	...	100–200	2000–4000	100–300	200–400
Resistivity, ohm·cm, log	11–17	11–17	15	12	11–14		
Dielect. const. at 1 kHz	3.0–3.5	3.0–3.5	2.8	7–9.5	5–8		
Diss. factor at 1 kHz	0.001–0.010	0.001–0.010	0.003	0.001–0.005	0.02–0.09		
Dielectric str., kV/in	100–600	100–600	500	250–600	350–525		
Reinforced stock:							
Tensile strength, psi	500–1200	500–1500	400–800	1300–1800	3000–10,000	1500–2500	2000–3000
Elong. at break, %	200–700	200–700	100–200	200–500	200–600	200	200
Hardness, Shore A	30–80	30–80	30–50	25–85	20–100	60–80	60–80
Resilience	Fair	Fair	Fair	Fair	Poor	Fair	Fair
Stiffening temp., °C	−50	−100	−50 to −100	−25	−25 to −35	−25	−45
Brittle temp., °C	−50	−120	−50 to −100	−50	−50 to −60	−25	
Cont. high temp. limit, °C	250	300	200–250	120	120	150	150

Resistance to:						
Acid	Fair	Fair	Fair	Fair	Good	Good
Alkali	Fair	Fair	Good	Poor to fair	Good	Good
Gasoline and oil	Poor	Poor	Excellent	Excellent	Good	Good
Aromatic hydrocarbons	Poor	Poor	Good	Good	Fair	Fair
Ketones	Excellent	Excellent	Good	Poor	Poor	Poor
Chlorinated solvents	Poor	Poor	Good	Poor	Poor	Poor
Oxidation	Excellent	Excellent	Excellent	Excellent	Excellent	Good
Ozone	Excellent	Excellent	Excellent	Excellent	Excellent	Excellent
Gamma radiation	Good	Good	Good	Good	Good	Excellent
Advantages	High temp. stability and low temp. flexibility; good electrical properties	Can be cast in place; good electrical stab.	Resists swelling in gasoline and oil	Excellent abrasion resistance; high strength; low coeff. of friction	Good solvent resistance; low air permeability; good resistance to oxygen and ozone	
Typical applications	Wire and cable insulation, tubing for aircraft systems, surgical implants	Caulking, seals, encapsulation, potting of elec. components	Gaskets, seals, diaphragms	Impellers, solid tires for fork-lift trucks, shoe heels	Seals, gaskets, wire and cable jackets, coatings	

*This abbreviation and all following are ASTM abbreviations, 1000 psi = 6.895 MPa; 1 kV/in = 0.0394 MV/m.

†1000 psi = 6.895 MPa; 1 kV/in = 0.0394 MV/m.

INDEX